Elements of Physiophilosophy
By Lorenz Oken

Elements of Physiophilosophy

"Of all truths relating to phenomena, the most valuable to us are those which relate to the order of their succession. On a knowledge of these is founded every reasonable anticipation of future facts, and whatever power we possess of influencing those facts to our advantage. Even the laws of geometry are chiefly of practical importance to us as being a portion of the premises from which the order of the succession of phenomena may be inferred."
John Stuart Mill.

Begun in the autumn of the year 1845, without the cognizance, or at the suggestion of a single human being, the present Translation is due to the fact of its original having encountered a somewhat kindred spirit, and aroused therein the desire to render others participant, if possible, in the large amount of instruction it is so well calculated to afford. And now that the work is done, what remains for the labourer at second-hand to say by way of preamble to his newly-dressed wares? Had the book been printed within the pale of a philosophical or physico-theological sect, the Translator's final duty would have been clearly enough prescribed. Already bound to the profession of "particular tenets," his main object would be to indulge in a laudatory but servile abstract of his author's doctrines, or, if having set out with the expressed intention of illustrating their bearings upon the state of science past, present, and to come, he would become so drunk beforehand with the large and unbridled potation of his creed, as to surprise the casual reader by informing him that such an intention is useless, for the two stand in direct antithesis to each other. Examples of this mode of procedure are not wanting at the present day, whether at home or abroad. They are the produce of that spirit, which, rife enough in the Middle Ages, has been so graphically described by Professor Whewell under the title of the "Commentatorial," and "whose professed object is to explain, to enforce, to illustrate doctrines assumed to be true, but not to obtain additional truths or new generalizations." While from dealings of this character, as being utterly opposed to the sacred cause of Truth, I turn away with feelings of repugnance, to which the lessons of some personal experience have lent their aid, it is not my business, upon the other hand, to enter the lists of controversy against those who, having neither the capacity, nor the desire of its cultivation, for the higher walks of science, delight to dismiss a work of the present kind with some idle anathema of mysticism or evasive outcry for more facts.

I refrain from essaying to give any condensed formula or outlineA of Professor Oken's Physio-philosophy: first, because its leading points have been already noted in his own prefaces to the German work and its translation; secondly, because the book will, I trust, best speak for itself to those who shall come with minds unprejudiced and duly prepared, each one in his particular department, to its study; and, lastly, because any such attempt would necessarily involve an amount of historical and critical details, which must be here superficially treated and so misplaced. Suffice it to observe, that the present work stands alone in Germany, as being the most practical application upon a systematic scale of the principles advanced by Schelling, more especially in the Mathesis and Ontology; for the concluding part or Biology stands almost "per se." As such it will form, apart from other and higher considerations, a readily available introduction to the writings upon similar subjects of Carus, Steffens, Hegel and others, and may induce further attempts to render, by translation or history, the English student familiar with much of what at present is known only by scattered fragments in journals, or through the medium of reviews. From what has been said, the reader will be at no loss to discern in what light the Translator humbly desires to be viewed in reference to the present work. He rests content with the confident hope that its pages will be, at least, found eminently suggestive, that new thoughts will be awakened by facts and their relations being here cast in a fresh mould, that shall stimulate others in the field of inquiry, and open paths hitherto untrod. In this he is but expressing the sentiments of the author himself, and acknowledging what the present time with its accumulating mass of knowledge presses upon us more and more—the necessity of work, wherein abstract science and experience, theory and facts shall advance together, the Ideal in part receiving and reflecting back with increased lustre the light which it has derived from the Real or outward semblance of things.

Meanwhile, it is with no small amount of diffidence and hesitation that the present Translation will quit my hands. Hemmed in by a rigid dialectic terminology upon all sides, I have had difficulties of no ordinary kind to contend with in adapting a language, composed of such varied elements as our own, to meet the requisites of general clearness and conciseness that form so prominent a feature of the German work. If errors and obscurities exist, the blame, it will be observed, attaches to myself, not to the distinguished author. Ill-health has conflicted much with the calmness and repose of mind so indispensable to an undertaking, at once novel in kind and character to the English reader; or otherwise, these (my last labours unto any extent as a Translator) might have been rendered more worthy of the Ray Society and the objects it has in view.

To those who have kindly afforded me assistance in the progress of the work, and to the latter body for undertaking it, I here return many grateful thanks. The Author himself in a letter to the Translator, dated Jan. 12, 1847, acknowledges the acceptance of his work by the Society in the following words:—"The intelligence of my Physio-philosophy having been deemed worthy of translation by so goodly and enlightened a Society, cannot be otherwise unto me than a source of delight."B

ALFRED TULK.

A For this the reader may be referred to the 3d vol. of Prof. Blainville's Hist. des Sciences de l'Organisation; Par. 1845; or better still, to the sketch (preceded by a view of Schelling's philosophy), which is given by M. Saint-Agy in the Tome Complémentaire of Cuvier's Hist. des Sciences Naturelles, 1845. He there rightly observes of Oken's work, that "pendant les quarante dernières années il n'a presque paru en Allemagne d'ouvrages d'anatomie, de physiologie, de physique et de chimie auxquelles elle n'ait servi de base." For what a master-mind like Oken's is capable of creating, I would especially refer to his theory of the Cranial Homologies, which has been in our own country so beautifully carried out, modified, and proved by the extensive researches of Professor Owen.

B "Die Nachricht dass meine Naturphilosophie von einer sociferigen und erleuchteten Gesellschaft der Uebersetzung für würdig erachtet worden ist, konnte nicht anders als mir Freude gewähren."

AUTHOR'S PREFACE TO THE TRANSLATION.

It is with no readiness or pleasure that I write introductions of any kind, and usually abstain from doing so, partly because they appear to me like a kind of apology or makeshift for the author, and partly because the contents of the book itself should indicate his status or position. With regard, however, to the history of the work, some few words are certainly requisite for its Translation. I wrote the first Edition of 1810 in a kind of inspiration, and on that account it was not so well arranged as a systematic work ought to be. Now, although this may appear to have been amended in the second and third edition, yet still it was not possible for me to completely attain the object held in view. The book has therefore remained essentially the same as regards its fundamental principles, such as those concerning the formation of matter, the protoplasmic substance (Schleim-Substanz) and vesicular form of the organic mass, the signification and function of the organs, as also the principles of classification in Minerals, Plants and Animals, so that all this is consequently as old as the first edition. It is only the empirical arrangement into series of plants and animals, that has been modified from time to time in accordance with the scientific elevation of their several departments, or just as discoveries and anatomical investigations have increased and rendered some other position of the objects a matter of necessity. This susceptibility to change will of course be persistent in the future, although the principles themselves should continue wholly unchanged; ay, the very stability of the latter will tend the more to invite the naturalist to the pursuit of empirical inquiries, by determining beforehand in what direction he is to extend his point of view, and thus spare himself the trouble of blindly and laboriously groping about in the dense labyrinth of facts. Such a work therefore as the present can only approximate completion through the progress made in science, and each new edition will supply some defect of its predecessor in the distribution or parcelling out of things.

In the first edition the principle was raised of individual bodies being alone the object of Natural History, and that in the next place they are to be arranged according to the combination of their organs or component parts, and by no means after the division or mere form of a single organ; that, for example, a special organ or anatomical system lies as the basis of each Vegetable and Animal class, and that there must be therefore as many classes, and no more, as there are cardinal organs present upon which to found them. On that account it was absolutely necessary first of all to find out these cardinal organs, and determine their rank; and, in so doing, it was shown that organs and classes are at bottom of one kind, and that the development by stages or degrees of the embryo is the antetype of that of the classes; furthermore, that each class takes its starting-point from below, and consequently that the classes do not stand simply one above the other, but fall into a series of mutually parallel ranks. Now it is this which, along with the doctrine of the infusorio-vesicular form of the organic mass, and that touching the signification of parts, as to how e. g. the blossom is the repetition of the vegetable axis or trunk, the cephalic bones that of the vertebræ, the feet of the branchiæ, and the maxillæ in turn of the feet, appears to me the cardinal point attained in my Philosophy of Nature; more especially, because it was these very doctrines which were first of all, i. e. before all the others, comprehended and almost universally adopted. The inorganic matters and activities pass, however, parallel also to the anatomical formations and functions; and that this is the case too with the spiritual or psychical functions the philosophy of the future will probably be in the condition to point out.

The reader will not expect to find that the serial arrangement of Plants and Animals, with their parallelism, has been in every instance thoroughly attained. The present is but a sample of how we are to proceed in our desire of obtaining a Natural system. With such an attempt one has something to change every year, and I have in the present translation made some alterations in respect to the Mollusca and Fishes. In this sense then it is my wish that the book may be

regarded, and accordingly received with its due amount of indulgence.
LORENZ OKEN.

PREFACE.

The first principles of the present work I laid down in my small pamphlet entitled Grundriss der Naturphilosophie, der Theorie der Sinne und der darauf gegrundeten Classification der Thiere; Frankfurt bey Eichenberg, 1802, 8vo (out of print). I still abide by the position there taken, namely, that the Animal Classes are virtually nothing else than a representation of the sense-organs, and that they must be arranged in accordance with them. Thus, strictly speaking, there are only 5 Animal Classes: Dermatozoa, or the Invertebrata; Glossozoa, or the Fishes, as being those animals in whom a true tongue makes for the first time its appearance; Rhinozoa, or the Reptiles, wherein the nose opens for the first time into the mouth and inhales air; Otozoa, or the Birds, in which the ear for the first time opens externally; Ophthalmozoa, or the Thricozoa, in whom all the organs of sense are present and complete, the eyes being moveable and covered with two palpebræ or lids. But since all vegetative systems are subordinated to the tegument or general sense of feeling, the Dermatozoa divide into just as many or corresponding divisions, which, on account of the quantity of their contents, may be for the sake of convenience also termed classes. Thereby 9 classes of the inferior animals originate, but which, when taken together, have only the worth or value of a single class. So much by way of explaining the apparent want of uniformity in the system.
I first advanced the doctrine, that all organic beings originate from and consist of vesicles or cells, in my book upon Generation. (Die Zeugung. Frankfurt bey Wesche, 1805, 8vo.) These vesicles, when singly detached and regarded in their original process of production, are the infusorial mass, or the protoplasma (Ur-Schleim) from whence all larger organisms fashion themselves or are evolved. Their production is therefore nothing else than a regular agglomeration of Infusoria; not of course of species already elaborated or perfect, but of mucous vesicles or points in general, which first form themselves by their union or combination into particular species. This doctrine concerning the primo-constituent parts of the organic mass is now generally admitted or recognised, and I need not, therefore, add anything by way of apology for it or defence.
In mine and Kieser's Beyträgen zur vergleichenden Zoologie, Anatomie und Physiologie; Frankfurt bey Wesche, 1806, 4to, I have shown that the intestines originate from the umbilical vesicle, and that this corresponds to the vitellus. It is true Friedrich Wolf had already discovered it in the chick, but his was only a single instance, and completely forgotten. I have also discovered it and without knowing anything about my being anticipated, since it was nowhere taught. But I have elevated this structure to the light of a general law, and it is that unto which I may fairly lay claim. In the same essay I have introduced into the Physiology the Corpora Wolfiana, or Primordial Kidneys, but, having failed to recognise their signification, any one who pleases may filch away the credit of their bare detection.
In my Essay: Ueber die Bedeutung der Schädelknochen, (Ein Programm beym Antritt der Professur an der Gesammt-Universität zu Jena; Jena gedruckt bey Göpfert, 1807, verlegt zu Frankfurt bey Wesche, 4to,) I have shown that the head is none other than a vertebral column, and that it consists of four vertebræ, which I have respectively named Auditory, Maxillary or Lingual, Ocular and Nasal vertebra; I have also pointed out that the maxillæ are nothing else but

repetitions of arms and feet, the teeth being their nails; all this is carried out more circumstantially and in detail in the Isis, 1817, S. 1204; 1818, S. 510., 1823. litt. Anzeigen S. 353 und 441. This doctrine was at first scoffed at and repulsed; finally, when it began to force its way, several barefaced persons came forward, who would have made out if they could, that the discovery was achieved long ago. The reader will not omit to notice that the above essay appeared as my Antritts-Programm, or Inaugural discourse, upon being appointed Professor at Jena.

In my Essay entitled Ueber das Universum als Fortsetzung des Sinnensystems; Jena bey Frommann, 1808, 4to, I showed that the Organism is none other than a combination of all the Universe's activities within a single individual body. This doctrine has led me to the conviction that World and Organism are one in kind, and do not stand merely in harmony with each other. From hence was developed my Mineral, Vegetable and Animal system, as also my philosophical Anatomy and Physiology.

In my Essay entitled Erste Ideen zur Theorie des Lichts, der Finsterniss, der Farben und der Wärme; Jena bey Frommann, 1808, 4to, I pointed out, that the Light could be nothing but a polar tension of the æther, evoked by a central body in antagonism with the planets; and that the Heat were none other than the motion of this æther. This doctrine appears to be still in a state of fermentation.

In my Essay entitled Grundzeichnung des natürlichen Systems der Erze; Jena bey Frommann, 1809, 4to, I arranged the Ores for the first time, not according to the Metals, but agreeably to their combinations with Oxygen, Acids, and Sulphur, and thus into Oxyden, Halden, Glanzen, and Gediegenen. This has imparted to the recent science of Mineralogy its present aspect or form.

In the first edition of my Lehrbuch der Naturphilosophie, 1810 and 1811, I sought to bring these different doctrines into mutual connexion, and to show, forsooth, that the Mineral, Vegetable, and Animal classes are not to be arbitrarily arranged in accordance with single or isolated characters, but to be based upon the cardinal organs or anatomical systems, from which a firmly established number of classes must of necessity result; moreover, that each of these classes commences or takes its starting-point from below, and consequently that all of them pass parallel to each other. This parallelism is now pretty generally adopted, at least in England and France, though with sundry modifications, which, from the principles being overlooked or neglected, are based at random, and are not therefore to be approved of. As in chemistry, where the combinations follow a definite numerical law, so also in Anatomy the organs, in Physiology the functions, and in Natural History the classes, families and even genera of Minerals, Plants and Animals, present a similar arithmetical ratio. The genera are indeed, on account of their great number and arbitrary erection to the rank whose title they bear, not to be circumscribed or limited in every case with due propriety, nor brought into their true scientific place in the system; it is nevertheless possible to render their parallelism with each other clear, and to prove that they by no means form a single ascending series. If once the genera of Minerals, Plants and Animals come to stand correctly opposite each other, a great advantage will accrue therefrom to the science of Materia Medica; for corresponding genera will act specifically upon each other. These principles, which I have now carried out into detail, were retained in the second, and have been also in the third or present edition of the Physio-philosophy, the arrangement and serial disposition of the natural objects having, with my increase of knowledge and concomitant views of things, been amended, enlarged or diminished, as the case might require, especially in the

Mineral, Vegetable and Animal systems. I am very well aware that there is many an object which does not stand in its right place; but where again is there a single system in which this is not still more strikingly the case? We have here dealt only with the restoration of the edifice, wherein, after years of long and oft-repeated attempts, the furniture may for the first time be properly distributed, without detriment to its general bearings or ground plan.

In my Lehrbuch der Naturgeschichte, the Mineralogical and Zoological portions of which are out of print, but the Botanical still to be had (Weimar, Industrie-Comptoir, 1826), I have arranged for the first time the genera and species in accordance with the above principles, and stated everything of vital importance respecting these matters. This was the first attempt to frame a scientific Natural History, and one unto which I have remained true in my last work, the Allgemeine Naturgeschichte, the principles whereof I have sought to develop more distinctly and in detail in the work now before the reader.

Thus then have I prosecuted throughout a long series of years one kind of principle, and worked hard to perfectionate it upon all sides. Yet, notwithstanding my endeavour to amass the manifold stores of knowledge so requisite to an undertaking like this, I could not acquire within the vast circuit that appertains thereunto, many things which might be necessary unto a system extending into all matters of detail. This it is to be hoped the reader will acknowledge, and have forbearance for the errors, against which every one will stumble who has busied himself throughout life with a single branch of the natural sciences. Natural History is not a closed department of human knowledge, but presupposes numerous other sciences, such as Anatomy, Physiology, Chemistry and Physics, with even Medicine, Geography and History; so that one must be content with knowing only the main facts of the same, and relinquishing the Singular to its special science. The gaps and errors in Natural History can therefore be filled up or removed only by numerous writers and in the lapse of time.

PHYSIO-PHILOSOPHY.

INTRODUCTION.CONCEPTION OF THE SCIENCE.

1. Philosophy, as the science which embraces the principles of the universe or world, is only a logical, which may perhaps conduct us to the real, conception.
2. The universe or world is the reality of mathematical ideas, or, in simpler language, of mathematics.
3. Philosophy is the recognition of mathematical ideas as constituting the world, or the repetition of the origin of the world in consciousness.
4, 5. Spirit is the motion of mathematical ideas. Nature, their manifestation.
6. The philosophy of Spirit is the representation of the movements of ideas in consciousness.
7. The philosophy of Nature that of the phenomena or manifestations of ideas in consciousness.
8. The world consists of two parts: of one apparent, real, or material; and one non-apparent, ideal, spiritual, in which the material is not present, or which is naught in relation to the material.
9. There are, accordingly, two parts or divisions of Philosophy, viz. Pneumato-and Physio-philosophy.
10. Physio-philosophy has to show how, and in accordance indeed with what laws, the Material

took its origin; and, therefore, how something derived its existence from nothing. It has to portray the first periods of the world's development from nothing; how the elements and heavenly bodies originated; in what method by self-evolution into higher and manifold forms, they separated into minerals, became finally organic, and in Man attained self-consciousness.

11. Physio-philosophy is, therefore, the generative history of the world, or, in general terms, the History of Creation, a name under which it was taught by the most ancient philosophers, viz. as Cosmogony. From its embracing the universe, it is plainly the Genesis of Moses.

12. Man is the summit, the crown of nature's development, and must comprehend everything that has preceded him, even as the fruit includes within itself all the earlier developed parts of the plant. In a word, Man must represent the whole world in miniature.

13. Now since in Man are manifested self-consciousness or spirit, Physio-philosophy has to show that the laws of spirit are not different from the laws of nature; but that both are transcripts or likenesses of each other.

14. Physio-and Pneumato-philosophy range, therefore, parallel to each other.

15. Physio-philosophy, however, holds the first rank, Pneumato-philosophy the second: the former, therefore, is the ground and foundation of the latter, for nature is antecedent to the human spirit.

16. Without Physio-philosophy, therefore, there is no Pneumato-philosophy, any more than a flower is present without a stem, or an edifice without foundation.

17. The whole of philosophy depends, consequently, upon the demonstration of the parallelism that exists between the activities of Nature and of Spirit.

DIVISION OF THE SCIENCE.

18. It will be shown in the sequel that the Spiritual is antecedent to nature. Physio-philosophy must, therefore, commence from the spirit.

19. It will also be shown in the sequel that the whole Animal Kingdom, e. g. is, none other than the representation of the several activities or organs of Man; naught else than Man disintegrated. In like manner nature is none other than the representation of the individual activities of the spirit. As, therefore, Zoology can be termed the Science of the Conversion of Man into the Animal Kingdom, so may Physio-philosophy be called the Science of the Conversion of Spirit into Nature.

20. Physio-philosophy is divisible, therefore, into three parts. The first of these treats of spirit and its activities; the second, of the individual phenomena, or things of the world; the third, of the continuous operation of spirit in the individual things.

The first division is the doctrine of the Whole (de Toto)—Mathesis.

The second, that of Singulars (de Entibus)—Ontology.

The third, that of the Whole in the Singulars (de Toto in Entibus)—Biology.

21. The Science of the Whole must divide into two doctrines; into that of immaterial totalities—Pneumatogeny; and into that of material totalities—Hylogeny.

Ontology teaches us the phenomenon of matter. The first phenomenon of this are the heavenly bodies comprehended by Cosmogony; these develop themselves further, and divide into the elements—Stochiogeny.

From these elements the Earth element develops itself still further, and divides into minerals—Mineralogy; these minerals unite into one collective body, and this is Geogeny.

The Whole in Singulars is the living or Organic, which again divides into plants and animals. Biology, therefore, divides into Organogeny, Phytosophy and Zoosophy.

After this division of the subject the question first of all arises, what is science, provided there is

one.
TRUTH.
22. Science is a series of necessarily inter-dependent and consecutive propositions, which rest upon a certain fundamental proposition.
23. Now, if anything be certain it can only be one in number. If, then, there be only one certainty, there can also be only one science, from which all the rest must be derived.

24. The Mathematical is certain, and, by virtue of this character, it stands also alone. Mathematics is the only true science, and thus the primary science, the Mathesis, or Knowledge simply, as it was called by the ancients. The fundamental propositions of mathematics must, therefore, be fundamental propositions for all other sciences also.
25. Physio-philosophy is only a science when it is reducible to, i. e. can be placed upon an equal footing with, mathematics. Mathematics is the universal science; so also is Physio-philosophy, although it is only a part, or rather but a condition of the universe; both are one, or mutually congruent.
26. Mathematics is, however, a science of mere forms without substance. Physio-philosophy is, therefore, Mathematics endowed with substance.
27. The substance of Physio-philosophy must be of one kind with the form of Mathematics.
28. The certainty of mathematical propositions depends upon no proposition being essentially different from another. Though there may be much that is diversified or heterogeneous, there is nothing new in Mathematics.
For to prove a mathematical proposition is to show (or demonstrate) that it is equivalent, i. e. of the same kind with another proposition. All mathematical propositions must, consequently, resemble a first proposition.
29. Physio-philosophy must also show that all its propositions, or that all things, resemble each other, and, finally, some first proposition or thing.
30. These natural propositions or natural things must, however, resemble also mathematical propositions, and depend, after all, upon the primary proposition of mathematics or the axiom. Now then comes the question, what is the first principle of Mathematics?

PART I. MATHESIS—OF THE WHOLE.

NOTHING.
31. The highest mathematical idea, or the fundamental principle of all mathematics is the zero = 0.
The whole science of mathematics depends upon zero. Zero alone determines the value in mathematics.
32. Zero is in itself nothing. Mathematics is based upon nothing, and, consequently, arises out of nothing.
33. Out of nothing, therefore, it is possible for something to arise, for mathematics, consisting of propositions, is a something in relation to 0. Mathematics itself were nothing if it had none other than its highest principle zero. In order, therefore, that mathematics may become a real science, it must, in addition to its highest principle, subdivide into a number of details, namely, first of all into numbers, and, finally, into propositions. What is tenable in regard to mathematics must be

equally so of all the sciences; they must all resemble mathematics.

34. The first act towards realization or the becoming something, is an origination of Many. All reality can, accordingly, manifest itself only in multiplicity.

That which belongs to the Many is a Definite; this again is a Limited; the Limited is a Finite. The Finite only is real.

The question now arises, how it happens that mathematics becomes a multiplicity, or, what is the same thing, a reality, a something.

35. The reality of mathematics consists in the universality of its quantities; viz. numbers or figures. Every number, and every thing which belongs to mathematics, can be derived from no other source than zero.

Mathematical multiplicity, or its reality, must have proceeded, therefore, out of zero.

36. Zero, however, contains no number and no figure really in itself; it contains, forsooth, neither 1 nor 2, neither a point nor a line within itself. The Singulars or details cannot, therefore, reside in a real, but only an ideal manner in zero; or, in other words, not actually, but only potentially. The conditions here are the same as with all mathematical ideas. We may conceive, e. g., an idea or definition of a triangle in so general a sense that it shall comprehend all triangles, without, however, a definite triangle being actually intended, or without even a triangle actually existing. In order that the idea of the triangle be realized, it must become a definite, in other words, an obtuse or an acute triangle. In short, the idea of the triangle must multiply itself, be self-evolved, or else it is as naught in reference to mathematics, or only a geometrical zero.

The individual objects or figures of mathematics thus attain existence, so far only as the idea comprising them emerges out of itself and assumes an individual character.

It is clear that all individual triangles taken together closely resemble the ideal triangle, or, to express the same in more general terms, that the Real is equivalent to the Ideal, that the former is but the latter which has become dissevered and finite, and that the aggregate of every Finite is equivalent to the Ideal. This will probably be rendered still more distinct by the example of ice and water. The crystals of ice are nothing else than water bounded by definite lines. So, also, are the Real and Ideal no more different from each other than ice and water; both of these, as is well known, are essentially one and the same, and yet are different, the diversity consisting only in the form. It will be shown in the sequel that everything which appears to be essentially different from another, is so only in the form.

The Real and Ideal are one and the same, only under two kinds of form. The latter is the same under an indefinite, eternal, single form; but the Real is also the same, yet under the form of quantity, and, as will be shown, of multiplicity. An infinity resides in both; in the Real an endlessness of individual forms; in the Ideal but one endless form; in the latter case an eternity, in the former an infinity. The quantity and multiplicity of the whole of mathematics is contained in the same manner in the 0, that the quantity and multiplicity of the triangles are in the ideal or primary triangle. Mathematics is a system of nullities or nothings, and this admits of being easily proved.

37. Zero is indeed the universality of mathematics, this, however, is not real, but only ideal. Every number issues out of zero, like the multiplicity of the real triangles out of the primary triangle. This progression of numbers out of zero takes place through a process of becoming determinate and limited; just as the real triangles are only definitions of the absolute triangle. The process of becoming determined is identical with becoming a Finite; becoming real is called becoming finite. Mathematical singulars or numbers can, therefore, be nothing else than zero

disintegrated, or rendered real by determination.
What zero is in infinite intensity, that are numbers in endless extensity. Zero is of two forms: under the ideal it is mere intensity; under the real mere extensity, or a series of numbers. The latter is only expanded intensity; the former, extensity concentrated on the point; both are, consequently, one and the same in toto. Numbers are identical with zero; they are zero in a state of extension, while zero is equivalent to numbers in a state of intensity. The sense in which numbers are said to come out of zero is, therefore, very clear; they have not issued forth from zero as if they had previously resided individually therein, but the zero has emerged out of itself, has itself become apparent, and then was it a finite zero, a number. So, also, does the idea of a circle become a real circle, not from the latter emerging from the former, but from this itself becoming manifest. The individual circle is a manifestation or phenomena of the spiritual circle.

38. All realization, therefore, is not the origin of a something that has not previously been; it is only a manifestation, a process of extension taking place in the idea.
Thus the Real does not arise out of the Ideal, but is the Ideal itself in a condition of definition and limitation, as are, e. g. the actual triangle or the actual circle. If, then, the Ideal and Real be one, everything is necessarily identical, and this identity dominates not merely between the Ideal and Real in a general sense, but between all individual members of the Real.
39. The identity of every Different, or of all things among themselves and with the highest unity, is the essence of things. The limitation or definition of the Ideal is their form. Limitation is the Impartient of form.
40. Limitation is originally only a quantitative relation, e. g. the size of the angle in a triangle; later on it becomes also a relation of direction or of position.
In both cases the limitation is only an ideal relation. Realization also takes place, therefore, only in an ideal manner; and the Real is therefore ideal, not simply as it regards its form, but also its essence. Every Plural resembles itself and the highest principle in essence; or, in other words, all Singulars are united through essence with the highest One. All diversity of the Plural resides merely in the form, limitation or manifestation. The one unchanging essence possesses one ideal form, which is that of pure unity, and the same essence has a limitation, a real form, which is that of subdivision. There is only one essence in all things, the 0, the highest identity; but there are infinitely numerous forms.
Numbers are naught else than different forms of the one unchangeable essence, namely, the 0.
If, then, all numbers are only zero in a state of extension, and are consequently identical with it, the question arises, what are the first finitings of zero, or as what does it appear when it is no longer merely ideal or indefinite; in short, what is the first form of the real zero, or of the essence in general?

ESSENCE OF NOTHING.
41. The ideal zero is absolute unity, or monas; it is not a singularity, such as one individual thing, or as the number 1; but an indivisibility, a numberlessness, in which neither 1 nor 2, neither a line nor a circle can be found; in short, an unity without distinction, an homogeneity, brightness, or translucency, a pure identity.
42. The mathematical monas is eternal. It succumbs to no definitions of time and space, is neither finite nor infinite, neither great nor small, neither quiescent nor moved; but it is and it is not all this. That is the conception of eternity.
Mathematics is thus in possession of an eternal principle.

43. Since all the sciences are equivalent to mathematics, nature must also possess an eternal principle.

The principle of nature, or of the universe, must be of one and the same kind with the principle of mathematics. For there cannot be two kinds of monades, nor of eternities, nor of certainties. The highest unity of the universe is thus the Eternal. The Eternal is one and the same with the zero of mathematics. The Eternal and zero are only denominations differing in accordance with their respective sciences, but which are essentially one.

44. The Eternal is the nothing of Nature.

As the whole of mathematics emerges out of zero, so must everything which is a Singular have emerged from the Eternal or nothing of Nature.

The origin of the Singular is nothing else than a manifestation of the Eternal. Thereby unity, brightness, homogeneity are lost, and converted into multiplicity, obscurity, diversity.

Unity posited manifoldly is an expansion without termination, but one that always remains the same.

Realization or manifestation is an expansion of the Eternal.

FORMS OF NOTHING.

45. The first form of the expansion or manifestation of the mathematical monas, or of 0 is + -.

The + - is nothing else than the definition of 0. 0 is the reduction of the positive and negative series of numbers, upon which the whole of arithmetic depends. A series of numbers is, however, nothing else than a repetition of a + 1 or a-1; consequently, the whole of arithmetic reduces itself to + 1-1.

What, however, is a + 1, or-1? Obviously nothing else than a single + or-. The figure is quite superfluous, and only indicates how often + or-has been assumed; instead, therefore, of + 1 we can posit +; instead of-1 simply-. The series + 1 + 1 + 1 is synonymous with + + +; or instead of 3 we may posit + + +, and so on for every figure ad libitum. The figures are nothing more than shorter signs for the two highest mathematical forms or ideas of numbers. Numbers are nothing different from the ideas of numbers; they are the latter themselves, only several times posited. Essentially numbers do not exist, but only their two ideas. These ideas, however, exist an infinite number of times.

Multiplicity or real infinity is, accordingly, nothing special or particular, but only an arbitrary repetition of the Ideal, an incessant positing of the idea. The idea posited is reality, non-posited it is = 0.

46. The first multiplicity is duality, + -. This duality alters nothing in the essence of the monas, for + - is = 0. It is the monas itself only under another form. In multiplication it is thus the form alone that changes.

There are many forms, but not many essences.

47. The first or primary duality is not, however, a double unity, both members of which are of equal rank, but an antagonism, disunion, or diversity. Many diversities are multiplicity. The Many is thus complex. The first form is not therefore a simple division of zero or the primary unity, but an antagonistic positing of itself, a becoming manifold.

48. Every Finite is in the same manner only the self-definition of the Eternal. The Eternal becomes, accordingly, real, by binary self-division. When the Eternal is manifested, it is either a positive or negative. The whole of arithmetic is nothing else than a ceaseless act of positing and negating, of affirming and denying.

All realization is nothing else than the act of positing and negating. The act of positing and negating of the Eternal is called realization.

49. Positing and negating is, however, an act or function. Arithmetic is, therefore, a ceaseless process of acting or performing. Numbers are acts of the primary idea, or, properly speaking, stationary points of its function, and hence proceeds a division into the two ideas + and -. If these remain always alone nothing is added to them. They alone produce the whole science of arithmetic, and simply because they are never exhausted by the act of positing themselves repeatedly, but capable after this of again becoming suppressed. Since + is in essence nothing else than a simple positing, a mere affirmation, and -a mere suppression of this affirmation, a negation; so is the positive unity = 1 nothing but an affirmation once declared, and the whole series of numbers is a reiterated affirmation. The act of affirmation alone gives the number, and the latter is thus the definite quantity devoid of intrinsic value. Bare affirmation alone without reference to any substance is unity, duality, &c.

SOMETHING.

50. Still, however, there must be something, which is posited and negatived. The form must have a substance.

This something is the primary idea, or the very Eternal of mathematics, the zero; for + - is = 0. The + is naught else than zero affirmed; the-naught else than this + 0 negatived = - 0. Now since an affirmation once declared is = 1, so are unity and zero identical. Zero differs only from finite unity in that it is not affirmed.

51. The - is not simply the want of affirmation, but its explicit abstraction. The + presupposes the 0; the - the + and 0; the 0, however, presupposes neither + nor -. Purely negative quantities are, as is known, a nonentity, because they can only bear reference to positive quantities. The - is, indeed, the retroversion of + into 0; yet alone, therefore, it is not perfectly equal to 0. It is a retrovertent, and consequently the second act, which presupposes the positive. By the - we know what is not; the 0 is, however, a nothing in every respect. The-is the copula between 0 and +.

52. If the + is the 0 posited, so is it a nothing posited or determined. This position is, however, a number, and therefore a mathematical something. The nothing thus becomes a something, a Finite, a Real, through the simple positing of itself, and the something becomes a nothing by the removal of this self-position. The nothing itself is, however, the mere neglect of its self-position. The something, the + -, has consequently not arisen or been evolved out of nothing, or been produced from it by addition; but it is nothing itself; the whole undivided nothing has become unity. The nothing once posited as nothing is = 1. We cannot speak of production or evolution in this case; but of the complete identity and uniformity of the nothing with the something; it is a virgin product or birth.

53. Zero must be endlessly positing itself, for in every respect it is indefinite or unlimited, eternal. The number of finite singularities must, therefore, pass into the Infinite.

54. The whole of Arithmetic is nothing but the endless repetition of nothing, an endless positing and suppressing of nothing.

We can become acquainted with nothing but the nothing, for the Original of our knowledge is the 0.

There is no other science than that which treats of nothing.

Every Real, if it were such in itself, could not be known, because the possibilities of its properties would pass into the Infinite. The nothing alone is cognizable, because it has only a single property, namely, that of having none; concerning which knowledge no doubt can be entertained.

A.—PNEUMATOGENY.

PRIMARY ACT.
55. The + - or, in other words, numbers are acts or functions. Zero is, consequently, the primary act. Zero is, therefore, no absolute nothing, but an act without substratum. Generally speaking there is, therefore, no nothing; the mathematical nothing is itself an act, consequently a something. The nothing is only postulate.
56. An act devoid of substratum is a spiritual act. Numbers are, accordingly, not positions and negations of an absolute nothing, but of a spiritual act.
57. The zero is an eternal act; numbers are repetitions of this eternal act, or its halting points, like the steps in progression. With zero the Eternal therefore originates directly, or both are only different expressions for one and the same act, according with the difference of the science wherein they are employed. Mathematics designates its primary act by the name of zero; Philosophy by that of Eternal. It is an error to believe that numbers were absolute nothings; they are acts and consequently realities. While numbers in a mathematical sense are positions and negations of Nothing, in the philosophical they are positions and negations of the Eternal. Everything which is real, posited, finite, has become this out of numbers; or, more strictly speaking, every Real is absolutely nothing else than a number. This must be the sense entertained of numbers in the Pythagorean doctrine, namely, that everything or the whole universe had arisen from numbers. This is not to be taken in merely a quantitative sense, as it has hitherto been erroneously, but in an intrinsic sense, as implying that all things are numbers themselves, or the acts of the Eternal. The essence in numbers is naught else than the Eternal. The Eternal only is or exists, and nothing else is when a number exists. There is, therefore, nothing real but the Eternal itself; for every Real, or everything that is, is only a number and only exists by virtue of a number. Every Singular is nothing for itself, but the Eternal is in it, or rather it is itself only the Eternal, though not the Eternal in itself, but affirmed or negatived. The existence of the Singular is not its own existence, but only that of the Eternal subjected to an arbitrary repetition; for the act of being and affirming are of one kind.
58. The continuance of Being is a continuous positing of the Eternal, or of nothing, a ceaseless process of becoming real in that which is not. There exists nothing but nothing, nothing but the Eternal, and all individual existence is only a fallacious existence. All individual things are monades, nothings, which have, however, become determined.
The Eternal must posit without cessation, for otherwise it would be an actual nothing, while in fact it is an act; but it must incessantly suppress also this position, else it would be only a finite act, or an act which had only one kind of direction, that of affirmation + + + +, and so on, which represents only the half of arithmetic. The totality of the Finite is, therefore, of eternal duration also: the Singular, however, issues forth and disappears like the numbers in arithmetic. The eternal duration of the Finite consists, however, only in ceaseless repetition. Such an Eternal is to be distinguished therefore from the Primary eternal, and is called the Infinite. The totality of finite things is not therefore eternal, but only infinite.
PRIMARY CONSCIOUSNESS.
59. Two tendencies are present in the primary act, both of which being inseparable are one in kind. It has the tendency to posit, and also to suppress, itself. The unity strives unto binary division or to antagonism, even as the 0 strives to produce + or -. While the primary act itself posits, it does this indeed out of its own strength, and that which it posits is also none other than itself; it itself posits i. e. actively; and is itself posited i. e. passively; it itself posits itself, is the self-position of itself; for + is nothing else than 0 self-posited. The positing and posited act are of one kind; the latter, however, is the Real, the Finite; the former the Ideal, the Eternal. Both are

distinguished from each other through this only, that the Real is the posited, numbered, and consequently determined act; the ideal, however, the positing, consequently numbering and thus undetermined act. While, however, the + is nothing else than 0, it must necessarily bear a relation to it, and thus retrograde into the 0. This retrogression is an act in the reverse direction, or what is indicated in mathematics by negation. The - has been therefore necessarily granted with the +, else the + would not be represented as = to 0. The act of positing is therefore at the same time also an act of negation. So soon as the 0 is or exists, it is = + -. The realization of the Eternal is accordingly a complete antagonism of itself. For 0 is equal to + -, not simply = + or = to-.

60. The being of the Eternal is therefore a self-manifestation. Every Singular is nothing but a self-manifestation; since all numbers are only positions of zero or of +, which can never be without-. In every essence there are two, but the two are the one essence itself, which posits itself by division. The Positing of the Eternal in the sense in which it has been hitherto adopted, namely, as a realization of the same, is not merely an act of positing, not an indeterminate Positing, but an antagonism of itself. The zero is simply the indeterminate Positing, or the negative Positing; but the number, or the real is the antagonism of zero, the + -, or the self-manifestation. The 0 cannot be thought of for itself alone without the +; the latter, however, not without 0, as well as the-also not without 0; for it is the suppression of the posited 0, namely, the +. Every act of self-manifestation is therefore twofold, a manifestation (= +), but a manifestation of itself, consequently a retrogression into 0 (= -). Through negation the Finite becomes united with the Eternal. Every disappearance of the Finite is a retrogression into the Eternal; for it must return to whence it came. It has arisen out of nothing, is itself the existing nothing; it must therefore retrograde again into the nothing.

GOD.

61. The self-manifestation of the primary act is self-consciousness. The eternal self-consciousness is God.

62. The continued act of self-consciousness, or becoming self-conscious repeated, is called representation. God is therefore comprehended in ceaseless representation. Representations are single acts of self-consciousness. Single acts, however, are real things. All real things, however, are the world. The world therefore originates with the representations of the Eternal.

63. The representations are, however, manifested or attain only reality through expression. The world is therefore the language of God; the creation of the world the speaking of God. "God spake, and it was." It is not merely said, God thought and it was. Thought belongs merely to spirit; in so far, however, as it becomes apparent, it is a word, and the sum of all apparent thoughts is speech. This is the created, realized system of thought. The thought is only the idea of the world, but speech is the idea actualized.

64. As thought differs from speaking, so does God from the world. Our world consists in our apparent thoughts, namely, the words. The universe is the language of God. So far as the thoughts lie at the foundation of the words, it can be said, that our world were the play of our thoughts, and the actual world that of God's. The word has become world. Worldly things have no more reality for God, than our words or our language for us. We carry a world within us while we think; we posit or create a world without us while we speak. Thus God carries the world within himself while he thinks; he posits the same without himself or creates it, while he speaks. In so far as thought necessarily precedes speech, it may be said, that there would have been no world, if God had not thought. In the same sense it may be also said, that all things are nothing but representations, thoughts, ideas of God. So soon as God thinks and speaks is there a real thing. To speak and to create are one. All, that we perceive, are words, thoughts of God; we are

ourselves nothing else than such words or thoughts of God, consequently his metatypes or images, in as far as we unite in ourselves the whole system of speech. There is therefore no being without self-consciousness. That only which thinks is (for itself); that which does not think is not for itself, but only for some other consciousness. The world differs from God as doth our speech from us. The self-consciousness of God is independent of the world, even as our self-consciousness is independent of our speech.

65. The divine laws are also the laws of the world; this has therefore been created and governed in accordance with eternal and immutable laws.

66. Physio-philosophy is the history of creation; the creation, however, is the language of God. The system of thought, however, lies necessarily at the foundation of the system of speech. Now the science of the laws of thought is called logic; physio-philosophy is therefore a divine doctrine of speech or a divine logic. The laws of speech instruct us in the genesis of language. Physio-philosophy is, therefore, the science of the genesis of the world, or Cosmogony.

FORM OF GOD—TRIUNITY.

67. As the complete principle of mathematics consists of three ideas, so also does the primary principle of nature, or the Eternal. The primary principle of mathematics is 0; so soon, however, as it is actual, is it + and-; or the primary idea resolves itself in being at once into two ideas, each of which resembles the other in essence, but differs from it in form. Thus it is here one and the same essence under three forms, or three are one. Now that which holds good of mathematical principles, must hold good also of the principles of nature. The primary act is manifested, or operates under three forms, which correspond to the 0, + and-.

These three ideas of the Eternal are all equivalent to each other, are the same primary act, each of them being whole and undivided, but each otherwise posited. The positing primary act is the whole Eternal; the posited is likewise the whole Eternal, and that which is subtractive, retrogressive, combining the two first, is also the whole Eternal. Although all three ideas are equivalent to each other, still the positing idea ranks first, the posited second, and the combining third; not as if they had first arisen successively (this is impossible, for they are coexistent, namely, before all time), nor as if they occupied different positions (for they are everywhere); but only according to their order and value. How one may be three and three one, is thus rendered comprehensible only by mathematics.

68. The first idea is the original, that therefore which is thoroughly independent, which having arisen from and being based upon itself, has consequently emerged from nothing else; in short, it is the eternal idea, like the mathematical $0 =$ Monas aoristos. Everything is possible with it; it can propose and solve all problems, knows therefore everything and creates everything. It is the generative, creative and paternal idea.

69. The two other ideas have emerged out of the first, although apparently equivalent to it; yea, they have themselves issued out of themselves. The second idea is, therefore, Dyas aoristos, and corresponds to the mathematical +; the third idea is the Trias aoristos, and corresponds to the mathematical-, so that by the three the primary trinity $0 + -$ is completed. The first idea labours or, what is more, rejoices from all eternity to convert itself into the two others. The action or the life of God consists in eternally manifesting itself, eternally contemplating itself in unity and duality, eternally dividing itself and still remaining one. The second idea has issued next from the first, and is therefore related to it as Son is to Father, when the ideas are viewed as personified. The third idea has emerged conjointly from the second and first, and forms therefore the spiritual union, the mutual love between both. It may be therefore simply called Ghost or Spirit, if it is thought of as personified.

70. Since every Singular, having been produced through the primary trinity, is only the expressed word of the primary trinity, so also must their qualities be recognizable in the same. The Singular is not simply therefore a position of one idea, but of all three. All things have issued out of the trinity. The essence of the universe consists in the trinity which is unity, and in the unity which is trinity; for it is a likeness of the primary trinity. Being, generally, is an act, and that, indeed, of a threefold nature. Apart from act or function there is no being. That, which is called nothing, is in itself an act, and there is, therefore, no absolute nothing. The nothing is only something relative to a particular being. Even the mathematical zero is not nothing, but an act. It is nothing only in reference to particular numbers. Numbering is a repetition of one and the same act. The forms or conditions of the primary act are Rest, Motion, and Extension or expansion.

a. PRIMARY REST. (First form of the Primary Act.)

71. The primary idea is the position simply without any relation, or any antagonism; it is the oscillating resting point in the universe, around which everything collects itself, and from which everything emerges; the Centrum ubique, circumferentia nusquam. The primary idea is the substratum of everything, which will come before us in the sequel of this work. Everything depends upon this primary essence; all action, motion, and form issues forth from it; or rather, in all phenomena naught else appears than the primary essence in different stages of position, just as in all numbers naught else appears but the zero. The primary idea is the absolute beginning. This primary idea is the non-representable, the never apparent and yet omnipresent, idea; but which is always withdrawing itself from our view when we imagine or believe that we gaze upon it; in short, the Spiritual, which declares itself in everything and yet always remains the same. The origin of all action may be termed the primary force.

b. MOTION, TIME. (Second form of the Primary Act.)

72. The primary idea operates only, while it posits; through positing, however, arises a succession of positing, or numbers; positing and successive positing are one. The function of the primary idea consists in an eternal repetition of the essence; the primary act is a continuous self-repeating act. Repetition of the primary act devoid of another substratum is Time. Time is none other than the eternal repetition of the positing of the Eternal, corresponding to the series of numbers $+ 1 + 1 + 1 + n$. Time has not been created, but has emerged directly out of the primary act and its position; it is the function of God himself. Something has thus already originated, which appears to conduct us into the universe. Time is the first portal through which the operation of God passes over into the world. Time is the infinite succession of numbers or the mathematical nothings. The mathematising, numbering act is Time. Numbers, however, are Singulars or finitudes, which constitute the world.

73. Time is infinite, for it is the totality of positing; it is only the points or numbers in it that are the Finite.

74. All things are created in time; for time is the totality of Singulars. Time is no stationary quantity, which is always changing itself into something new during its progressive flux. It is not a continuous stream, but a repetition of one and the same act, namely, the primary act, like as it were to a rolling ball, which constantly returns upon itself. There is no endless, still less an eternal thing; for things are only positions of time. Time itself is, however, only repetition, and thus also a suppression of these positions. The vicissitude of things is in fact time; if there be no change, there is also no time. Time is an universal property of things. Exemption from time is only in the Eternal.

75. Time, not being itself the Finite, but creating it, is not itself a Real, but still an Ideal, a form only of the primary act, an idea, with which finite things have been directly posited. Time is the

act of numbering; numbering is thinking; thinking is time. Our thinking is our time. In sleep there is no time for us. God's thought is God's time; God's time, however, is all time, consequently time of the world. Time is not of earthly but heavenly descent or origin. In so far a divine quality belongs to all finite things. They are divine, in so far as they are time; terrestrial, in so far as they are evanescent moments of time.

POLARITY.

76. Time is an action of the primary power; and all things are active only in so far as they are filled or inspired with the idea of time. The whole activity of things, all their forces arise out of the primary act or primary power, are only moments of the same. There are, however, no positive without negative numbers, consequently also no moments of time without suppression of the same. There is, therefore, no single force, but each is the position of + and -. A force consisting of two principles is called Polarity. Time is, therefore, the primary polarity, and polarity is manifested at the very instant in which the creation of the world is stirring.

77. Polarity is the first force which appears in the world. If time is eternal, polarity must also be eternal. There is no world, and in general nothing at all without polar force.

78. Every single thing is a duplicity.

79. The law of causality is a law of polarity. Causality is valid only in time, is only a series of numbers. Time itself has no causality. Causality is an act of generation. The sex is rooted in the first movement of the world.

MOTION.

80. Polarity may be viewed as a single positing of + -; if, however, this positing repeats itself, Motion originates, viz. when many + - + - are consecutively posited, and thus the principal poles separate from each other, as in an iron bar when magnetizing. Time is a polar positing of the primary act, and an endless repetition of this positing; through this originate individual things, whose succession is motion.

81. Primary motion is the result of primary polarity. All motion has originated from duplicity; consequently from the idea in a dynamic not a mechanical manner. A mechanical motion, which might be produced ad infinitum by mechanical impulses, is an absurdity. There is nowhere a purely mechanical motion; nothing, as it is at present in the word, has become so by impulse; an internal act, a polar tension lies at the bottom of all motion.

82. Motion itself, however, is not twofold of character; it is unity, but the result of duality. In time we have to distinguish the polar act of position, and the act of repeating this position, which is motion. Motion is the simple repetition of the polar, twofold act, or the ceaseless separation of poles; but, as in every polar line the two poles are in all cases together, so even is this mutual separation of poles only a repetition of polarity.

83. Motion also is not created, but has emerged directly from the Eternal, is the primary function itself repeated. Motion is the ever self-manifesting, consequently progressive God.

84. Motion is thought, which is manifested as speech. Thought polarizes the fingers. If the thought be powerful it moves them, and through them other bodies. Speech is only a thought that has passed over into motion. The world is the thought of God that has been translated into motion, the moved thought of God—thought spoken. It is here evident that the world is not simply the thought but the language of God; for there is no action without motion; consequently no thought without speech, and vice versâ.

85. There is no thing which were without motion, just as there is none without time. A Finite without everlasting motion is a contradiction. All rest in the world is only relative, is but a

combined motion. There is only rest in the Eternal, in the nothing of nature.
86. The primary motion is only possible in a circle, because it fills every thing.
87. The motion of finite things by polarity may, in a wider sense, be called life; for life is motion in the circle. Polarity, however, is a constant retrogression into itself. Without life there is no being. Nothing is, simply by virtue of being, e. g. by its mere presence; but everything of which a being can be declared, is only, or manifests itself, by its polar motion or by life. Being and life are inseparable ideas. While God acts, he creates life.
88. Life is nothing new, that came first into the world, after it was created, but an Original, an idea, a moved thought of God, the primary act itself with all its consequences.
89. There is in the universe no vital force of its own; the individual things lie not there some time and await the polarizing breath, but they first become through the breath of God. The Causa existentiæ is life.
90. There is nothing properly dead in the world; that only is dead which is not, only the nothing. Something can only cease to live, when its motion ceases; this, however, ceases only when deprived of its polarity; the polarity dissolved, however, is zero. Thus individual things retreat into the Absolute, if they cease to live. Everything in the world is endowed with life; the world itself is alive, and continues only, maintains itself, by virtue of its life; just as an organic body maintains itself, only while it is constantly being generated anew by the vital process.
91. Every living thing is twofold in character. It is one persistent in itself, and one immersed in the universe. In everything, therefore, are two processes, one individualizing, vitalizing, and one universalizing, destructive. By the process of destruction, the finite thing seeks to become the universe itself; by the vitalizing process, however, the variety of the universe, and yet with that to remain a Singular. That only is truly living which represents the Eternal, and the whole multiplicity of the universe in the Singular.
92. The whole in the singular is called Individual. The individual is an example of computation, which admits only of being developed, from its comprehending the whole of arithmetic in itself. Nothing individual can persist eternally; it must eternally move itself, consequently fill up everything, displace everything, must become itself the universe.

MAN.

93. Time consists of single acts; i. e. the life or the absolute act does not work with one stroke, but an infinite number of times. All acts, therefore, taken together, all finite things in time, are equal to the primary act or the Eternal.
94. There are two totalities, a primary totality 0 + -, and a secondary, or the summing up of all numbers 0 + n-n; the former the eternal, the latter the finite totality, or the one the eternity, the other the infinity.
95. The more a thing has adopted into itself of the Manifold of the universe, by so much the more is it animated, by so much the more does it resemble the Eternal. It is conceivable, for a finite or living essence to unite all numbers or acts in itself, without, however, its being the very Eternal. It would, however, be obviously the most perfect finite essence, and, as a secondary totality, be the likeness of the primitive; the former the compound universality, the latter the identical.
96. Such an essence would be necessarily the highest and last, whereunto creation could attain; for more than the universe cannot be represented in one thing. With such an essence creation would be closed or would terminate.
97. Since the realization of the Eternal is a becoming self-conscious, so is the highest creature also a Self-conscious, but a Singular. Such a creature is the finite God, or God become corporeal.

God is Monas indeterminata, the highest creature is Monas determinata, Totum determinatum. A finite self-conscious being we call Man. Man is an idea of God, but that in which God wholly, and in every single act becomes an object unto himself. Man is God represented by God in the infinity of time. God is a Man representing God in one act of self-consciousness, without time.

98. Man is God wholly manifested. God has become Man, zero has become + -. Man is the whole of arithmetic, compacted, however, out of all numbers; he can therefore produce numbers out of himself. Man is a complex of all that surrounds him, namely, of element, mineral, plant and animal.

99. The other things below man are also ideas of God, but none of these ideas is the whole representation of arithmetic. They are only parts of the divine conscience posited in time; but man is God, planted or posited uninjured in time. Man is the object in the self-consciousness of God; the creatures below man are, however, the objects only of the consciousness of God. Thus, if God places before and from himself only single qualities, there are worldly things; if, however, God in this crowd of representations attains to his own entire representation, then arises Man. God is = + 0-, Man = + [oo] 0-[oo], the animal is = + n 0-n. The animals are only represented in part. The subject of self-consciousness is = + 0-; the objects, however, are the numbers which are equivalent to this, being = [oo] + 3 + 2 + 1 + 0-1-2-3-[oo]. Thus if all numbers, all world-elements, together with their perfections, occur in consciousness = + 0-, there is a Man; if only single, and perhaps but few things, such as food, stones, &c. (with the entire exception of the celestial bodies), enter consciousness, there is an animal. They are represented only partly or in a portion of the universe, but man is represented wholly or in all its parts. Animals are fragments of man.

100. No creatures below Man can possess self-consciousness. They have, indeed, consciousness of their several acts and of their sensations, and possess memory; but as these several acts are only parts of the world, or of the great consciousness, and are not the Whole, they can never become objective unto themselves, never imagine. Animals are men, who never imagine. They are imaginative, but never of themselves wholly; they are therefore beings who never attain to consciousness concerning themselves. They are single accounts; Man is the whole of mathematics.

FREEDOM.

101. An action, which is not determined by some other action, is free. God is free, because apart from him there is none other action.

102. Man, as being an image of God, is likewise free; as being an image of the world he is devoid of freedom. Man is, therefore, in his primary commencement or principle free, but not in his end or object to be attained. In the resolution Man is free, in the execution he is not free. The mathematician can select at pleasure any proposition; but having selected it, must solve it in accordance with necessary laws and with definite numbers and figures. Man is a twofold being, compounded of freedom and necessity.

RETROSPECT.

103. Hitherto we have considered simply the arithmetical relations of the primary act and of the universe. We have shown, to wit, that all ideas fluctuate simply under the forms of numbers; that everything was comprised in the 0 + -. Time was only the active series of numbers; motion was the actual arithmetical calculation, namely, the process of reducing numbers to absolute identity, to zero.

104. Life is moreover only a mathematical problem, which, the higher it ascends, approaches so much the nearer to absolute zero in its attainment of the infinity of numbers, becomes so much

the more endowed with life.

105. Arithmetic is the science of the second idea, or that of time and motion, or of life; it is, therefore, the first science; mathematics not only begins with it, but creation also, with the becoming of time and of life. Arithmetic is, accordingly, the truly absolute or divine science, and therefore everything in it is also directly certain, because everything in it resembles the Divine. Theology is arithmetic personified.

106. Hence it follows in the most perfect manner, that every science, if it would possess certainty, must resemble arithmetic. Now a science always implies a science treating of certain objects; all certain objects must, therefore, resemble the objects of arithmetic; or all objects, of whatever denomination, whether natural or spiritual, must correspond to arithmetical objects, consequently in idea be numbers, an actual arithmetical problem, as it were the numbers of motion, of life.

107. A natural thing is nothing but a self-moving number; an organic or living thing is a number moving itself out of itself, or spontaneously; an inorganic thing, however, is a number moved by another thing; now, as this other thing is also a real number, so then is every inorganic thing a number moved by another number, and thus ad infinitum. The movements in nature are only movements of numbers by numbers; even as arithmetical computation is none other than a movement of numbers by numbers, but with this difference, that in the latter this operates in an ideal manner, in the former after a real.

c. FORM, SPACE. (Third form, of the Primary Act.)

108. Viewed arithmetically every position is a number, geometrically, however, it is a point. What the 0 is in arithmetic, the point is in geometry; the one the arithmetical, the other the geometrical nothing. Both sciences commence with nothing and are only different views of nothing. The 0 is a temporal nothing (a number), the point a spatial nothing (a figure).

109. The first motion of numbers or of points is the motion of the primary number, the 0, or the primary act; and this motion depends upon the multiplicity of numbers or points, upon the disintegration of the identical primary number, upon the + -. The first motion of the primary act is an expansion of itself into multiplicity, whereby not merely sequence but an addition also is posited. The primary act is not simply positing, but also posited; as the former it is time, as the latter it is time posited universally. Time remaining stationary is Space. Space is not different in essence from time, but only according to position; it is only time resting, while this is moved, active space.

110. Space has first arisen out of time, as the third idea out of the second, but only ideally. It has arisen out of it, while, time being the act of positing, it is the posited; now as time posits from eternity, so is space also from eternity and in eternity. The eternity of space, however, depends not upon duration, but upon extension; it is unlimited.

111-112. Space is everywhere, as time is ever. Two spaces can no more exist than two times. There is only one Eternal. Time and space are, however, nothing special that has attained unto the Eternal, but the Eternal itself. They are also not two kinds of qualities subsisting near each other, but are one in kind. The series of numbers is infinite, thus universal; space is consequently universal.

113. Space is an idea like time, a form of God like time; it is the passive form, the extended $0 = + 0-$.

114. All temporal things are also in space and limited. An unlimited thing extended through the whole of space is an absurdity. God's operation only is extended through the whole of space; it is space itself; when he willed to act, he became time; but when he was time, he became space.

115. Space has not been created, but has emerged out of the Eternal; it is nothing new in the universe, nothing next to God and present with him, but coexistent with God.

116. Single things must be both in space and in time; or a real thing first originates, where time and space cross each other at one point; they cross, however, everywhere, and therefore things are everywhere.

117. There is no void or empty space, no time and no place, were a Finite could not be; for time and space are virtually the manifesting primary act, the zero that has become thing.

POINT.

118. Time has begun with number, space with the point, with the spatial nothing, with the zero of space. This point necessarily posits itself "ad infinitum;" it extends itself also in all directions, and necessarily in equal distances. Such an extended point is the Sphere.

119. The sphere is nothing peculiar, nothing new in the thoughts of God, but only the point expanded, while this again is but a contracted sphere, just as the totality of numbers is an expanded 0, and this their contracted sphere.

120. Space is spherical, and, indeed, an infinite sphere. The sphere has been posited with space, and consequently from eternity; it is also an idea, and that, indeed, the total idea; for time and space have in it been posited together.

121. For God to become real, he must appear under the form of the sphere. There is no other form for God. God manifesting is an infinite sphere.

122. The sphere is, therefore, the most perfect form; for it is the primary, the divine form. Angular forms are imperfect. The more spherical a thing is in form, by so much the more perfect and divine is it. The Inorganic is angular, the Organic spherical.

123. The universe is a globe, and everything, which is a Total in the universe, is a globe.

LINE, LIGHT, MAGNETISM.

124. While the point expands, it is active; this active expansion is a simple repetition of the point, and this is a Line, which in the sphere, however, is a Radius. With time originates not merely a series of numbers, but together with it also the line. The line and time are of one kind, repeated positions of the nothing, of the point. It is consequently clear, how that time were a repeated positing of the Eternal itself: for the line is only a repeated self-positing of the point, of the nothing. God fluctuating in his eternity, and the point, are one in kind; but God acting is a line, being or existing a sphere, i. e. the point in the act of being.

125. The line is nothing new in creation, but time itself, when regarded more closely. God creates the line as little as he does time; but this originates unto him, while he moves, while he thinks. It is impossible to think without producing a line. The line is therefore from eternity, is a series of numbers.

126. The essence of the line does not consist in its two extremities being continued with equal significance into the Infinite; but in its radiality, i. e. that one extremity turned towards the centre has become central, converging, absolute; but the other turned towards the periphery has become divergent, finite, multiplicity. The primary line is a line produced with two antagonized characters. The central extremity is 0, the peripheral is the bisected zero = ±. This radial line gives us the antetype of a new polarity. The two extremities are not related as + and-towards each other, but as 0 and + -. At the instant, when a line originates in the universe, it is not a line merely, or an indefinite line that originates; but one that is definite at both extremities, polar, indeed, but after a determinate fashion. Nothing, not even a finite thing, exists in an indefinite manner.

127. There is no mathematically straight line in the world: all real lines are polar; they are all rooted in God by one extremity, by the other in finitude. The primary act becomes in its first operation not merely a posited nothing, a numerical series; not merely time, not merely an aoristic line, but a Linea determinata; in short, God can step forth into time only as radius. The Monas determinata is a Monas radialis, or a centroperipheric monas.

128. The essence of the primary antagonism is a centroperipheric antagonism. As centre is related to periphery, so is here one pole related to the other. Polar existence and central or peripheral existence are one. Primary polarity is centroperiphery. The primary line is constantly in a state of polar action, which is called tension; for it is always converging and diverging, at once central and peripheric. Every line originates, therefore, only by tension, and is only by it, yea, every line is nothing else than this tension.

129. A line, one extremity whereof strives towards the centre, the other to the periphery, the one to identity, the other to duality, will exhibit itself in the world as a line of Light, in the planet as a Magnetic line. Magnetism is centroperipheric antagonism, a radial line, $0—\pm$, the action of the line being cleft at one extremity. Magnetism has its root in the beginning of creation. It is prophesied with time.

SURFACE, ELECTRICITY, OXYDATION.

130. The periphery is the boundary of the sphere, and is, consequently, a superficies or Surface. This, therefore, originates also directly with the positing of the Eternal.

131. As the primary line is not a purely polar, but a radial line, so is the primary surface not a level, but a curved or convex surface.

132. There is no level surface in the universe, no pure surfaces any more than pure lines. All surfaces are curved. For example, those of drops, of the heavenly bodies, of animals. The surface of a sphere is no Continuum; but consists properly of the divided peripheric and upright extremities of the radii; it is a $\pm$.

133. The surface of a globe has no centre, no 0, like the radius; but is an absolute Dualized, a $\pm$ without 0.

134. This mode of operating of the primary act is manifested as electricity. Electricity is a merely peripheric antagonism, without centre, thus without union; an eternally Dissevered without rest. Electricity is thus also a special form, under which polarity makes its appearance, and is likewise rooted in the primary creation. There is, consequently, no thing which were not magnetic and electric.

135. The idea of a surface is constantly that of surrounding. It is not generated by a section of a globe, but by the completion, the circumferential limitation of the sphere. The essence of the sphere is boundary. Every surface is finite, is convex. In the divine position a surface never occurs, save on the boundary of the primary sphere.

136. As no thing can exist without a line, without a radius, so also none can be without surface, without circumscription. The single surface is identical with the Locus of the old philosophers. Every Finite is a closed whole, and that thing is of the most perfect kind which has the most perfect closure, surface, periphery (or skin).

137. The surface is also not different from the primary act, but a form of the primary act itself; or a boundary, which, however, nowhere remains stationary, but is always displaced by means of the eternal act. Therefore the world is at once unlimited and limited; the latter in reference to the closure of the surface, the former to the endless expansion of the same.

138. The periphery is the object in divine consciousness, the point which, posited without the

centre, is thus one and the same, centre (subject) and periphery (object). It is everywhere the same point, the same 0, wherever it be posited. Hence the profound saying, "Mundus est Sphæra, cujus centrum ubique, circumferentia nusquam."

139. The surface stands in antagonism to the line, like periphery and centre; it stands perpendicular upon the radius, and can never pass parallel to linear action. Electricity ranks in eternal antagonism to magnetism.

SPHERE, HEAT, CHEMISTRY.

140. The line and surface are density, the representation of time and space; they have therefore like these originated out of nothing, namely, out of the point. The sphere is the expanded nothing. Nothing thus extended, or nothing posited, becomes a something, viz. line, surface, density, polarity. The line is a long nothing, the surface a hollow nothing, the sphere a dense nothing; in short, the something is a nothing which has received only predicates. All things are nothings with different forms. The point is $= 0$, the line $= +$, the surface $= + -$, the sphere $= + 0-$.

141. The internal motion of the globe, or the becoming of the globe, is manifested in the universe as Heat, in the planet as Chemistry.

ROTATION.

142. The primary sphere is rotating, for it has originated through motion; the motion of the sphere cannot, however, be progressive, for it fills everything. God is a rotating globe. The world is God rotating. All motion is circular, and there is everywhere no straight motion any more than there is a single line or straight surface. Everything is comprehended in ceaseless rotation. Without rotation there is no being and no life; for without it, there is no sphere, no space and no time.

143. The more perfectly circular the motion of a thing is, so much the more perfect is it. Straight motion is only the mechanical; such, however, exists not through itself. The more a body moves in a straight direction, the more mechanical and ignoble is it. Straight motion too yields only straight form.

GEOMETRY.

144. The sphere with its attributes is the totality of numbers, is thus a rotating number. The universe is the same. In arithmetic the quantity of divine positions is regarded; in the sphere, however, the direction of these positions, or of series of numbers.

145. The doctrine of the sphere is Geometry; for all forms are contained in the sphere. All geometrical proofs admit of being conducted through the sphere. Geometry has originated directly from arithmetic, or is arithmetic itself, with this difference, that the latter regards series of numbers as individualities, the former, however, as a whole. Arithmetic is a geometry seriebus discretis; geometry is an arithmetic seriebus continuis, a solidified arithmetic.

146. Geometry is a science of equal value with arithmetic; it is even as certain, because it has no other propositions; it is equally eternal, is the same realization of the primary act, the Deus geometrizans of the Pythagoreans. Everything to be certain must therefore resemble geometry, must be itself a position of geometry, only under other relations.

147. Geometry is more real, more finite, therefore also more apparent, and, as it were, more material than arithmetic. The ideas in it have become something determinate, have assumed form, while before they still fluctuated formless in arithmetic; here were they mere ghosts without veils, but in geometry they have received these veils. Time has received for its form, its body, the line; space, the surface; life, the globe, consequently the rotation for its form or body. It is to be here remarked, that ideas always become more real and more finite, always approximate

nearer to actual manifestation, the lower they descend or the more they are considered individually. Geometry has not originated later than arithmetic, but is only a more individual view of ideas, arithmetic being more universal. Geometry is arithmetic with stationary numbers, = points. The Divine thus approximates to manifestation, to materiality, the more individual it becomes; and this is very natural, for it verily limits itself more and obtains always more predicates. The more a thing obtains predicates, by so much the more perfect is its finiteness. By geometry we are actually transferred into the universe, but only into the formal, in which it has, like a skeleton, been sketched for us solely upon a general plan; namely, as infinite extension, in which line and periphery, central and peripheric action, magnetism, electricity, and rotation, &c., have been prefigured.

B.—HYLOGENY.

a. GRAVITY. (First form of the World. Rest.)

148. In arithmetic the divine acts are only undetermined = numbers. In geometry the numbers obtain determinate or finite directions, become figures. All figures have, however, an especial direction to the centre. Figures are nought but centres manifoldly posited.

149. The direction to a centre is, however, an act, which never ceases to operate. The primary act strives therefore to posit ad infinitum nought else than a centre, i. e. points.

150. If there are points without the centre, it so happens only because the succeeding points have been displaced by the points that were first posited. The peripheric points are only with reluctance out of the centre. The globe only exists in an uneasy state, because it has no place in the centre.

151. Every finite thing strives towards the centre. The finite is only something, in so far as it is posited in the centre, and it maintains its value according to its distance from the centre. This exertion or endeavour, by virtue of which things would be in the centre, is Gravity.

152. What the retrogression of numbers into 0 is, that is the gravity in the sphere. The gravity is a geometrical reduction of position unto nothing. The sphere is only produced by action, and that indeed by the centroperipheric. This action must therefore manifest itself in two ways, as centrifugality and centripetality. The first is the dispersion of the primary act or of points, the second is the collection of the primary acts or points into the unity, and this is gravity. Centrifugality originates only in a constrained manner or with reluctance, for the primary act always seeks the centre, and only moves towards the periphery, because it has no longer any place there. If centripetality be regarded as a force, then is centrifugality no force, but only centripetality itself retreating from the centre; even as cold and darkness are probably no particular forces in themselves, but only weaker degrees of heat or light.

153. Gravity is not motion simply, but motion unto a centre, unto rest.

154. That which is itself in the very centre is therefore not heavy. The primary act is not heavy.

155. As all finite things are positions of the primary act in the sphere out of the centre, so are all of them heavy. Gravity is the force that strives unto the centre, and which is there impeded by other forces already present therein. A finite thing, that is not heavy, is a contradiction. The gravity of the single thing is weight. The world itself has no weight, or else it must be heavy in relation to something else without it. The ideas of gravity and weight, as we speak of them in reference to individual things, are not applicable to the world, still less to God.

156. Gravity is also nothing new in the world, but it is only the positing of the centre in space. As necessarily as the Eternal, when it manifests itself, must appear under a definite form, so also must it be necessary with the eternal effort, to return into itself or appear as gravity. Gravity is nothing different from the primary act, nothing specially created; but the spherical position of the

same tending unto the centre.

157. Now, as the sphere has originated out of nothing, so also has gravity originated out of the same. The form is a formed nothing: the form is, however, no form without internal forming force, and to this gravity belongs. The being of form and the being of gravity are one. Gravity is a weighty nothing, a heavy essence, striving towards centre, a realization of the first divine idea. Gravity cannot, therefore, be perceived in the universe as a whole, but only in its parts.

158. If gravity is the primary act that has become real, so must everything originate out of gravity, or everything must acknowledge gravity to be the common mother of the finite. It is in all cases, or in every individual thing, only the gravity, the Ponderose, which exists, otherwise nothing exists; for verily nothing exists without the divine primary, which is incessantly a central, act.

MATTER.

159. Points, which strive towards the centre, are compressed, because they would all occupy one and the same spot. These points, however, are forces, which take up space and therefore exclude other points. A space that excludes another is Matter. Everything which has been said of gravity holds good in respect to matter; for matter is only another word for gravity. A heavy thing is a material thing.

160. To the totality of a thing belongs not merely its figure nor its tension or motion simply, but also its gravity. This is, however, a whole sphere. Matter is, consequently, a total position of the primary act, a trinity of ideas.

161. Matter has been imparted with space and time. All space is material; ay, matter is itself the space and the time, the form and the motion; for space is nothing special, but only extended or formed force. It is here also shown that the nothing does not exist. There is as little nothing in the universe as there is an 0 in mathematics. So soon as the nothing is, there is something. The whole universe is material, is nothing but matter; for there is the primary act eternally repeating itself in the centre. The universe is a rotating globe of matter.

162. But the universe is an acting gravity, a matter, in which the centroperipheric antagonism is active; it is therefore everywhere matter only, which acts. There is no activity without matter, but also no matter without activity, both being one; for gravity is itself the activity, and itself the matter. Matter is only limited activity. A matter which does not move is not; it can only subsist through continuous origination, through life. There is no dead matter; it is alive through its being, through the eternal that is in it. Matter has no existence in itself, but it is the Eternal only that exists in it. Everything is God, that is there, and without God there is absolutely nothing.

163. It is an illusion to believe that matter were an actual something subsisting in itself. It is even so with numbers, upon which a reality also is bestowed, when they are still demonstrable nothings. A number is nothing truly than an affirmation several times repeated, a reiterated deposition of what is nothing, what is no number. This deposition happens likewise in the universe, where it is the primary act, that is deposited. Where, however, this is, no other station can occur. This exclusive property is usually called the Impenetrable, the Material. It cannot be said in what spot matter arises, so secretly and suddenly does it start into existence. It is matter properly at the first manifestation of being, of time and of space; for at the same instant also the line, surface, density and gravity have been given. The line does not exist if it does not act; the sphere does not exist if it be not inert, i. e. if its forces do not strive towards the centre, and consequently to connexion. Nothing exists if it is not material. Matter is accordingly coexistent with the presence of God.

164. The Immaterial does not exist; for even the Material which is not, is the Immaterial.

Everything that is, is material. Now, however, there is nothing that is not; consequently, there is everywhere nothing immaterial. Immateriality is only a postulate principle, by which to get at matter, like the 0 in mathematics, which is nothing in itself, does not even exist, but that still must be posited, in order that numbers may bear a reference to it.

165. God only is immaterial; he is the only permanent immaterial invention, and the axiom being the Formless, Polarless, Timeless. A spirit with form is a contradiction. But the matter also does not exist, because matter is nothing, because it is only a sphere of central actions, which is gravity.

166. The material universe is called nature. There can be only one nature, as well according to time as to space and to divine animation. There is only one God, whose operations expressed, or materially posited, are nature. Nature has originated out of nothing like time and space; or with these has nature also been. God has made heaven and earth out of nothing.

167. God has not found matter co-eternal with himself, and, like an architect, arranged this to his fancy; but he has, out of his own eternal omnipotence, by his will simply, evoked the world out of nothing unto existence. He has thought and spoken, and it was.

168. The doctrine of matter is the Philosophy of Nature. It is therefore the science also of all Singulars, like geometry and arithmetic; thus at bottom is only the third part of mathematics, and is even as certain and demonstrable as this.

ÆTHER.

169. The matter, which is the direct position of God, which fills the whole universe, which is the time in a state of tension and motion, the formed space, the heavy primary essence, I call the Primary matter, the matter of the world, cosmic matter, Æther. The æther is the first realization of God, the eternal position of the same. It is the first matter of creation. Everything has consequently originated out of it. It is the highest, divine element, the divine body, the primary substance $= 0 + -$.

170. The æther fills out the whole universe, and is, consequently, a sphere, yea, the world's sphere itself; the world is a rotating globe of æther.

171. The sphere of æther, not as yet individualized, I call chaos. From the beginning was chaos, and this was æther, and unto the end will chaos become æther. The æther is the apparent nothing, and thus it is the chaos. This was not the latter and not the former, but an existing nothing.

172. The æther is the imponderable matter, because it is gravity and totality itself, because it is the infinite matter.

173. The æther has no life; it is the only mortuum, because it is the heavy 0. In æther, however, reside all the principles of life, all numbers; it is the substratum, the essence of life. There is only one universal substratum in nature.

HEAVENLY BODY, POINT, CENTRE, GRAVITY.

174. Everything which emerges out of æther and is posited as a finite matter, can be nought else than a repetition of the sphere.

175. The æther subdivides into an endless number of subordinate rotating spheres, and so it must, because the world is not a whole devoid of parts, but only a whole in the parts, only a repetition of positions. The chaotic sphere of æther consists essentially at one and the same time of an infinity of spheres.

176. A chaos has never existed. The General never exists, but only the Particular. Chaos was from eternity a multiplicity of ætherial globes. Chaos is only postulate.

177. Every sphere of æther is complete in itself and closed, and therefore rotating around its axis and around the universal axis of the æther.

178. The new rotation in the heavenly bodies condensed in the periphery of the æther, follows as a necessary result, on account of the unequal velocity of its outwardly and inwardly lying points.
179. Every individual sphere has two motions in itself; the one depends upon the representation of the primary act in itself by the special rotation; the other re-attempts to regain the primary centre, through the general rotation around the universal axis.

180. Such a sphere rotating for itself is called a Heavenly body. A heavenly body again is the image or metatype of the Eternal; it is a whole, it is alive; everything, even the highest, can originate out of it; everything develop itself out of the coagulated, individualized æther. The heavenly body has a double life, an individual and an universal, since it is for itself and at the same time in the general centre. Every Individual must have a double life.
181. The heavenly bodies are as old as the æther, consequently they are from the beginning, and endure also without end. As they are only coagulated æther, so are they susceptible of being resolved into the same, such as are probably the comets.
b. LIGHT, LINE. (Second form of the World. Motion.)
182. The æther is from eternity, not merely monas, but also dyas; from eternity it stands in a state of tension with itself, when, as the image of the existing primary act, it has emerged out of itself into two poles. This self-egression or self-manifestation of æther, or of substance simply, is the self-egression of the point into the periphery. As dyas, æther exists under the form of polarity, of central and peripheric effort; the æther in a state of tension is a centroperipheric antagonism.
183. The æther is separated from eternity into a central and peripheric substance, and that indeed through its simple position as a globe. The universe is a duplicity in the form of æther; it is both indifferent and different æther, both central and peripheric. The central mass of æther may be called sun, the peripheric planet. Only one sun can originate in a globe of æther, but many planets.
184. Between the central and peripheral mass of the æther, between the sun and planets, there is tension. Through this solar-planetary tension the æther fluctuating between the two becomes polarized.
185. The tension of the æther proceeds from the centre and thus from the sun. Were the sun to be removed, the polarity of the æther would be annulled; it would be again the indifferent chaotic æther, the null and void matter. For the absolute substance to exist it needed not simply itself, but an identical centre and a dissevered periphery. Is there no peripheric mass, no planet, so also is the tension annulled; centre cannot be without periphery, sun not without planet, nor vice versâ. The tension of æther is thus excited by the sun and conditioned by the planet. The planet is not the principle, but the Redintegrant of the tension of æther by the opposition.
186. At that part of the universe where no periphery stands opposed to centre, no planet to the sun, the æther is not tense, but indifferent, annulled. There can thus only be columns of æther, which are rendered tense, namely, those columns of æther which are found between the sun and the planets. Near to the planets the æther is void of action, indifferent, non-apparent. There are consequently as many apparent æther-columns, as there are heavenly bodies, that stand in the process of polarity opposed to each other. These columns move with the planets around the suns. The indifferent æther of the world-space is, therefore, successively rendered tense, as the planets move around suns, and becomes again indifferent behind the planets.
187. But besides this, that columns of tension only exist, and therefore that the æther is nowhere active as a sphere, there is still no spot in the world-space where there would be only indifferent æther, nothing; for the æther consists of many globes of æther. There is thus nowhere an

indifferent æther, consequently nowhere an empty space. The idea of the repletion of space is not that of the sphere, but of the columns of tension, which by their crossing in every direction form a sphere only externally.
188. That which is thought of as originally filling space is not the quiescent but only the moved, tensed æther. The former is the void space, the nothing.

189. The tension of æther is an action, which operates according to the line. This linear activity, which makes its exit from the central mass and is excited hence to the peripheral mass, is Light, or in short, light is tension of æther.
190. Light is a traction of lines or radial action; consequently an antetype of magnetism. A ray of light is a radius. The ray of light has two extremities different from each other; that turned towards the sun is 0, that which comes in contact with the planets is ±. Light is, therefore, a splitting, rending action.
191. Light is the life of the æther. Hitherto the æther was an inactive nothing, mere substratum for a future. This nothing, when it becomes centroperipheric, seeks to tear the mathematical point into radii and circumferences, appears; and this centroperipheric manifestation we call light.
192. The untensed, indifferent æther is, therefore, darkness, and this is the essence, the rest of æther. Chaos was thus darkness; the world arose out of darkness when light became. Light has originated out of darkness when the chaos was moved. Were it possible therefore for all light to vanish, the world would again return into its old nothing; for darkness and nothing are one. God has separated the light from the darkness.
193. If light be only a column of æther in a state of tension, light is or exists only between the planet and the sun; near to the planet and behind it is darkness. The primary sphere is a dark sphere, transpierced only by single rays of light. Each star, however, stands in a state of tension with another; thus many thousand rays of light stream forth from each, and fill out the world-space in all directions. There is therefore no absolute darkness, because there are infinitely numerous rays of light. In the night also there is always as much light present as is necessary to maintain the heavenly bodies in their action. For the world there is no night, but only for the planets. It will be shown that the air maintains its existence simply through the operation of light; were it therefore always dark, were night to endure for ever, the air must soon assume another composition or mixture, and everything that lives in it must fall to ruin. This is shown also by the diseases and crises of the same.
194. Light is from eternity, for the tense æther was from eternity. The dark chaos exists only as postulate. Light is time that has become real, the first manifestation of God; is God himself positing, is the dyadic God. The dyas is not merely radiality but light; or both are one, time and light are one, motion and light are one. When God numbers, when he draws lines, he thus creates light. God becoming self-conscious is light. Light is God illuminating. Darkness has accordingly never existed, although the light is derived from the darkness, like numbers and figures are out of nothing.
195. Light is no matter. There is no substance called light, but the æther is illuminant through its binary division. The sun does not, therefore, stream forth when it illumines the planets, and loses nothing of its magnitude; it is not to be feared that we shall ever lose it. That the sun is an undulating sea of flame, that it is throughout a volcano, that combustions or electrical processes of light, appearing to us as light, occur in its atmosphere; that the velocity of rotation hurls about the light-particles, and that these particles scattered in the world-space are, by an unknown route, or by means of comets, again brought back to the sun, are opinions unworthy the inquirer into

nature. The sun gives out nothing but the impulse, not, however, the mechanical, which makes the space of heaven tremble upon which it shines; but the purely spiritual, as the nerves rule the muscle. The sun can never be extinguished, never become dark; for it gives out light, not as a fire, but simply by reason of its standing in the midst; its simple position, its enchainment to the planets is light. A fire upon the sun would not be perceived by us; it would not lighten nor warm us, because of its having no relation to us. The central relation of the sun toward us cannot, however, remain unobserved, and this observation is even that of light.

196. Matter has become by means of light, is a child of light, is but illuminating æther. Every binary division of matter manifests itself as light.

197. The whole universe is transparent, because everything has issued out of the tension of æther. Everything which is matter is light, and without light there is nothing. Without light the universe is not only dark but it is even not. Light is the universe, and every Finite is only a different position of light. The world is a thoroughly illuminating globe, a rotating globe of light. The solar system must have been created according to the laws of light. The phenomena of the world are only representations of optics, consequently of living geometry. What we see is nothing but optical construction or figuration. (Vid. Oken's 'Essay über das Licht;' Jena. Fromann. 1808.)

c. HEAT, DENSITY. (Third form of the World. Shape.)

198. Light is not simply a motion in itself, a mere continuous excitation of polarity in the æther, but it is also the æther itself set in motion thereby. All polar actions resolve themselves finally into motion of the polarized masses. The end of electricity, galvanism, magnetism, is motion. It will be, however, shown that all these polar functions are only repetitions of the primary polarity; this must therefore produce what the former did, namely, ætherial motion.

199. Every point of the æther becomes polar, every one attracts and repels the other; whereby motion arises in the innermost parts of the æther itself. Not a portion of the æther is moved on, but motion originates in the mass of æther itself. The æther-atoms quit each other.

200. The æther is, however, that which is filling space, is space itself, the Expansissimum of the world, the Formless and therefore that which adopts all forms. The formless æther, when it moves itself, must be connected with a phenomenon, that depends upon its expansion and identification, which is polarized by light. This action of the æther does not therefore depend upon the tension of æther, not upon production from differences in the same, but upon dissolution of the tension, therefore upon extension, upon the indifferent representation of space. This action, which is at the same time universal, can only be heat. Moved æther is Heat.

201. Heat is the contest of the indifferent æther with light; light alone produces heat. Without light the world were not only dark, but also absolutely cold. Cold is untensed and quiescent æther, death, nothing; dark and cold are one. Heat is therefore the result of light, but equally eternal with it; it is space represented really, as that is real time.

202. Heat is not moved indifferent ether, which is = nothing; but moved and tensed or the moved light.

203. Heat penetrates into thickness as an extending function, but does not oscillate between two poles like light. It is only the function of density, and depends upon nothing else, not upon lines or mere surfaces, but upon the living sphere.

204. Heat and light, although characters of one substance, yet stand in an antagonism like thickness and line, or as indifference and difference. Heat is properly the first perfected position of the primary act, while light is only the act of positing; the latter therefore is +, the former -. Or also, gravity is the absolute position, simply = 0; light is the commencing egression of this

position out of itself + -, heat the completion + 0-, and therefore the position everywhere; it will everywhere deposit, therefore it is motion, repletion of space, expansion. Light is gravity become real, or 0 become real; 0 however, rendered real, is + -. Heat is as-at the same time + - and 0, or light and gravity, material light, light that is filling. Both will also assert the antagonism of their genesis through all forms of the world. The heat seeks to destroy the line, which the light endeavours to establish; heat seeks to produce homogeneity in the Dissimilar, light to effect the reverse. Heat is slow in its motion; with it the mass of æther must continually move, or move whither it will operate; light, however, acts spiritually and rapidly, without motion of the mass, but only glides continually with the latter. Heat is not created, it is no special matter different from æther. There is no body of heat.

205. Heat is everywhere where the æther is, and must consequently be regarded as a sphere. Heat is not present merely in the columns of æther between the heavenly bodies, but everywhere. Therefore heat does not move itself in the direction of the line, but it extends itself on all sides, as real space.

FIRE.

206. Light and heat were the first phenomena of the world. Heat with light, however, are Fire. Fire is the totality of æther, is God manifested in his totality. God, previous to his determination to create a world, was darkness; in the first act of creation, however, he appeared as fire. There is no higher, more perfect symbol of divinity than fire. God's whole consciousness, apart from individual thoughts, is fire. The Holy Scriptures therefore usually admit of God appearing under the form of a fire, as a fiery bush, a flame. The world is none other than a rotating globe of fire.

207. Everything that is, has originated out of fire; everything is only cooled, rigidified fire. As everything has become out of fire, so must everything to be annihilated have recourse to fire. If finite things be only fire singly posited, so must every change occurring in the same be an igneous change. Nothing changes in the world but fire. The essential change of things take place only by fire.

RETROSPECT.

208. The Triplicity of the primary act in the universe has now been completely demonstrated. The first manifestation of God is monas; to this corresponds Gravity, Æther, darkness, the cold of chaos. The second manifestation of God is the dyas; to this corresponds the æther in a state of tension, the Light. The third manifestation of God is the trias; to this corresponds the want of form, Heat. God being in himself is Gravity; acting, self-emergent, Light; both together, or returning into himself, Heat. These are the three Primals in the world, and equal to the three which were prior to the world. They are the manifested triunity = Fire.

PART II. ONTOLOGY—OF SINGULARS.

A.—COSMOGENY.
a. REST, CENTRE.

209. Through light duplicity originates in the æther, by virtue of which the æther divides into central and peripheric æther. The peripheric necessarily rotates around the central. Every part of the æther is a sphere; the æther therefore is separated by the light into infinitely numerous central and peripheric spheres. Creation is an endless position of centres. The primary centre is

inventive.
210. There cannot be therefore only a single central mass; otherwise the universe would be a finite.
211. The central spheres are characterized by absoluteness, the peripheric, however, by finiteness, division; the former are something in themselves, but the latter are so only by opposition; yet the two could not be without each other.
212. Every central body must be surrounded by several peripheric bodies. The peripheric spheres rotate around the central, the images of the primary centre. A Whole, consisting of a central body and several peripheric bodies, is called solar system.
213. Chaos is not conceivable, without being at the same time solar system. The solar systems are nothing specially created, but have been given with chaos or with light, are indeed only the æther separated by light. The primary matter appearing as light must appear at the same time as sun and planet. Primary act, sun and planet are of one kind, and differ only in this, that the former is posited individually in the latter, while in itself it is non-posited.
214. There is no general central body, no central sun, about which all suns and planets gravitate. The essence of the æther consists in its complete dissipation. There exists only an infinity of solar systems, which taken together form the central body. All solar systems pursue a course to and fro through each other, like the blood-globules in the vessels. The general central body is only inventive. That the general central body may be dark (that it must be, if present, from its being invisible) is an assertion which betrays an ignorance of the essence of light. A dark central body is an absurdity.
b. MOTION, LINE.
215. Sun and planet, as individual spheres, have also their own individual gravity. The æther therefore must exist otherwise than in the universal sphere. The next change of the æther is condensation, more intense gravity, because it becomes more individual, centre and periphery approximate more closely to each other. The heavenly bodies must contain more matter, more æther in an equal space than the terrestrial globe.
216. The heavenly bodies have obtained their matter nowhere else than out of the primary matter, the æther; they are condensed æther. The heavenly bodies of a solar system have derived their mass out of the æther, which is found within the confines of this solar system. The matter of the heavenly bodies was thus previous to its coagulation strewn in the space of the solar system, and has been by so much the rarer, as the space of the solar system is larger than the volume of all the planets together with the sun. It admits therefore of being calculated how much rarer the æther is than e. g. water.
217. The æther is therefore not absolutely imponderable, but only so in relation to the heavenly bodies. Light and heat are therefore ponderose substances, though they are not ponderable.

218. The separation of the æther into central and peripheric mass has happened according to the laws of light, and thus according to the centroperipheric primary antagonism. As a consequence of this, only one central body can originate in a solar system; the mass of the periphery can, however, divide into several, and must divide into as many as the light has moments of operation; of this we shall speak for the first time in treating of colours.
219. The matter of the periphery can be condensed by light into no other form than that of a hollow globe around the sun. The planets are originally concentric hollow globes, in the midst of which the sun is formed. There are several hollow globes, because the light has several points of contraction at certain distances from the sun.

220. The number of hollow planetary globes is a definite one, and it is not an arbitrary matter how many of them originate.

221. The matter of such a hollow globe of æther is still, however, rarer by so much than the present planetary mass, as that of our earth would be rarer if it were to form a hollow globe around the sun, about as thick only as from the earth to the moon.

222. This hollow globe rotates with the sun, because the whole globe of æther, which fills out the space of the subsequent solar system, rotates; therefore everything necessarily tends in one direction.

223. These hollow planetary globes, on account of the rarity of their mass, their rotation, and the greater tension of light, could not subsist in the equatorial plane of the solar system, but coagulate together in equatorial rings about the centre of the whole system. The planetary fœtuses are only solar rings, which rotate with the sun.

224. If the whole coagulated æther of the solar system be so small in quantity, that when extended around the sun in a planetary track or course, it still does not become solid; so also can the orbitar ring not persist, but it contracts itself through light, rotation and the peculiar excited gravity into a globe. This globe continues to rotate, as it did when under the conditions of orbitar ring, of hollow globe and as æther; i. e. it pursues a course around the sun. The peripheric globe travels necessarily in the same plane in which the sun rotates. This is therefore called the zodiac. This globe rotates also around its own axis and virtually in the same direction, according to which it performs its course or the sun rotates. A globe coursing and rotating around the sun in its equatorial plane and in its direction is called planet.

225. At the first aggregation of the mass of the planetary ring into a planetary globe, the latter was still very much extended, the earth extending beyond the moon. The mass was thus gaseous. What happened in the great globe of æther, of which the sun has become the centre, happens also here. An opposition of centre to periphery again originates; and a subordinate sun and new orbitar rings are formed. If the mass of the planetary equatorial ring be only small and consequently rare, it rolls into a globe and together with this into moons.

226. If it be much, consequently so dense, that it coheres, it remains stationary, and is Saturn's ring.

227. This is the genesis of the planetary system, but everything has become, and remained as it became, at one stroke. The moon can never have existed as an orbitar ring around the earth in time, or else it had been solid. Being once solid, it can no more coagulate into a globe. Still less, however, have the planets originated from conjoined moons. From whence then have the moons come? The solar system has not arisen mechanically, but dynamically; it has not become what it is by being projected or hurled from the hand of God, nor by impulses and aberrations; but by polarization according to eternal laws, according to the laws of light.

228. As a necessary number of planetary productions exists, so also is their magnitude, distance and velocity a determinate one. No planet, whatever its situation, has attained that by chance. Were the earth larger, it must also have occupied some other place, and have had another velocity, another density of mass, &c.

229. The coagulating matter of æther must collect into a larger mass in the centre than in the periphery. The centre will exist everywhere, and the periphery comes only to its behalf as if it were a scaffold or prop only to existence. The sun can only be the principle of the determination of the planets by the preponderance of its mass. Our sun comprises above 700 planetary systems in itself.

230. Sun and planet are mutually conditionated; both have originated at the same time, the

former as the positive pole, the latter as the negative, as the necessary counterpoint, or the one as 0, the other as ±. The hypothesis, which surmises that the planets have come from another solar system, is not maturely considered. For how have they there originated? Such explanations are mere child's play. Sun and planet are in idea but one piece, only one line with two different extremities. The same act which polarizes the sun polarizes also the planets out of chaos. One and the same æther that has become positive, is called sun, when negative, it is called planet. Both are only a single globe of æther, the centre of which is called sun, the periphery, planet. The latter belongs to the sun, like a stone though detached from, belongs to, the earth; its rotation is therefore similar, but retarded.

c. FORM

231. The sun cannot be in the absolute middle of the solar system, on account of its antagonism with the planets, which would likewise become the centre. The collective mass of the planets is the secession of the sun from the centre. The situation of the sun or the degree of its excentricity bears relation to the polar force of the planets. The form, under which the solar system really exists, cannot therefore be the sphere, but the ellipse, i. e. the duplicity of the centre.

232. The sphere is only the type of the universe, of the æther, but not of the solar system nor the Finite. No Finite is absolutely spherical. As the real universe can only exist in a bicentral condition, so is there in this respect also no universal central body. It is there, but under the form of bicentrality, as sun and planet. God only is monocentral. The world is the bicentral God, God the monocentral world, which is the same with monas and dyas. The primary polarity, the dyas, the radiality, the light establishes itself in nature as bicentrality, which is the cosmogenic expression for self-manifestation or self-consciousness. Self-consciousness is a living ellipse.

233. The bicentrality determines the distance of the planets from the sun. If the sun as the active pole be strong or energetic, the planets will occupy a remote situation; if it be feeble or weak, one that is near. The strength of the polar energy depends, however, upon the quantity of the mass. Were the mass of the sun less, all the planets would range nearer to it; were it greater, they would be all driven further off, as electricity repels the pith-balls of elder-wood; more than this the planets are not towards the sun, but even less. The energy of the solar polarization depends not merely upon its magnitude, but also upon the velocity of rotation, which harmonizes with the former; the latter, however, depends upon the original velocity of rotation of the æther. The velocity of the rotation of æther being assumed as definite, that of the sun must be definite also, and with this everything accords.

234. The circumvolution of the planets around the sun is a polar process of attraction and repulsion, by virtue of the primary law in the solar system, by virtue of the light. The planet then can only be repelled in the neighbourhood of the sun from the sun, when it has the same solar pole in itself, when it has become positive; and can only attract it at a distance from the sun, when it has received the opposite pole to the sun, or has become negative.

235. This is only conceivable in that the planet, while it draws nearer to the sun, extinguishes in itself by its own power the negative pole, and produces on the contrary the positive pole, or becomes a sun; and that, as it removes itself from the sun, it again extinguishes the positive solar pole, and generates the negative planetary pole within itself. This substantial production of alternating poles upon the planet takes place through the diversity of its surface as water and land, through the oblique position of its axis, whereby summer and winter are produced, through the processes, or through the life that is upon it, through the processes of decomposition and combination effected by water, through the revival and death of vegetation, and even the white colour of snow. The planet discharges its pole in the neighbourhood of the sun, like a cork pellet,

and reloads of itself at a distance from the sun; and thus oscillates to and fro, like the hammer in an electrical bell. The course of the planets takes place with the greatest ease. It is everywhere no force of weight, of impulse, but of the easiest self-motion. The planet revolves by its own force to and from the sun, like the blood circulates to and from the heart.

236. The planet cannot, however, be diverted from its course, because the other heavenly bodies, probably the comets, do not act mechanically but only polarly upon it. By means of this polarity they maintain themselves always at a distance, even as the sun keeps itself at a distance from the planets. In addition to this, the polar tension between the comet and the sun is stronger than between the comet and the planets. The perturbations of the planets depend upon their polar relations to each other. Although the planets have a centrifugal tendency, they are not thrown by a prodigious mechanical force in the direction of the tangent, and then drawn back by an attractive force of the sun, that has no import or meaning; but they course in a playful manner round the sun. A theory of attraction of this kind has no physical sense. Such an attraction is a Qualitas occulta, an angel which flies before the planets. It does not create the world by impulses and strokes, but only by vivification.

237. Were the planet dead, it could not be attracted or repelled by the sun; it would have from the very beginning always maintained a similar pole in itself, and it could therefore only move in a circular manner around the sun. The circular motion or course around the sun is not generally conditioned by the polarity of the planets, but depends upon the primary rotation. Proportionably to the mutual interchange of polar operation between the sun and planet, the latter would only approach the sun in the line of the apsides, and thus remove from it; but by the primary rotation it is conducted around it. The elliptical path is consequently the result of rotation and of the polar or linear interchange of operation between the two heavenly bodies.

238. The moon would keep a wholly circular path around the sun, if it were not disturbed by the earth, were it not through the difference of the earth's poles to passively retain also different polarities; for the moon is in itself dead.

239. The moon is not attracted more forcibly by the earth than the sun; and therefore it remains not by the earth. The sun exercises more polar action, more photal action upon it than the earth, and yet it falls not into the sun, for the very same reason that the earth itself does not fall in. The moon is forsooth to be regarded as itself a planet with a definite charge of electricity, which is always equably maintained by light; as such it rotates circularly about the sun. But it rotates in the same path wherein the earth rotates; therefore the latter operates upon it and draws it in its strange serpentine line around the sun.

240. The more living a planet is, by so much the more excentric is its path, because it enters into great opposition with the light.

241. If polarization by light be the cause of the attraction and repulsion of the planets from the sun, so is it also the cause of the distance of the planetary masses generally. The individual distance of the several planets is determined by the energy of their own polar excitation. Planets, which possess a strong polar energy, must range further than the others from the sun. This polar energy is, however, dependent upon the magnitude and density of the mass, upon the level state or unevenness of the surface, upon the capacity for heat, upon the quantity of water, upon the position of the axis in regard to the path, upon the possible processes of vegetation; it is thus not to be determined. Before vegetation was upon the earth, there were other processes, e. g. the aqueous precipitations, that changed the polarity; so that the path might formerly have been different to what it subsequently was or is now.

242. Planets are consequently those bodies which possess in themselves a peculiar degree of

polarity and a substantial change of the same, whereby their individual distance and the nature of their paths are determined.

COMETS.

243. The comets are heavenly bodies, devoid of a persistent grade of polarity, and without any substantial change in the same. They obtain their polarity solely from the sun, like the cork-pellet from the electrical machine. The comet is therefore repelled as far from the sun as there is still an action between it and the polarity that has been imparted to the comet.

244. At the point where all antagonism between comet and sun ceases, the former must remain stationary, and resolve itself again into æther. This is the case with those comets that never return. These comets are temporal coagulations of æther by light, and thus continued creation.

245. The æther coagulates where the light, already polarized in part by the operations of other heavenly bodies, encounters it. This depends upon fortuitous constellations.

246. These comets originate like the planets; they are æther condensed in the form of an orbitar ring. This dissevered orbitar ring is the tail, which is only a more gaseous æther, through which, or even through the nucleus itself, the stars are seen. The tail follows the comet not really but only ideally. Around the nucleus, so far as it is prolonged, the light concentrates the æther. New æther is constantly emitting light, while that which was before illuminating as tail again becomes dark and again sinks into a state of indifference. The tail is only an optical spectrum. For how can the tail be really a part of the comet, since it is always turned backwards from the sun, since it therefore follows and precedes the nucleus? The nucleus is only the lamp which kindles the æther surrounding it for some time. The light suffers a modification through the nucleus; it therefore polarizes only the æther behind it. The tail is the evident example of what is antecedent in the origin of the heavenly bodies. It is the heavenly body conceived in the act of becoming, but unto which polarization is wanting; it cannot therefore concentrate itself, but again dissolves when the nucleus is gone. Every heavenly body is a mass of æther in the world-space, which is materialized by light, and separated out of its indifference into difference, into more solid masses. Finally, the tail becomes dense æther, a nucleus.

247. These comets are thus true meteors; as they originate, so originate the globes of fire, by polarization occurring in the atmosphere, or even too above the limits of the atmosphere.

248. Meteoric stones are terrestrial comets. The opinion that they come from the moon has no foundation. There is probably as little metal as water upon the moon.

249. Returning comets are probably polarized by two suns.

250. A comet can never come into collision with a planet; the fear of such an event is equally absurd with the hypothesis, that a comet had produced the deluge or displaced the earth's axis.

251. Two planets also can never come into mutual collision, not even those that have been recently discovered, although their paths intersect each other.

252. The planets are returning comets, which, however, before they have come to the second sun, have produced within themselves the opposite pole to the sun. What happens to the comets through the influence of the second sun, the planets effect of themselves.

B.—STÖCHIOGENY.

CONDENSATION.

253. Through its separation into polar masses the æther becomes condensed, heavy and material.

254. This condensation is the result of the fixation of a definite pole on a definite mass of æther. The essence of the æther consists in its having no fixed pole, but that all the poles oscillate to and fro with the greatest facility from one particle of æther to the other. This is what is meant by

indifference, by equivalency of poles; no part of the æther differs from another, because none retains permanently a definite pole, but each of them all the poles. The formation of the heavenly bodies is none other than an union of poles to a definite mass of æther.

255. A mass of æther with a fixed pole is a dense matter; such a mass of æther I call terrestrial matter, but the æther itself the cosmic. Sun and planet must be terrestrial matters, for the essence of both consists in the difference of their poles.

256. The cause of the fixation of poles resides in light.

257. The heavenly bodies go to ruin by removal of the fixation of the pole abiding on the mass, substratum or substance, not by mechanical demolition. The destruction of the heavenly bodies is a retrogression of their mass into æther by means of fire. Heat does not drive the bodies, after the manner of a wedge, from each other, but only suppresses their polarity, and then the atoms must withdraw from each other. Heat depends only on the destruction of poles, not upon extension. The heavenly bodies are ruined in the same way that they have originated, namely, through the primary act in its retrogression.

258. It is only the pole, no other concealed property, which maintains the being of the mass. The mass is not a terrestrial mass subsisting simply by its own rest. Nothing material is the cause of the form of matter, but the Spiritual. Matter has therefore no quality, no consistence of itself, but is nothing, is æther. Mass cannot supplant mass, nor mechanism destroy anything material. The destruction must proceed from within.

259. The fixation of poles in the substance is the impenetrability of matter. It is only the spirit in matter which renders this impenetrable, not the mass itself.

260. The æther is penetrable, and therefore also penetrating. Heat is penetrating; light, as æther in a state of tension, is only partially penetrating.

261. All the diversity of matter depends upon the fixation of poles in the substance. For there is no diversity in the universe without poles, without binary division. The substance always remains the same, it is only the poles that change. The substance is the Indestructible, the Persistent, the æther, the nothing.

262. The fixation is the perquisite, but the necessary one, of the substance. The diversity of things resides only in the perquisite. In the substance all are alike. There is only one substance, only one essence.

ELEMENTAL BODIES. (How many kinds of Æther-condensations may exist?)

263. The æther has three forms, and can therefore condense itself after a ternary manner, or in other words, there can be only three kinds of fixations of poles.

264. The condensations of the several forms of æther must be simple matters or Elemental bodies, as they are called.

265. There can therefore be only three simple bodies, a body of gravity = 0, one of light = +, and one of heat = -.

266. If the heat of the æther becomes fixed, the rarest, most mobile and lightest body must originate. The body of heat is Hydrogen.

267. If the light of the æther becomes fixed, a less dense, and thus a less heavy matter, must originate, and one whose atoms are moveable against each other. The body of light must be the most active in nature; it must determine the changes of all other elemental bodies. The body of light is Oxygen.

268. If the gravity of the æther become fixed, the greatest condensation must originate. The densest matter is necessarily the heaviest. The dense matter must be immoveable in its atoms, i.

e. endowed with form. The body of gravity is Carbon (as basis of the metals).

269. Besides these 3 elemental bodies, hydrogen, oxygen and carbon, there can be no other simple bodies. All other bodies must be only different degrees of fixation of the above-mentioned bodies, or combinations of the same. Different degrees of carbon are without doubt the metals. Different degrees of oxygen are probably chlorine, iodine, bromine. Different degrees of hydrogen are probably sulphur. Nitrogen is probably peroxydised hydrogen, or an oxyd of hydrogen; this is indicated by its medium weight, and its perfectly azotic character.

ELEMENTS.

270. Simple bodies cannot exist for themselves, for there can nowhere be an æther, which merely belongs to gravity, or merely to light or to heat.

271. An elemental body is never a Total, but invariably a Polar, a something not whole, properly a half or rather but a third essence, a fraction. One-sidedness is therefore the character of the elemental body.

272. One pole is nowhere produced, but all are invariably present together. The terrestrial matter completed must therefore consist of the three primary bodies, but occurring in diverse proportions. As light and heat can never subsist without the substance of the æther, so also can no body of light and no body of heat subsist per se without the body of gravity or carbon, and vice versâ. The general materials of nature are therefore combinations of the three primary bodies.

273. The æther is the totality of the primary bodies in equal proportion, where thus no pole is fixed, but all are comprehended in fixation, i. e. in constant change.

274. All other general matters must be also combinations of the three primary bodies, but with different fixation, or in unequal proportion. There can consequently be only four general matters.

275. The first general matters are called Elements. There are only four elements, one general and three particular.

1.   Element,   Fire.
2.   - - - -   Heat.
3.   - - - -   Light.
4.   - - - -   Gravity.

276. Each element is a total representation of the æther.

277. An element is not that which is chemically inseparable, but it is only the Whole, which has first originated. But the elemental bodies are chemically non-decomposible, because they are already separate, being moieties or fractions.

278. The heat element is the hydrogen element—Air.

279. The light element is the oxygen element—Water.

280. The gravity element is the carbon element—Earth.

281. In each element, beside the basic or combustible elemental body, there is also oxygen; for they are verily naught else than the æther fixed by light, æther that has become heavy by means of light.

AIR.

282. The first condensation of the æther must be that which corresponds to its condition as heat. This ele-

ment, as being that in which the atoms have no connexion, must be therefore the lightest and rarest. In this element the poles must be fixed in the least degree, and therefore change with the slightest operation. This element is therefore moveable in all directions, is the most unstable, and in form most similar to the æther.

283. Active freedom from form predominates in it, i. e. its atoms are constantly striving to withdraw from each other, or the mass to extend. This endeavour is called elasticity. Elasticity is none other than the endeavour to become a greatest or interminable globe. The terrestrial matter, with this striving towards an universal globe, is called gas.

284. The formless internally moveable element, constantly extending itself and changing its pole, is the Air.

285. The air is the first terrestrial element, the first degree of ætherial condensation associated with the feeblest fixation of poles, the constant change of which is manifested in its electric relations. It corresponds in every respect, in mobility, extension, general penetration, &c., to heat. The air consists of a preponderance of the body of heat or hydrogen (oxydulated as nitrogen in the proportion of 79 by volume), and of a fair quantity of the body of light or oxygen (21); also of a very small amount of the body of gravity or carbon, as evidenced in the carbonic acid.

286. The air is a maximum of air, a medium of water, and a minimum of earth.

287. As heat is not merely indifferent æther, nor merely its motion or extension, but is the æther moved by the polarity of light, so is the hydrogen gas in the air not in a pure state, but converted by oxygen into nitrogen. The air is in every respect therefore an element that has undergone combustion, an oxyd of hydrogen and carbon.

288. The oxygen is that which is everywhere active, exciting, moving, and vivifying everything; it is the light in the Terrestrial. The nitrogen is inert, as it were mortified, and therefore mortifying or causing death; the former the +, the latter the-. The greatest activity among all terrestrial elements resides in the air, since all polarizations issue from it.

289. The changes in the air are accompanied by constant changes of temperature, for they are verily in themselves nothing else than changes of caloric-æther.

290. All subsequent elements must originate from or be condensations of air, even as this has arisen out of, and been a condensation of, the æther.

291. Condensations, however, are fixations of poles; the other elements differ therefore only from air by having other poles fixed in them.

292. Since the poles are at the same time fixed more internally on these elements, they can no longer have the gaseous form.

293. They must on this account contain more bulk and be therefore heavier.

WATER.

294. If the polarity of light becomes fixed in a certain quantity of the mass of æther, or the oxygen of the air obtains the preponderance, a less changing element originates possessing a more definite character, and the atoms of which adhere more strongly to each other than those of air.

295. This element has, in addition to the gaseous effort towards a general globe or periphery, the effort at the same time also to a centre, or to an individual globe. It is therefore neither elastic or gaseous. The effort of a mass to a special and general globe is a conflict betwixt form and want of form. This effort is called fluidity.

296. The fluid element must contain a preponderance of oxygen (85), and less hydrogen (15). There is also some carbon present in it. The carbon of water is to be sought in the slime of the sea, for the sea, and not fresh water, is the primary water.

297. The fluid element oxygen is the Water. Water in large as well as in small quantities, seeks to represent the globe, namely, to form drops. It possesses therefore the effort unto form, while it is always relapsing into formlessness. This oscillation between form and formlessness is the

conception of fluidity, which is therefore essentially different from that of gasidity; it might be said that the latter were the arithmetic or constant change of numbers; but that fluidity were the combination of arithmetic with geometry.

298. If the essence of water consists in the contest between form and formlessness, it must thus seek to produce fluidity everywhere. Liquefaction is, however, called solution, namely, globules are formed, both on a large and small scale. The function of water is therefore solution. It dissolves the air, (imbibes it) like the earth.

299. Water is more difficult to analyze than air, because its poles are more fixed.

300. In the analysis of water, the body of heat emerges in a pure state as hydrogen, because the antagonism here subsists in an abrupt manner; in the air it is constantly changing. Hydrogen is therefore nitrogen wholly deoxydised.

301. If water is the oxygen-element, so is it the light-element or condensed light-æther; thus it is as little something absolutely new as the air.

302. Terrestrial life originates out of water, as does the cosmic life out of light. All form originates from water; for it is the general fluid, or that which strives towards form. Without water, there would be no life, no Solid and no Organic.

EARTH.

303. If the gravity of æther condenses itself, or the action of gravity be fixed in a quantity of æther, there originates immobility of the atoms, i. e. an effort upon their part towards a single direction, namely, simply towards the centre. The effort towards a single direction or towards the centre, is cohesion or rigidity.

304. The mass with fixed gravity is carbon. If therefore the carbonic acid of air, or the carbon of water, obtain the preponderance over the other elemental bodies, there thus originates the rigid centripetal element.

305. The heavy, rigid, carbon-element is the Earth. The earth is neither gaseous nor fluid. The earth contains a preponderance of carbon, with a tolerable quantity of oxygen, and a slight amount of hydrogen and nitrogen. The earth is an oxyd of carbon.

306. If fire is indicated by + 0-, the air then corresponds to the-, the water to +, the earth to the 0. The earth is therefore the Identical, water the Indifferent, air the Different; or the first the centre, the second the radius, the last the periphery of the general globe or of fire. The earth is naught but an accumulation of points. If radii occur in it, it happens only because all points have not place in the middle point.

307. The capacity for analysis of the elements comports with the serial order of their origin. The air is most easily analysed, the water with difficulty, the earth scarcely or not at all. The æther is occupied in eternal analysis, and therefore appears only when it is momentarily polarized unto light or heat, i. e. obtains the disposition to fixation.

308. If air represent arithmetic, so does earth the geometry or universality of forms. Water is the synthesis of both, the algebra; æther the analysis.

309. The geometrical figures of the earthy are called crystals; the geometry of the earth is Crystallography.

310. In the creation the three primary ideas attained only by degrees to reality. First of all the trias becomes real in the air, then the dyas in water, and lastly the monas in the earth. The creation of the elements is none other than a representation of the three divine ideas in a finite sphere. Creation is a process of formation of the nothing.

311. Creation ceases with the production of the fixed or stable form; for all ideas have parted from each other, and settled down into the most Individual, with which separation all further

formation of new matters necessarily ceases. Creation is a constant analysis of the æther, of air, and finally, of water.

312. The element that is correspondent to gravity necessarily occupies the centre upon the planet. It is surrounded by the element corresponding to light, the water, like the centre is by the radii; both are enveloped by the heat-element or air, which forms the periphery of the globe, the integument of the planet.

313. The forms of the elements are the following; water is spherical in its greatest as well as least parts; for it is the point merging out of itself, and can therefore nowhere acquire form. The earth is everywhere nothing but point; it is therefore concrete, and every part self-subsistent or individual, while in water no part subsists for itself, but at every opportunity is confluent with the other, and therefore arrives nowhere at individuality. Finally, air is the eternal flight of the smallest parts to the periphery. In the earth the Finite or Singular is for itself; in the water it is so only through the Whole; in the air it is not indeed for itself, but is there only the Whole without individualized parts.

314. The world is twofold, an ætherial and a terrestrial; both are transcripts or copies of each other, and both ultimately of God. The terrestrial world has originated out of the æther; it is therefore further removed from God than the æther; this is the discharged, purified Terrestrial.

315. God is a threefold Trinity; at first the Eternal, then the Ætherial, and finally the Terrestrial, where it is completely divided.

316. The holy primary number is 3; the second is 9. The æther is 1 in 3; the other elements are simply the 3 of the æther, together 4. 2×3, however, or 6, lies at the bottom of this 4. The symbolic numbers are consequently 1, 3, 4, 6, 9, which fundamentally are one and the same, but in different combinations. With this, however, all formation does not yet terminate; to the 4 elements is added the vegetable and animal kingdom. The number of the days of creation is 6.

C.—STÖCHIOLOGY.
FUNCTIONS OF THE ELEMENTS.
1. Functions of the Æther.

317. The spiritual activity of æther manifests itself as a process of combustion, which appears as fire. The combination of the substance with light and heat is fire; the elements have therefore originated by a process of combustion. All matter has undergone combustion, and indeed æther has been submitted to the same process. The process of fixation of the æther or the process of condensation, is consequently a process of combustion.

318. Each heavenly body has originally burnt, was nothing but fire; the Vulcanic has preceded the Neptunian agency. It was, however, the primary process of combustion, through which the at present combustible matters first originated. The present volcanoes must not be regarded as synonymous with the primary combustion. There it was not Vesuvii that burnt, but æther.

319. What has not been burnt upon the planet, is again reduced. A something that has not originally suffered combustion is a contradiction. The metal is therefore not the Original, as for obvious reasons also, the solid cannot proceed directly out of the æther, without having been first gaseous and fluid. The processes of combustion upon the planet are all secondary, are imitations of the primary combustion in matter, in the elements. Fire consists of the combinations of the three activities, gravity, light, and heat, which are now to be separately considered.

a. GRAVITY.

320. The functions of gravity are exhibited principally in the motion of the heavenly bodies, and are so completely unfolded in works upon Physics and Astronomy, that they may here be passed

over in silence.
b. LIGHT.
321. The æther and the terrestrial matter are correlative, like Higher and Lower, unity and multiplicity, and therefore stand in the same antagonism with each other, as the air with the two inferior elements. The æther is constantly seeking to convert the matter into itself, to render it indifferent by depolarization, this, however, to condense it. Matter is only condensed æther. This condensation proceeds, however, from the polarization of light, and it is consequently the activity of light by which the æther hardens into matter.
322. The activity of the æther or the light dies or becomes obscured in matter. The next obscuration of light, or its immediate transition into matter is the polar primary body, oxygen. Oxygen is the corporeal light. It is the spirit of light to posit every thing with an internal polarity, to convert everything into oxygen, to render everywhere free the oxygen pole; for the process of fixation can only happen through activity = light, and fixation is a production of primary bodies.
323. Light is the manifestation of positive tension, of the tension of oxygen. When this reaches its maximum, light issues forth. This is evident upon the planet. Every generation of the Similar takes place through the similar principles; the cosmic generation of light must be therefore imparted also by positive tension, by that of oxygen. No reference has been here made to the negative relation of oxygen in electrical tension; it does not come under the denomination.
324. The sun is the body of oxygen, the water in the world-space; the planet, however, is the basic body, the earth in world-space. The æther is diffused between the two as the air of heaven. The sun appears to have only the density of water; for it is four times less dense than the earth, and is thus pretty nearly in the condition of water.

325. The sun must be water, even because it is a body of oxygen. It must be denser than water, because it is in the centre, is central water.
326. It gives out light only, because it is water; for as such it is in eternal motion.
327. The solar water is moved by the planets, like ebb and flow. At every point of the sun opposite to which a planet stands, there is flow; there the illumination is stronger, in other situations weaker. There must be several seas of light upon the sun, as many as there are planets placed opposite to it. There is nowhere a perfectly quiescent point in the sun; therefore it is nowhere wholly solid.
328. The shining is an ebb and flow of the sun. Query Do not the spots and flashes of the sun depend upon this? The ebb and flow also of the sea gives out light; every motion in water shines. The sun does not simply shine by external motion, but because it is by this motion polarized unto the Innermost. It is a true gelatinous animal, a body trembling through its whole mass, and thereupon phosphorescent.
329. The sun is not inhabited. It has no firm ground.
330. The contest of the primary principles of the elemental bodies upon the planet appears as light.
331. Light is now more closely characterized. It is no longer merely the tension of æther generally, but tension of material elemental bodies. Thereby the light has been torn from heaven and given to the earth. Light has a chemical relation, and admits therefore of comparison with terrestrial matters.
332. Through light the negative, its opposite or the basic, pole has been evoked in matter. The sun is self-posited as oxygen against the planet as Azotic or phlogiston; hydrogen and carbon. Light therefore deoxydises the bodies; it converts them into itself, into a polar principle. Acids

placed in the light become deoxydised; nitric acid, muriate of silver. Oxygen is developed in light out of water. In like manner the constituent parts of the air continue separated only through light.

INFLEXION.
333. The light tends from the sun to the earth, not merely because the polarization in accordance with its nature streams forth from the centre to the periphery, because the light is radiality; but because the earth is the basic pole of the sun. It is thus polarity, and not simply the straight line, which light obeys. Light tends to the centre of the earth, because between this and the sun the tension oscillates. The line of tension is only between the two centres.
334. In whatever direction light may fall upon the earth, it must strive towards the middle point. Those bodies that have in themselves the earthy nature, attract the light, not by virtue of an hidden quality, but out of the antagonism with the sun; now it may proceed either from greater density or from true basic import.
335. Rays of light, which pass close to the earth, as rays of tension between the sun and another star, become, from the polarity of the earth being stronger than that of the star, diverted from their straight direction and drawn towards the middle point. This diversion of light from its direct course is called Inflexion.
336. It is chiefly basic bodies that inflect the light towards themselves. No such body has an infinite shadow. In other respects, all bodies inflect because they are much denser than light.
TRANSLUCENCY AND REFRACTION.
337. Light, as an Ætherial, permeates matter, and must on that very account pass toward the middle point of the earth, because it is virtually none other than the tension of both middle points, the earth and the sun. Originally therefore the light must have gone through the earth.
338. This permeation is not, however, mechanical but dynamic, and is indeed necessarily a propagation of the tension of æther through the matter.
339. Matter is susceptible of the same polarization of which the æther is susceptible, because it is on no account different from the æther; the polarization only takes place more slowly.
340. The transmission of light is called Translucency.
341. The æther is transparent, because it is everywhere the propagation of the tension of light, because it is itself, or becomes, light everywhere. Bodies can also be transparent, only in so far as they are the light itself, i. e. in so far as the same polarization can be excited in them which is constantly excited in æther by the sun. But this is not only possible, but necessary; for matter is surely the æther itself, only condensed. The polarity æther must therefore be capable also of being excited in the condensed æther, although in a much less degree. The transparency of matter is a tension of æther continually ringing through matter. The whole universe originally was transparent; it has only originated through tension of light.
342. Matter is a tension of light that has become central. This continuous tension of light in matter in relation to the centre, and thus with curvation, is called Refraction.
343. All transparent bodies must refract light. The bodies are, however, denser than the æther; therefore the light, which passes from a rarer into a denser medium, must be refracted towards the centre (plummet of incidence), and in the reverse case turned from it.
344. Materiality is not the only determinant of refraction, but also the density of the element, the earth refracting light more than water, and this more than air.
345. The density also is not the only determinant, but the quality also of the matter; the Basic or Planetary must refract more than the Oxygenic or Solar.

346. As translucency is not a dead transmission of light, but a continued propagation of tension; so must it be viewed as a process of light in matter, but one excited from without. Translucency is a co-illumination, like the concord of equally attuned instruments.

347. Co-illuminating bodies are thus analogous still to the æther. If there are non-transparent bodies, they can only be found in the earth-element, which, being furthest removed from the æther, has perfected itself independently, and they must indeed be wholly deoxydised.

348. The co-illumination of bodies or their transparency is an effort of deoxydation. Bodies that cannot at all be deoxydised are non-transparent.

REFLEXION.

349. Transparency belongs only to those bodies which have in themselves a twofold character. It will be shown that the metals are absolutely identical matters, and therefore non-transparent. The metals are the only non-transparent bodies. Metaleity = non-transparency.

350. The light falls upon a non-transparent only through a transparent body, and thus one in which the tension of light propagates itself. This co-illumination of the matter placed in front of the non-transparent body cannot cease to co-illuminate; the tension must thus abide in it, and turn back from the non-transparent body, in a straight direction if the tension fell direct upon it, at a certain angle, if obliquely. This phenomenon is called Reflexion.

351. Reflexion is no repulsion of light, but only its tension continued into the medium, in which the tension has been.

352. A non-transparent body indicates nothing for the tension of light but the limit of the co-illuminating matter; it does not at all operate itself upon the light, it is as it were a void space.

353. Transparent bodies also reflect partly, because they are only relative æther, because they only co-illuminate, are not themselves tensed; or because the basis in all is the metal. Every other medium is, however, an æther differently fixed; in every one therefore the tension has been changed; every medium is thus a limit for the tension, and therefore the transparent bodies also reflect. Since the tension becomes altered, when it passes into another medium, it always remains by preference in the neutral medium; therefore reflexion originates also by the air, when the light passes out of glass very obliquely into it.

Operation of the terrestrial Elements upon Light.

DECOMPOSITION OF LIGHT—COLOURS.

354. It results from all this, that light cannot enter unchanged into mutual operation with matter. The tension of æther changes itself in matter. This change of light through the influence of matter is a debilitation of the tension of æther and lastly its complete cessation. There can be therefore no absolutely transparent matter; the æther only is this absolutely transparent matter. The denser a material is, by so much the more will it be capable of suppressing in itself the tension of light. The most transparent bodies must also become with a greater density adiaphanous or opaque, because the Metallic increases in them.

355. This suppression or expiration of the tension of light in bodies has received, as likewise proceeding from the mechanical theory of light, the name of absorption. The absorption is not a mechanical adherence of the particles of light in the pores of bodies. There are no pores for light, and this requires none.

356. The absorption or decrescence of light is a retrogression of light into the indifference of æther, into darkness. Light in conflict with matter does not continue light, but becomes a mean condition between light and darkness.

357. The substratum of light, the æther, has two extreme conditions, and only two, the tensed and the non-tensed; the one is the light, the other the dark. Between the two, however, are the

mediate conditions of twilight or "clare-obscure." The light æther emits rays, the dark does not; the mediate conditions are half the two. The light condition is the clear unsullied light, the absolute translucency; the darkness is the absolute want of translucency; the mediate members are offuscated light, mediate tensions of æther.

358. The mean tension of æther, or light mingled with darkness, is called Colour. Colour is a finite, fixed light, the actual transmission of light into matter.

359. No matter can be uncoloured. An uncoloured matter is a nonentity.

360. Since matter is rigidified light, even so must it be posited in reference to colour, like light. Pure light materially substantiated or posited is White. The untensed æther materially posited is Black.

361. The mediate tensions of æther, or the mixture of Light and Dark, are mean conditions of White and Black, mixtures of the two extremes or androgynisms of White and Black. If we do not call White and Black colours, colours are then partial positions of light in matter, or in the dark.

362. Colour originates only in the confinity of Light and Dark, or in the limit between White and Black. They are therefore microscopic.

363. Darkness is the cause of colours.

364. There is nothing visible but colour, but the coloured matter. The Non-corporeal itself is invisible. Darkness is the cause of all visibility. Were there no darkness, there would be no world for the eye. Colours are only illuminated darkness.

365. In the limit between Light and Dark there is neither White nor Black, but their possible mediate conditions, or the proper colours, the material tensions of æther. If the shadow-line of light be viewed under a magnifying glass, colours will be seen to reside in it. They are invisible only before on account of their minuteness. The prism and the lens do nothing else than magnify the shadow-line of light. They only show the colours that already exist therein, but do not create them.

366. There is properly only one colour between White and Black; it is the transmission of light into matter generally. If we look through a prism with the refracting angle presented downwards, at an horizontal fissure in the shutter of a dark chamber, the red colour is then exhibited upon the upper and lower borders of the spectrum so formed within the eye; thus, in both instances, where the Dark is above and the White beneath, as also where the latter is above and the former beneath, as on the inferior border of the opening. Upon the lower border of the upper Red, and thus in the Clare, Yellow appears, which is consequently a mixture of Red and White, as seen through the thinner part of the prism. Yellow is thus brighter Red. Upon the upper border of the lower Red, thus also in the Clare, Blue appears, which is consequently a mixture also of Red and White, but the latter seen through the thicker part of the prism. Blue is thus offuscated Red. If Yellow and Blue be mixed, Green then originates. There can be therefore only four colours, whereof the Red is a mixture of Black and White, Yellow of Red and White, Blue of Red and Black, Green of Blue and Yellow. The first three are simple or mixed colours, the last a compound colour or a medley. These colours are parallel to the gradations in nature, or the latter are none other than the materializations of colours or the gradations of light. All other colours must be contained in the Red; it must serve as the basis or groundwork of all; it must be the noblest, most total, fullest and purest colour. This colour is the first position of the æther as matter, and thus of fire. Fire-colour is the first-born, the noblest, highest, fullest, purest; it is the ætherial, cosmic colour. In fire the light is offuscated by gravity, and thereby coloured.

367. The light is not, however, perfected by its position as fire, but is posited also terrestrially.

There are therefore terrestrial colours also.

368. There can be only three terrestrial colours, neither more nor less; for there are only three different material or offuscated positions of light.

369. The first terrestrial offuscation of light is the air. The colour of the air is thus second in the rank of colours. As the fire-colour plays the chief part in the cosmic and in all colours, so does the air-colour among the terrestrial. It is the highest colour of the planet.

370. The second offuscation of light is water. The colour of water is the third colour.

371. The third offuscation of light is the earth; and this colour is the last, the most ignoble. The colours part into two series, the cosmic or solar, and the terrestrial or planetary. The cosmic is the Red. The first terrestrial is Blue. The second is Green. The third is Yellow. Red alone is worth as much as all the three others taken together. It is the identification of all numbers. Green is merely their synthesis, the terrestrial, finite totality.

372. The genesis of colours is thus the genesis of the elements, or that of matter. It cannot be otherwise; for the becoming of matter is verily an offuscation of light, a coloration. Colour agrees essentially with the elements, and is itself nothing different from element. Fire is in its essence red, as being the impartient of light and heat; air is in its essence nothing else than the blue æther by virtue of its being gaseous; water is the green æther, earth the yellow. If the æther is tensed, it then becomes red or fire; if it attains its blue stage, it becomes air; at the green stage, water, upon the yellow, earth.

373. The elements are only gradations of light, colours. They have therefore been formed according to the laws of light; for colours are without doubt the legitimate developments of light.

374. Red, as being the solar or fire-colour, ranks parallel with oxygen; the more powerful indeed the combustion, the more powerful is the oxydation, and by so much redder the flame. Matters also become red through oxydation. The Red vanishes lastly into White, and thus the highest oxydation is white.

375. The next interchange of Red is with Blue; it becomes red by oxydation, this again blue by desoxydation, but by excessive alkalization and terrification, yellow. It is Red that imparts oxydation, Blue that resolves it and reduces the poles to indifference.

376. From the same cause Red warms, but Blue on the contrary does not. The calefaction given out in prismatic spectra of colours is an impure work, in which refractions, diversions and convergences of light, as well as demi-foci, cooperate.

377. The sun in the firmament may be viewed as the bright opening in the darkened chamber. Colours are therefore nothing but images of the sun in darkness, self-manifestations of the sun in dark matter. A point of light thrown into darkness is colour. This is the case around the sun, which is therefore surrounded by a hollow globe of colours, by its own refulgence. The rainbow is a ring around the sun consisting of infinite positions of the solar spectra in darkness.

378. The symbolical doctrine of the colours is correct according to the philosophy of nature. Red is fire, love—Father. Blue is air, truth and belief—Son. Green is water, formation, hope—Ghost. These are the three cardinal virtues. Yellow is earth, the Immoveable, Inexorable, falsity the only vice—Satan. There are three virtues, but only one vice. A result obtained by Physio-philosophy, whereof Pneumato-philosophy as yet augurs nothing.

COLOURS AND PLANETS.

379. Every condensation of æther by light is consequently a production of colour, and inversely, every production of colour is a condensation of the æther. The laws of coloration run parallel with those of materialization, or, what is more, are the same. The planets are thus produced

according to the laws which light exercises in the production of colours.
380. There are as many productions of heavenly bodies as there are of colours, and thus there are four.
381. The sun is the incorporation of Red or fire, the planets are that of the three terrestrial colours. The comets belong to the kingdom of darkness.
382. The planets are only suns reflected in darkness as in a mirror; they have originally been hollow globes of colour, then orbitar rings of colour (solar rainbows), then points of colour. The planets are coagulated colours, for they are coagulated light. At that very distance from the sun, where light begins to grow dim, where, to speak in the Newtonian sense, it begins to refract, there planetary mass originates. The mass of the planet thus coagulates together around the sun, but not in an uniform manner like a mass of pulp, but in pauses of colours, exactly like a rainbow.
383. These planetary chromatic arcs or bows of colour are related to the sun like the three terrestrial colours to the cosmic, or as the three terrestrial elements to fire. Three planetary productions must have thus formed around the sun, because the light condenses, materializes itself in three moments. Therefore the planets range themselves in groups at three great distances. To the first production belong Mercury, Venus, the Earth, Mars, Vesta, Juno, Ceres and Pallas. They are the first digression from Red, the Yellow; the Earthy preponderates in them. They are all placed close together. To the second production belong Jupiter and Saturn; water rules them, and fluctuates visibly upon them; they are the Green. They range at a great distance from the former group. To the third and most remote productions belongs Uranus; it is the Blue, in which the air preponderates. It again ranges at a great distance from the former group. The comets are naught but æther, which is about to become air.
384. The production of the earth-planets or of the first group is dispersed into so many as it is on account of the proximity of the sun, on account of the energy of the rays of light, as well as the import of the Earthy, which exists essentially under many forms. The planetary rings have been arranged like scales behind each other, like clouds through electrical pauses; but these repeat themselves more rapidly, in the neighbourhood of the sun.

c. HEAT.
385. While the æther falls into a state of tension or shines, it is thrown into motion. This motion of æther manifests itself as the conatus or effort to extension. The extension, however, considered as a world-phenomenon, is Heat.
386. Light, when it operates upon terrestrial matter, excites this to special polarization, whereby the Ætherial that is in it is set in motion, i. e. heat is generated.
387. Light never moves directly the mass itself, but only the Ætherial that is in it. Through this motion of the æther it becomes separated from matter; and this separation is manifested as free heat.
388. Heat is not matter itself any more than light is, but it is only the act of motion in the primary matter. In heat, as well as in light, there certainly resides a material substratum; yet this substratum does not give out heat and light; but the motion only of the substratum gives out heat, and the tension only of the substratum light. There is no body of heat; hydrogen is the body of heat, just as oxygen may be called the body of fire.
389. Heat is real space; into it all forms have been resolved, as all materiality has been resolved into gravity, and all activity, all polarity into light. Heat is the universal form, consequently the want of form.

390. Light properly develops heat out of matter through separation of the fixed poles from the substance, whereby the latter again passes over into æther.

391. The development of heat in a body is not an extrusion of a matter adherent, and as it were foreign, to it; but an ascent of the matter itself into heat. The matter does not develop, or give out heat, but becomes heat, namely æther.

392. The loss which a body sustains by the radiation of heat is as slight as the æther is subtile or rare; thus it is infinitely small, where the æther is infinitely light. We cannot speak of the loss of matter, while it is hot; although a true loss is present, if the point in question be philosophically regarded, it can come, however, as little under consideration as the weight of the æther.

393. The fusion of bodies is a diminution of the fixity of their poles, their further evaporation, and thus an approximation to indifference or the apolarity of æther. Heat is the actual retrogression of matter into æther; light is only the efficient of this transition.

394. As heat becomes originally excited, so must it be always excited; two causes for one effect are impossible. The excitation of heat by oxydation takes place in the same way as by light, namely by polarization and separation therefore of the Indifferent. The generation of heat by chemical processes is based upon the same principle. Lastly, the generation also of heat by compression and simple friction is wholly similar to that which is caused by light. In every case they are only polar, and by no means mechanical operations upon matter, whereby the fixity of the poles becomes changed.

395. It is not a change of cohesion, which the friction mechanically effects, but the act is purely dynamical. The essence of friction consists in the constant renewal of polar change, because thereby an infinite number of projecting points or apices are alternately brought into contact. There is verily no smooth body.

396. Heat is the transition of light unto darkness; for it is indifferent æther, only moved. Colours are thus also a conflict of light with heat, and out of this conflict issues the most beautiful, the highest colour, the Red of fire. In fire, the contest between light and darkness has risen to the highest pitch; the æther therefore is also moved to the greatest degree, becomes hot. If the indifference becomes the maximum, the vital tension then relaxes, the fire is extinguished; finally motion ceases, it is cold and dark.

397. In matter also light and heat operate against each other. Light deoxydizes, heat oxydizes. If light appears to oxydize, it is only by evoking heat.

398. Heat is the function of expansion for matter. Every body has a definite degree or amount of expansion, therefore a definite fixity of æther; this æther is latent heat.

399. Heat operates spherically in matter or in all directions, not in the linear direction like light. The propagation of heat can only take place slowly, because it is not a polar action, but only the result of such an one, only motion. Heat does not penetrate bodies mechanically, but dynamically like light, yet without decomposing them, as light does.

400. Heat is related as indifferent æther to the matter as to a Polar. This relation imparts the process of conduction. Light, however, is itself polar and disturbs matter, while it passes through it.

401. With the exception of their cohesion matters are not directly changed by heat.

402. During every process of decomposition, during every process of light heat must be produced, but not light also during every process of heat.

403. Dense materials must conduct heat because they are most opposed to it. It is only the formed element therefore that can possess capacity for conducting heat.

404. Absence of form is the character of isolators of heat, form that of the conductors, apart from

every remaining quality. Solid bodies, which easily pass into the formless condition are isolators.
405. The densest bodies among the solids must be the best conductors. Regard may be first paid to the nature of their constituent parts in the sequel of the present section of this work.
406. The conduction of heat is a continuous excitation from one part striving against or resisting the other; the earths (as metals) are the best conductors.
407. Matters, which are images of heat, do not conduct, because they enter only as minimum into conflict with it, and while they expand at once convert themselves into the same. Such is the air. The Heterogeneous only conducts. The heat expands in the air only by continuous motion of the aerial particles. The air is an isolator. Water ranks in the middle between air and earth.

408. With respect also to conduction, light is opposed to heat. Light is conducted by those very bodies, which isolate heat, and isolated, not admitted to permeate or absorbed, by those which conduct heat. The air conducts light, isolates heat; the metal, the earth conducts heat, isolates light; water holds a mediate relation towards the two, yet towards light that of a conductor, because it is deoxydizable.
409. The conduction of light is likewise a process of deoxydation, or a disintegration of matter. By the conduction of light the bodies are chemically analysed, and finally resolve themselves into their principles; such after all is the case with every glass and crystal. As the process of conduction of light may be called a process of deoxydation, so also may it be called a process of the generation of colour. The conduction is an offuscation of light, a colouring; the deoxydation is a solution of the material bonds, an elevation unto colour.
2. Functions of the Air.
ELECTRICITY.
410. The air is the slightest combination of the primary bodies, and stands in opposition with the two other elements, as more solid combustions. In this antagonism the air lays claim upon the other elements to analyse them; these, however, upon the air, to combine, and undergo more vivid combustion.
411. This antagonism is on a large scale an antagonism of periphery and centre, like the primary antagonism, by which planets and suns have withdrawn from each other. The tension of air with the other elements is called Electricity. The centroperipheric antagonism between the sun and the planet, between light and colour, represented in an elementary manner is electricity.
412. Sun and planet are electrically related to each other, and the circumrotation of the latter may be viewed as conditioned by the change of electrical poles. Colours also are only electrical productions. Light itself is similar to an electrical tension of the æther.
413. Electricity is an action of the periphery or limit and thus of the surface of the globe. The surface of the globe is, however, everywhere + - without centre. The principles of electricity are therefore eternally separated without having a middle point, as occurs in magnetism. The electrical poles live in eternal animosity, because they have no point of union. Such is then the essence of electricity. Electricity is therefore only a function of surfaces without any line. It clings only to the upper surface of the bodies, and does not penetrate into their thickness. It is only the tension of the surfaces of bodies against each other, of the apices of divided radii.
414. The air is the periphery, the limit or boundary of the earth. Electricity is therefore the spirit of air. It is in its most active state in that stratum of air which is in contact with the earth, because there the limits are situated. Upon this lightning depends.
415. Electricity, as an aerial function, is terrestrial heat. Both are therefore conducted by the same rigid linear bodies and isolated by the same. The isolation of electricity coalesces with

absence of form, or with the transition of denser bodies into air.

416. Electricity is an antagonism between air on the one side and water and earth upon the other. By these therefore two kinds of operation are posited in the air.

417. While electricity is the tension of air with the other elements, it is also the tension of the principles themselves of air. Electricity is a twofold character appertaining to the two principles of air. The tension of æther and of substance, thus the tension of fire repeated upon the two elementary bodies of air, is electricity. Electricity exists under two forms, as the electricity of the substance or body of fire, and as that of the planet. These two conditions are perhaps incorrectly named + E and -E, or positive and negative electricity.

418. The + E is the more energetic, active in itself, polar; it is the electricity of fire represented in oxygen. The -E is the weaker, that which has only been evoked, the basic; it is the electricity of the planet represented in nitrogen, hydrogen, carbon, sulphur. It may be said that -E is related to + E as planet is to sun, as periphery to centre. The sun is + E, the planet -E, the one the electricity of oxygen, the other of hydrogen.

419. These electrical conditions must always be changing in the air, according as the slightest influence is brought to bear upon them from without. The air consists in this change of electrical poles. Were the poles not to change, the air would be a solid element. The earth is an air with fixed electricity, the water with extinguished or neutralized electricity.

420. The twofold character of the aerial principles is increased by every polarizing action from without, and therefore principally by the surface of the earth, which consists of two elements. Were there no surface to the earth, were simply air present, then there would be no electrical change of poles. The surface of the earth itself, however, produces no change of poles in the air, for it remains always neutral; but it becomes unequal or polar from the change wrought upon it by water and earth, by light, by heating and by chemical processes.

421. The air also changes the electricities while it roams over the earth. This wandering motion is a contact of differently polarized tracts of the earth. Every mountain, valley, and river, every meadow is differently polarized; from each the air derives another electricity. Through this ceaseless alternation of polar exchange, its activity becomes so elevated that at last the electricity makes its appearance in a manner cognizable by the senses. The production of electricity by friction admits of a similar explanation. Friction is in miniature, what the sweeping of air is over the earth. Were the earth quite level, and composed of homogeneous matter, the air would not become electric by motion.

422. All terrestrial electricity has been evoked by a change of poles analogous to friction. Through light nothing foreign is posited in the air, any more than by friction, and by both the electricity is attained in a similar dynamical manner.

423. The elements of the air, polarized to the highest degree by electricity, must combine, and this combination is the process of combustion. The final result of electrical tension is combustion of air. It is only, however, the two primary bodies that undergo combustion; the two constituent parts of the air must, therefore, be driven by electricity to their last extreme, even to the most perfect element of fire and the basic or terrestrial substance. The internal combination of both these primary or elemental bodies, or the product of this aerial combustion is next of all water. The termination of electrical tension in the air is rain. All rain is the extinguished function, the dying spirit of air. The two hostile principles are reconciled in water. Water accompanies every process of combustion.

424. By electricity the air was separated into the two inferior elements, into water and earth. At present, where the whole water and the whole earth has been precipitated from air, rain is

certainly as a rule only water condensed and held in solution in the air.

425. The nitrogen gas is the residue of the primary rain. After the air has become sea by the combination of oxygen and hydrogen, it no longer creates water out of its own bulk, but now rederives it from the sea, and still but imitates its primary process in rain.

426. Were there merely solid land upon the planet, the oxygen pole would be evoked in the air, and the latter would be precipitated wholly as water. Were there merely water or sea upon the planet, the nitrogen pole would be evoked, and the air be wholly precipitated as earth. But since the two elements are constantly and alternately operating upon the air, so must at one time water, at another earth, be generated and precipitated from it. The earths in the act of falling down, or in precipitation, are the meteoric stones. They are only the after-births of the primary decomposition of air.

427. It can indeed be none other than carbonic acid, which is converted by decomposition into meteoric stones, or at least into the metals and earths which they contain. Sulphur must be regarded as the melting down of carbon and hydrogen. In the upper lighter regions of the air carburetted hydrogen gas may very well be present, and condense itself into meteoric stones.

428. The condensation must take place by means of electricity. The greater condensation or solidification, as well as the calefaction, results indeed first through the fall.

429. Their bursting is only conceivable when some hydrogen gas is in the meteoric stone, which gas, when submitted to the highest degree of heat, becomes finally fulminating gas and bursts the stone.

430. Meteoric stones are children of our planet, and not of the world-space. They are the antagonism of the water-formation or of rain.

431. The falling stars are simply indeed condensed carbon or hydrogen gas, but which by reason of its small mass does not become solid. It is probably converted only into sulphur. Thus falling stars would be the other halves of meteoric stones; the latter consisting of a preponderance of carbon, being therefore earths and metals; the former of a preponderance of hydrogen, being therefore carburetted hydrogen or sulphur. A greater number of meteoric stones must fall over the sea than over the solid land.

3. Function of the Water.
SOLUTION.

432. The function of water is necessarily homologous with the process of combustion, because the main bulk or proportion of water is oxygen. In conformity with its spiritual activity does water seek to convert the two other elements into water, to impart to them its form, to fluidize them. This happens with the air when it has been absorbed; but upon the earth also the water exercises the same action.

433. The function of water is the formation of globes or the Process of Solution; it directs itself chiefly against the solid; for the solid element is the redintegrant factor of water. Solution is a positing of the solid under the internal polar form, but the poles of which have not yet separated. Every solid formation has come out of water, as water has out of air; every new formation must also return out of water, by fluidization, by relaxation of the poles. By solution, solid matters are again reduced to their primary condition, and are then capable of reassuming new fixities. The process of solution is a process of becoming water, not by agglutination, but by liberations of fixity; a Solution in the strongest sense of the term.

434. No process of solution is conceivable without oxydation. The dissolved body, while it obtains the aqueous nature, is taken up in the sense of oxygen. No solution occurs without

oxygen, as well as no combustion is possible without water. The solvent character of water is based upon the preponderance of the oxygen over the hydrogen.

435. During every solution the two principles of water enter into a state of tension with each other, as the two aerial principles do in electricity. This tension is established by that which is to undergo solution; for everything so circumstanced is polar towards water. During every solution the oxygen is elevated in its pole, and the hydrogen likewise. If the solution be very heterogeneous, they separate, the water is decomposed. In the pure process of solution the water simply abides in a state of tension; if each aqueous principle is actually and independently self-evolved, chemistry then originates; but of this we shall discourse in the sequel. The process of solution may be characterized as the equilibrium of the process of tension between the object that is to undergo solution and the solvent, and between the two principles of the latter, whereby separation is not thus attained. As electricity finally strikes out into the process of combustion, so does that of solution into the chemical.

436. Solution is in essence like electricity. Solution is an electrical tension between oxygen and carbon; electricity is a process of solution between oxygen and nitrogen, a process of tension without separation of principles. Air and water are in a state of constant tension towards each other; and hence therefore result the constant evaporation and the clouds.

437. What lightning is in the air, namely formation of water, that is chemistry in water. The salt in the sea is what the clouds are in the air. Two electrical clouds are what two salts are in the water. Rain is the imitation of the creation of water. Precipitation of salt is the imitation of the creation of earth.

4. Functions of the Earth.

CRYSTALLIZATION.

438. The earth-element is the highest result of combustion in creation, the highest fixation of æther. The earth is the æther represented as centre in the Material, or it is the identification of all polar binary division in the Terrestrial, independently of oxydation; therefore its parts are motionless, be they dust or compact masses. The earth is the corporeal gravity, the substance as a perfectly simple position without emergence out of itself, the 0, the terrestrial monas. The earth is consequently the heaviest and densest element, and is that which must include the middle point of our own and every planet.

439. In the middle of the planet there is only earth, and nothing else; the middle is not hollow, does not contain any central fire as has been imagined, nor air, and the science of Geogeny will show that no metal also could be contained in the interior of the earth.

440. In the air both material principles are only associated with each other, in water they are mixed, but in the earth identified, blended together.

441. The earth is to the other elements what the sun is to the planets, namely, the basis, the centre and that the mathematical as well as the dynamic centre. That this is the part played by the earth-element is proved not only by its character, but also by its volume. The earth-element exceeds the other elements in mass as much as the sun does the planets; the water is only the vascular system in the flesh of the earth-element; while the air is only the expression of the limit or in other words its integument.

442. Everything therefore that now occurs upon the planet develops itself out of the earth, the water and the air being only the auxiliaries of generation. The Earthy is developed in the water by the air. As creation has been closed with the earth, so may the solid materials, which are now and then found upon the planet apart from the earth, not be products of the first creation, but only

developments of the planet when created and cosmically completed.

443. The earth as material gravity is solid. It has, however, originated out of the fluid, therefore by a process of cohesion; this is called the Process of Crystallization.

444. The process of crystallization is perfectly equivalent to the process of fixation of the æther, and is only the termination of the same. As light at any spot in the world-space creates a central point, the nucleus of a comet, around which more of the mass of æther is constantly accumulating till it finally coagulates into the solid condition; so does the process of crystallization evoke some particular spot, point or nucleus in the water, wherein the crystallizing forces are excited, which attract the mass that is susceptible of fixation and fashion it into a crystal. The process of crystallization is a process of fixation, and with it also is furnished the theory of crystallization. The process of crystallization is a process of polarization, and one indeed that proceeds from a centre; or, properly speaking, the point, from which the polarizing process emerges in a fluidity that is fixable or in other words susceptible of crystallization, becomes a central point, a middle point and virtually the middle point of the crystal.

445. The process of polarization does not originate absolutely in the fluid, any more than light has the power of concentrating or crystallizing itself in any given part of the æther; but by an external determination. This is a granule, a projecting point in the vessel or in the hollow of the earth, in which the crystals originate. The crystal never begins in the middle of the fluid, but only on its walls or on the surface. The point of polarization or of crystallization has been granted; now this is polar towards the fluid, and works therefore by polarizing upon it, and through this water also passes over into the crystal, forming what is called water of crystallization.

446. This polarization of the fluid passes in every direction; for every polar point is polar all around. Thus a spherical portion of the fluid is polarized round about the point. The fixable parts are spheroidally attracted and gather together from all sides around the point. For were the polarization not to traverse the whole mass, but only according to individual lines, the crystal must then indeed be jagged or indented.

447. In this manner the crystal would have been a globe, from the fixable particles lying together in distinct points, after the manner of pulp. But this is impossible, because the point of starting or departure is differently polarized to the fluidity, being negative according to observation. Every polar process does not operate in continuity, so that one end of the line should be purely positive, the other, however, purely negative; but every polar line is an infinity of poles, where, however, at one end the positive character only, at the other the negative preponderates. Such a line is e. g. as follows, + - + - + -, which begins with + and ends with-; it therefore has a preponderance of + at one end, of-at the other, and yet is both everywhere. By this infinity of polar change the fixable particles range themselves behind each other, while they separate from each other to an infinitely minute degree; these parts polarized behind each other are lines or fibres. Every crystal must accordingly consist of fibres; none possesses an homogeneous or pultaceous structure.

448. In the crystal one principal direction of polarization originates, which is effected by the antagonism of the point of crystallization with the fluid mass. It gives the direction of the crystal and its energy gives the length. This principal line consists of two poles that recede from each other, and these determine the two ends of the crystal, which are always similar, provided no mechanical obstacle be interposed.

449. From each of the mutually seceding poles lines of polarization issue at definite angles, which (like elliptical radii on the periphery) meet at the sides of the newly produced crystal. Then again between these radii tension arises, so that the fibres become lamellæ. The main line between the two mutually seceding poles is the central line or polar axis of the crystal; the

angular lines which determine the position of the lamellæ, are the polar radii. The polar radii determine the nucleus of the crystal and are therefore nuclear lines; the polar axis determines the whole of the crystal, is the crystal, the central-line, and determines the form in general, or what has been called the secondary form.

450. Since all polar activities operate only in a straight line, there can thus be no globular crystal. Water is only susceptible of assuming the globular form upon a large as well as a small scale, because there are no fixed poles in it. The nucleus does not originate previous to the secondary form; since it is verily impossible for the polar rays to originate without a polar axis.

451. There are no actual degradations in the genesis of the crystal; they are only a mathematical expression for the finished form of the crystal.

452. The number of possible or actual nuclei is definite. They are based upon the combination of the laws of the globe with those of the polarity.

453. The simplest angular body must be circumscribed by at least four surfaces, and thus be a tetrahedron.

454. The fundamental nucleus of crystals is, however, the double tetrahedron or the hexahedron, namely the trilateral double-pyramid; for radii do not proceed simply from the point of commencement, but also from the extremities of the axis. When the superior and inferior radii meet, they must form a double-tetrahedron. The disposition to this form has been implanted in all crystals. If the nucleus becomes no such hexahedron, the aberration from, still admits of being referred to, the hexahedron.

455. There is no prismatic nucleus. The columns and parallelopiped nuclei are only mutilations.

456. The tetrahedron is also only a mutilated nucleus. To the essence of a nucleus belong two tetrahedra, with their bases joined to each other.

457. The six-sided double-pyramid is a duplication of the hexahedron. The octahedric nuclei, are things intermediate between the three-and six-sided interruptions of the natural type, like quadrinumeral or tetrapetalous corollæ in flowers.

458. Columns originate only between the two tetrahedra, without doubt owing to deficiency in bulk.

459. If the three-sided double-pyramid be the primary form of crystals, so must the six-sided column with trilateral terminal pyramids be the ultimate form. The rhomboidal-dodecahedron is therefore the most perfect crystal. It is the most perfect representation of the globe in the angular form.

460. The crystal can not commence with the nucleus and then for the first time continue to grow or even change into the perfected crystal, because it becomes only crystal in the conflict of the linear and spherical action. As little as the sun can be produced without the planet, or vice versâ, so little can a nucleus subsist without what is called the external shell or crust. The nucleus is in fact determined by the shell of the crystal, namely by the polar axis. But inversely also the shell is determined by the nucleus, by the polar radii. A nucleus alone would be called a centre of a circle without circumference. Microscopic crystals therefore have at once the same perfected form as the largest in size. A crystal is an entire heavenly body; it is determined by central forces, which have, however, been roused and conducted by dualizing forces, forces of light. Everywhere do we meet with the same laws of the fixation of æther, upon a small as well as large scale.

461. Every solid matter and thus the Terrestrial generally, is crystallized upon a small as well as large scale. There cannot be an atom that were not crystallized, not arranged according to central

and polar forces. Every crystal is therefore, and especially by reason of the infinity of the subordinate poles, crystallized again upon an infinitely small scale, or in other words it consists of infinitely numerous crystals. Every lamella or every particle of the lamella of a crystal consists again of crystals. These are what are called the integral parts, properly integral forms of the crystal. They are all probably hexahedra. The metals usually form but very small crystals, probably because the atoms are too heavy and cannot therefore be attracted from any distance.

462. Crystallography has been incorrectly made the principle of division in Mineralogy. A single character can never become a principle of division. If also it were actually true, that the form is always disposed according to the Interior, yet the form could never be the principle of division, but the Interior itself. The form is only the sign, but not the spirit, the essence of the mass.

D.—KINGDOMS OF NATURE.
INDIVIDUALS.

463. All the matters that have hitherto originated have done so only in a general not a particular manner. They are constituent parts of the universe, in which as yet no distinctions reside. So soon as distinctions occur in the elements, they cease to be general matters, and become particular or individual things. The sum of the individuals is the Kingdoms of Nature.

464. The kingdoms of nature are the repetition of the world upon the planet. This repetition in consciousness is Natural History.

465. Acts of the world repeated upon the planet are combinations of the elements. Creation, which has hitherto advanced, now retrogrades, and thus by the combinations of general elements that have been already created.

466. Combinations of the elements, in accordance with the laws of the world, are upon the finite planet particular or individual bodies. The kingdoms of nature are the totality of particular bodies.

467. That which is not a Particular belongs not to the kingdoms of nature, and thus also does not come within the province of Natural History, but of Physics.

468. The earth-element lies at the basis of all the combinations of the elements. These combinations are therefore ascensions or retrogressions in creation. Only three such combinations are therefore possible, viz. 1. Of the earth with water, or air, or fire—binary combination. 2. Of the earth with water and air, without fire—ternary combination. 3. Of the earth with water, air, and fire—quaternary combination.

469. Out of the binary combination the quiescent bodies originate, for they are only a part of the planet—Minerals, Earths.

470. Out of the ternary combination originate bodies that are internally moved, for they are a whole planet in Particulars—Plants.

471. Out of the quaternary combination particular bodies, moved throughout and rotating around themselves, originate; for they are representations of the whole universe—Animals. Individual bodies that are moved internally are called organic.

472. There can be therefore only three Kingdoms of Nature. The first consists simply of individualities, because it is not the equal proportion of all the elements. The two other kingdoms, however, are combinations of the individualities of the earth-element with two or three elements, and are thus equivalent to the planet or to the whole universe. The organic bodies are thus combinations of the Singular with the Whole, and supply the third part of the Philosophy of Nature, the Organology.

FIRST KINGDOM.MINERAL OR EARTH KINGDOM.

473. Uni-or bin-elementary terrestrial bodies are minerals or earths; their development is Mineralogy in the general sense of that term. The earths regarded individually gives us the science of Mineralogy proper; combined to form a whole, that of Geology.

I.—MINERALOGY.
474. Mineralogy teaches us the development of the earth-element.
475. The earth-element does not exist universally, but only in particular bodies or individuals. There is no general earth, but it is either silicious earth or common salt, sulphur or iron, and so on.
476. The earth-element or the earth can only sustain changes, which are permanent or abiding; for in it alone fixation has become formation, in which the atoms do not move, or whereby at least a constant individual character of body, or one that is chemical, becomes apparent. The changes undergone by the other three elements are not constant, because of the atoms ceaselessly moving and balancing themselves. They do not exist individually, but only universally. There is only one water, one air, and only one fire; there are therefore no igneous, aerial, and aqueous individuals.
477. The changes of the earth-element can only take place upon its fundamental or characteristic body, thus on carbon.
478. Nothing can, however, change of itself. All change must proceed from external influence. All things can be changed therefore by such influences only as are already antecedent to or before them. The two other bodies, however, are prior to carbon; before the earth-element only the three other elements.
479. The earth can therefore be changed in only two ways; either the carbon by the other elemental bodies, or the total earth-element by the other elements.
480. The changes wrought by the influence of these bodies are, however, only partial or fractional changes. Therefore partial or chemical diversities only originate, and with them other different bodies or degrees of such. The changes effected by the elements are, however, total changes, which bear not only reference to the carbon, but to all the constituent parts of the earth-element.
481. Total changes or different conditions of the earth-element are called Minerals, or earths.
482. The genesis of minerals, thus their collective character, as differently posited fixations of earth, determines the classes, orders, and genera.
483. The genus is the product of a genetic moment, and is therefore always a definite, chemical mixture, which alone consequently expresses the essential character. Hitherto there has been no definition of mineral genera.
484. Species of minerals are successive developments of the genetic moment, thus stöchiometric subdivisions of the genetic mixture, e. g. the different degrees of oxydation of nitrogen, in the oxyde, binoxyde gases and nitric acid. Hitherto it was not known what a mineral species might be; Physio-philosophy has been the first to introduce clearness to these conceptions.
485. A stöchiometric mixture in the earth-element is an individual.
486. Individuals only are the object of natural history, and thus neither water, air, nor fire. This

also was not known previous to Physio-philosophy; it is, however, gradually acknowledged also by empirics.

487. The crystalline form is merely an external character for the species, and therefore the same nuclei may occur in the different orders.

488. Kinds or varieties are different conditions of cohesion. They are therefore not determined by the form of the secondary crystal, since the aberration of forms results only from a stoppage upon their part half way or from the quantitative energy of the polar radii or polar axes.

489. While æther, air, and water, as being general matters, do not belong to the mineral system, what have been called artificial salts must on the contrary be admitted therein, because they are no works of art. The chemist only brings bodies together which do not come together accidentally in nature. It is a true misapprehension of nature's products if those substances only, that adhere to the earth, are recognized as such; surely this definition is perfectly ridiculous.

490. Two kinds or modes of division are possible, a chemical, and a genetic or philosophical.

491. The chemical principle of division of the earth are the elemental bodies. The philosophical or naturo-historical principles of division are the elements.

492. In reference to the chemical bodies four combinations only are possible. 1. Carbon, represented as perfectly pure, may be regarded as Metal. 2. Carbon united with hydrogen, is manifested in the Inflammables. 3. Carbon with oxygen makes its appearance in the Earths. 4. Carbon with oxygen and hydrogen in the Salts. According to this view, the classes would succeed each other thus:—

1. Ores.
2. Inflammables.
3. Earths.
4. Salts.

Now, as the earths here intervene between the Inflammables and salts, it is at once seen that the series is incorrect; for the earths form by far the largest mass, and must therefore constitute the groundwork or basis of Mineralogy, and thus stand at the commencement. If all metals, Inflammables and salts were to be deducted, the globe of the earth would still lose but little of its magnitude.

493. This chemical division admits thus of no strict arrangement, since what are called minerals follow each other unnaturally. Meantime the chemical view admits also of a philosophical treatment and amelioration of the serial order. It may be said that the earth consists of much carbon, little oxygen and very little hydrogen, without any other element. Salt, of little carbon, much oxygen and little hydrogen, together with water. The Inflammables of little carbon and oxygen, much hydrogen, besides air. Ore, of much carbon, little hydrogen, and still less oxygen with fire. As the fire or the æther is imponderable, so do the three elemental bodies appear blended together into one apparently simple body, with which gravity, light as lustre, heat as spirit and the conduction of heat, are only spiritually combined.

494. But this view leads directly to the genetic division, as the only true one, to that, namely, which has been based upon the mutual influence of all the elements. It is itself the ultimate cause or foundation of chemical division.

495. There can accordingly, as there are only four elements, be only four kinds of minerals. The Earthy either continues unchanged, or it is changed by water, air and fire.

496. When the earth-element originates or separates itself from the water, in order to free itself from all the properties of the latter as well as from those of the air and fire, and to become stiff and solid, the remaining elements exert an incessant influence upon it, and draw a portion of it

into their circle, i. e. they confer upon it their properties.
a.   The Earth-element can be changed by fire   Fire-minerals.
b.   Or changed by air   Air-minerals.
c.   Or changed by water   Water-minerals.
d.   Or lastly, it is severed wholly and substantially free   Earth-minerals.
497. Through the influx of fire upon the formation of the Earthy it becomes an identical, homogeneous mass, in which the possibility resides, as in the æther itself, of undergoing all changes. This developmental stage of the earth-element is represented by the metal. The homogeneous mass of the metal can become earthy by oxydation, aqueous or saline by acidification, aerial or combustible by being hydrogenized.
498. The metal is unanalysable, as is the æther, although it consists of three forms. The metal is easily restored or brought back from its combinations.
499. Besides, however, the identical, homogeneous or simple character, the metal has still also the three characters of fire or of the æther. It is therefore a triplicity in identity.
a. In so far as gravity is represented in it, it has the identical or homogeneous mass already indicated, and is heavier than all other bodies. It is central mass. It must be regarded as pure carbon. Metal and the body of gravity are one.
b. In so far as light is represented in it, it has the peculiar lustre, which stands again also in intimate connexion with the homogeneous mass. The usual colour of metals is white, the colour of unsullied light. The lustre is properly a self-illumination, and thereupon depends their repulsion of light, or opacity. Metals are therefore adiaphanous or opaque, because they are noncombustible by light. As soon as they become decomposible, namely oxydes, they become also transparent. The metals are the only opaque bodies, because they alone are non-decomposible. All matters become only opaque by admixture with metal, or in so far as the metallic body resides at the bottom of all. The visibility of the world is based upon its metallic character. Without metal we would see nothing.
c. In so far as heat is represented in metal is it extensible, fusible and fluidifiable. Metal is water that has become dense.
500. In so far as the air has acted upon the Earthy during its origin, it has imparted to it electrical and combustible properties; the metal has combined with hydrogen, has become an Inflammable, as in sulphur or pit-coal. Sulphur may be regarded as the intimate fusion of hydrogen with metal; coal probably a combination of the same probably elicited by means of oxygen. Inflammables are idioelectric and combustible, because they are rigidified air. That matter belongs only to the Inflammables, which, being once kindled in exposure to the air, continues to burn of itself. The Inflammables are volatile, since they undergo combustion, i. e. they take on the condition of their antetype, the air. They have from metal the opacity and the colours, but they do not preserve the lustre or self-illumination. They become transparent simply by crystallization or oxydation.
501. With the generation of the Earthy water imparts also to a portion of the same its properties, dissolubility and transparency. To the metal and hydrogen oxygen is next added. An hydrated Earthy originates. The Aqueo-earthy is fluid in water; it is salt. Salt changes its form in the readiest manner, because it is the metatype or likeness of water; and hence its susceptibility to crystallization. It is not combustible by itself, because it is essentially an oxyde and hydroid. Salt is a metal or Inflammable that has undergone combustion, and can therefore never be simple.
502. Now that part of the earth-element, which remains after the salt, the Inflammable and the metal have been separated, is plainly the Earthy or the earth. It has therefore no aqueous properties, is not soluble; it has no aerial properties, is not electric and combustible; has no

metallic properties, is not heavy, nor opaque and glittering, not fusible and malleable or extensible. The pure Earthy is always fixed or firm, and therefore figurate. The Earthy is a metal, with which the oxygen has been intimately melted down; for it is the identification of all elements.

503. The Earthy is the principal mass, because it represents the earth-element itself. Salt, Inflammable and metal are only subordinate masses, because they are only displacements of the earth-element by the other elements. Therefore a small part only of the Earthy has become salt, a yet smaller Inflammable, and the smallest, metal.

504. Although the metal is simple, it can by no means correspond, as might otherwise appear, to the earth-element; for every element is a totality of elemental bodies, and therefore those minerals, which represent the pure earth-element, must be compound, without, however, exhibiting the characters of the other elements. This is found only in the earths.

505. There are, accordingly, in a genetic point of view, four, and only four, mineral classes. They originate in an ascending direction, from the earth-element by water and air up to fire. The Classes are—

I.   Earth-minerals Earths.
II. Water-   "   Salts.
III. Air-   "   Inflammables.
IV. Fire-   "   Ores.

a. Earths are those minerals, which admit of being changed neither by water, nor air, nor by fire; i. e. which are neither soluble, combustible, fusible, neither yield colour, nor are particularly heavy. Such minerals have been properly called earths, as silicious, sacrilegious earth, &c.

b. Salts are those which have aqueous properties, i. e. are soluble.

c. Inflammables are those which have aerial properties, i. e. are inflammable and volatile.

d. Metallic ores are those which have the three properties of fire, are superlatively heavy, yielding light or colour, and fusible.

506. The Earths are to be regarded as the proper total earth-element, namely as carbon neutralized by oxygen. The Salts are to be regarded as combinations of the earth-and water-elements; therefore as combinations of carbon with oxygen and hydrogen. The Inflammables are to be regarded as combinations of the earth-element with the air-element, thus of carbon with hydrogen, which supplies the place of nitrogen. The Metals are to be regarded as combinations of the earth-element with the fire-element; therefore as carbon without any other body, only combined with spiritual actions, namely gravity, light and heat. Hence the apparent simplicity of metals, and the great number of special properties, which are absent in the other classes.

DEVELOPMENT OF THE CLASSES.

507. As there is not simply a single earth, salt, Inflammable and metal, but in each class many of them; we have to inquire what is the groundwork of their further distinctions or of their systematic division. Here also may we go to work again chemically and philosophically.

a. Chemical Division.

The Metals, chemically regarded, do not admit of being separated into constituent parts. They exhibit only physical differences in gravity, colour, hardness, malleability, conducting power, tension or their mutual polarity. If it be endeavoured to arrange them according to these respects, nothing but disorder results. The same is the case in reference to their affinity for oxygen, sulphur, the acids and other metals. Rather more order is at once displayed if their philosophical composition, namely as carbon and fire, be submitted to our consideration.

508. In consequence of this view, the Metals must divide into Earth-metals and Fire-metals; and the latter again into three subdivisions, nearly as follows:—
A. Earth-metals—difficultly fusible and invariably oxydized—Sidereometalla, e. g. Iron—Manganese, Wolfram, Uranium, Titanium, Chromium, &c.
B. Fire-metals.
a. Heavy metals; difficultly fusible, unoxydized or noble metals, e. g. Platinum, Nickel, Cobalt.
b. Light metals; the easily fused noble metals—e. g. Gold, Silver, &c.
c. Heat-metals; the easily fused, ignoble and frequently volatile metals, e. g. Lead, Tin, Antimony, Zinc, Arsenic, &c.
509. The Inflammables divide under a chemical point of view into two groups—into the varieties of Coal and Sulphur, whereof the Earthy lies at the basis of the former which is non-fusible; the Aerial at that of the latter. They do not admit of being divided, unless a mean betwixt the two be taken, the combinations of carbon and hydrogen in the resins.
510. The Salts admit of a better dismemberment. Their constituent parts are alkalies and acids, the former the Earthy, the latter the Aqueous. The mean condition is exhibited by the neutral salts, so that three orders are the result.
511. Now by the philosophical view we have first attained to the very remarkable import of the acids. They are forsooth nothing else than oxydized elements and mineral classes. In the nitric acid it is evident that, as the acid of nitrogen, it is the aerial acid; as sulphuric acid is the inflammable acid; arsenic acid the metallic acid. Upon this ground we may expect that the other acids also have a similar origin. Without much hesitation hydrochloric acid may be viewed as the aqueous acid, which is associated with the sea; the carbonic acid as æther- or igneous-acid, as well on account of its constituent parts and gaseous character, as chiefly on account of its general diffusion. There remain then only two that have been called mineral acids, the fluoric and boracic acids, the first of which, as conqueress of the earths is the earth acid, the last being thus the acid of the salts. We have accordingly—
a. Elemental acids.
1. The oxydized ætheris   Carbonic acid.
2. The oxydized air   Nitric acid.
3. The oxydized water   Hydrochloric acid.
b. Mineral acids.
4. The oxydized earth is   Fluoric acid.
5. The oxydized salt   Boracic acid.
6. The oxydized inflammable   Sulphuric acid.
7. The oxydized metal   Arsenic acid.
512. The vegetable and animal acids are none other than repetitions of the elemental and mineral acids. They may perhaps be parallelized in the following manner.
Fire-acid   (Carbonic acid)   Acetic acid   Hæmatosine.
Air-acid   (Nitric acid)   Malic acid   Lactic acid.
Water-acid   (Hydrochloric acid)   Saccharine acid   Mucic acid.
Earth-acid   (Fluoric acid)   Tartaric acid   Phosphoric ac.
Salt-acid   (Boracic acid)   Tannic acid   Uric acid.
Inflammable-acid   (Sulphuric acid)   Succinic acid   Sebacic acid.
Ore-acid   (Arsenic acid)   Indic acid   Formic acid.
All the remaining acids must be viewed as subordinate to, or as kinds of these.

513. The alkalies appear to follow the same course, though it does not admit of being so completely demonstrated.

Fire-alkali	Ammonia	Vegetable,	and	Animal alkalies.
Air-alkali	Potash	Alcaloids		Alcaloids.
Water	Soda			Urea.
Earth	Lithium			Bile, &c.
Salt				
Inflammable				
Ore-alkali				

514. The earths proper do not consist of two principles, and do not, therefore admit of being chemically divided.

515. This division is only incorrect in a naturo-historical sense, because it has no reference to the totality. Inasmuch as every mineral class is viewed as having originated out of only one or two elements, it divides by the chemical method only into constituent parts or fractions, as the acids and alkalies, which are obviously only moieties, and taken in a strict sense are not true minerals.

b. Genetic Division of the Classes.

516. The total division only is genetic and consequently correct.

517. As the classes have originated through that which directly preceded them, namely, the elements; so must the divisions of the classes be determined by the other classes. Such divisions are called orders. Every class necessarily divides into four orders.

Order 1.   Earths.
2.   Salts.
3.   Inflammables.
4.   Ores.

CLASS I.

EARTHS.

518. There must be therefore pure earths, haloid or salt-earths, inflammable earths and metallic earths or ores.

1. The Earth-earths must have neither saline, nor inflammable nor metallic properties, and thus also be insoluble in acids. Such is the case with the Silicious earths.

2. The Haloid-earths must have saline properties, dissolve in acids, but not fall to pieces when exposed to air and fire. Such is the behaviour of Argillaceous earth; it admits besides of combining with water, that antetype of the salts.

3. The Inflammable earths must be soluble in acids and exhibit electric or aerial properties. Such is the behaviour of the Talcose earths; its minerals are unctuous, fall when exposed to the air into electric lamellæ, and burn brittle.

4. The Metallic earths must undergo change in acids, air and fire. The calcareous earth dissolves in all acids, burns corrosive and becomes almost a metallic calx. The Orders of earth are consequently:

1. Earth-earths   Silica; Quartz, &c.
2. Haloid-earths   Clays; Felspar, &c.
3. Inflammable-earths   Talcs; Mica, &c.
4. Metallic-earths or Ores   Calx.

519. Nature does not produce any so-called pure calcareous earth, but only this earth in an oxydated condition. Carbonic acid is the oxygen of the earth that has become free, and the corroding calx is the Metallic, the other constituent part of the Earthy, which has obtained some

oxygen, but lost the Aqueous by the carbonic acid and thereby has become corrosive.
520. The carbonate calcareous earth is the whole earth, not the corrosive. This is only the half of the earth-element, only its basic or phlogistic principle. What has been called pure calcareous earth is a half earth; the perfect or naturo-historical earth is just that which is chemically impure.
521. The calcareous earth is not, however, perfected with one position. It still exhibits several stages of development which appear to be approximations to the salt, e. g. Strontian and Baryta.
522. The silicious earth, which principally represents the Earthy, holds its principles more firmly together. No separation occurs there in the carbonic acid and the basic or corroding body of earth; no association with water, no great activity, no direct participation in the highest evolutions of the planet; but it continues to lie in an extreme state of contraction, and in a state of indifference in the non-differencing darkness.
523. This pure earth is the basis, the pedestal of all the other earths, and the foundation of the planet; for it alone is the earth proper, the earth-abiding earth-element, while the other masses of earth, divided in their principles, have pitched themselves in outward opposition to the sun and other elements. The silicious earth is in every respect the centre of all earthy productions, these being only digressions from it. The Zircon earth is only a removal or displacement of the silicious toward the argillaceous earth.
524. The argillaceous earth also is not dissevered into its principles; it is not found as a carbonate. On the contrary, it is at once shown to be far more pliable by its capacity for being kneaded and moulded in water, and by its hardening when exposed to air and fire. It is also seized upon and dissolved, i. e. reduced to the aqueous condition, by all acids. Its kindred earths are the Glucine and Yttria, verging towards the talcose earths.
525. The first dismemberment of principles is shown by the talcose earth. Where it appears uncombined with the former earths it is carbonate, yet still feebly corrosive.
526. These three principal earths together make up the body of the earth, while the calcareous earth is only spread over them like a mantle or crust.
527. As no earth is in its totality corrosive, and none such occurs in nature or has at least not been originally produced from it, so may the insolubility of the earths in water be set up as an essential and thoroughly valid characteristic of the earths. Their distinctive characters have been sedulously rendered fluctuating, by having been drawn not from nature, but the products of art. That the corrosive chalk is soluble in water, and may therefore be a salt is true; but it has not issued thus out of the womb of nature. Mineralogy knows nothing of a corrosive calcareous earth. The earths are sufficiently separated from the salts by their insolubility in water. They are separated from the ores by their incombustibleness, or, if they have been already burnt, by their incapacity for reduction. As both of these qualities are imparted by fire, so the earths are distinguished by their immutability in fire, whereby is naturally understood not the scoriation, but change of the earthy character. They differ also in the same manner from the Inflammables. Nature does not undertake the artificial reductions of earths to Metalloids, at least not so, that they may become again of themselves earths. The metals are permanent reductions.

528. Earth is thus the body, which is mutable neither in water, air, nor fire. Earth is a water-, air-, and fire-proof body. This is the brief, rigid, wholly exclusive, and significantly expressive definition, which a so-called empirical science could never, but philosophy alone, bestow.
529. The Ore is not soluble in water, nor mutable in air; on the contrary, it is fusible, oxydizable, or reducible in fire. Ore is a water-and air-, but not fire-proof body.
530. The Inflammable is immutable in water, but mutable in air and fire. The Inflammable is a

water-proof, but not air-and fire-proof body.

531. The Salt is soluble in water, and decomposible in fire, but immutable in air. Salt is an air-, but not water-and fire-proof body. The legitimate series of gradations comprised in the above four definitions cannot escape the attentive reader, nor moreover that the properties of the earths are all affirmative. Nature has not employed such insignificant means of distinction as our mineralogy has done; has nowhere used an acid in order to distinguish the metals from the earths, nor savour to separate the salts from the earths; but she selects universal reagents which are the elements themselves. So simple is Nature, if we do not violate her by art.

DIVISION OF THE EARTHS.

532. There is not merely a single silicious mineral, but many such, just as in clay, talc, and calx. How, then, do differences occur in these earths? When we survey the science of Mineralogy we remark that most minerals are composed of several earths; with them also metals, coal, sulphur, alkalies and acids are frequently associated. It follows thereupon that the further distinctions are no longer of an internal kind, namely, alterations of substance; but proceed from combinations and thus indicate stöchiometric bodies. The next division of the orders I call Families.

Order 1. Silicious minerals. *

533. With how many bodies now can the silicious earth combine? It will first of all appear in a pure condition, as in quartz; then, in the next place, combine with the other earths, thus with clay, talc and calx. We have thus four families of Earth-silices.

Fam. 1.   Pure-Silex   Quartz.
2.   Argillaceo-Silex   Zircon.
3.   Talco-Silex   Emerald.
4.   Calcareo-Silex   Leucite.

534. Thus the hardest minerals or the silicious precious stones are here placed. But these are obviously not exhausted with the above four combinations, but more of the latter must still be sought for. Those bodies which rank next to the earths, and can therefore enter into the following combinations are the other mineral classes, such as the salts, Inflammables, and metals; and we accordingly obtain the following silicious minerals, as constituting classes.

Fam. 5.   Salt-Silex   Topaz.
6.   Inflammable-Silex Diamond.
7.   Ore-Silex   Garnet.

535. Still all the silicious minerals are not exhausted with these combinations. But now the silicious earth can combine with nothing more than the elements, whence three families originate.

Fam. 8.   Water-Silex   Hornstone, Silicious schist, Jasper, Flint, Opal.
9.   Air-Silex   Silicious sinter, as Tripoli and Polierschiefer.
10.   Fire-Silex   Obsidian with Pitchstone, Pearlstone and Pumice.

536. Upon casting a glance at this series, it is shown, that the first seven families occur in a crystalline, but the last three only in a compact or structureless condition. The latter occur at the same time in large masses, the former, on the contrary, but scantily dispersed. The first family or the quartz, occurs as well in a compact and massive state as crystallized; the others, on the contrary, taken collectively, are only crystallized, and scarcely form small rocks here and there, but never mountain-chains. They are the precious stones proper, both on account of their hardness, as also their rarity. Precious stones are thus only combinations of silex with other earths, and with the classes; on the contrary, the elemental silices only, viz. the earth-, water-,

air-and fire-silices, are massive.

537. It is here shown, that freedom finds a place also in dead nature. Quartz only is necessary as the earth in general. Its marriages with the other earths, &c., to form precious stones are not necessary, but free or accidental, and may therefore happen for the first time in the laboratory.

538. If we now proceed to the arrangement of the Clay, we find exactly the same law to prevail in the genesis of its minerals i. e. stöchiometric combinations with other orders, classes, and elements. We have likewise—

A.—Earth-Clays.
Fam. 1. Silicious clays Felspar.
2. Argillaceous clays Sapphire.
3. Talcose clays Ruby.
4. Calcareous clays Epidote.

B.—Class-Clays.
5. Salt-clays Schorl.
6. Inflammable clays Azurite.
7. Ore-clays Harmotome.

C.—Elemental-Clays.
8. Water-clays Clay-slate.
9. Air-clays Potter's-clay, Clay-stone.
10. Fire-clays Lavas, Phonolite, Toad-stone.

The water-clays are hydrates; the air-clays volatilized hydrates; the fire-clays are clay fused or transmuted by heat. Here also the first 7 families only are crystallized; the 3 last, on the contrary, as well as the first in part, occur only in a compact state and in large masses.

539. The Talcs follow the same laws, and we have—

A.—Earth-Talcs.
Fam. 1. Silicious talcs Mica.
2. Argillaceous talcs Sapphirine.
3. Talcose talcs Talc, Chlorite.
4. Calcareous talcs Augite.

B.—Class-Talcs.
5. Salt-talcs Hornblende.
6. Inflammable-talcs Asbestus.
7. Ore-talcs Olivine.

C.—Elemental-Talcs.
Fam. 8. Water-talcs Serpentin, Steatite.
9. Air-talcs Lithomarge, Fuller's-earth, Bole.
10. Fire-talcs Basalt.

Here also the first 7 families only are crystallized and occur for the most part in a scattered manner; but the aqueous, aerial, and igneous families, as well as the first family in part, are merely compact and mountainous masses.

540. The fourth order or that of the Calcareous earths is developed likewise according to the same laws. As, however, it approximates the salts, and therefore combines with acids, it presents

many anomalous varieties, of which account cannot be taken in every instance. These minerals are soft throughout, change by fire and admit of being wholly or partially dissolved in acids. Here belong the zeolites, or combinations of the calcareous earths with the other earths.

A.—Calcareous earths, Zeolites.
1. Silicio-calcareous earths   Lapis lasuli, Scapolite.
2. Argillaceo-   "   Mesotype, Analcime, Stilbite.
3. Talco-   "   Stellite.
4. Calcareo-   "   Tabular spar.

B.—Classes of Calcareous earths.
5. Halo-calcareous earths   Boracite.
6. Inflammable-   "   Phosphate of lime, Fluorspar?
7. Ore-   "   Titanite, Tungsten.

C.—Elemental-Calcareous earths.
8. Water calcareous earth   Hydrophyllite? Wavellite.
9. Air-   "   Gypsum, Heavy-spar, Celestine.
10. Fire-   "   Limestone.

Here also the first 7 tribes only occur crystallized, the 3 last, in a great measure, compact, and as mountainous masses.

CLASS II.
WATER-EARTHS. SALTS.
541. The chief distinctions of the salts consist also in their combination with the other classes, and we have therefore 4 orders—
1. Earth-Salts   Double-salts.
2. Saline-Salts Neutral-salts.
3. Inflammable-Salts Saponaceous-compounds.
4. Metallic-Salts   Vitriols.

The same will hold good without doubt of the orders, as in the case of the earths. They form as many families as there are principal masses of them present, with which they may combine. As the acids, from being the children or offspring of water, play the chief part in the water minerals, and are themselves nothing else than oxydized and outwardly lying masses, they carry consequently within themselves the number and import of the families; thus it is they which determine indeed the division. If the bases were to be taken as the groundwork of arrangement, there would be only earths and alkalies, and on the other hand numerous metals, by which step the mineralogist would fall into the unprincipled method of classification adopted by empirics. Here also the philosophy of nature shows, and that indeed upon sound reasoning, that the acids and not the bases afford the principle of a natural classification. A somewhat different opinion is held by the chemist, who must characterize the salts according to both series; but this is by no means the course taken by the historian of nature.

Order I.
Earth-Salts—Double Salts.
(Combinations of acids with earths.)
Fam. 1-4. Earthy-acids or Fluoric-acid earths; here also belong Bromic, Iodic, and Cyanic acids.
5. Salt or Boracic-acid.

6. Inflammable or Sulphuric-acid—Alum, Sulphate of magnesia.
7. Metallic or Arsenic-acid.
8. Water or Hydrochloric-acid—the earths Barytes, Strontian, and Lime; Murias ammoniæ, Chloride of Calcium.
9. Air or Nitric-acid—Strontium, Nitrate of lime.
10. Fire or Carbonic-acid—Vegetable-acid earths.
Order II.
Salt-Salts—Neutral Salts.
(Combinations of acids with alkalies.)
1-4. Fluoric-acid
5.   Boracic-acid   Borax.
6.   Sulphuric-acid Sulphate of soda, Sulphate of potash.
7.   Arsenic-acid.
8.   Hydrochloric-acid   Rock-salt, Muriate of soda, Sal-ammoniacum or salmiac.
9.   Nitric-acid   Nitrate of potash or saltpetre, Cream of tartar.
10.   Carbonic-acid Soda, Subcarbonate of potash, Binoxalate of potash, Acetate of potash.

Order III.
Inflammable-Salts—Saponaceous-compounds.
(Soluble, and at the same time combustible bodies.)
1-4. Earth-Soaps   Calcareo-sulphuret of potash.
5.   Salt-Soaps   Common sulphuret of potash.
6.   Inflammable-Soaps   Fatty or soft soaps.
7.   Metallic-Soaps   Metallic Soaps.
8.   Water-Soaps   Animal mucilages.
9.   Air-Soaps   Saccharum or sugar.
10.   Fire-Soaps   Vegetable extracts.

Order IV.
Ore-Salts—Vitriols.
(Combinations of acids with metals.)
1-4. Fluoric-acid
5.   Boracic-acid
6.   Sulphuric-acid Iron, Copper, Zinc, Vitriol.
7.   Arsenic-acid   White arsenic.
8.   Hydrochloric-acid   Calomel, Corrosive sublimate.
9.   Nitric-acid   Nitrate of silver.
10.   Carbonic-acid Sugar of lead.
CLASS III.
AIR-EARTHS—INFLAMMABLES.
542. It is very difficult to arrange this class, because it has been wholly neglected by mineralogists, and is, properly speaking, quite unknown to them, because they have had recourse only to those combustible bodies which occur accidentally in the earth, while according to philosophical principles everything belongs to the province of natural history, that has originated or may originate in nature, so that its situation is a matter of complete indifference. If we follow the same laws according to which the earths and salts have been so excellently arranged, we

must here also adopt four orders, namely, combustible things which bear a resemblance to earths, others to salts, others to metals; finally, others which represent combustibility in a pure state, and thus we obtain—
1. Earth-Inflammables Coals.
2. Salt-Inflammables Fats.
3. Inflammable-Inflammables Resins.
4. Ore-Inflammables Colouring matters.

The Earth-inflammables will be such as are solid and burn, without becoming fluid, e. g. Common Coal. The Salt-inflammables will either be or become fluid before they undergo combustion, and are readily converted of themselves into acids, e. g. Animal and Vegetable Fats. The Inflammable-inflammables will be of a sulphurous character, solid or fluid, fragile, electric, fetid and fluid before they burn. These properties are found in the Resins. The Ore-inflammables are those which, independently of their combustibility, possess pre-eminently one property of metals, namely their non-transparency, or coloration, e. g. the Pigments or colouring matters from the organic kingdoms.

Order I.
Earth-Inflammables—Coals.
1-4.	Earth-Coals	Common-coal, a mixture of coal and earths.
5.	Salt-Coals	Gunpowder, viz. a combination of charcoal with a salt.
6.	Inflammable-Coals	Glance-coal, viz. carbon without earths.
7.	Ore-Coals	Black-lead, or carburet of iron.
8.	Water-Coals	Peat-bog and brown or Common coal?
9.	Air-Coals	Lignite or wood-coal?
10.	Fire-Coals	Animal carbon, fibrine.

Order II.
Salt-Inflammables—Adipaceous or Unctuous bodies.
1-4.	Earth-fats	Spermaceti? Tallow.
5.	Salt-fats	Lard and Train-oil?
6.	Inflammable-fats	Butter?
7.	Ore-fats	Wax?
8.	Water-fats	Vegetable oils?
9.	Air-fats	Desiccative or drying oils.
10.	Fire-fats	Greasy oils.

Order III.
Inflammable-Inflammables—Resins.
1-4.	Earth-resins	Sulphur, Phosphorus.
5.	Saline-resins	Chloride of sulphur, Chlorate of Sulphur.
6.	Inflammable-resins	Mineral-pitch, Amber, Turpentine.
7.	Ore-resins	Balsams.
8.	Water-resins	Gum-resins.
9.	Air-resins	Ætherial oils.
10.	Fire-resins	Alcohol, Æther.

Order IV.

Ore-Inflammables—Pigments.
1-4. Earth-pigments   Ochre-pigments.
5. Salt-pigments Soluble-pigments from roots and wood, such as Krapp and Dier's-weed.
6. Inflammable-pigments   Retinoid-pigments from roots and wood, such as Dragon's-blood, Turmeric.
7. Ore-pigments Indigo or devil's-dye.
8. Water-pigments   Sap-colours, such as Sap-green, Oak-gall.
9. Air-pigments Flower colours, such as Saffranon and Saffron.
10. Fire-pigments Animal colours, as Scarlet and Blood-red.

CLASS IV.
FIRE-EARTHS—ORES.
543. The metals are again easier of arrangement, because they have a resemblance to earths and have been better worked out both in chemistry and mineralogy. They divide very naturally into—
1. Earth-Ores Ochres, combinations of metals with oxygen.
2. Salt-Ores   Haloids, insoluble combinations of metals with acids.
3. Inflammable-Ores Blendes, combinations of metals with Sulphur, Phosphorus, and Selenium.
4. Metallic-Ores   Pure metals.

The principles of this arrangement which has been at present pretty generally followed, were first published in my essay 'Das naturliche System der Erze,' 1809. In order to gain a proper insight into the serial gradation of all the families we must first regard the 4th order.

Order I.
Earth-Ores—Ochres.
1. Silicious-Ochres   Metallic calces with silicious earth, as Lierite, Dioptase, Electric calamine.
2. Argillaceous-Ochres   Clay-iron-stone.
3. Talcose-Ochres   Blue-iron-stone.
4. Calcareous-Ochres   Black oxyde of manganese.
5. Salt-Ochres   Calces not peroxydized. Bog-iron.
6. Inflammable-Ochres   Pure calces without metallic lustre, as Wolfram, Protoxide of Uranium, Rutile, Tin-stone.
7. Ore-Ochres   Oxydulated, as Iron-glance, Red oxyd of Copper.
8. Water-Ochres Hydrates, as Brown-iron-stone-ore, Gray ore of manganese.
9. Air-Ochres   Malm-rocks; volatilized Ochres of the difficultly fusible metals, as Umbra, Yellow earth, Earthy manganese, Black cobalt.
10. Fire-Ochres   Slags, volatilized calces of the difficultly fusible metals, as oxyde of or White antimony, Protoxide of Arsenic.

Order II.
Salt-Ores—Haloids.
4. Earth-Haloids Fluoric-acid.
5. Salt-Haloids   Boracic-acid.
6. Inflammable-Haloids Sulphuric-acid, as Sulphate of lead, Phosphoric-acid as Green and Blue phosphates of iron, Diarsenate of iron, Uran-glance, Green phosphate of Lead.
7. Ore-Haloids   Chromic-acid, as Chromate of lead, Arsenic-acid, as Cube-ore, Arseniate of iron, Olivenite, Cobalt-bloom.

8. Water-Haloids Hydrochloric-acid, as Muriate of copper, Horn-silver.
9. Air-Haloids Nitric-acid.
10. Fire-Haloids Carbonic-acid, as Iron-spar, Red manganese-ore, Earthy blue-copper, Malachite, Carbonate of lead.

Order III.
Inflammable-Ores—Blendes.
1. Silicious-Blendes Zinc-blendes, Cinnabar, Red antimony and Ruby-silver-ore.
2. Argillaceous-Blendes Iron and Copper pyrites.
3. Talcose-Blendes Sulphuret of Titanium, Chrome, Uranium.
4. Calcareous-Blendes Sulphuret of Molybdena.
5. Salt-Blendes Copper-glance, Gray copper.
6. Inflammable-Blendes Nickel-glance, Cobalt-glance.
7. Ore-Blendes Sulphuret of Platinum.
8. Water-Blendes Gray antimony, Galena.
9. Air-Blendes Bismuth-glance, Arsenical pyrites.
10. Fire-Blendes Silver-glance-ore.

Order IV.
Metallic-Ores.—Metals.
(Pure or reduced metals.)

544. The classification of metals is one of the most difficult, because no natural arrangement of them has been as yet attempted, and their signification is also so mysterious that we can only get at it, by clinging fast to the laws of development. Thus assuming, that they likewise arrange themselves according to the elements, classes and orders of the earths, we have at once the Elemental metals. The earth-metals are without doubt the difficultly fusible, ignoble or oxydized, such as iron with its congeners. Then the air-metals present themselves with their peculiar character of volatility, as Arsenic with its congeners. These being once rendered solid, the easily fusible but non-volatile will correspond to water, such as Lead with its congeners. The noble metals consequently as Gold, Silver, &c., must doubtless be regarded as fire-metals. Having once separated these 4 groups, the Class-metals admit of being more readily brought into their place. There is one metal, which subjected to moisture is readily converted into a salt, namely copper. This is consequently the representative of the salts among the metals. Ore-metals are, without doubt, those resembling iron, which do not, however, occur in an oxydized condition and are therefore noble Irons. Of this kind are Platinum with its retinue. Between Copper and Platinum nothing else can be introduced but Nickel and Cobalt, as they are likewise difficultly fusible and tolerably noble. They are thus the Inflammable metals. After all these separations a great group is still left of the earth-metals or the difficultly fusible and ignoble. They divide therefore without doubt according to the 4 earths. If now iron approximates the argillaceous-earths, so will those metals whose oxydes are distinguished by a striking colour be regarded as talc-metals. Of this kind are Titanium, Chromium, Uranium, which crystallize moreover into spiculæ like hornblende, or into lamellæ like mica. These again being separated the silicious and calcareous-metals remain for investigation. The former are those which scarcely admit of being reduced; the latter, on the contrary, those which approximate to the noble, difficultly fusible metals, namely to platinum. It can hardly be doubted that Tantalum is the silicious-metal. For the calcareous-metals Sulphuret of Molybdenum is left, to which Osmium seems to approximate. We have accordingly the following genetic arrangement:

A.—Earth-Metals.
(Difficultly fusible and ignoble.)
1. Silicious-Metals   Tantalium.
2. Argillaceous-Metals Wolfram, Cerium, Manganese, Iron.
3. Talcose-Metals   Titanium, Chromium, Uranium, Vanadium.
4. Calcareous-Metals   Molybdenum, Osmium.

B.—Class-Metals.
(Difficultly fusible and noble.)
5. Salt-Metals   Copper.
6. Inflammable-Metals Nickel, Cobalt.
7. Ore-Metals   Platinum, Palladium, Iridium, Rhodium.

C.—Element-Metals.
(Easily fusible or noble.)
8. Water-Metals   Antimony, Lead, Tin
9. Air-Metals   Zinc, Cadmium, Bismuth, Arsenic.
10. Fire-Metals   Tellurium, Mercury, Silver, Gold.

Every one will easily see that these groups of metals agree with their antetypes in properties, as also that this arrangement is more natural than any that has hitherto been advanced. Glancing at it, it must strike the reader that in several of the families 4 metals are present, and that none exceed this number. There are 4 sideroid, 4 titanoid, 4 platinoid, 4 arsenicoid and 4 argyroid metals. Now, as they are to be viewed as deoxydized earths, it must be assumed that in each family they are mindful of their origin, and everywhere represent the 4 earths along with the character of their family. They are earths divided unto the last members, or reduced in a primo-chemical manner. In order to recognize the parallelism of the classes, orders and families, we have only to compare the adjoining table. We cannot expect to find all the minerals ranging there in their proper place. For we are treating the subject at present only as regards its principles.

Table A
FIRST CLASS.
EARTH-EARTHS—EARTHS.
ORDER I.   ORDER II.   ORDER III.   ORDER IV.
Earth-Earths. Salt-Earths. Inflammable Earths. Ore-earths.
Silex. Clays. Talcs. Calcareous-earths.
Fam. 1. Pure Silex.   Fam. 1. Silicious Clays.   Fam. 1. Silicious Talcs.   Fam. 1. Silicious Kalke.
1. Quartz.   1. Feldspar.   1. Mica.   1. Lapis-lazuli.
2. Iron Flint.   2. Anorthite.   2. Pinite.   2. Hauyne.
3. Petalite.   3. Holmesite.   3. Sodalite.
4. Oligoclase.   4. Margarite.   4. Scapolite.
5. Triphane.   5. Nepheline.
6. Andalusite.
7. Crucite.
F. 2. Argillaceo-Silex. F. 2. Argillaceous Clays.   F. 2. Argillaceous Talcs.   F. 2. Argillaceo-Kalke.
1. Zircon.   1. Sapphire.   1. Sapphirine.   1. Fugenstein.

2. Œrstedite.   2. Chrysoberyl.   2. Seybertite.   2. Frehnite.
3. Cyanite.   3. Chabazite.
4. Sillimanite.   4. Laumontite.
5. Stilbite. 6. Desmin. 7. Analcime. 8. Mesotype.
F. 3. Talco-Silex.   F. 3. Talcose Clays.   F. 3. Talcose Talcs.   F. 3. Talco-Kalke.
1. Emerald.   1. Spinel.   1. Talc.   1. Adelforsite.
2. Davidsonite.   2. Automolite. 2. Pyrophillite. 2. Stellite.
3. Euclase.   3. Dichroite.   3. Chlorite.   3. Mellilite.
4. Phenacite.   4. Humboldtite.
F. 4. Calcareo-silex.   F. 4. Calcareous-Clays.   F. 4. Calcareous Talcs.   F. 4. Calcareo-Kalke.
1. Leucite.   1. Epidote.   1. Augite.   1. Tubular spar.
2. Glaucolite.   2. Mangan-epidote.   2. Diopside.   2. Apophyllite.
3. Sahlite. 4. Pyroxene. 5. Coccolite. 6. Hedenbergite. 7. Diallage. 8. Bronzite. 9. Hypersthene.
F. 5. Salt-Silex.   F. 5. Salt-Clays.   F. 5. Salt-Talcs.   Fam. 5. Salt-Kalke.
1. Topaz.   1. Yttro-cerite. 1. Grammatite.   1. Boracite.
2. Physalite.   2. Schorl.   2. Strahlstein. 2. Datholite.
3. Pycnite.   3. Axinite.   3. Hornblende.
4. Anthophyllite.
F. 6. Inflammable Silex.   F. 6. Inflammable-Clays.   F. 6. Inflammable-Talcs.   F. 6. Inflammable-Kalke.
1. Diamond.   1. Lazulite.   1. Asbestus.   1. Cryolite.
2. Turquois.   2. Fluorspar.
3. Amblygonite.   3. Wagnerite.
4. Apatite.
F. 7. Ore-Silex.   F. 7. Ore-Clays.   F. 7. Ore-Talcs.   F. 7. Ore-Kalke.
1. Garnet.   1. Harmotome.   1. Chrysolite.   1. Titanite.
2. Vesuvian.   2. Gadolinite. 2. Hyalosiderite.   2. Tungsten.
3. Acmite.   3. Orthite.   3. Pharmacolite.
F. 8. Water-Silex.   F. 8. Water-Clays.   F. 8. Water-Talcs.   F. 8. Water-Kalke.
1. Flint-stone. 1. Wærthite.   1. Schiller-spar.   1. Diaspore.
2. Jasper.   2. Allophane.   2. Serpentine.   2. Wavellite.
3. Hornstone.   3. Pyrophyllite.   3. Steatite.   3. Hydrophyllite.
4. Opal.   4. Clay-slate.   4. Meerschaum.
F. 9. Air-Silex.   F. 9. Air-Clays.   F. 9. Air-Talcs.   F. 9. Air-Kalke.
1. Tripoli.   1. Potter's-clay.   1. Agalmatolite.   1. Aluminite.
2. Polierschiefer.   2. Clay-stone. 2. Lithomarge. 2. Heavy-spar.
3. Silicious-sinter.   3. Porcelain-clay.   3. Fuller's-earth.   3. Celestine.
4. Cimolite.   4. Bole.   4. Gypsum.
F. 10. Fire-Silex.   F. 10. Fire-Clays.   F. 10. Fire-Talcs.   F. 10. Fire-Kalke.
1. Pitchstone. 1. Clay Iron-stone.   1. Basalt.   1. Mellilite.
2. Pearlstone.   2. Toad-stone. 2. Magnesite.
3. Obsidian.   3. Phonolite.   3. Strontianite.
4. Pumice.   4. Lavas.   4. Limestone.
SECOND CLASS.

WATER-EARTHS—SALTS.

ORDER I.	ORDER II.	ORDER III.	ORDER IV.
Earth-Salts.	Salt-Salts.	Inflammable Salts.	Ore-Salts.
Double Salts.	Neutral Salts.	Soaps.	Vitriols.
Fam. 1. Silicic Acid.	Fam. 1. Fluoric Acid.	Fam. 1. Silicio-sulphuret of Potash.	Fam. 1. Fluoric Acid.
Fluoric-acid earths.	Alkalis.	Metals.	
F. 2. Albuminic Acid.	F. 2. Bromic Acid.	F. 2. Argillaceo-sulphuret of Potash.	F. 2. Bromic Acid.
Bromic acid.			
F. 3. Oxide of Magnesia.	F. 3. Iodic Acid.	F. 3. Magnesio-sulphuret of Potash.	F. 3. Iodic Acid.
F. 4. Oxide of Calcium.	F. 4. Cyanic Acid.	F. 4. Calcareo-sulphuret of Potash.	F. 4. Cyanic Acid.
Cyanic acid.			
F. 5. Hydrochloric Acid.	F. 5. Hydrochloric Acid.	F. 5. Salt-Soaps.	F. 5. Hydrochloric Acid.
Boracic acid.	1. Sassoline.	1. Potassio-sulphuret of Potash.	
2. Borax.	2. Ammoniaco-sulphuret of Potash.		
F. 6. Pyric Acid.	F. 6. Inflammable-Acid.	F. 6. Inflammable-Soaps.	F. 6. Inflammable-Acid.
1. Alum.	1. Sulphate of Soda.	1. Hard or Soda-soap	1. Green or Iron-vitriol.
2. Magnesia.	2. Sulphate of Potash.	2. Soft or Potash-soap	2. Blue or Copper-vitriol.
3. Mascagnine.	3. Linimentum volatile		3. White or Zinc-vitriol.
4. Sub-nitrate of Bismuth.			
F. 7. Metallic Oxide.	F. 7. Ore-Acid.	F. 7. Metallic Soaps.	F. 7. Metallic Acid.
Arsenic acid.	1. Antimonium diaphoreticum.	1. Carbonate of Lead.	1. White Arsenic.
2. Liquor arsenicalis Fowleri.			
F. 8. Water-Acid.	F. 8. Water-Acid.	F. 8. Water-Soaps.	F. 8. Water-Acid.
1. Sulphate of Magnesia.	1. Rock-salt.	1. Mucus.	1. Butter of Antimony.
2. Sulphate of Baryta.	2. Muriate of Soda.	2. Gelatine.	2. Calomel.
3. Murias Ammoniæ.	3. Salmiac.	3. Albumen.	
4. Chloride of Calcium.		4. Coagulable lymph.	
F. 9. Air-Acid.	F. 9. Air-Acid.	F. 9. Air-Soaps.	F. 9. Air-Acid.
Nitric acid.	Nitrate of Potash.	1. Sugar.	1. Nitrate of Silver.
1. Nitrate of Strontian.		2. Manna.	
2. Nitrate of Lime.		3. Honey.	
F. 10. Fire-Acid.	F. 10. Fire-Acid.	F. 10. Fire-Soaps.	F. 10. Fire-Acid.
Carbonic acid.	1. Soda.	1. Extracts.	1. Tartar emetic.
Super-carbonate of Lime.	2. Subcarbonate of Potash.	2. Sugar of Lead.	
3. Binoxalate of Potash	3. Fulminating Silver.		
Acetate of Potash.	4. Acetate of Potash.		

THIRD CLASS.
AIR-EARTHS—INFLAMMABLES.

ORDER I.	ORDER II.	ORDER III.	ORDER IV.
Earth-Inflammables	Salt-Inflammables.	Inflammable-Inflammables.	Ore-Inflammables.

Coals.	Fats.	Resins.	Pigments.
Fam. 1. Silicious Coals.	Fam. 1. Spermaceti.	Fam. 1. Sulphur.	Fm. 1. Silicious-Pigments.
1. Common Coal.	1. Alcohol of Sulphur.	1. Litmus.	
2. Orpiment.			
F. 2. Argillaceous Coals.	F. 2. Adipocire.	F. 2. Boron.	F. 2. Argillaceous Pigments.
1. Common Coal.			
F. 3. Talcose Coals.	F. 3. Oleine.	F. 3. Selenium.	F. 3. Talc Pigments.
1. Common Coal.			
F. 4. Calcareous Coals.	F. 4. Tallow.	F. 4. Phosphorus.	F. 4. Calcareous Pigments.
1. Common Coal.	1. Stearine.		
F. 5. Salt-Coals.	F. 5. Salt-Fats.	F. 5. Salt-Resins.	F. 5. Salt-Pigments.
1. Gunpowder.	1. Lard.	1. Chloride of Sulphur.	1. Krapp.
2. Train-oil.	2. Chloride of Phosphorus.	2. Dier's-weed.	
F. 6. Inflammable-Coals.	F. 6. Inflammable-Fats.	F. 6. Inflammable-Resins.	F. 6. Inflammable-Pigments.
1. Anthracite.	1. Butter.	1. Mineral-pitch.	1. Sandal-wood.
2. Cream.	2. Amber.	2. Log-wood.	
3. Turpentine.	3. Curcuma.		
4. Caoutchouc.	4. Chlorophyle.		
F. 7. Ore-Coals.	F. 7. Ore-Fats.	F. 7. Ore-Resins.	F. 7. Ore-Pigments.
1. Black Lead.	1. Wax.	1. Turpentine.	1. Succory.
2. Pyrorthite.	2. Balsam of Peru.		2. Quercitron.
3. Mecca-balsam	3. Woad.		
F. 8. Water-Coals.	F. 8. Water-Fats.	F. 8. Water-Resins.	F. 8. Water-Pigments.
1. Brown Coals.	1. Cocoa-butter.	1. Assafœtida.	1. Sap-green.
2. Peat.	2. Palm-oil.	2. Gumboge.	2. Oak-gall.
3. Nutmeg.	3. Myrrh.		
4. Laurel-oil.	4. Opium.		
F. 9. Air-Coals.	F. 9. Air-Fats.	F. 9. Air-Resins.	F. 9. Air-Pigments.
1. Lignite.	1. Linseed-oil.	1. Petroleum.	1. Saffranon.
2. Nut-oil.	2. Dippel's Oil.		2. Saffron.
3. Hemp-oil.	3. Camphor.		3. Anotto.
4. Poppy-oil.	4. Oil of Turpentine.		
F. 10. Fire-Coals.	F. 10. Fire-Fats.	F. 10. Fire-Resins.	F. 10. Fire-Pigments.
1. Fibrine.	1. Rape-oil.	1. Spirits of Wine.	1. Scarlet.
2. Olive-oil.	2. Sulphuric Ether.		2. Blood-red.
3. Oil of Almonds.	3. Acetic Ether.		
4. Formic Spirit.			

FOURTH CLASS.
FIRE-EARTHS—ORES.

ORDER I.	ORDER II.	ORDER III.	ORDER IV.
Earth-Ores.	Salt-Ores.	Inflammable-Ores.	Ore-Ores.
Ochres.	Halde.	Inflammables.	Metals.
Fam. 1. Silicious Ochres.	Fam. 1. Silicious Halde.	F. 1. Silic. Inflammables.	F. 1. Silicious Metals.

1. Lierite.   1. Fluor-Cererium.   1. Zinc-blende.   1. Tantalium.
2. Dioptase.   2. Cinnabar.
3. Antimonial Silver. 3. Red Antimony.
4. Electric Calamine. 4. Ruby Silver-ore.
F. 2. Argillaceous Ochres.   F. 2. Argillaceous Halde.   F. 2. Argillaceous Inflammables.
    F. 2. Argillaceous Metals.
1. Clay Iron-stone.   Bromic acid.   1. Iron Pyrites. 1. Wolfram.
2. Polymignite.   2. Copper Pyrites.   2. Iron.
3. Yttro-tantalite.   3. Tin Pyrites. 3. Cerium.
4. Manganese.
F. 3. Talc Ochres.   F. 3. Talc-Halde.   F. 3. Talc Inflammables.   F. 3. Talc-Metals.
1. Blue Iron-stone.   Iodic acid.   1. Vanadium.
2. Uranium.
3. Titanium.
4. Chromium.
F. 4. Calcareous Ochres.   F. 4. Calcareous Halde.   F. 4. Calcareous Inflammables.
    F. 4. Calcareous Metals.
1. Black oxide of Manganese. Cyanic acid.   1. Sulphuret of Molybdena.   1. Molybdenum.
2. Pyrochlore. 2. Osmium.
F. 5. Salt-Ochres.   F. 5. Salt-Halde.   F. 5. Salt-Inflammables.   F. 5. Salt-Metals.
1. Bog-iron.   Boracic acid.   1. Copper-glance.   1. Copper.
2. Gray-copper.
F. 6. Inflammable-Ochres.   F. 6. Inflammable-Halde.   F. 6. Inflammable-Inflammables.
    F. 6. Inflammable-Metals.
1. Wolfram.   1. Sulphate of Lead.   1. Nickel-glance.   1. Nickel.
2. Protoxide of Uranium.   2. Blue phosphate of Iron.   2. Cobalt-glance.   2. Cobalt.
3. Rutile.   3. Uran-glance.
4. Tin-stone.   4. Green phosphate of Lead.
F. 7. Ore-Ochres.   F. 7. Ore-Halde.   F. 7. Ore-Inflammables.   F. 7. Ore-Metals.
1. Iron-glance. 1. Chromate of Lead.   1. Sulphur-Platinum.   1. Rhodium.
2. Titanate of Iron.   2. Cube-ore.   2. Iridium.
3. Chromate of Iron.   3. Olivenite.   3. Palladium.
4. Red Copper ore.   4. Cobalt-bloom.   4. Platinum.
F. 8. Water-Ochres.   F. 8. Water-Halde.   F. 8. Water-Inflammables.   F. 8. Water-Metals.
1. Thraulite.   1. Muriate of Copper. 1. Antimonial Nickel. 1. Antimony.
2. Brown Iron-stone ore.   2. Horn-lead.   2. Gray Antimony   2. Tin.
3. Gray ore of Manganese.   3. Horn-silver. 3. Lead.
3. Galena.
F. 9. Air-Ochres.   F. 9. Air-Halde.   F. 9. Air-Inflammables.   F. 9. Air-Metals.
1. Umbra.   Nitric acid.   1. Acicular Bismuth.   1. Zinc.
2. Yellow-earth.   2. Cadmium.
3. Black Copper.   2. Bismuth-glance.   3. Bismuth.
4. Black Cobalt.   3. Arsenical Pyrites.   4. Arsenic.
F. 10. Fire-Ochres.   F. 10. Fire-Halde.   F. 10. Fire-Inflammables.   F. 10. Fire-Metals.
1. White Antimony   1. Iron-spar.   1. Foliated Tellurium. 1. Tellurium.
2. Mennige.   2. Earthy blue copper. 2. Silver-glance.   2. Mercury.

3. Bismuth Ochre.   3. Carbon of Lead.   3. Silver.
4. Protoxide of Arsenic.   4. Calamine.   4. Gold.

II.—GEOLOGY.

545. Geology is the history of the formation of the planet. It is the doctrine which comprises the structure and thus the form, with the organs or members of the planet, if we would compare the latter with an organic body.

1. Form of the Planet.

546. Crystallization belongs to the essence of the earth as the globular form does to that of water. The life of the earth consists in the formation of crystals. The being of the earth and of the crystal are identical. The solid planet earth, has originated also according to the laws of crystallization.

547. It is not, however, a single large crystal, the structure of which appears to be homogeneous; but it is crystallized in its smallest parts; it is an accumulation of crystals, which its atoms, integral parts, or constituent forms present for our examination. If a schorl or feldspar were extended to the size of the whole earth, its integral parts, though undiscoverable before by the microscope, would then become visible. Crystals would be exhibited therein of silicious, argillaceous, talcose, and calcareous earths, of iron, boracic acid, &c. In short it would prove a complex kind of rock or mountain. The fundamental or principal mass of the planet is thus a granular kind of rock or mountain, probably like granite. Each of these constituent forms is crystallized for itself out of the fluid-mass according to the laws developed in the theory of crystallization, in every point of the fluid a globe of crystallizing forces being constituted, that generated the constituent forms.

548. The earth (regarded as planet, not as element) has during its coagulation into a solid nucleus, generated an infinity of polar spheres, as every polar line consists of an infinite vicissitude of poles.

549. These integral crystals have originated only in drops of water; for then only was an infinite multitude of polar axes and polar radii separated from each other. Water in infinitely numerous drops is rain. In the primary rain each drop crystallized, and each fell towards the centre, because the primary water ranged over a vast extent. The granular rock has originated in and out of rain. It is crystallized rain.

550. It does not follow from this, that the earth should be an accidental accumulation of small crystals, which, by the rotation of the planet, formed themselves mechanically into a spheroid. As in small crystals the infinity of poles reunites to constitute some principal polarities, so also is this the case with the globe of the earth; this results from its genesis whereby it is present in a definite space, and hence coheres or hangs together as one piece. The earth is only one small punctule of contraction wrought in æther-space by the agency of light. Again, it was without doubt a single central tension, which, occurring upon a large scale, attracted together all the particles of æther, and from that circumstance arranged them also. It was probably magnetism, which is so intimately connected with the rotation of the earth's axis, or the conflict of magnetism in its interior with the electricity upon its surface, which was active in the condensation and arrangement of the masses.

551. This arrangement of the parts of the earth, upon a large scale, is a regulation of its constituent forms. The adjusting forces are, however, those that operate through the whole sphere, and are thus linear and spherical at one and the same time. By these the laminæ of the nucleus were determined. The constituent forms of the earth are consequently arranged in laminæ. What in the crystal is called the cleavage of the laminæ, is in the earth stratification. The strike of the strata combined with their dip determines the crystal nucleus of the earth.

552. The strike and dip of the strata happened without doubt according to definite laws of crystallization, and has by no means been resigned to elevating force, mechanical dislocation, or even to chance.

553. On this very account the two directions of the earth's laminæ cannot have a similar bearing over the whole earth. They can only have a long tract extending in the same direction in individual mountain-chains. This does not, however, exclude a parallel strike and dip occurring in some wholly different quarter of the world; it must indeed occur, and in such instances we meet with the opposing sides or edges of the earth's nucleus.

554. The earth has without doubt originated according to the laws of the polyhedron, which represents in the nearest manner the globe. The polyhedron of the globe is the rhomboidal dodecahedron.

555. The land cannot therefore have an equal elevation everywhere above the water, because the crystal consists of edges, angles, and surfaces or sides. The mountain tops are probably the angles, the mountain ridges or chains the edges, the plains the lateral surfaces of the crystal.

556. Several mountain-chains run parallel, but interrupted, with the equator. This parallelism extends to the most temperate zones. Then follow oblique mountain-chains, as the Carpathian, Alps, Pyrenees. Lastly, mountain-chains pass from the poles to the equator, as the Sewo, Ural and Altai mountains. Subordinate mountain-chains unite the latter transversely and the former in the direction of the meridian. The earth is probably a regular net of crystal edges and angles, and thus of crystal surfaces also.

PRIMARY VALLEYS.

557. Although the earth may be regarded as originally a crystal, that consists of level surfaces, edges and angles, wide fissures may still have originated between its laminæ, such as we see in large crystals of felspar. These fissures or gaps are the primary valleys.

558. There must be therefore valleys or parallel valleys, which probably extend for hundreds of miles, and are many miles deep—Longitudinal valleys.

559. The laminæ of the earth had without doubt transverse fissures, which have been called hidden passages. These transverse fissures are the transverse valleys, which are consequently less long and deep.

560. The mountains originate of themselves. They do not properly originate, but valleys only originate, and the ridges of the crystal laminæ afford the mountains. The mountains have not been originally upheaved above the surface of the earth, nor the valleys depressed. A valley, which is several miles broad, must originally have been several miles deep, and the mountain wall consequently several miles high. The earth at its origin was a cloven and jagged polyhedron, a polyhedric star, such as the moon is still.

561. The mountains are not therefore large crystals, which crystallized above the surface of the earth. They are only crystal laminæ, and may be as irregular as possible in form; for they are ruptured crystals.

562. The water, which from the beginning had covered the polyhedron, now sunk into the primary valleys. From the water resulted new and final crystallizations, and these deposited themselves in the valleys, upon the level ground and the flanks of the mountains; and thus the fathomless primary valleys have been in part filled up. There are no longer any primary valleys upon the earth.

563. After the water was once confined in narrow canals, it must begin to flow, and by the force of its current many a steep primary wall must have fallen in, been crumbled into ruins, and either been left upon the spot or washed away—Diluvial drift rocks (Trümmersteine), Nagelfluhe,

Stratified rocks.

564. The principal direction of the water was formerly, as it is even now, determined by the rotation of the earth; it flowed therefore from east to west under the equator, from north-east to south-west in our temperate zones, and pretty well from the north or from the poles toward the equator in the frigid zones.

565. The primary valleys, which had originated in these directions, were more excavated by aqueous agency than those which ran in other directions, new valleys being also produced; the mountain chains therefore upon the earth agree in the main with the water courses, and, though not generated from, have been certainly changed in character by, the latter. Such must be our conclusion, if in the formation of the earth nothing but crystallization be taken into consideration. Condensation alone brings yet other phenomena along with it.

566. The first and most important of these is the elevation of temperature. We cannot think otherwise than at the first precipitation of the earthy a number of huge cavities remained in the interior of the earth, which were filled with water. This being heated, was converted into steam, which thrust up the superincumbent rock, and converted it into new mountains or mountain-chains. These agencies of heat may be called primary volcanoes, although they are not to be confounded with volcanoes proper.

567. The igneous cavities were probably placed in a certain order, according with the original edges and angles of the earth's body. They may have therefore thrown up mountain tops and chains.

568. But the cavities, besides raising mountain chains, could just as well depress or allow them to sink in, and thus produce valleys. By these means the seas have probably been formed. Like the seas so also have the inland seas or lakes originated.

569. Lastly, a similar origin must be ascribed to the world-sea or ocean. Many earthquakes arose, with all their concomitant phenomena, by partial and sudden falling in of the deep descending cavities of the earth. Frequently that which fell in was again driven out by the expansive force of the suddenly compressed air.

2. Organs of the Planet.

570. The principal mass or body of the planet is formed by the earths proper, as being the typical or genuine representatives of the earth-element. The other classes, as the metallic minerals or ores, the Inflammables, and salts, are to be regarded only as viscera of this body. If we take a circumspective glance at the planet, we cannot fail to recognize this remarkable relation, that it is properly the elemental families only of the earths that form the mountain masses, while the other families are only scattered within them, like the glands in an animal body. Let us again place the Families in tabular contrast with each other.

SILICIOUS,	ARGILLACEOUS,	TALCOSE and	CALCAREOUS-EARTHS.
A. Earth-silices.	Earth-clays.	Earth-talcs.	Earth-calces.
Fam. 1. Quartz.	1. Felspar.	1. Mica.	1. Lapis-lazuli.
2. Zircon.	2. Sapphire.	2. Sappharine.	2. Mesotype.
3. Emerald.	3. Ruby.	3. Talc.	3. Stellite.
4. Leucite.	4. Epidote.	4. Angite.	4. Tabular-spar.
5. Topaz.	5. Schorl.	5. Hornblende.	5. Boracite.
6. Diamond.	6. Azurite.	6. Asbestus.	6. Phosphorite.
7. Garnet.	7. Harmotone.	7. Olivine.	7. Titanite.
B. Water-silices.	Water-clays.	Water-talcs.	Water-calces.
8. Hornstone.	8. Clay-slate.	8. Serpentine.	8. Hydrophyllite.

C. Air-silices. Air-clays. Air-talcs. Air-calces.
9. Tripoli. 9. Clay-stone. 9. Lithomarge. 9. Gypsum.
D. Fire-silices. Fire-clays. Fire-talcs. Fire-calces.
10. Obsidian. 10. Lava. 10. Basalt. 10. Limestone.

571. Now we here observe that only Quartz, Hornstone, Tripoli and Obsidian, occur as mineral aggregates in large masses, while all the others are only rare precious stones. Among the Clays, only Felspar, Clay-Slate, Hornstone and Lava. Among the Talcs, only Mica, Serpentine, Lithomarge, and Basalt. Among the Calcareous-earths, which incline towards the nature of salts, all the first families are of rare occurrence, and the latter only appear as mineral masses. This regularity speaks moreover retrospectively in favour of the correctness of the classification. The families have thus the same import or value in Geology as in Oryctognosy.

A.—EARTHS.

572. The proper organs of the planet are the mineral aggregates or rocks. These are either presented to us as they have originally been precipitated from water by chemical process, or as they have been changed by the co-operation of the other elements. There are thus genetic and metamorphosed kinds of rock. The former were crystallized previous to the current of the water—Primary rocks. These rocks are changed either—
1. By water—Transition-rocks, or
2. By air—Trappean-rocks.
3. By fire—Volcanic-rocks.
There are properly four kinds of rock formations—
1. Earth-formation, masses that have originated through the crystallizing force of the earth-element itself—Primary-rocks.
2. Aqueous-formation—Transition-rocks.
3. Aerial-formation—Trappean-rocks.
4. Igneous-formation—Volcanic-rocks.
The stratified rocks, having originated for the most part mechanically, are only the object of Physio-philosophy, in so far as chemical precipitations partly occur amongst them.

573. The earth-formation is represented by the crystallized primary rocks.

a. Earth-Formation.

574. The earths can only be precipitations from the element that immediately precedes them, and thus from water. They have been enveloped in the water, as this has been in the air, and the latter again in the æther, but not mechanically, as we at present dissolve clay in water, nor chemically either, as lime dissolves by corrosion in water, but dynamically. The water did not exist from the beginning as water in the air, but only its principles, which became for the first time water, when they had combined by electricity.

575. Even so was it the case with the earths. The primary water, which was present prior to the solid nucleus of the earth, is not the water, with which we are at present acquainted; it had still the earthy principles, the basic in itself, which being separated by a differencing act from the oxygenous of the water, was deposited as carbon specially produced. No calcareous, argillaceous, and silicious earth was imbedded in the primary water. How then could they have been dissolved in it? Some say, by a great quantity of acids which prevailed at that time; but from whence then did these acids come? The largest quantity of them resides in lime, gypsum, and common salt; but how could this quantity, which has not so much as acidified all the calcareous earth, have held the rest of the earths in solution? And then again are all the earths

soluble by the process of acidification? We cannot reason in geogeny, as we would of a common chemical precipitation, where we precipitate earths that have been dissolved by elective affinities.

576. The earths originated for the first time, when they were separated from the primary water. The instant of their precipitation was also the instant of their generation, even as rain is, or rather was, a production of the water. It is easy to say that the earths had been dissolved in the water, and were then precipitated by elective affinities. But it is absurd to rest content with such an assertion. The chief question still remains; how did the earthy originate, before the earth was? Every one can say how that which had already originated, and was only suspended in water, was precipitated. But we are in the habit of distinguishing the several earths, before having taken a survey of matters and inquired if earths then existed.

577. It must not, however, be thought, that all had happened in a gradual manner; that air was first converted into water, and lastly, after a lapse of some thousand years, the latter again into earth, without with the one element the principles conducting to the others being already and necessarily imparted. No, everything has been granted and determined at one stroke, even as with the impregnation of the ovum all the organs of the future embryo are determined, although they first develop themselves gradually. The same agent, which creates the air solicits also its two principles unto combustion in the water; and the same act, which separates the oxygen gas out of the air as water, separates also the carbon from it to constitute earth. Not one can be posited, without the other being codetermined.

578. The analysing principle cannot be a something internal residing in the planet itself. But all that it is, it is through antagonism with the sun. This antagonism is light. This it is that divorces or separates the elements from their matrix; it is light, which has sundered the æther into the twofold air, and the air again into the denser elements, water and earth, separating the oxygen from the nitrogen and later on from the carbon, which must be taken up in water as an oxide.

579. At the commencement of the separation the fluid must be the first to emerge, because the cohesion or fixation of the poles is possible only in a successive manner, yet not as if water were according to its determination to be the first and the Earthy the second in the order of production. As the oxygen separated from the air to create water, so also must the carbon that was left have been precipitated from the water and metamorphosed into earth. Thence and thence only may we venture to say, that the earths originated out of water; for, properly speaking, they arose as absolutely for themselves as air and water; but as air succeeded æther, and water air, so did the earth the water.

580. Whatever be the bulk of the mass of water, to the same extent is the Earthy, or the germ of the Earthy, diffused throughout it. The whole is a fluid mass of earth. Light, however, penetrates the mass of water as a transparent body. As translucency is not a mechanical but a dynamic act, or process of differentialization in matter; so does this earth-water become separated into a mass of oxygen and carbon, or into ordinary water and into the earth-element.

581. The genesis of the earths is a process of conduction of light. All transparency is a formation of earth; for it is a separation of the Aqueous from the body of gravity. Where non-transparency exists, there has earth been already formed.

582. As the conducting process of light is an act of deoxydation, so are the earths at the same time also deoxydized by precipitation, and this in four stages which indicate the elements whereby light operates upon the Basic in water; or it might be said, by the four colours of, or by coloured, light. We already know that the earth, which presents in itself the most dismembered

character is the calcareous earth; but that those which have preserved a more identical character in themselves are the silicious, argillaceous, and talcose earths. These earths may be viewed as those in which the lime has absorbed a proportion of oxygen, which has in it become carbonic acid. There is indeed only one earth-substance in water. In this substance, which is neither silex nor calx, the polar principle has distributed itself, and that very portion, which has obtained the most thereof, has become calcareous earth.

583. The calcareous earth has originated in the upper parts of the aqueous globe, the other earths, however, in the depths, in the middle of that globe; for in the upper regions of the water, the light can exert a greater polarizing influence, and therefore that very earth is generated which stands nearest to the æther or to the light, viz. the different calcareous earth. But in the depths of the water, the light loses its energy, and is no longer in a condition to elicit the Oxygenous in the Basic; thereby identical and more fixed earths originate.

GRANULAR ROCK OR GRANITE.

584. The differenced calcareous earth has been associated with the differential water, remained for a longer period identical with and dissolved in it, and was therefore the last to be precipitated from it. The silicious earth with its neighbours must necessarily have been the first to separate from the water, as it is in a proper sense, that earth from which all water has been withdrawn. Two periods of precipitation exist therefore in geogeny, one that of the identical or fundamental earths, and one of the internally subdivided calcareous earths.

585. The silicious, argillaceous, and talcose earths must occupy, from their having been first precipitated, the middle of the planet. The water being earthy everywhere, had everywhere the capacity to become earth; but different earths originated, where the light was different in the fluid mass of earth. In pure light, or that upon the surface, the Earthy dualized itself into calcareous earth; in situations, where the light exercised less influence, the Earthy became talcose, still deeper argillaceous earth; lastly, at such a depth where the light could scarcely reach, the Earthy became a pure Earthy or silicious earth.

586. Precipitation, both upon a large and small scale, is a process of crystallization; the integral parts of the planet originate, like the nucleus and the perfected crystal, through central action and polar action. The integral forms of the planet impart the crystalline granule or its joints.

587. The integral forms of the three fundamental earths crystallized with and through each other into one mass, or, in other words, the nucleus of the earth consists of a crystalline mass of the three fundamental earths. The solid nucleus of the earth consists of crystals upon a small scale of silex, clay, and talc.

588. As, however, no pole is produced of perfect purity in nature, so are the integral forms not perfectly pure fundamental earths, but other and later factors enter also into their composition, e. g. calcareous earth, with even ores and salts. The silicious earth is crystallized as quartz, the argillaceous as feldspar, the talc as mica.

589. The mixture of the three crystallized fundamental earths, which composes the nucleus of the earth, and upon which the polarized masses of soil have been supported, is therefore a definite species of rock, which has all the properties of Granite, although that which has been extruded from the earth may have been altered by heat.

590. As the mass of the earth is about five times denser than water; so must the planet, before the Earthy was separated, have been much denser than at present. With its separation the fluid must have suddenly diminished, and moved towards the middle point of the planet. During the descent of water for many miles in extent, it must separate in drops. The separation of the earths was

combined with rain.

591. In every falling drop the three fundamental earths, which are insoluble in water, crystallized. The first crystals therefore are only of the size of drops. It is only by this mode of origin that the crystallization into one another of the three constituent parts of granite without any cementing substance can be comprehended. Granite is an earthy hail-storm. The hail-stones crystallized during their fall in a similar manner with each other.

592. Upon the whole the silicious character predominates in granite, as the primary formation of the planet, and must do so, for silicious earth is the primary earth, the principal earth of this precipitation, and one from which clay and talc subsequently proceed, being higher heterogynisms produced by light. Properly speaking only two perfectly different characters of earth exist, the silicious earth as the pure separation of the Earthy, and the calcareous earth, as the last separation of the water from the Earthy. It may also be said that in the calcareous earth fire, in the talcose earths the air, in the argillaceous earth water has exercised its influence and displaced the Earthy. Not only quartz is silicious earth, but feldspar and mica consist in a great measure of the same. Still, however, these last two constituent forms are wholly different from quartz, and very far removed from the nature of silex; the feldspar obviously forming the transition to clay by its resolution into porcellanic earths, but the mica passing over into talc.

593. There does not exist in nature an order of silex, clay, or talc, so soon as we take up these earths in a simply chemical manner. Nature instead of silex produces quartz, instead of clay feldspar, instead of talc mica; and these must be the earths, that determine this part of the mineral system; they are the characters of orders; there is thus properly no order of silex but of quartz, none of clay but of feldspar, none of talc but of mica, at least according to their signification.

594. The points of origin for this division of the mineral system or that formed by the fundamental earths are the three fundamental crystallizations, and upon these everything that does not belong to the calcareous earth must and does naturally arrange itself. The mineral system is only the developed and separated granite. The fact cannot rest unobserved, that mica has only one laminar cleavage, feldspar two, quartz three, this being for the first time a perfect crystal, a double tetrahedron with one column.

595. With granite and with it alone the solidity, the body and form of the earth has been given; it is the homogeneous fundamental mass of the planet, and is therefore crystallized throughout to the finest degree in its three constituent forms.

GNEISS AND MICA-SCHIST.

596. If, in addition to granite, several formations are displayed which bear traces of the same mode of origin, of the same component parts and the same aggregation, it may be inferred that they are only metamorphoses of the same precipitate. Granite is the basis of geogeny.

597. Granite is a totality for the earth, it is a representation of the three terrestrial elements under the form of identical earth; it is the earth represented in silex, water in clay, and air in talc; it is an universe represented individually in the earth-elements.

598. The granite can undergo metamorphosis in only three ways; for its essence is indeed only trinity. Nothing can individualize itself from granite but quartz, feldspar, and mica. All the formations of this period are thus quartz, feldspar, or mica formations. The granite appears under a threefold form as quartz, feldspar, and mica-granite.

599. The first quartz formation is properly the granite itself, and its character will be also the determinant for all the metamorphoses of quartz—quartz-granite.

600. The first structure, in which the character of feldspar as regards its laminated form and its more argillaceous nature obtains the preponderance, is Gneiss—Feldspar-granite.

601. The first structure in which mica is the predominant character, is the Mica-schist—Mica-granite.

602. All rocks that do not belong to the calcareous formation come under these three forms. The progressive formation of the earth takes a threefold course, since it begins in a threefold and yet single manner in granite.

603. The gneiss and mica-schist precipitations followed subsequently to the completion of granite. For all precipitation is a true process, in which water has a certain tension peculiar to this process, by virtue of which this form of earth and no other has been produced.

604. By granite the silicious principally came out of water, but what was argillaceous and talcose remained behind.

605. Gneiss and mica-schist are indeed subdivided granite, but not in the sense of the already perfected granite having been again stirred up and dissolved, but as already separated in principles, in the primary water.

606. Gneiss and mica-schists are products of a more mighty operation of light than granite. They are nearer to the upper layers of water.

607. After the granitic rain, or after the formation of granite, the sphere of water was no longer perfectly transparent; but water was now found also in the primary valleys, in which the light had more power, and thus greater capacity for splitting. During the formation of granite the water has only become polar by light; but, as it was there wholly transparent without opposition, it could never attain to a perfect dualization. In the second earth-rain and in the valleys, the light on the contrary produces dualization of the fundamental earths, since the sides of the mountains afford opposition to the light, become themselves polar against the water, and at the same time heat is produced.

608. Gneiss and mica-schist enter into a polar relation with gravity, and that indeed as a Different to an Identical, as periphery to centre or as light to gravity.

609. Now the falling granite had in part lost its quartz and obtained a predominance of feldspar. Furthermore as the water-globe was already greatly collapsed, the crystals thus originated in larger drops, and occurred besides in water already stagnant or flowing. The schistose gefüge must have emerged as well through the preponderance of the laminar felspar, as by the flowing of water and the attraction of the granite-walls. This schistose granite is gneiss.

610. When the gneiss was thrown down the talc predominated in the water; it now fell in the same manner with less quartz and feldspar, and was deposited in a still more schistose form than mica-schist.

611. Granite, gneiss and mica-schist are the first that together form a Whole, each factor whereof has been evolved in an equally perfect manner.

LAMINATION.

612. By this active antagonism of granite to gneiss and mica-schist the Lamination of the latter is determined. Every particle of gneiss is attracted from the granite-wall and placed in a definite direction, corresponding to the polar operation of granite; the particles of gneiss already deposited attract the coming ones, and so on. The parts of the gneiss and mica-schist are not deposited upon the granite by virtue of their dead or inert gravity, but by virtue of living polar attraction. They are not therefore deposited in the depth of the primary valleys, nor do they fill up the latter; but they are attracted by the granite-walls and deposited to a greater or less extent like laminæ of crystallization, in large perpendicular layers.

613. By these two precipitations the primary valleys were in part filled up, and partly narrowed

by the polar attraction of the walls. The primary valleys therefore are no longer present upon the earth, unless everything be called a primary valley that has not arisen or been excavated by the current of water.

614. Gneiss and mica-schist have indeed taken part in the primary crystallization, yet are, however, only its last movements, as the water had already met with resistance, and was partly stagnant; their mass therefore is not so purely crystallized as granite. They are not parts of the earth's nucleus, but lie only like a crust upon it like hollow crystals.

615. It may be said that gneiss and mica-schist originated only because there were fathomless valleys in the granite, in which the dissevering actions were inclosed, and extended themselves from one mountain-wall to another, while the light could be reflected in them and heat the earth.

616. So long as the granite was devoid of valleys, so long also did no other formation originate. This is proved by the fact that upon the highest mountains the granite is bare and uncovered, while this is not the case upon its lateral walls. It is therefore the second and third earth-rain that first originated, after the earth was heated. Gneiss and mica-schist are, so to speak, precipitated by reflected light.

617. The principal valleys of gneiss and mica-schist have not originated by themselves, but have been modelled according to the form of the granitic valleys. The valleys of those earth-precipitates are properly only subsequent valleys.

618. The lamination is not everywhere a mechanical phenomenon, but without doubt also a polar. It is exactly the same law, which determines the lamination of crystals, that does that of the strata of the earth, and operates also in producing their lamination.

PRIMARY LIMESTONE.

619. The earths must be viewed as one mass, the component parts of which observe a mutual relation. Now, after the silex, clay and talc had been precipitated, a proportionate abundance of lime became free, which was then thrown down at the end of this period of precipitation, as Primary limestone.

620. The calcareous mass may be viewed as corrosive earth, from which a certain quantity of carbon has been set free. As such, and combined with the oxygen developed out of the water by the influence of light to form carbonic acid, the lime was insoluble and was precipitated.

621. The primary limestone has also not been mechanically deposited. It has a crystalline texture, is a calcareous granite, and generally succeeds the mica-schist formation.

b. Water-Formation—Transition-Rocks.

METAMORPHOSIS OF THE PRIMARY PRECIPITATIONS.

622. The metamorphosis of the granite did not terminate with these precipitations. They are only the starting points for the metamorphosis, in which the effort was first manifested to free themselves from the primary combination. In the metamorphosis of earths, the fundamental earths, or the constituent parts of granite, strove to become each for itself a particular rock. They were all identified in quartz; in granite they for the first time parted company with each other, yet still formed a common sphere; lastly, in gneiss and mica-schist they all separated into three spheres, but which still did not differ in mass from granite.

623. Individualization had not yet been attained. The gneiss had still all the constituent parts of the granite, as well as the mica-schist; both are only a more peripheric, slaty granite, the one having a preponderance of the Argillaceous, the other of the Talcose. With these formations, however, Geogeny cannot remain stationary; for verily the law of the development of the world is individualization. Instead of granite, simply quartz, instead of gneiss simply clay, instead of mica-schist simply talc must be separated; and then the termination of this period is attained, the

trinity of the simple earths has been completely represented.

624. All the Earthy could not have been separated from the water by the first storm of precipitation. For the water now rested very much collapsed upon the earth's nucleus, and could no longer therefore assume the form of rain. That therefore which was now precipitated could no longer be thoroughly crystallized, but must follow the current and disquietude of the water, and thus emerge from it in a slaty or massive condition.

625. The quartz of the granite endeavours to set itself free from the clay and talc, or it becomes freed by the latter removing from it in virtue of its polar behaviour. There is therefore one series of rocks, in which the granite is constantly rejecting the feldspar and mica more and more, and at last subsists as quartz simply, which quartzose rocks as forming entire mountains are for intelligible reasons not of frequent occurrence.

626. The completion of the gneiss in its entire separation from the granite, and the evolution of feldspar upon a large scale constitutes the clay-slate, and finally clay-stone and clay-porphyry. This yields us a new series of formations, in which the gneiss gradually attains to being divested of quartz and mica and to a pure position as feldspar. The clay-slate is a true gneiss, that has lost the definite particles of quartz and mica.

627. The position of mica-schist constitutes in its purity the talc-formation, talc-slate, chlorite-slate, hornblende-slate.

628. After these several precipitations, the calcareous mass remains behind in the water, and now, as in the first periods of crystallization, is charged with carbonic acid, and is precipitated as transitionary calx under the form of mountain limestone.

629. These formations are found upon the whole to be arranged on the earth, in the order of time at which they were precipitated from the water. In the middle of the loftiest mountains is granite, then gneiss and mica-schist; then follow quartzose rocks, clay-slate or porphyry, talcose rocks, and lastly on the edge of all these runs the chain of alpine or mountain limestone. In the last of these formations are found fossil remains of corals and molluscous animals. For these formations fell first of all after the water had a solid bottom, and the granitic mountains projected above it.

SEDIMENTARY OR STRATIFIED ROCKS.

630. The period had now arrived in which the fundamental earths, being upon the whole completed, predominate. That which was separated in a chemical way from the Earthy out of water, has been in great part precipitated. This period, however, although the first and most extensive, indicates but the half of Geogeny, or as yet only one pole in the genesis, which requires the other. In the beginning both poles were in the water, that of the fundamental earths as well as that of the calcareous earth; the light shone upon, disseverred them, and earths, the most heterogeneous in respect to water, were first of all precipitated.

631. While the fundamental earths were precipitated, the calcareous earth was repelled and retained, on account of its homogeneity, in water, because the acid half continued longer fluid than the basic. The water was thus after its separation from these substances a true limewater.

632. Through this separation, however, the great antagonism in the water ceased; and subordinate antagonisms now made their appearance, which were kept united by the former. The calcareous earth is now occupied no more as one pole, but is the whole water itself, upon which, as it is less deep, the light acts anew and with greater force.

633. The dispersions of earths began just at this period to multiply themselves, from the only fettering agent, namely gravity, having betaken itself to rest; every earthy now emerges from its connexions, the factors falling wholly asunder into alkalies and acids, which combine in a

multifarious manner.
634. These dispersions associated with the torrents of water that were now everywhere present prevented crystallization from taking place upon a large scale; they moreover mingled with the mechanically water-borne and crumbling débris of the earlier species of rocks; their laminations therefore resemble rather a mechanical deposit from water. They are the Stratified rocks.
635. As the first period must include the calcareous earth, so also in the period of strata or in the dualized period this earth is not without a slight antagonism of the fundamental earths; and this it is which for the first time becomes distinct, but always with a preponderance of the calcareous over the fundamental earths, while in the primary periods this relative proportion of the two was the reverse.
636. The primary period repeats itself again in the second, and thus strata consisting of the fundamental earths originate, as we have seen exemplified in the primary and transition formations of limestone. The precipitation of strata is divided also into four formations, into silicious, argillaceous, talcose and calcareous strata, close to which range also the strata of ores, Inflammables and salts.
637. In other respects the chemical deposits of this period are so blended with the mechanical, that their mode of origin seems for the most part to have happened in both ways.
638. The silicious formation returning in the stratified periods is chiefly under the condition of sandstone. Apart from that, which has originated through the detritus of the older kinds of rocks, it may be assumed, that the prevailing lime still held some silicious earth in a state of moisture within itself, and that this during its separation was precipitated as a fine alcohol, namely, as sand. If, however, sand fell, so also must a proportionate quantity of lime fall, by combining itself with an acid. Sand and lime therefore usually accompany each other. If the two be regarded also as only floated freely and suspended in water, still the chemical antagonism manifests itself between them as if they were in a mortar, and they have been precipitated in layers alternating with each other. The sandstone is as a rule therefore imbedded in the lime; it is a mortar containing but little lime. The mechanical silicious deposits are exemplified in the Nagelfluh, old red sandstone, Grauwacke, sandstone and drift-sand.
639. The stratified clay appears to have been deposited as clay-stone; it passes over into slate and potters' clay. The talcose strata pass by serpentine and potters' stone into steatite and meerschaum.

STRATIFIED LIMESTONE.
640. The pole which had operated continuously from the fundamental earths contained in the Earthy has now separated from it, and the tension is again extinguished. The Earthy is now contained in a pure state in water without continuance of the silicious pole; the influx of water has now obtained the preponderance. So soon, however, as the antagonism of this water to the stratified silex, clay and talc ceases, the more internal, hitherto restrained by the feebler antagonism, becomes awakened, as it did after the precipitation of the fundamental earths.
641. The principles are necessarily combined more firmly in the fundamental than in the calcareous earths. In this both the oxygenic and basic earth-principle must each attain for itself completion, and represent the two primary bodies in the earth with the same capacity for separation and activity.
642. The production of earths results from a constant antagonism subsisting between them and water. The more the Basic is thrown down, by so much does the Oxygenous preponderate to a greater degree in water. The water becomes oxydized and seeks to divide into its two principles, into oxygen and hydrogen.

643. By this contest tension is also excited in the earth-principles, namely, oxygen and carbon, and they begin in themselves to separate. The metallic basis of the earths strives to become free. During the separation, however, the oxygen snatches as it were some carbon along with it and appears as carbonic acid; but the carbon of the earthy lays claim to some of the hydrogen and oxygen to combine with it and appears as corrosive or calcareous earth.

644. The acid is therefore a half of the earth, which passes over into water, and the corrosive earth is also a half of the earth that has lost its Aqueous. The former is the Aqueous in the earths, the latter the Earthy itself separated from the former.

645. The corrosive force is therefore no peculiar action in nature, but only the effort made by the earths to complete themselves and imbibe water or acid. The corrosive force is no synthesis, but a moiety.

646. A total earth may be therefore regarded as a combination of acid and corrosive body. These two component parts are separable in the calcareous and talcose earths; but in the argillaceous and silicious earths they are so intimately dissolved, that they can not be separated from each other.

647. Carbon, hydrogen and nitrogen, but not oxygen, fall or range upon the corrosive side of the earth in question.

648. Regarded in this general manner the corrosive principle stands opposite to the principle of combustion, and what is combustible, is in idea corrosive. The corrosive power is, however, but feeble in the gaseous nitrogen, stronger in hydrogen and finally strongest in the body of earth. The earthy carbon is the proper corrosive principle. The direct antagonism is not therefore between the corrosive body and oxygen, but between it and the carbonic acid, and therefore between it and all the acids, or acidity generally. The last earthy antagonism is that between the corrosive body and acid.

649. The corrosive body regarded specially as simply a pure earthy body, must stand opposite to the two moveable elements. The corrodent is therefore constantly striving to draw water and air into itself, and upon this depend also the effects produced by the corroding matters. The corrosive calcareous earth acts in a destructive manner, it abstracts water and air from bodies. The action of the corrosive body is a deprivation of water and air, and hence the elevated temperature of burnt lime in water. If again the corrosive lime be full of water and air, it is neutralized. It is now forsooth again a total earth, in a mechanical sense being again provided with water and acid. All earths are an equal or identical mixture. Acids and alkalies are thus to be regarded in this respect as moieties, and thereupon their chemical relation appears to depend. The elemental bodies are desirous to complete themselves. If therefore a base stands in corresponding import with a certain acid, it will thus have a greater affinity for the latter, tending to separate it from some other combination. Upon this principle, which has indeed been hitherto unknown to exist, the grades of affinity appear to depend.

650. As regards the mode of occurrence of the calcareous earth, it also is not of so mechanical a nature as is generally supposed. Its legitimate relation to sandstone and other precipitations, speak against that. But crystallization has for the most part disappeared in it; and it is only in cavities that crystals shoot out, like the ores in metallic veins. In granite the commencement is crystal, but in lime it is the termination; crystallization determines the character in granite, in calx or lime, however, the crystals are only blossoms.

651. The calcareous earth multiplies itself as a reduction of the earth of gravity and that indeed three times. There exists, so to speak, a corrosive silicious as also argillaceous and talcose earth. The three corrosive earths are calcareous earth, strontian and baryta, or it may be said that the

first would be salt, the second Inflammable, and the third, metal.

652. Still a polar separation emerges in the stratified calx, while the two earth-principles become more individualized. The carbonate of lime ranks on the lowest stage. In this, however, the differencing process of light had not remained stationary, but elevated the carbon to a higher grade; carburetted hydrogen and sulphur originate in the calcareous earth combined as gypsum, with oxygen.

653. It may be said, that calx were decomposed into alkali and carbon; water into oxygen and acid. Carbon and oxygen become carbonic acid in limestone. Carbons and hydrogen become sulphur, combined with oxygen sulphuric acid in gypsum. Hydrogen and oxygen become hydrochloric acid in common salt.

654. Gypsum is to be regarded as a calcareous earth, which is inflammable in character, as the fundamental earths were metallic. The philosophical essence of gypsum consists not in its oxydation by sulphuric acid, but in the combination of calcareous earth with sulphur, as occurs in that of iron-spar with iron; in this combination, however, the carbonic acid has still remained, whereby the sulphur became acid. The gypsum was therefore a carbonaceo-sulphate of lime, an oxydized metal with a very large proportion of calcareous earth.

655. Gypsum and chalk are related polarwise to each other, separate during the general precipitation, and are deposited opposite to, or alternately upon, each other.

656. In fluor-spar, apatite and boracite the last differentialization of lime and carbonic acid is lost. The principal masses are the carbonate and sulphate of lime.

657. The strata of the Inflammables, such as of pit-coal, and of the ores, as of iron, calamine, appear to have originated in simply a mechanical manner.

REPEATED SEDIMENTARY PRECIPITATIONS.

658. The precipitating process is a process of polarization, which comprises several stages. In it there are moments of time.

659. If the fluid mass be large, this polarization will then require considerable time to penetrate throughout it.

660. The polarizing process will issue from a definite point, which is different from water, and thus form the point upon which the light operates with greatest force.

661. This is solid ground. During the calcareous precipitation, there was no other ground present save that of the mountain tops. It was thus from these that the calcareous precipitation set out.

662. With each precipitation a greater number of mountain tops made their appearance, because the water sank; for the calcareous earth is about three times denser than water.

663. Such being the conditions belonging to the sedimentary periods, several consecutive centuries characterized by precipitations with repeated recessions and elevations of the water, elapsed and have left evidence of the time thus consumed.

664. The recession of the water was not always an ebbing or sinking in, but a diminution or disappearance of the same, like the water in a glass lessens in quantity, or becomes thoroughly solid, if salt crystallizes therein.

665. By such precipitations whole basins of land became dry or freed from sea-water. Streams therefore of fresh water originated, and with them corresponding organizations.

666. These streams gradually filled the basins and formed seas. As the seas coalesced by the constantly descending water, an inland sea arose.

667. This inland sea became again salidified, and that indeed of itself by the influence of light and the dissolution of the salt-banks, which were not overflown. Such is still the Caspian sea and others of that kind. In these marine plants and animals could again originate.

668. Thereupon new precipitations of salt succeeded, the gypsum and calx being again dissolved, and consequently diminutions also of water, whereby rivers again originated. One and the same basin of land was alternately covered with fresh and saline water.

669. Marine and fresh-water animals could therefore originate and perish alternately. And this is the explanation of the fact why banks inclosing both kinds of animals are found above and below each other.

670. An alternating ingress of the sea is not therefore necessary in all cases to explain the occurrence of marine fossil remains. Such an assumption is also wholly inconceivable. Nor is alternating elevation and depression of the soil necessary to the explanation of this phenomenon.

671. During the time of precipitation the temperature of the water and consequently of the earth and air also was necessarily raised. All creatures, which then originated, must correspond therefore to those of warmer climates.

672. The fossil remains do not require the assumption of a change having taken place in the inclination or bearing of the earth's axis; nor of a heating of the surface by a fiery interior.

673. With every later precipitation other animals and plants must originate, because the temperature and also the mixture of water was changed. The fossils therefore indicate the age of the sedimentary strata.

674. During the last precipitations the creatures of colder climates must have originated.

675. Land animals cannot, or but rarely, be found in the sedimentary strata, if even they had already been in existence prior to their formation. For the inundations did not break in suddenly, but the water rose by degrees. They had time therefore to retire to the high grounds.

676. Land plants may, on the contrary, lie in the sedimentary strata, because of their inability to escape.

677. The bones of birds and men must be found least of all fossilized, because a retreat by them was most easily effected. It does not follow, from our not finding them, that they have not existed.

678. The different fossil remains have therefore not simply lived, where they are found, but originated there also. Some of course may have been floated also to these localities.

679. The inundations of water were in general necessary, because basins of land and precipitations were everywhere present; but not all on that account at the same time.

680. In this sense there was a general flood, a deluge, namely, for every land.

c. Air-formation.
TRAP-ROCKS.

681. Vapours and gases of different kinds may be contained in the interior of the earth in two ways, either chemically combined as carbonic acid, or mechanically inclosed in cavities of the earth. Both may be developed, or expanded by calefaction, and the latter by diminution of pressure.

682. If those that are chemically combined be developed by calefaction, they then form vesicular spaces in the masses of earth whereby the latter are extended and raised above the surface; such as amygdaloid, basalt and others of this kind.

683. Warm springs may originate by subterraneous processes, chemical or volcanic; probably, too, by the compression of air that has forced its way into these situations.

684. Earthquakes may indeed originate in different ways; as in addition to the falling in of cavities, by the chemical development also of gases, by their subjection to heat, by aqueous vapours, and also by the sudden diminution of pressure upon these incarcerated gases.

685. This diminution of pressure proceeds from sudden rarefaction of the atmosphere—due probably to the disappearance of air in a particular place, or resulting from a change of the wind, or the formation of heavy rain. Earthquakes can therefore extend through many countries, without requiring to depend upon each other or on a common focus of action.
686. The silicious trap is silicious sinter, tripoli and polierschiefer.
687. The argillaceous trap, amygdaloid, clinkstone, several porphyries.
688. The talcose trap, basalt.
689. The calcareous trap, probably chalk.

d. Fire-formation.
VOLCANIC-ROCKS.
690. Volcanoes are secondary combustions of masses that have originated through the primary combustion, and are therefore only of local occurrence. Such combustible masses are without doubt bodies belonging to the class of Inflammables, and thus carbons, sulphur, sulphuretted metals. Simply burning gases would throw up on high the masses of earth, but not heat them to the degree of fusion.
691. By the heat of these combustions the masses of earth have been fused, forming lavas. The silicious lavas are obsidian, pitch-stone. The argillaceous lavas are the kinds usually met with. Next come the talcose lavas. The calcareous lavas are probably dolomite.
B.—METALLIC ORES AND INFLAMMABLES.
692. Metallic-ores and Inflammables are products of the planet, when completed, and have not originated along with the origin of the latter, like the earths. The question accordingly arises, what have been the forces by which the metals and Inflammables were produced.
a. METALLIC VEINS.
693. Fissures in rocks, so narrow that they cannot be illuminated by the sun, are called passages or veins. They are rarely found in granite, appear generally for the first time in gneiss, more rarely in the later kinds of rocks, and almost cease to be met with in the stratified chain of mountains. They are found principally in mountains, and thus in masses of earth which project above the level land. We must thus arrive at the conclusion that they have there originated by actual fissure, and that indeed for this reason, that masses which project or stand freely out would admit of yielding asunder more easily than the masses of the plains. This fissure may take place by mechanical disruption, by land-slips, or even also by desiccation.
694. The veins are not prolonged into the kind of rock that underlies them, as e. g. gneiss veins into granite, and so on; they have hence originated from above.
695. They are open and wider above and strike out below; they have not therefore originated by a force acting from beneath.
696. In the schistose rock they form generally transverse fissures.
697. There was a time in which the veins stood empty, as well as a time in which the primary valleys were empty, namely unreplenished with gneiss, mica-schist, and such like minerals.
b. PRODUCTION OF ORE.
698. Geogeny takes two directions; the one passes upon the periphery into the splitting action of light, the other into the abyss, where darkness reigns.
699. The valleys were the condition that conduced to the differentialization of the earths, because in them light had power to produce the highest polarity. By the valleys the Earthy has been separated into its principles; silex has separated into clay and talc, to which finally carbonate calcareous earths and salts succeeded.

700. The Earthy cannot subsist in its identity in the broad valleys; the earth cannot be represented as the pure symbol of gravity. All bodies that have originated upon the surface of the planet are oxydes or salts.

701. If the earth-difference be generated in the illumined valleys, so must the earth-identity be produced in the dark valleys; for it is the absence of light alone that allows the purely Basic to subsist. This earth generated out of gravity is the ore.

702. The ore is a child and a treasure of darkness; where light is, it must vanish; it cannot endure its gaze. Metal when exposed to day is given up to annihilation, to oxydation.

703. Darkness is, however, no power, and can consequently be only the opportunity, not the cause of anything's happening. Other forces, instead of that of light, must have therefore operated in the production of ores. In order to discover these forces the relations of the ores must be carefully weighed or considered.

704. The ore is in a philosophic sense a reduced earth, and so reduced indeed that the basic principle has obtained the preponderance over the oxygenic or supporter of combustion, and attained unto substantiality.

705. In light, in the water forsooth when illuminated, the two earth-principles were already divided internally, but not completely separated; salt only originated, namely an acid and alkali.

706. The ore is, however, a salt wholly reduced, and indeed the reduced alkali has become metal, the reduced acid with the basis of hydrogen, Inflammable, namely, coal or sulphur.

707. Now as light was not able to produce such a separation in the free or full sense of the term, forces must thus have been present in the dark passages, which completed this separation.

708. Ore and Inflammable are the total salt dissevered, and this is the dissolution of the two; the former are blue and yellow, the latter is the compound green.

709. The processes of the formation of salt and of ore are both indeed processes of separation, but yet they stand opposite to, or rather transcend each other. Both mutually conditionate each other.

710. While the earths submitted to the action of light upon the surface of the planet are converted into salt, the process of the formation of ore takes place in the dark or under the earth; or while above the oxygen is predominant, the basic body is that below. The ore imparts upon a large scale its oxygen to the salt, and the salt bestows its basic body on the ores.

711. No ores could originate in the middle of the earth were light even to have no access thereunto. For not merely do earth and darkness belong to the genesis of the ore, but earthy water like as unto salt.

712. The ore is not a conversion of earths that have already existed, or been actually separated; but it originates first of all during the process of separation. Where ore is, fluid has thus been, and polarity, which is not directly derived from light. The ore is a mere child of the planet, a pure terrestrial essence generated without the joint assistance of the heaven, and therefore the highest substantiality of the planet, the spirit of the earth.

713. What gneiss and mica-schist, calx and salt, are in the bright valleys, such is the ore in the dark valleys; the former are the differenced ore, and this is their identification.

714. There is no peculiar metallic body or seed, which had already existed in the primary creation as something special or peculiar, and which by one process only, as perhaps even by its gravity, was precipitated or posited from the fluid mass. One and the same substance, if found exposed in a valley to the light, becomes earth; but ore when it is in a dark passage.

715. Of a certainty neither clay, sand, talc, or calx become metal. For these are at once

definitions of the spirit, words that have been already completed and expressed, and cannot be recalled; so also the ore will not again become clay, even if it be submitted to light. The indeterminate substance only, which under other conditions might have become clay, becomes in the darkness ore.

716. What is not in idea, prior to the adjustment or fixation of a pole, reduced to ore, can on no account become that out of an already finished mass of earth. Conversions of earth into ore by chemical arts are labours bestowed in vain.

717. Yet if ores do originate, they originate only out of the indefinite Basic, which is still in the water, just as trap stones originate not from a stony dust that pre-existed, but out of the pure indifferent substance of the air.

718. The veins and the formation of ore are one, as are the valleys with the calcareous and salt formations; and he who asks how has the ore originated, must forthwith inquire as to what is the essence of the veins.

719. The ore has not originated externally to the veins and been at some time or other conducted into them by means of water. For how should it thus originate? The reply is, a specific action must have been at work in the fluid, which determined it to separate ore and nothing else. But where is this ore-forming action in the free space of water? Nowhere. And if also the ore had been separated or diffused throughout the whole mass of water, what a world-wonder is it that it merely flowed into the veins and some stockworks? What prevented it from filling in large masses the broader valleys? The mechanical theorist upon metallic veins must assume an attraction on the part of the veins for the ore-particles in water; but how could this attraction have drawn these particles for miles in extent out of the water? and were this action strong enough, it must still be gifted with greater power to produce or at least separate the ore out of the water, that is found in the empty spaces formed by the veins.

720. Since ore has on a large scale separated from the calcareous and salt formations like the Identical from the Different, it has done so also upon a small scale in the veins. There the same process of separation has preceded it.

721. Now, however, nothing can be separated, i. e. nothing be reduced, without the oxygen accumulating upon something else. The Earthy must therefore separate during the formation of metal into the Reduced and Peroxydized.

722. The reduced earthy is the metal or inflammable; the peroxydized, however, an earth proper. This earth is called vein-stone.

723. The ore has only originated in opposition to the vein-stone, and when this has taken the Different of the Earthy into itself. Therefore the vein-stones are also different from the kind of rock in which they occur, and that through greater differentialization; they have even receded for the most part into acid and alkaline poles, as calc-spar, fluor-spar, heavy-spar, which are the usual vein-stones. All vein-stones are oxyds and as a rule, those in which the oxygen is freely manifested, namely as acid. The vein-stones were the sheath of the ore, which could first appear when this sheath had withdrawn itself.

724. The metal stands in relation to the vein-stone. Thus in argillaceous vein-stones we commonly meet with iron, manganese; in the quartzoze with gold; in the calcareous with lead, &c. There are here also extremes. There are vein-stones, called sterile veins, in whose antagonism no metal has been formed; and there are veins, that are merely filled with metal, such as the Stockwerke, Lager.

725. Since the ore and the vein-stones thus originate together, and in such a manner that they conditionate each other; their fundamental mass must have been one, and a separating force,

which is not light, must have operated upon them.

726. Moreover, as the ores occur only in narrow spaces with their vein-stones and both form alternating tables upon the walls of the veins, they must have been attracted by the latter.

727. The walls of the vein consequently exert a polar influence upon the ore and vein-stone. Now, if this be their mode of action, they must be in a condition to separate the fundamental mass.

728. It is thus the vein themselves, which, by a vital force, produce the metals; they are thus a living womb, or matrix as it has been emphatically termed.

729. Two walls in close juxtaposition are requisite for the production of metal. Upon a freely exposed wall or face of rock no metals are found.

730. By this separation, however, two kinds of minerals originate, Inflammables, and ores proper or metals. The action of the walls must be therefore of a twofold nature.

731. But two cases also are conceivable, in accordance wherewith this polarity of the vein admits of being divided. It subsists either quite alone between the two walls, as surface-polarity or electricity; or it subsists between the mediate point of the earth and the walls, constituting centroperipheric polarity or magnetism.

732. The product of the surface polarity are the Inflammables, of the radial polarity the metals.

733. Since no more metals and Inflammables originate at the present day, although magnetism and electricity are in continual operation; a third influence moreover must have been in constant activity. Now this cannot be thought of as any other than heat. The metals must thus have originated while the earth was still in a glowing state, and when thus also magnetism and electricity could operate more powerfully towards effecting a reduction of the mass. By heat the mass in the veins was probably converted into gas, through which its separation by means of magnetism and electricity into ore and transition earth could more easily take place. The metals are thus sublimations, which were first deposited, when the earths or rather the mountain-stock began to cool. Thus also zeolithic crystals were deposited in the upper stores of mines. Metals are thus the children of heat, of magnetism and electricity; the heat renders the mass in a fit state for being separated in the next place by the polar forces.

734. Metal is carbon completely reduced, which contains nothing more of the other elements in itself, namely, neither hydrogen nor oxygen. It is consequently the Basic of the earth-element without material admixture, and thus is earth with the properties of its prototype, of fire or of gravity, of light and of heat.

735. In accordance with this interpretation metal can be produced by no other process than the centro-peripheric. Its occurrence in fissures of the earth that are mostly perpendicular likewise proves this.

736. In the deeply situated veins, therefore, the more identical or purer formations of ore must occur; while, on the contrary, in those situated higher up or exposed to the day, i. e. in closer proximity to the water, air and light, the more different or compound formations of ore must be produced.

737. The four classes of ores occupy a position in the veins tolerably accordant with these relations; an additional proof that they have originated in the vein and not been floated into it by the action of water.

738. The production of ore which occurs in the upper parts of the veins, furnishes us with the saline ores; for here the water, air and earth are principally active. In the upper depths are found most frequently the oxygenized metals, or as they are called mineral spars, as spathic-iron,

malachite, calamine, lead-spar, pyromorphite, &c. Finally, the metallic or ore-spars actually pass over into salts, and by the horn-silver ores into the vitriols. They are for the most part crystallized.

739. The salt-formation of the ores has always decreased in proportion to the light having less access to it, and hence in greater depths, or in wholly mountainous masses. There the ore has not been deposited as a light-difference, but only as a Terrestrial, just as the earths, especially the calcareous earth, have obtained the same. It has not attained to a complete evolution of the polar body so as to constitute an acid, but the two principles have only emerged opposite to each other. They are oxydized ores, consequently those among the ores, that represent the character of earths, namely, the ochres. They are frequently uncrystallized.

740. So far the appearance of these ores is thoroughly earthy and devoid for the most part of metallic lustre; finally, the genesis of the ore turns upon the side of the identity, the principle of gravity having secured itself wholly in the depths, previous to the adverse accession of water and light. In such situations nothing more than heat and the centro-peripheric polarity of the vein-wall operates. The Metallic recedes wholly from the salt and from the earth. The oxygen disappears, but in its place comes sulphur, and the sulphuretted ores originate, e. g. the bi-sulphurets of iron, blendes, glance-ores.

741. Lastly, every combination or influence of the other elements and mineral-classes vanishes; acids, oxygen and sulphur are no longer generated in the perfectly dark depths, and the ore stands there in its entire identity, homogeneous, resplendent and heavy as the sterling metal. The same results from sublimation. The heavier metals remain below, the lighter ones and the sulphur ascend.

742. As the orders of earth are placed in the veins, so also are the different metals themselves; for the cause of genesis is alike in both.

743. The earth-like metals, as iron, manganese, &c. which occur constantly oxydized, are usually found upon the surface of the planet; the hydroid or water-like, as lead, tin, lie usually deeper; the aeroid or air-like, as arsenic, zinc, exhibit pretty nearly the same relations as the sulphur metals; lastly, the pyroid or fire-like, e. g. the noble metals such as gold and silver occur frequently in great depths and not unfrequently in granite; the two preceding metals generally occur in gneiss, but the first even in younger or more recently formed rocks.

744. As a certain regularity prevails in the arrangements of ores and of metals in the veins, so also does this hold good in respect to their distribution over the planet.

745. The more the polar earths are separated and deposited in the broad valleys exposed to light, by so much the more in quantity and of greater purity is the ore produced in its own valleys of darkness. The first effect, however, happens through the power of the sun; the more powerful therefore its influence, although indirect, so much the greater in quantity and purity is ore produced in the depths. The greatest quantity of ore and that of the noblest kind was inevitably generated beneath the equator. There also more lime as well as more salt are probably found upon the surface of the water; towards the north the oxydized metals, or the whole series of iron-metals, were produced. In the Temperate Zones we find more lead, zinc, bismuth and arsenic.

746. Thus the theory itself of ores furnishes us with a proof, such as the theory of the earth that has been hitherto entertained could not afford, that the equator forsooth has since the formation of the metals not been displaced. The metals formed themselves cotemporaneously with the gneiss, and have thus begun to form prior to the existence of the organic world. It is therefore a vague opinion, devoid also of foundation, that since the earth has been inhabited by animals, even by the higher or hair-clothed vertebrata, its axis has been changed.

747. No earthly phenomenon speaks so clearly and loudly against the mechanical theories in the natural sciences as ore. Not only has the whole planet been included in the ore, but also the whole of science, the whole of philosophy.

748. The first transition of the Earthy into the metallic character is indicated by iron. Iron ranks next to earth, especially to the argillaceous earth, is everywhere associated with it and is most generally distributed, being mixed with almost every earth and even all organic bodies, to their very elements.

749. The whole series of metals has but one root. What the primary earth is for the metamorphosis of earths, that is iron for the metals: it is their silicious earth.

750. The iron as being the first transition from the Earthy into the Metallic has the highest grade of fusion, and all metals which approximate to this, belong to the retinue of iron. This grade may be set down as 20,000F.

751. As iron is the root of all metals, so every division or group of ore has a principal metal, which occurs in more considerable quantity than the rest and characterizes the division. Among the saline ores, copper under the form of malachite is the principal metal. Its fusing point is 6000F. Among the sulphurous ores, lead is the chief metal; it has with its neighbours the lowest grade of fusion, which may be set down in round numbers as 600F. Among the volatile, arsenic is the principal metal.

752. Among the standard metals, silver is the principal metal; it has with its affinities a fusing point, which ranges midway between that of iron and lead, being probably about 5000F.

753. There are four grades of fusion of the metals, which are removed from each other by very wide intervals, between which no metal is situated. Quicksilver is fluid at the temperature of the air and becomes volatile like arsenic. Moreover the artificial metals are associated with the alkalies and acids. Lead, with its congeners, melts at 500F.; silver with gold and copper at 5000; iron, platinum and such like at 20,000.

754. There are four metallic characters, which are shown to be peculiar in all their relations, in their affinities for oxygen, acid, and sulphur, in respect to specific gravity, fusibility, extensibility, in their electric relation, in mode of fracture, in occurrence, age, and geographical distribution, &c.

c. POISON.

755. It is a remarkable fact that the principal metal of the elemental metals, obtains mostly by oxydation or acidification poisonous properties, while the proper earth-metal, iron, acts beneficially upon the animal organization. Among the water-metals lead becomes poisonous by acidification; among the fire-metals mercury. Among the air-metals arsenic ranks highest, becoming poisonous by mere oxydation. Among the earth-metals, only one among the salt-metals is poisonous, and that is copper.

756. The metals thus appear to become poisonous, when they enter into or put on the character of salt, or of water.

757. The above are also those very metals, which unite most readily with the others, to form alloys, amalgams or metallic compounds. Copper is very readily alloyed; lead almost always contains silver; mercury is susceptible of amalgamation; arsenic metallizes the others almost like sulphur.

758. The air-metals appear to have lost for the most part the metallic character; arsenic therefore destroys also magnetism.

759. The essence of metallic poison thus appears to reside in the endeavour on the part of the metals to suppress the metallic character and convert themselves into the formless elements. The

metallic poison is the direct opponent of the metals themselves, and through this, of everything that has form, and thus of the Organic also.

d. MAGNETISM.

760. Two actions are necessarily manifested in iron, one clearing or dividing in so far as it is earth, and one to be identifying, in so far as it is metal. Iron is the fluctuation between oxydation and reduction, between light and gravity, and this conflict of the two latter is Magnetism. Magnetism is the spiritual function of the metals.

761. Magnetism belongs essentially to the metals only. What is not metal, is magnetic only according to idea or signification; it may be therefore aptly said, that such a body hath no magnetism, and that what has it, were metallic.

762. Magnetism is the direct property only of iron; this alone is the hybrid or heterogynous metal.

763. Magnetism appertains only to the other metals in as far as they are positions of iron; and is the more powerful, the nearer they stand in relation to iron. All metals are magnetic in idea, whether magnetism be manifested in them or not.

764. All metals have originated through magnetism, through the radial polarity, or the conflict of light and gravity. Magnetism is the action betwixt light and darkness, periphery and centre. Magnetism as being a metalgenerating action tends towards the centre of darkness of gravity. What in the earths and salts is the duplex tendency of crystallization, is in the metals the identifying magnetism, as an everlasting operation of attraction.

765. Magnetism is still, however, not identical with gravity. Gravity is the centre abstractedly from the periphery; but magnetism is the centre only in relation to the periphery or light.

766. Polarity belongs to the essence as well as the genesis of magnetism; the metal subsists only through a constant resistance against the universal process of oxydation, against the developmental process of the earth-principles, which the metal is always striving to conceal. The metal is altogether the most mysterious essence of the planet. This resistance to the disclosure of the Innermost of the earth is magnetism. Where magnetism has wholly attained this concealment, it renders the metal perfectly free and disappears, because it has become completely embodied. In the noble metals it has attained what it is still seeking in iron.

767. Magnetism is a linear action with two different extremities, like the primary radius. By one extremity magnetism runs towards the identical centre, by the other towards the partite, electrical periphery, towards the oxydized earths. One extremity will reduce, the other oxydize; one will become metal, the other earth. This is the difference between north and south pole, the former centre, the latter periphery.

768. There is no peculiar magnetic fluid, any more than there is a matter of light, heat and electricity. In magnetism the spirit only of the earth appears, as in light the spirit of heaven.

769. Magnetism is a constant process of excitation. This process of excitement is the process of imparting and of propagation. Magnetism has not been given, but excited. It breathes life into the iron bar, whereby the latter awakes, and that is magnetism.

770. Every action, which induces differences in a line of iron, renders it therefore magnetic; thus electricity, unequal calefaction and a blow, whereby it is thrown into a state of vibration. An iron bar planted perpendicularly becomes magnetic, because it is then a radius to the earth. From the same reason it becomes magnetic, if placed in the magnetic meridian.

e. TERRESTRIAL MAGNETISM.

771. If all metals are in idea magnetic; so must the metallic veins, as products of magnetism, be

magnetic lines. Every vein has a north and a south pole.

772. As every vein is a magnetic line or magnetic needle upon a large scale; so must two veins abutting against each other represent likewise a magnetic tension. A mountain of ores is a net of numerous magnets interlacing each other. As one vein is related to the other, so must one metalliferous mountain be related to the other; and thus two mountains of this character stand in magnetic polarity with each other. The whole earth is surrounded by a magnetic net.

773. As every metal, every vein, every rock is in miniature a magnet, so must the earth be a magnet upon a large scale. There is a Terrestrial magnetism.

774. This magnetism belongs only to the earth only in so far as it possesses a metallic quality, for magnetism is only the spirit of the metals, not of the other terrestrial bodies, as the earths, Inflammables, and salts.

775. Magnetism is no general character of the earth, still less of the whole solar system, except in so far as the metallic principle lies at the basis of every thing earthy. Magnetism does not operate outwardly over the earth.

776. The determinants of terrestrial magnetism are the metallic veins, or the metallic beds in the crust of the earth.

777. Terrestrial magnetism has not been produced or determined by a magnetic nucleus; since a metal in the middle of the earth is a contradiction. All determinations of terrestrial magnetism depend upon the nature, character, distribution, number and direction of the metallic veins or beds.

778. Thus the direction of the magnetic axis of the earth, its mutability, the declination of the magnetic meridian, the inclination or dip of the needle, in short every phenomenon without distinction that concerns magnetism, must be derived from the nature of the metallic veins. Another momentum does not indeed exist for magnetism.

779. Terrestrial magnetism can only be based upon the polarity of metallic veins, and this upon the ores they contain. It must therefore accommodate itself to the distribution of the chief masses of metals, especially of the idiomagnetic metal. Now it has been ascertained, that the noble metals are accumulated about the equator, such as gold, silver, copper, which may be regarded as non-magnetic. In the northern temperate zones, the mountains contain for the most part metals that are non-magnetic and semi-noble, as lead, zinc, antimony, which are usually combined as ores with sulphur. Iron, on the contrary, being the only magnetic metal, (a fact which is also determined chiefly by the magnetic needle in terrestrial magnetism) is accumulated in greater quantity towards the north-pole and becomes always rarer in occurrence towards the equator. The southern hemisphere of the earth, is indeed less known in this respect; but, that towards the south-pole iron reincreases in quantity, is rendered evident by the magnetic needle when conveyed beyond the equator, inclining to the south-pole. If the cause of the inclination upon the northern hemisphere be sought for in the presence of iron, the same must be done in regard to the southern. The arrangement of the metallic groups from north-to south-pole is thus; iron, lead, silver, lead, iron.

780. Iron has been deposited at both poles and this in accordance with the genesis of metals, for iron is a half reduced metal, and must consequently be subjected to the demi-action only of light. At or beneath the equator the Ferrogeneous is wholly reduced, and becomes noble.

781. There is obviously, however, less iron at the south-than the north-pole, because the greatest part of the planet is there covered with water, and thus in general with less earth less metal also is to be found. From this alone it may be explained why the magnetic needle conveyed beyond the equator still remains horizontal and first inclines about the tenth degree of southern latitude

towards the south-pole.

782. The north-pole is thus more energetic than the south-pole, so that the two poles must be also inversely related to each other. It is only from this antagonism of the two poles of the earth that we comprehend, why upon the southern hemisphere the south-pole of the needle dips towards the earth, and why the needle does not turn completely round. For if there was an equal quantity of iron in both hemispheres of the earth, the southern must have the same magnetic pole, and thus attract the north-pole of the needle; and under the equator the needle must have actually no direction. The cause of the direction and dip of the needle rests thus in the antagonism of the two poles of the earth, and this again in the unequal masses of iron.

783. The action of terrestrial magnetism does not reside in the iron formation alone, but in its antagonism to the reduced ores. The terrestrial magnetism is a tension between iron and silver. The lead imparts the poles.

784. As the earth is a globe, so are the two masses of iron situated nearer to its axis than the masses of silver and lead. If therefore the latter exert also any influence upon the direction of the magnetic needle, yet nevertheless the poles of magnetism must coincide nearly with the poles of the earth's axis, because the exciters of magnetism operate in this direction.

785. The magnetic meridian runs indeed in general from pole to pole, but as it is not the earth's axis, which determines the magnetism, but the metallic masses, so the direction of the meridian deviates from the earth's axis in accordance with these masses.

786. The magnetic needle can therefore assume a different direction on every part of the earth, according as its relative position is varied between two principal masses of metals, and even as is indicated by the movement of the compass with the change of geographical longitudes, it is removed from one metallic mass, approaches another, and by this becomes more powerfully attracted. Upon the whole, however, the direction must tend toward the poles. Such are the phenomena of declination of the magnetic needle; and we can now comprehend why this is present.

787. But there must be also spots upon the earth, where the needle points straight to the north, probably, when it is between two metallic masses, or stands at a certain angle to them. These are the lines without declination, of which there are many as is well known, but which have not as yet been reduced to any law. Nor can they ever be so, because we shall never become acquainted with the metallic beds.

788. As the solid land, so far as it projects out of the water, forms a horse-shoe figure, of which the two Americas represent one leg, Europe with Asia and Africa the other leg, the sea being interposed between them; so must the lines without declination fall principally in the ocean, between the earth's crura. Moreover as both these crura of the earth are unequal in size, so must the one influence the needle more than the other, and in this respect also there can be therefore no regularity in the lines without declination. Thus everything co-operates, to the effect of rendering the direction of the magnetic needle unequal; such as the distribution of metalliferous mountains, of masses of iron, of the earth's hemispheres, of the earth's crura, and inversely the earth has probably obtained this horse-shoe form through magnetism. The cavities of the earth probably do not fall in there, because the ground, namely the mountains, is supported by metallic plates. On this account the earth probably maintains an oblique position in her course. In accordance with this the earth's axis had first changed, when the metals were generated. According to this also, the sea had rushed in first, when the metallic veins were present. Lastly, in accordance with this, the metals would be generated, when the whole earth was still covered, and the veins filled with water. Unto such conclusions the philosophy of nature can alone

conduct us. The cause of a horse-shoe magnet acting more powerfully than any other shaped instrument, resides probably in the form of the earth's crura.

789. The magnetic meridian varies, however, not only according to places, but times also. This is explicable from the mutation of metals under the poles, as well as under the equator and in the temperate zones. Processes of oxydation and reduction are always occurring, the more too if the water recedes and tracts of land become dry. To this, culture, the clearing of woods, the draining of swamps and probably mining operations themselves conduce.

790. It must besides be clear to every one, who has learnt to look upon nature as a whole, that the numerous metallic masses of metal upon the earth are not indifferent to the genesis of metal, that they are not foreign to iron; but that they stand invariably in one relation to it, which can only be magnetic; for therein only are they metals.

791. Magnetism is an infinity of tensions spread over the whole earth, of which the tension towards the axis is but the principal, not the sole, tension. The expression hereof is the net of metallic veins.

792. Every magnetic line consists of an endless number of shifting poles; for every magnetic line can be but the metatype of terrestrial magnetism.

C.—INFLAMMABLES.
Electrism.

793. Coal and sulphur may be regarded as the representatives of the Inflammables, making their appearance in the carbonic acid of lime, and the sulphuric acid of gypsum, just as the metals did among the alkalies.

794. The Inflammables are accordingly associated with the acids or the salts, the ores with the earths. It may be said that the former are reduced acids, the latter reduced earths.

795. The Inflammables are consequently those that succeed next to the salts or water-minerals. Their determining element is in this respect also the air; that of the ores is therefore the fire.

796. The Inflammable, as being the reduced acid, must have the strongest affinity for oxygen. A body, which by its own force, attracts the oxygen from the air, so that it appears luminous, is called combustible.

797. The generating spirit of the Inflammables coincides with the spirit of air, and thus with electricity. The generating spirit of metals coincides with the light; it is the radial action in the Massive, or magnetism.

798. Electricity has become embodied in the Inflammable, i. e. idioelectric; in metal, light has become embodied, i. e. idiomagnetic.

799. Now, as the Inflammable exists under two forms, with the preponderance of the earth-nature as coal, and with that of the air-nature as sulphur, so must the electricity appear fixed chiefly in the latter. This fixation is the idioelectricity.

800. As electricity is in its essence a constantly dualized agent, so can only one pole belonging to it become fixed. In sulphur this is what has been called the negative pole.

a. SULPHUR.

801. As the air stands opposed to the earth, so must sulphur to coal. The latter is thus endowed with positive electricity.

802. Coal is, however, the fundamental body of the metals. The metals are consequently related as positive electrics to sulphur. Sulphur is air-metal or idio-negative; metal is earth-or idiopositive sulphur. Sulphur therefore occurs almost solely with metals, as iron, pyrites, glance; yet frequently with arsenic, the metal that resembles it, e. g. in realgar.

803. Sulphur is the basis of all idioelectrism, and this property occurs only in bodies, in so far as they are positions of sulphur.
804. Magnetism and electrism are correlated, as iron and sulphur, as gravity and light, as centre and periphery. The same spirit, which when ruling in the dark, exhibits itself as magnetic, is manifested when it has attained to light in sulphur as electrical. Magnetism is only the electricity identified.
805. We may therefore speak of idiomagnetic metals as well as idioelectric bodies.
806. Magnetism therefore stands in accordance with these relations in opposition to electrism; they mutually change or annihilate each other.
807. Electrism has, in accordance with its signification, the power of manifesting itself with one pole accumulated or set free from the other, as e. g. the negative in a cake of resin; in magnetism, on the contrary, both poles are always together and inseparable. The radius is divided into two in every part of its length.
808. As the functions of metal and of sulphur are correlated, so also are their substances; they are opposed, and hence the metallization by means of sulphur with all its results. This antagonism is, however, dormant or concealed; that of the functions manifests itself much more clearly.
809. The metals, as being dense, central, and linear masses, must fall into a state of tension with electricity as with heat; this is called conduction. The metals are therefore conductors of electrism. In antagonism to the conducting power of the metals sulphur is naturally an isolator; for what is idioactive is virtually also isolating. Iron may be likewise called an isolator of magnetism. There is only one series of bodies in nature, belonging to the peripheric and expansive functions, that conducts; the metals only are conductors. Isolation belongs to the essence of electricity. Isolating action and Electricity are one; for electricity is the surface-function, wherein the line, which is the only conductor, disappears.
810. Electrism does not tend towards the metals, and can therefore have no definite direction in the earth; there is neither an electric meridian, nor an electric equator. There is only an electrical surface to the earth, and this is alike in all regions of the world.
811. The metals must accordingly stand opposed to sulphur as positive bodies, if not as idiopositive, yet as such when brought into collision with sulphur. The metals, when rubbed with sulphur, constantly become positive, and the sulphur remains negative.

812. The earths also become positive when rubbed with sulphur; in short, everything which, in the genesis of the earth, ranks below sulphur, is positive. Heated bodies rubbed with cold, and rough bodies with smooth, must become negative.
813. Bodies become positive with sulphur, simply because the essence of sulphur is of a negative character, or because, in other words, it is nothing else but negativity; the persistency of one pole and the counter-resistance to every other, is called isolation. The metals are conductors, because they stand opposed to sulphur.
814. Positive isolation only is evolved opposite to sulphur, in zinc, probably because this belongs to the air-metals.
815. What sulphur is in its series, that is zinc in the metallic series; the isolating electric rod, with which the other bodies are associated; the one the positive, the other the negative isolator; in so far forsooth as one can isolate bodies that have arisen through linear action. With zinc the other metals become negative, because it can be nothing else than positive, as sulphur can be none other than negative. (That this does not hold good absolutely, we need scarcely be reminded).
816. Two fixations of electricity thus exist, and from these the electric phenomena must be

derived. So long as we imagine that electrical proportions run in a continuous line, so long shall we never be able to avoid contradictions. Two rods stand firmly, and from out and around these two heaps of bodies form, which in reference to their electrical relation (according to the experiments hitherto performed) are naturally exhibited as only one series.

817. Sulphur does not stand alone, but is associated with a series, especially of the higher Inflammables, bitumens or mineral-resins, æthereal oils and hydrogen gas. The higher the inflammability ascends, by so much the more energetic is the negative character, so that, finally, the sulphur itself becomes positive towards such matters.

818. If in every polar action it can be proved, that each polar line consists of infinitely numerous poles, and that each point in it can be alternately changing both polarities, in accordance with the mutation of the principal poles that exert their influence; so is it in electricity. There is scarcely a single body which cannot be positive as well as negative, if it only becomes displaced in its own series, or is transported into the other.

b. COAL.

819. During the electrical separation of the Basic of the earth, or during the communication of the aerial character to the Earthy, a body remains behind with positive character, or the Coal.

820. Coal may be regarded as volatilized metal, as a metal which can change by the action of water or acid upon it into air. Black-lead is a coal, which is directly associated with the metals.

821. Coal appears therefore less in particular places, than as expanded into entire rocky masses, as e. g. in the clay-slate and as carbonic-acid in lime.

822. The coal was, during the earth-formation, separated from the sea, yet not, or only rarely, by itself, but along with other masses of earth, while the sulphur rather accompanies the metals. Coal passes over into the earths, the sulphur into the metals.

823. The volatilized earth or coal, i. e. the earth that has ascended through water or salt unto air, is associated with a higher kingdom, and that indeed the general mass of the vegetable kingdom, as is the case in the pit-coals, which are reversions of plants.

824. As the earths and metals extend into pit-coals, so does sulphur lose itself in idioelectric, inflammable substances, which are likewise reversions of a sulphur that has escaped into a higher kingdom. Here belong the amber, mineral resins and naphthas.

825. There are thus two ways, by which the reduced earthy seeks to mount aloft; by the carbon, as belonging to the more inert earth; and the Resinous, as belonging to the more active air. The vegetable kingdom has its root in the simple earths, especially the hydroid argillaceous earths; the animal kingdom in the divided calcareous earths.

826. Sulphur is yellow, because it is the earthy that has come to light, the carbon is black, because it is sulphur volatilized, moistened in the gloom or darkness of the earth.

D.—SALTS.

Salt-periods.

827. So long as the basis of the acid is an earthy, such as carbon or sulphur in the carbonic and sulphuric acids, does the earthy also obtain the preponderance, and the lime as well as the gypsum or sulphate of lime are precipitated as insoluble bodies.

828. It is only through the influence of light constantly becoming more powerful by reason of the solid land under the water that the oxydation of water rises to the highest degree, so that this element finally converts itself into an acid, or hydro-oxide. This process must be regarded as a decomposition of water, whereby a portion of the hydrogen forms sulphur with the carbon, the rest with the oxygen an hydro-oxide.

829. The hydro-oxide is hydrochloric acid. This acid must be regarded as peroxydized hydrogen.

Its signification is thus that of being water itself, or a whole element with a preponderance of oxygen. It attains this rank by its constituent parts, namely, the two general gaseous primary bodies, by its distribution as a whole element around the earth; by its occurrence as an earth-formation in rock-salt; finally, by its presence in all vegetable and animal juices. Hydrochloric acid is the type of all acids, as the iron is of all metals. All acids are but imitations of the hydrochloric. All abide by the signification of water, or are conversions of elements or earths by oxydation into an hydroid condition.

830. Between the acidified water and the Earthy a higher antagonism now emerges. The Earthy separates a part of its carbon from the carbonic acid and sulphur, so that the rest remains behind peroxydized, and makes its appearance as an alkali.

831. The alkali is to be regarded as the last conversion of calcareous earth in respect to water. It is an earth, whose oxygen has converted itself with a portion of its carbon into acid and been set free; a salt halved upon the basic side. This general alkali, that has originated in water, is soda or natrom.

832. Alkali and acid are the last antagonism in the earthy, moieties, which can never subsist without each other.

833. The alkali is corrosive, because it seeks water and acid, in order to perfect itself; acid is pungent to the taste, because it seeks earth or alkali.

834. Their antagonism is the highest antagonism between water and earth; it is also the representation of the antagonism between fire and the terrestrial elements, or also between light and gravity. Therefore this antagonism has a cosmic or universal signification.

835. The combination of this antagonism is the sea-or common salt.

836. Sea-salt is the universal salt. All other salts are to be regarded only as metamorphoses of it, as well as the acids only conversions of the acid of common salt, and the alkalies of soda.

837. The sea-salt is essential to water. It is the product of geogeny, has not entered the water from without, but been generated in it, and is constantly being regenerated, so long as light shines upon the sea. Properly speaking, sea-salt has been in the water from the beginning; but it was previously shrouded in the other earths, and could act substantially for the first time, when they had been separated from it. It has become salt, or water and earth-element by the agency of light.

838. The sea-salt has also been generated in opposition to the calcareous earths, and during its separation been rendered polar towards the latter. The salt mines are therefore associated with the last calcareous formation, the gypsum, and this it is also that determines their lamination.

839. As it may be said, that the metals separate into coal and sulphur, namely, pass over at their iron-extremity into coal, at their arsenical, into sulphur; so may it be said, that the earths separate into acids and alkalies; the one by the conversion of silica into fluoric acid, the other by that of calcareous earth into soda. Carbonic and sulphuric acid take possession of the calcareous earth; the hydro-oxide of the alkali.

840. The sea-salt has been the last to be separated from, because it was last generated in, the water. The salt beds belong to the last precipitation, by not having been mechanically thrown down, but as already observed and as their occurrence proves, by an alternating process of separation from the acidified lime. It is absurd to wish to explain the presence of common salt in the sea by a solution of saline beds. For where have the latter come from?

841. With the separation of the ore and the Inflammable out of the primary water and the confluence of the Earthy into marine salt, its metamorphosis upon the surface or when exposed to light is at an end. All forms of the planet have been successively developed out of the earth-

element. It can attain no further to anything new, and if nature had not yet been concluded, that which in the sequel is still dynamically developed upon our earth, must be thus a product, which extends beyond the mineral kingdom.

842. As the fundamental earths lose themselves in pure carbon, in resin, and so in the vegetable world, so the series of corrosive earths resolves itself likewise in remnants of an organic, and that indeed the animal, kingdom. As the pit-coals and resins are associated with the metals and Inflammables, so are the fossil animal remains with the calcareous earth; and thus the voice of the organic world speaks already with force and clearness to those that hearken, from out the stones.

843. For the metamorphosis of the earths, nature has twice prepared herself, has planted at the same time two great points of origin, according to which all her action is directed, and which remain in constant correspondence with each other.

844. The main pole is granite. It is at once the primary pole, to which the second main pole, the lime, is directed. The granite brings the series through gneiss and mica-schist down to clay-and talc-slate, then makes a sudden transition to the ores, and terminates at a boundary, where pit-coals and resins conduct us into a new kingdom.

845. The lime rejects from itself the sand and sedimentary clay, progresses through barytes and strontian up to gypsum, makes a sudden bound or transition from thence to the salts, and terminates at a limit, where corals and molluscous animals conduct us into a new kingdom.

846. Salt concludes the growth of the earths; it is the eruption or breaking out of the soul, as the metal was the body of earths completed. Both finally pass into a higher world, the metal into the corporeal, the salt into the psychical.

CHEMISM.

847. The spiritual activity, the soul of the earth has declared itself in crystallization, the spirit of the metals in magnetism, that of the Inflammables in electrism. The calcareous epoch is also the manifestation of a peculiar activity, whose ultimate product is salt. It has been already shown what the functions of the latter are, but the signification of its acts has not yet been mentioned.

848. By the influence of light the water becomes elevated in its oxygen-action, enters thus into tension with itself, and this constitutes the tendency to solution which is the function of water. The oxygenous water solicits the basic principle in the earth; this issues forth, but still combined with its oxygen that has become free; the formerly identical earth is a calcareous earth in a state of tension with itself. What does not admit of being brought into a state of tension, is thrown down as fundamental earth. Hitherto this process was a mere process of solution, i. e. it had attained in the solid and fluid only to tension, not to disseverment of the poles.

849. But the light always renders the water more oxygenous, and therefore the earth always more basic; finally, the one portion of the basic in the water, namely the hydrogen, separates itself, and becomes peroxydized or the hydrochloric acid. A portion of the Basic in the Earthy, namely the metallic body or carbon, separates itself also, and the remainder becomes peroxydized or soda. As the calcareous earth was at first dissolved in, and therefore one mass with, water, so at present is the alkali also combined with the acid water; while both dissolved in each other constitute salt.

850. In this process therefore whole elements have been taken only as one primary body, and they have combined with each other like the two primary bodies. The water has no longer become an element or Equiponderant, but an Oxygenous, a true light-body; the earth has no longer continued a total earth of equal specific gravity, but by relative peroxydation has become an Aqueous, a Soluble. The salt has thus from the union of the two lower elements, seeing that

each was of no more value than one primary body, become a new element.

851. This process converts the elements again into their primary condition, creates new elements and thus actually new matters. It is therefore a struggle of the elements with their primary bodies, a separation and interchange of the same. Such a process is called Chemism. This is the essence or interpretation of chemism, viz. the creation of new elements out of the old, by the reduction of these to the nature of the primary bodies.

852. Chemism, which separates or combines, ranks a step higher than the process of solution, which has the power only of heterogynizing, but not of separating. Thus the electrical spark separates and combines in combustion, while the tension of air enforces only evaporation. Chemistry drives the elements to their utmost. In water the oxygen is the predominating; it becomes, however, perfect first in the acid of salt. In the earth the body of gravity prevails; but it becomes first of all predominant in soda.

853. The opposition of the two primary bodies has been represented in the two inferior elements as chemical tension, and the combination of these primary elements is a chemical product.

854. Chemism is moreover a process of combustion, in which, however, a whole element supplies the place of oxygen, and a whole element the place of the base. It is an elementary process of combustion.

855. Chemism is the metatype of primary creation, both from its being a material process of combustion, as from its creating new elements. It is the union of the antagonism between æther and terrestrial matter occurring within the circle of the terrestrial elements. Chemism is a true conversion of substances according to their fixation.

856. All chemism takes place only in water; not only because the particles can move therein, but because chemism is a process of combustion of the elements themselves. The inferior elements, however, such as water and earth, can only undergo combustion with each other, because the two are moreover related to each other as æther and mass; or as oxygen and base, for the two, so to speak, have become unipolar. Without Fluid and Solid we cannot think of chemism.

857. The chemism of air is in the beginning electricity and then the true process of combustion; both are similar, but different in position. In the process of combustion both elements are unipolar in the air, thus moieties; but in chemism two elements unite so as to constitute a Whole. The product of air-chemism is water, as the product of earth-chemism is salt. Water and salt fall into one position, but transcend each other, even as the process of combustion invades the province of chemism. The relation of electrism to chemism has now been expressed in the clearest manner. The one is chemical tension of air, but chemism is the electrical tension of earth and water.

858. Chemism is related to magnetism, as salt is to metal, as the sedimentary to the primary periods. The whole sedimentary period is a product of chemism, as the whole primary period is a product of magnetism; salt and metal are only the last evolution of these periods, and the products for whose sake all the preceding actions and formations have taken place; granite and lime with their ramifications, are but the stems, upon which metal and salt are borne as blossoms.

859. Magnetism and chemism are thus the creating agents for the solid nucleus of the earth, and through both is it completed. The process of earth-formation is a magneto-chemism.

860. Regarding the earth as an entire crystal, magnetism is the Determinant of its polar axes and polar radii, while chemism is the same in respect to its integral parts.

861. All terrestrial action is an interchange of these two functions or souls, which are none other than the living gravity and the living light upon the planet. The electricity, like the heat, only maintains them in eternal tension or extension.

862. Chemism is the process of space, density, quiescent heat; therefore the latent heat or the temperature must change in every chemical process. Chemism is related to magnetism, as heat is to gravity, to electricity like as to light. Crystallization is point, magnetism line, electrism surface, chemism cube, or expressed according to their powers: O0, O1, O2, O3.

863. Nothing can become solid without taking water into admixture with it. This water is the water of admixture. Nothing also can assume form, without taking water into itself—water of crystallization.

864. In so far as magnetism is active in crystallization, it renders the water identical, basic, and this therefore becomes solid; the water is not as water in a crystal, but it first becomes so by separation.

865. All chemical processes are based on the union of bodies, which are elements, but which, like acids and alkalies, have assumed the nature of the primary bodies.

866. The elective affinities are based upon the polarizibility or transmutability of the Fluid and the Solid into the primary bodies. That is decomposed and combined, which during admixture maintains the animation in the strongest degree unto the origin. What cannot be so reduced, is precipitated, as is silex. The chemism is a bin-elementary process, and therefore constitutes the termination of this period of creation, or of the mineral kingdom. So soon as a tri-elementary process originates, the products pass over into a new kingdom.

PART III. BIOLOGY—OF THE WHOLE IN SINGULARS.

A.—ORGANOSOPHY.
I.—ORGANOGENY.
A. Galvanism.

867. If we take a retrospective glance at the development of the planet, we find that it commenced with the simplest actions, and then assumed a more elevated character by gradually drawing together several actions and letting them work in common. In magnetism the earth-element alone was active, and this having freed itself from the other elements by crystallization, asserted itself as a particular form upon the planet. By this single act of the planet, an extensive series of positions or numbers originated, which may be called mineral individuals.

868. Up to the formation of the solar system or of the planet, the character of creation is analytic. The three primary ideas emerged from each other as gravity, light and heat, and appeared as fire. These three united actions emerged again from each other and became air, water and earth, which together make up the planet. This was therefore the descending creation. But from and after this period, the character of the development of the planet becomes synthetic, for the divided elements again united with each other. By synthesis only or by combination of the elements does the planet progressively advance, and by it only does it divide into lesser planetary masses or bodies, called individuals. This is the ascending creation.

869. To the earth-element, in which the active magnetism is isolated, comes the element-water; and by the identification of both into one body, a new process is evolved, which we have recognized by the term chemism—salt. Then the earth-element combines with air and becomes an Inflammable, in which the process is likewise extinguished and only a dead product is left—

the Inflammable. Then it combines with fire and is converted into ore. There never originates therefore from the twofold combination of the earth-element with any other element, but one product, that namely in which the Earthy obtains the preponderance, or a mineral.

870. Two elements only belong to the essence of chemism, and they indeed are the two lowest, the elements carbon and oxygen, both being reduced to their primary condition, i. e. to that of an alkali in natron, and an acid in the hydrochloric or that of common salt.

871. As the principle or rationale of chemical action consists only in the potentiality of two elements to revert to their polar, or the oxygenous, condition, this action must thus become extinct, so soon as the creation of the new or secondary element has resulted. For if the tension equalizes itself in the two, and the two only be coexistent, so in accordance with the compensation no new tension can originate, and yet such is the groundwork of all chemical action. The result of the chemical process is consequently death; and furthermore, because it is also a simple bin-elementary process, it cannot be the ultimate goal or limit attained by the development of the planet.

872. The next stage to which the genesis of the planet ascends consists in the bin-elementary processes being associated with the third terrestrial element. In this manner a process originates in which the powers of earth and of water marry or conjoin with the power of air, and thus originates a chemical power or chemism, influenced by the air.

873. The chemism, when influenced by air, is one of a perpetual character; for this power dies only because the tension of its two elements is balanced or equalized; the influence of the air is, however, none other than the constant renewal of the tension.

874. Now, the process of tension in the air is electrism or that action in which the two poles being devoid of indifference range opposite to each other, can therefore never unite, and the end attained by which is oxydation. The new process is consequently a chemical power constantly excited by electrism—it is an electro-chemism. (Ed. 1st, 1810.) This composite process is known under the name of Galvanism.

875. Hereby the galvanism has been most rigorously and characteristically separated from the chemism, and the succession of stages been exactly indicated. By the accession of a single but higher nature-factor, namely, the air, chemism advances one and only one stage higher. We have consequently made no leap or abrupt transition in tracing out our genesis of nature. Magnetism is the uni-elementary, chemism the bin-elementary, galvanism the tri-elementary process of the planet, in so far as it is occupied with its own evolution, or that of the Solid.

876. Considered in relation to the result and also the internal nature of the process, galvanism is in no wise different from chemism, but only in reference to the continuance of the tension. The fluid and solid are in both the co-equal media or means; the decompositions, separations and combinations also, are alike in both. The air has no other office than to sustain the opposition, which in chemism proceeds through the difference of the two unipolar elements, acid and alkali.

877. The air maintains this animosity of the elements only by oxydation, and so far takes part in the contest like a fellow-combatant; yet this invariably happens only while the water is preserved by these means in its primary condition, that of the acid. The air breathes life only into the chemical body, without being body itself. No galvanism therefore continues, if it be denied the access of air. The chain or column enters it is true into tension also without air, but remains only for a short time in that state, or so long only as there is a trace of oxygen in the water.

878. Galvanism, as a tri-elementary process, represents the planet in its totality. The galvanic column is an entire planet, a planet upon the planet, the planet individualized.

879. The individual, taken in a strong sense, is an entire planet taken up into Singulars, a

triplicity of the elements in the particular or special unity. In galvanism there consequently issues forth for the first time an individual, which is equivalent to a cosmic totality. Galvanism is the metatype of the planet. All other and profounder processes are not total in character, nor metatypes of a whole system, but only moieties thereof.

880. The planet, regarded in itself, in its three elements, apart from its relation to the sun, is a galvanic body, a column, just as inversely this is a planet.

881. The attributes, which consequently belong to the planet, abstractedly from the sun, must belong to every galvanic process, or to such a body. The planet is a Whole included in itself, and thus is galvanism. The latter acts only in a closed chain, or only by its own body or its materiality forming one circle that returns into itself. The three elements are mutually self-excited and moved, and that indeed from internal causes, though not apart from external conditions. Thus galvanism is like an individual planet.

B. Primary Organism.

882. An individual (total, self-included) body, excited and moved by itself, is called Organism. Organism is what individual planet is. The metatype of the planet is organism; or a planet upon the planet is organism. The planet is not itself an organism, because it is not individual or galvanic in every point.

883. The self-excitation of the individualized elements, is called life.

884. Galvanism is the principle of life. There is no other vital force than the galvanic polarity. The heterogeneity of the three terrestrial elements in a circumscribed individual body is the vital force. The galvanic process is one with the vital process.

885. Organism is galvanism residing in a thoroughly homogeneous mass. The galvanic column is no organism, because it only admits the galvanic process just as the planet does, in individual places. A body only, which is zinc-pole, silver-pole and moist pulp at every conceivable point, is an organism. A galvanic pile, pounded into atoms must become alive. In this manner nature brings forth organic bodies.

886. Electrism has a basis; it is the air. Magnetism has a basis; it is the metal. Chemism has a basis; it is the salt. So has galvanism a basis; it is the organic mass.

887. Accordingly, what would be organic, must be galvanic; what would be alive, must be galvanic. Life is not different from organism, nor also from galvanism. For life is verily the vital process. But the vital process is an organic, galvanic process. Galvanism lies at the basis of all the processes of the organic world. They are either modifications of it, or only its combinations with other and still higher actions. A living thing, which is not galvanic, is a nonentity.

888. With galvanism consequently the first step has been made out of the inorganic into the organic kingdom. Every aught of nature, which has hitherto originated, is inorganic. These, however, were mere individualities. The character of the Inorganic consists consequently in something being a Singular, a moiety, or a metatype of a Singular; the character of the Organic in its being the metatype of a whole or round number. Organic things are internal self-exciting numbers; the inorganic things are fractions.

889. Every fraction is dead. No moiety can attain to life, for it does not receive its complement. What is simply fluid, cannot be organic, because it is not the totality of the planet. What is simply solid, cannot be organic. It is only a third of the organism. Every organism is produced according to the laws of galvanism, according to the law of the triplicity.

890. As the terrestrial magnetism is indeed only one, but includes an infinity of magnets, which are rendered manifest in the progress of the earth's life; so also in the great galvanism of the earth

an infinite number of subordinate galvanic triplicities reside inclosed, which become gradually detached, and, instead of the universal galvanism, represent an infinity of individual galvanisms. The universal galvanism cannot exist, without establishing itself as an infinity of individual galvanisms. As magnetism is only associated with the net of metallic veins, so is the absolute only, with the universality of its finite positions. The number of organisms is infinite, both in coexistence as also in consecutive existence.

891. An organism is an individual in the rigid sense of the word, because it is ruined, so soon as one of its three members parts from the rest. In this sense only are there properly speaking organic individuals.

892. If we do not confine indivisibility to what is mechanical, but extend it also to the chemical; individuals may be likewise granted to the mineral kingdom. The minerals are chemical individuals; for by separation they are likewise annihilated as such, and moreover the relation of mixture of chemical bodies is not one of an arbitrary kind. The gray ores are a definite mixture of sulphur and antimony, and are thereby individuals. The silver, lead, and copper, that are accidentally mixed in this compound, do not at all alter the individuality, and by no means prove a capacity residing in the matters for mixture in all conceivable numbers. Such a mixture would be a medley only. In plants and animals casual component parts occur also frequently. Thus the individuality of the ruby-silver appears to consist in the definite mixture of sulphur and arsenic, with the addition perhaps of antimony. The silver is only mingled with it, and therefore present in all numbers. The same holds good of the bi-sulphurets of iron, and the glance-ores. (Ed. 1st, 1810.)

Creation of the Organic.

893. It has been demonstrated from the genesis of the Organic, that its essence consists in the universality of the planetary processes. Every organic individual has essentially three processes in itself, which must be regarded as its fundamental processes, whereof no one can ever be wanting. Seeing, that if this be the case, the body is only a chemical or magnetic, a crystallized carcase.

894. The first three planetary processes, namely, the earth-process, water-process, and air-process, or the forming, chemicalizing, and electrifying or oxydizing processes, are also the first three vital processes.

895. It has been shown, that with every new process and with every new combination of processes, the materials also of the same were altered, ennobled, rendered more composite, and thereupon also more decomposible. Herein also nature advances consecutively and creates new materials for the organic world.

896. In the metamorphosis of the earths, when the chemism was added to the process of formation, not only the alkalinity and acidity issued forth in the calcareous earth and the salts, but the pure Earthy also became free from fixity, and manifested itself as carbon in the carbonic acid.

897. The last product of an antecedent stage is always the basis of that which is subsequent. The fundamental matter of the organic world is consequently the carbon.

a. ELEMENTARY BODY—PRIMARY MUCUS.

898. If in this carbon the three processes of the planet, namely, the formative or its special, the chemicalizing or fluidizing, and the electrifying or oxydizing, process, concentrate themselves, and are present with all their energy in every atom of the organic body; so must the mass of carbon be at the same time solid, fluid, and aerial, oxydizable in every spot, and thus also soft. Now a carbon mixed identically with water and air is Mucus.

899. Mucus is oxydized, hydrated carbon; or expressed in purely philosophical language, mucus is the universality of the minerals and elements, or the synthesis of earth, salt, Inflammable, and ore in water and air.
900. Every Organic has issued out of mucus, is naught but mucus under different forms. Every Organic is again soluble into mucus; by which naught else is meant, than that the formed mucus becomes one devoid of form.
901. The primary mucus, out of which every thing organic has been created, is the sea-mucus.
902. Mucus belongs originally and essentially to the sea, and has not been mixed with the latter through the dissolution in it of putrefying substances.
903. The sea-mucus has originated in the progress of planetary development, like the calcareous earth has with the carbon and like the sea-salt. As little as this could have entered the sea originally through solution of rock-salt; so little could the mucus through the perishing of animals and plants, for none of these were yet present, but could be first developed, only with the production of this mucus.
904. The sea-mucus was originally generated through the influence of light and by the denudation of the crude masses, especially of the earths and salts, which was thereby effected; while with the metals and Inflammables ranging opposite to these, the carbon thus became free, and betook itself as carbonic acid to the water and air. Thus also has salt been produced.
905. The sea-mucus, as well as the salt, is still produced by the light. Everything takes place through the differentialization, or by the absolution of fixed poles on the earth-element. Light shines upon the water, and it is salted. Light shines upon the salted sea, and it lives.

906. All life is from the sea, none from the continent.
907. All mucus is endowed with life.
908. The whole sea is alive. It is a fluctuating, ever self-elevating and ever self-depressing organism.
909. Where the sea-organism by self-elevation succeeds in attaining unto form, there issues forth from it a higher organism. Love arose out of the sea-foam.
910. The primary mucus was and is still generated in those very parts of the sea where the water is in contact with earth and air, and thus upon the shores.
911. The first creation of the Organic took place, where the first mountain summits projected out of the water; and thus indeed without doubt in India, if the Himalaya be the highest mountain.
912. The first organic forms, whether plants or animals, emerged from the shallow parts of the sea.
913. Man also is a child of the warm and shallow parts of the sea in the neighbourhood of the land.
914. It is possible, that Man has only originated on one spot, and that indeed the highest mountain in India. It is even possible, that only one favorable moment was granted, in which Men could arise. A definite mixture of water, definite blood-heat, and definite influence of light must concur to his production; and this has probably been the case only in a certain spot and at a certain time.
915. The first men were the littoral and mountainous inhabitants of warmer countries, and found therefore at once reptiles, fishes, fruit, and game for food.
CHANGE.
916. The number of individual organisms is not persistent. For they are verily only products of a ceaseless polarization or a constant evocation of poles in the great galvanism, positions of the

general galvanism in time. Thus, as the poles change, so also do the organic individuals. The kingdom of organisms is an iron bar, in which the magnetic poles originate and vanish or change, according as the polarizing magnet is removed. Organisms change, because they are numbers, thoughts of God.

917. The process of change in organic individuals is that of their destruction.

918. But this destruction is as nothing for nature. There originate again in the same moment other organisms in other situations. The process that destroys the poles is only one that effects their change.

919. The world-organism only is eternal, and devoid of change, with the exception of that which is within its poles. It can itself change with no other, because it is only one.

920. No individual organism is eternal, because it is only a changing pole of the world-organism.

921. There is no constancy in the individualities. Change only is persistent.

922. The world only is persistent. Nothing in it is constant. Were individuals not to perish, but live for ever, the world must then die; for the life of the world, like every life, consists only in the change of poles. Individuals could in no way therefore continue alive, if the world were to remain alive, because this is only possible through change of the individuals, which are its organs; nor could they, were the world to die, because the totality of individuals is the world itself.

923. Death is no annihilation, but only a change. One individual emerges out of another. Death is only a transition to another life, not unto death.

924. This transition from one life to another takes place through the primary condition of the Organic, or the mucus.

925. If new individuals originate, they could not therefore originate directly from others; but they must be redissolved into mucus. Every generation is a new creation.

b. FORM—GLOBE.

926. The organism is a metatype of the planet and must also have the corresponding form. It is the Sphere. This results also from the combination of the three actions, which being in equiponderance could only produce the globe.

927. The sphere must commence with the idea of the point. For the idea of the sphere is the idea of the centre, which is a point. The point, however, is not different from the sphere. It is only the infinitely small or minute sphere.

928. The primary mucus is globular in form. The primary mucus does not swell into a single sphere, but it divides into infinitely numerous spheres. For were it only one sphere, it would be the planet itself. But it is an individual, or only one sphere in the great sphere. The idea of the great sphere consists however of an infinity of small spheres.

929. The primary mucus consists of an infinity of points. This admits of being proved by its mode of origin. It is formed on the limit between water and earth, consequently in a line. This line, however, becomes constantly dissevered by disquietude, and divides therefore necessarily into infinitely numerous points.

930. The primary Organic is a mucous point.

931. The organic world commences not merely with one point, but with infinitely numerous points. Where earth, water, and air are found in one spot, there also is an organic point.

932. The organic points originate upon the surface of the earth, not in it, and not in the air. For only between earth and air do all three elements enter into collision.

PRIMARY VESICLE.

933. Through the oxydation of the air an opposition of the component parts, or of the Fluid and

Solid, issues forth in the organic point, and these mutually conditionate each other. The Fluid and Solid cannot, however, be otherwise conditioned, seeing that the former is the Contained, the latter the Containing. The Solid is only a precipitate from the Fluid wrought by the influence of the air. The air, however, is externally related to the mucus-point. The Solid can therefore originate nowhere else than between the Fluid and the air. It consequently surrounds in accordance with its genesis the Fluid. The physical cause thereof is naturally the oxydation of mucus upon its periphery. A globe, the middle of which is fluid, but the periphery solid, is called a bladder or cyst.

934. The first organic points are vesicles. The organic world has for its basis an infinity of vesicles. (Ed. 1st, 1813. §. 922.)

INFUSORIA.

935. The mucous primary vesicle may in a philosophical sense be aptly called infusorium, like as we designate the primary condition of the embryo, by the word vitellus. Now are we making use of definite expressions.

936. Everywhere, where the three elements cooperate, are infusoria present—thus upon the sea-shore, the tide-mark or strand, and shallow watery places.

937. The infusorium is a galvanic point, a galvanic vesicle, a galvanic column or chain.

938. In every infusorium there is triplicity of the poles, or properly speaking, of the processes. Each one maintains itself by the nutritive, digestive, and respiratory process, or what amounts to the same, the infusorial globule of mucus assumes a figure, its peculiar fluidity is formed in its interior, and it becomes oxydized. As is well known, no infusorium can live without moisture, and none if the access of air having been prevented, or the water boiled, it is freed from the air and the Earthy.

939. If the organic fundamental substance consist of infusoria, so must the whole organic world originate from infusoria. Plants and animals can only be metamorphoses of infusoria.

940. This being granted, so also must all organizations consist of infusoria, and during their destruction dissolve into the same. Every plant, every animal is converted by maceration into a mucous mass; this putrefies, and the moisture is stocked with infusoria.

941. Putrefaction is nothing else than a division of organisms into infusoria, a reduction of the higher to the primary, life.

942. Organisms are a synthesis of infusoria. Their generation is none other than an accumulation of infinitely numerous mucous points, infusoria. In these the organisms have not forsooth been at once wholly and perfectly depicted as on the smallest scale, nor contained in a state of preformation; but they are only infusorial vesicles, that by different combinations assume different forms, and grow up into higher organisms.

THEORY OF GENERATION.

943. The theory of generation is in this sense a synthetical and epigenetic, not an analytic.

944. The theory of preformation contradicts the laws of nature's development.

945. Generation is a successive formation, both in relation to the quantity as well as the quality, and the specific organs. It having been preposited, that an organism has several organic systems, so must these range according to their importance, and like the systems of nature, behind each other, and be also developed in this order. As the whole of nature has been a successive fixation of æther, so is the organic world a successive fixation of infusorial mucus-vesicles. The mucus is the æther, the chaos for the organic world. The semen of all animals consists also of infusoria; the same may be said of the vitellus. The pollen of flowers consists in like manner of microscopic vesicles with globules, which have a life of their own and move themselves in

water. Many confervæ indeed divide evidently into a multitude of living, self-moving globules, which, after they have swam about for some time, again unite to form a stem of conferva.
946. Every generation consequently commences à priori or from the beginning. The organic substance must again be dissolved into the original chaos, if any thing new should reoriginate.
947. Out of an organic menstruum only can a new organism proceed, but not one organism out of the other. A finished or perfect organism cannot gradually transform itself into another.
948. The generative juices, or semen and vitellus, are none other than the total organism reduced to the primary menstruum.
949. Physically regarded also every individual originates only from the Absolute, but no one out of the other. The history of generation is a retrogression into the Absolute of the Organic, or the organic chaos—mucus, and a new evocation from the same.
950. This development from mucus is only applicable however to the generation of the perfect organisms, but not to the origin of the organic body, or the infusorial mass. The former originate only from an organic mass that has been already formed; but the infusorial mass, as constituting the organic primary bodies, cannot have originated in the same way. It does and must originate directly from the Inorganic. For whence can the organic matter have otherwise proceeded?
951. The infusorial mucus-mass originated, as has been already remarked, at the moment when the earth's metamorphosis was at an end; at the moment, when the planet succeeded in so bringing together and identifying all the elementary processes, that they were all together or at one and the same time in every point.
952. Hence the organic primary body originated also by synthesis, not by analysis, if regard be paid to its factors. But do we consider this substance as first emerging into view, when the coarse, abundant, isolated materials, such as earth, metals, Inflammables, and salts had separated themselves from it; that this organic primary body then remained behind as it were for the first time: it has then originated through analysis, or was preformed; but so preformed as are also the metals, and as is everything. It need scarcely be observed, that this last separation from out the Earthy is the carbon, the dissolution of the earths into atoms, and thus again into points or globules susceptible of form.
953. Everything is preformed in æther, like as every Mathematic is preformed in zero, every Active in God; yet for that very reason nothing individual is preformed therein; but it originates first through fixation of poles on the substance. This is the true meaning of the original generation of the Organic.
954. This origin of the organic primary bodies I designate Generatio originaria, Creation.
955. But infusorial vesicles can also originate by mere division of larger organic carcases, and these can again originate as well through the combination of these secondary as of the primitive vesicles, or as it were by coagulation only, such being the case indeed in the intestinal worms also. I nominate this generation, Generatio æquivoca.
956. All generation is Generatio æquivoca; whether imparted by sexes or not. For the generative juices of the sexual organs themselves are naught else than organic primary mass, and have originated by division.
957. There are only two kinds of generation in the world. The creation proper and the propagation that is sequent thereupon, or the Generatio originaria and secundaria.
958. No organism has been consequently created of larger size than an infusorial point. No organism is, nor has one ever been, created, which is not microscopic.
959. Whatever is larger, has not been created, but developed.
960. Man has not been created, but developed. So the Bible itself teaches us. God did not make

man out of nothing; but took an elemental body then existing, an earth-clod or carbon; moulded it into form, thus making use of water; and breathed into it life, namely air, whereby galvanism, or the vital process arose.

961. The original origin of organization has been imparted by the co-operating influence of heat and light. By the heat, because without this no galvanic and no chemical process is possible; further, because heat is the totality of æther, the moved æther, the ætherial air, and thus the menstruum of all action. But the heat is not sufficient to animate the three terrestrial elements, because it imparts only the possibility of procedure and of action; while it does not differentialize, nor posit tension, but maintains everything in identity, fluidity. In the heat alone every thing must become fluid and finally decompose. Unto heat therefore the accession is yet necessary of the Cosmic-differencing, or the light. The light inspires the body prepared by heat with life, antagonism, polarity.

962. The æther imparts the substance, the heat the form, the light the life. (Oken first started this opinion in his work, 'Die Zeugung,' Frankfurt, Wesche, 1805.)

c. Processes of the Organic.

963. The life of the organic body is not a single but a threefold action, consisting of the actions of the three terrestrial elements, which become the three fundamental processes of the body, or of life, and in which three processes galvanism consists.

1. EARTH-PROCESS, NUTRITIVE PROCESS.

964. The magnetic earth-process is virtually the formative; and in organic bodies is called the nutrient process.

965. The process of nutrition is the principal process in the organic world. Its product or its basis is the fundamental mass of the body itself. As crystal and the process of crystallization are related to each other, so are the body and the process of nutrition.

966. The process of nutrition is the sustaining, and proper fundamental process of the organism.

967. It is present entire and indivisible in every part of the body. Whencesoever it is ablated or withdrawn, there is death.

968. It operates according to the laws of crystallization.

969. Its very forms are crystals modified by the organic mass or, what amounts to the same, by the other processes combined therewith. The organic body is an accumulation of an infinity of (organic) crystals (cells).

970. In the planetary process of formation, which is a process of crystallization, the organism is continually comprehended. It is the planetary body ever becoming; the latter is an organic body, which has ceased to become.

2. WATER-PROCESS, DIGESTIVE PROCESS.

971. Another action, that helps to constitute the organic body, is the chemism, which is not only the process of liquefaction, but the process also of the formation or creation of new organic matter. It is known to us under the name of Digestive process.

972. The digestive process elevates the Inorganic up to the organic mass, like the chemism has converted the Earthy into carbonate of lime and finally into muriatic natron. The digestive process is the process of the formation of mucus. Regarded philosophically the nutritive juice (or chyle) is naught else than mucus. This is also correct in a physiological point of view.

973. The digestive process is the second organic process, in so far as it has been fashioned after the type of the water; but the first, in so far as every Organic has originated out of water.

974. As forming only the mucus it is not directly distributed in every part of the body like the

nutritive process, which is the body or planet itself. But it interposes, or is mediate, everywhere.
975. As the water of the planet is related to the continent or earth's nucleus, so are the digestive matters or the mucus of nutrition to the body. The Earthy, however, is the principal mass of the planet, upon which the others have been supported. So is the nutritive body the principal mass, upon which the digestive body has been supported.
976. No organism is conceivable without a digestive process.

3. AIR PROCESS, RESPIRATORY PROCESS.

977. The action of the air finally settles down also in mucus. It is that which sustains the constant heterogeneity of the organic factors, the electrical tension. The electrical tension has, however, oxydation for its result. The organic process of electricity is thus at the same time a process of oxydation. It is called Respiratory process.
978. Without respiratory process, no organism is conceivable. By its influence difference has been induced in the chyle, and by this difference only does the latter become decomposible or serviceable for the process of nutrition.
979. The respiratory process is also present not immediately in every part of the body, but only mediately. It is the atmosphere of the body.
980. The mutation, which the juices undergo through the process of respiration, is none other than an emergence from their state of indifference. Thereby each point of the juice becomes polar towards every other; all are mutually attracted, all repelled, whereby a decisive vortication originates.
981. As every globule of sap or mucus is indifferent, it has thus naturally an affinity for air. The air itself is comprehended, like the water and the earth, in the organism. And thus it may with full force be said; that the organism is elevated by respiration to the element air, by digestion to the element water, by nutrition to the element earth. So that respiratory process = air-process, digestive process = water-process, and nutritive process = earth-process.
982. The first three organic processes are consequently true synotypes of the planetary processes—are planet-forming processes in miniature in individuals. The fundamental organism has thus been shown to be in its apparent processes a synotype or likeness of the planet; in other words, a microscopic planet.
983. These three processes constitute the galvanic process. Making use of the expressions applied to the inorganic kingdom, we found the organism to be a combination of magnetism, chemism, and electrism; while in organic parlance it is a nutritive process, maintained by respiration and digestion. The processus nutritorius, digestivus, respiratorius, together constituting galvanism.

4. MOTION.

984. Motion is no peculiar or self-persistent process, but the necessary manifestation of galvanism. Motion has been established with the three organic fundamental processes.
985. Every motion depends upon the galvanic process. Taken in a strict sense, there is no process of motion, but motion only. For motion is verily but the phenomenon of galvanism. The process of motion is synonymous with the galvanic process.
986. The galvanic process is a process of motion effected in circles, in its own factors, in its planet, but not from without; it is consequently an actual vital process.
987. As the process of motion is the phenomenon common to all three organic fundamental processes, so is the whole organism characterized by it. The essence of the Organic depends consequently upon its automatic or self-motion.
988. Self-motion is the only, but essential and ultimate, distinction between the Organic and

Inorganic. (Ed. 1st, 1810. § 904.) All other distinctions that have been advanced do not suffice; because they do not comprehend the totality of the organism, nor the three fundamental processes in one phenomenon, but only individual attributes.

989. A circumscribed, closed mass, which moves itself, is an organism. The perpetuum mobile is only the organism.

990. Every Inorganic moves not itself, but is only moved by external influences; because every Inorganic is only a part of a whole.

991. The organic motion is present and possible in every point of a body. A mass that is automatic, or thoroughly moved by itself, is an organism.

992. The Inorganic consists in motion having vanished from it, and in being simply mass. But the Organic consists exactly in this alone, namely, that the Massive has disappeared, or that the mass is in constant motion. The Organic becomes destroyed, so soon as motion disappears in it; the Inorganic is destroyed, so soon as motion enters it. Motion is therefore the soul, whereby the Organic is elevated above the Inorganic.

2. ORGANOGNOSY.

Division of the Organism.

PLANETARY AND COSMIC ORGANISM.

993. Hitherto we have regarded the organism merely in a general point of view; namely, as regards the substance, form, and processes, which must indiscriminately occur in every organism. We have seen that it is composed of at least three elements, the earth, water and air. There is still, however, one combination that is possible and therefore also actual, namely, with the æther or the fire.

994. The organic world has two stages in its development. Upon each stage, however, it is the totality or synotype of nature, yet is different in each.

995. The organism represents the whole solar system; but this divides into two stages. The lowest of these is the Planetary, or totality of the Earthy, Aqueous and Aerial; the higher is the Solar or Cosmic, namely, the totality of earth, water, air and fire. Thus there is a tri-elementary and a quadri-elementary totality. As the first is already an organism, so much the more too must be the second. Thus there must be one organism, which comprehends indeed all systems in itself, but with the preponderance of the Planetary; and one with the preponderance of the Solar.

996. In the planetary organism the æther-systems will be either wanting, or only indicated as mere projections; they are there only, in so far as the planet itself is not without light. In the solar organism, however, the planetary systems are subordinate to the æther processes; the former are only there, because the sun cannot be without planets. In the planetary organism the æther-system has only been taken up into the Terrestrial; but in the Solar the Terrestrial has been taken up into the æther.

997. In the planetary organism all the processes launch out into production or alterations of the matters; it is a chemical organism; in the solar organism there are processes, which neither change, nor produce matters, it may be therefore styled light organism.

998. The chemical organism is associated with the earth, the spiritual with the water and air; the former must therefore consist principally of carbon, the latter of the combination of oxygen with hydrogen, and thus of nitrogen.

999. The carbon-organism must moreover in accordance with its import be associated with the Inflammables and metals, and through these with the silicious earth.

1000. The nitrogen-organism on the contrary with the salts and calcareous earths; thus we have silicious organisms and calcareous organisms; Inflammable organisms and salt organisms. From this it is already clear, that the planetary or primary organism is not general, nor indefinite, but the plant; for no General or Indefinite has existence. The solar or quadri-elementary organism admits in like manner of being recognized as animal. This is the philosophical deduction. But there is also a physiological, which conducts to the same result.

1001. The mucus-vesicle can feasibly pass into two kinds of condition only. It either remains in the water, or is cast upon the shore, or in the mud. In the last case it continues to lie, and is only supplied with light, and oxydized by the air upon the upper side; in the first case, on the contrary, it rolls about constantly in the water, and is alternately illumed and oxydized upon all sides. The first vesicle thus obtains a single axis from above downwards, between light and darkness; the second on the contrary gains a multitude of axes from without inwards, where it is alone dark and deoxydized. The first is thus devoid of any middle point, and finds its centre of gravity only in the middle of the earth, while the latter acquires its centre of gravity in its Interior, and this renders the Interior polar towards all points of the circumference. The plant is only one axis, or from having no middle point, is properly only one radius, which has its centre in the centre of the earth; the animal is an infinity of axes or radii, which concur or converge in the creature itself; the plant is an inverted cone, the animal an infinity of cones or a globe.

1002. Thus the planetary organism originates, if the primary vesicle having been taken out of the water is given up to the earth, to immobility and to darkness. But the light-organism arises, if the primary vesicle continue in water, or in the Moveable and Diaphanous. Here then in their genesis an essential difference is declared between the two organic worlds. Planetary organism originates, if the vesicle develops itself apart from the water, in which case it is withdrawn on one side from the light; but the light-organism originates if it remains in water itself, where it can be supplied on all sides with light. The essence of both is expressed by the names darkness-organism and light-organism.

1003. The basis of both kingdoms is therefore exactly similar; the vesicle and the mucus lie at the foundation of both. It depends solely upon the surrounding element, whether out of one and the same mass this or that organism should arise, or rather upon the active influence of the light, this being conditioned only by the elements. Not a word can accordingly be spoken about preformation. In darkness-organism the water-vesicle has been placed between earth and air, and thus fettered to the earth; in light-organism, however, the vesicle has been placed in the water and so freed from the earth.

1004. The planetary organism has, in accordance with its situation and import, been bound to the earth. It must originate like the metal in the earth, in the darkness and, as it were, in a vein. But it is at the same time a light-product; it must rise from out of the earth into the air and towards the light. It is a mucus, living metallic vein, which elevates itself from out the earth into the air.

1005. This organism, which originates in the darkness of the earth, and grows therefrom into the air so as to meet the light, is plant.

1006. The solar-organism is, in accordance with its import, void of connexion with the earth; like a planet it revolves freely about the earth, and everywhere receives its image or likeness in the influence of all four elements.

1007. The organism, which, free from the earth, has originated in water, or properly speaking, in the transparency, is animal.

1008. The vegetable and animal are the only organic, kingdoms. In both, nature has exhausted herself, and has in the last kingdom, as in a mirror, been wholly reflected. They are together planet and sun, or thus solar system. But since the animal comprehends all elements in itself, so it contains also the plant, and is therefore for itself vegetable and animal kingdom, or the whole solar-system.

1009. The plant hath no free system of motion, because motion is wanting to it; bound to the elements, by these it is determined. The element of motion, the æther, lies apart from it. It has only motion, if and while the elements act upon, or solicit it, thereunto.

1010. It moves itself only by an external or foreign stimulus. If no foreign stimulus be present, it does not move itself. A root grows, moves itself towards one spot, not because it there seeks for moisture, but because it is affected by the moisture which is there found. Were the moisture not to act upon it, it would wither.

1011. The animal has independent motion. For it has indeed taken up the centre, the earth-and the light-system which is the principle of motion, into itself.

1012. Thus the animal moves itself independently of external stimuli. The animal can move itself from want of stimulus. It moves itself to seek for, and thus from want of, nourishment, which consequently does not act upon it; the plant cannot, however, move itself owing to want of food, but only die.

1013. This is the essential and only conclusive distinction between animal and plant. All others that have been advanced are not sufficient.

Processes of the Cosmic Organism.

ÆTHER PROCESSES.

1014. The primary or planetary organism cannot be the last product of nature's development: for it is only indeed the metatype of the three terrestrial elements, and consequently not of the totality of nature. The mucus-organism ascends to a higher stage, since it superadds to its three elements the primary element, or ascends itself to primary elements. It becomes an ætherial globe of mucus.

1015. Hitherto there were merely three processes in the organism; to these consequently the fourth is added, which is the æther-process. It may be called the fire-process.

1016. With the fire-process the development of the organic world has been carried to the highest pitch and therewith closed.

1017. The highest organism is a quadri-elementary individual, or a quadri-elementary mucus.

1018. The four elements are, however, the universe. The higher organism is consequently not merely a synotype of the planet, but of it and the sun, or of the whole universe. The higher organism is an universe in miniature; in the profoundest, truest sense of the word is it small world, microcosm.

1019. The planetary, terrestrial organisms are related to the solar or cosmic, as the planet is to solar-system.

1020. The cosmic organism has besides the systems of nutrition, digestion, and respiration, those of the æther in itself, and thus of the gravity, light and heat. These are immaterial, spiritual processes, which produce no more matters.

1. Process of Gravity.

1021. The organ of gravity is that of quiescence or rest, the organ constituting the basis of the organic body, or rather of the other æther-organs, the centralization.

1022. It imparts form to the higher organism.

1023. The quiescent, sustaining, form-imparting system is the rigid earth-system, and appears as the osseous system.

2. Process of Heat.

1024. As the heat is the motion of the æther, so is there a system of motion in the organism.

1025. The system, which has no other function to perform than to move, is the muscular system.
1026. The osseous system is related to the muscular system as mass to motion, as Passive to Active. The former maintains the form, the latter changes the same exactly like gravitation and heat.
3. Process of Light.
1027. The light-system must be related to all other spiritual systems and the three fundamental systems of the organism, or to the simply organic systems as light is to matter, being thus polarizing or dominant.
1028. Now the domination of light consists in the sustenance of polarity in all matter. The air itself is preserved in its duplicity only by light. All the points of the organism are polarized by the light-system. The light-system acts consequently through the whole body.
1029. The light-system is not capable of producing matter, like the terrestrial systems. This is self-evident.
1030. It does not polarize by effecting chemical changes. If nevertheless these are present, they are thus only results, occurring while the terrestrial processes are set in action by the process of light.
1031. Light polarizes the Material by mere fixation or discharging of poles, and thus in a spiritual manner. So also does the light-system of the organism. It governs the organism not by mechanical power, not by mass, but by a spiritual breath or aura.
1032. The organic light-system is the Animative of the organism. In it the spirit exercises its power over the mucous mass. It is the nervous system.
LIFE OF THE NERVOUS SYSTEM.
1033. The light-polarity can bring about no other tension in the mass than that which is peculiar to it, and thus the galvanic tension, whose highest and purest phenomenon is motion. The light-system principally causes motion in the mass, like as in æther.
1034. The nervous system has, however, a life also in itself, or the internal light-polarity that is without any relation to the organic mass. This action of the nerves is called sensibility, and its phenomenon, feeling or sensation. The system of sensation is the nervous system as sun in itself; in the motor-system it is as sun in a centre of planets.
1035. The organism, like the elementary nature, is completed by four systems.
1. By the Nutrient,
2. "   Digestive,
3. "   Respiring, and
4. "   Motive, unto which the nerves, muscles, and bones belong.
1036. It is impossible for more than these to be developed in an organism; impossible, for any thing else but what is in nature to originate therein; impossible, that any thing new be born by it. Everything in nature is only repetition of an Antecedent or something that has gone before. How could the organism be aught else, how aught else than the focus of the four elements!

SECOND KINGDOM. VEGETABLE KINGDOM.

1037. The Vegetable Kingdom is the individual development of the three planetary elements.
I.—PHYTOGENY.

1038. Phytogeny represents the developmental history of individual plants, or, properly speaking, the idea of the plant.

1039. To the plant belong all the definitions that have been hitherto deduced. It is an organism fettered to the earth, is developed only apart from water, only in the dark, in the earth; is associated with the metal, with the carbon; is a magnetic needle that has been attracted out of the earth into the air towards the light. Seeds germinate better if they have been protected beforehand from the access of light; the radicle sinks, indeed, into the earth, because it obeys the gravity, the quiescence or rest; but it is therein maintained, because it is there humid and dark. This is a reason that has not yet been connoted, for the plant having been fettered to the earth. There are indeed plants which also take root in water, but the water is still darker than the air. The root has, in this respect, completely the character of the metal, that is a child of darkness.

1040. Consisting for the greatest part of carbon, plants are associated with the pit-coals, through these pass over into the carbon of the clay-slate rocks; finally, through the black-lead unto the iron. In like manner, through their hydrogenous import, they pass over into the inflammable asphalts, and through these unto sulphur. Metal and sulphur have, in the Geogeny, announced themselves as the precursors, or harbingers, of the vegetable world. In this respect, also, can the vegetable kingdom be regarded as the mineral kingdom, that, having continued to grow, has become alive. The ore, which becomes organic, becomes carbon or plant.

PARTS OF THE PLANT.

1041. The character of each development consists in the separation of the Indifferent or Chaotic into its ideas or actions, i. e. the development of every system is first completed, when it is divided into as many substantial systems as it numbers factors, or has processes in itself.

1042. Although the plant is only essentially a planetary-organism, it must yet be developed unto an æther- or light-organism; and it therefore divides into planetary- and solar-or light-organs.

1043. The planetary organs are those that have the earth-, water-, and air-process above them, and which are made known in the root, the stalk and foliage, which together constitute the vegetable stem.

1044. The light-organs begin to be stirring in the blossom, and are divulged as sexual organs. They are a repetition of the trunk.

1045. The vegetable body divides therefore into two great principal parts, which are synotypes of each other, into trunk and blossom or inflorescence. If we regard the vegetable trunk empirically; it is then divisible into three stages, whereof each consists of the organs of the three fundamental processes, which seek to separate from each other.

a. The first stage is that of the three tissues namely of the parenchyma, medulla or pith; of the cells, ducts, and tracheæ or spiral vessels.

b. The second stage is that of the shaft or main axis, where these three have separated concentrically into cortex or bark, liber, and wood, constituting the anatomical systems or sheaths.

c. The third stage is that of the caudex proper or the trunk, in which the three tissues have separated in the direction of the longitudinal axis into root, stalk or stem, and leaves, these making up the organs proper or members. The inflorescence divides into two stages, into flower and fruit.

d. The fourth stage or that of the flower repeats root, stalk and leaves, in seeds, pistil and in the corolla.

e. The fifth stage or that of the fruit is a further repetition of these three parts of the flower in the nut, plum and berry, unto which, as synthesis, comes the apple.

A. Vegetable-trunk.

1046. The vegetable-trunk is the development of the three fundamental processes up to their complete separation or substantial representation. It divides itself into the tissues or the pith (parenchyma), into the shaft and into the trunk.

1047. The plant is a galvanic water-vesicle, and as such earth, water and air. Upon this vesicle it is, however, the earth-element that chiefly acts. While the earth seeks to encroach upon the vesicle, the magnetic process becomes active therein, and it enters into opposition with the air. The vesicle becomes now determined by two elements, by the earth and by the air; it stands itself in the category of the water.

1048. The plant may be characterized as organic water which is polarized upon two sides, towards the earth and the air. The vegetable vesicle must therefore maintain two poles. While it would represent in itself the magnetic pole, it endeavours to identify itself, to obey gravity and merge into the darkness towards the mediate point of the earth; but that it may remain a galvanic pole, it becomes excited by the air, strives to become a Different and to attain the light.

1049. The vegetable vesicle receives two opposed extremities, an identical earth-and a dyadic air-extremity; and thus the plant must be regarded as the organism, which manifests a constant endeavour, upon the one side to become earth, on the other air, upon the one side identical metal, on the other duplex air.

1050. The plant is a radius, that towards the centre becomes identical, towards the periphery divides or starts asunder. The plant is not therefore an entire circle or globe, but only a section of such, a cone, whose apex has been turned towards the centre of the earth, or would become earth-centre. It can therefore have no middle-point. It will on the contrary demonstrate that the animal is the totality of radii, is consequently diameter, and has therefore a centre of its own, or is entire globe. As the whole earth is surrounded by plants, and all their roots turn towards the centre; the whole vegetable kingdom only forms a sphere, composed of infinitely numerous cones. On the contrary every individual animal forms a sphere for itself alone, and is therefore worth as much as all plants taken together. Animals are entire heavenly bodies, satellites or moons, which circulate independently about the earth; all plants, on the contrary, taken together are only equivalent to one heavenly body. An animal is an infinity of plants.

1051. In so far as an organism strives unto identity or to gravity, it seeks to produce the Metallic, the carbon, the Alkaline. The indifferent and alkaline character appears in the earth-extremity of the plant. Mucus and acid bodies are evidenced for the most part in the root. In so far as the organism strives unto duplicity, it will produce the salt, the acid and the Inflammable. Acids and electric bodies are manifested in the air-extremity of the plant.

1052. The two vegetable extremities are accordingly related to each other as alkali and acid, and as carbon and hydrogen. In the air the water is divided into oxygen and hydrogen, acids and oils; in the earth it hardens into earths and carbon.

1053. The earth-end or the alkaline extremity of the plant is the root; the air-end, or the acid and oily, is the entire stem-fabric, or body. The plant has first of all two cardinal organs, viz. root and stem-fabric. Both together represent the water divided into earth- and air-mucus. The root is the central extremity of the plant, and is therefore prolonged or runs out into magnetic points; the stem-fabric is the peripheric and therefore expands into branches and electric surfaces.

1054. But besides the air, the light also operates upon the plant and stimulates it to grow aloft and produce a light-organ. This light-organ can thus originate only upon the apex or summit. It is the flower. The flower can therefore stand nowhere else than on the summit or end of the plant. The light however acts upon many points of the upper surface of the vegetable trunk and elongates the same. One plant can therefore support numerous flowers, but all of these must stand upon an extremity. Wherever therefore a flower may happen to stand, that spot must be regarded as a summit or end. There is thus also, according to the physiological view of the matter, a light-organ in the plant, which is its animal pre-affection. The chief antagonism in the plant is in this respect therefore between trunk and inflorescence the former is related to the latter as plant to animal. Were the plant to attain unto animal functions; they could thus only take place in the flower.

I. Tissues.

1055. The tissues are the unseparated organs of the three fundamental processes, the earth-, water-, and air-process.

1. WATER-ORGAN, CELLULAR TISSUE.

1056. If a mucus-vesicle lie upon the ground, it thus continues indifferent upon the lower or dark side, and is only affected by the gravity and the water; the upper side, on the contrary, by the differencing air and light. It is consequently prolonged into the earth and into the air. It must pass over from the round into a linear form. The elongation is not a mere protraction of the vesicle, but an apposition of new vesicles. For it happens through polarization, and thus by infinite repetition of the primary vesicle. The plant is thus a body of infinitely numerous vesicles.

1057. In so far as the plant is a multiplication of the primary vesicle, it consists of Cellular tissue. The anatomy of plants informs us, that there is nothing originally in the plant but cellular tissue, and that other forms first emerge or make their appearance in the sequel.

1058. The cellular tissue indicates the Indifferent in the plant, for it is only an accumulation of the indifferent primary vesicles. In so far as the plant consists thereof, is it indifference—water-plant.

1059. The cellular tissue is only oxydized, desiccated mucus. Chemistry has proved, that the wood is only oxydized mucus.

1060. The cellular tissue being the water organized and saturated with earth, or the organized mucus, has consequently the chemical function in itself of solution, homogeneous production, or formation of mucus. As therefore the plant originates, so does it enlarge. It originates as vesicle, and its growth is a constant origination of vesicles; from the Indifferent, which is the water. The sap contained in the cells consists of water and starch-granules, which constantly circulate therein in a circle.

1061. The fundamental form of the cells is the rhomboidal-dodecahedron (Kieser's Phytotomie); for around a globe only 6 others of equal size can be placed, whereby its 6 lateral surfaces are pressed in, which during the induration impart to it the form of a six-sided column. Above and below these 7 globes only 3 others admit of being placed, whereby 3 point-converging surfaces originate, which thus complete the middle globe as rhomboidal-dodecahedron.

2. EARTH-ORGAN, VASCULAR TISSUE.

1062. So long as the vesicles or cells lie as globes upon each other, triangular interspaces are found between them, which stand in conjunction with each other on all sides. As water is found in these interspaces, it is plain that they do not entirely disappear with the transformation of the cells into rhomboidal-dodecahedra. These spaces are called intercellular passages or sap-tubes, Vessels. In many plants, such as those which contain a milky juice, particular sap-tubes run through a part of these intercellular passages, and are probably formed by condensation of the sap. Both are therefore in a physiological respect of one kind. At bottom also the blood-vessels of animals are naught else but passages in felted cellular tissue.

1063. As the principal polarity of plants has been directed upwards, and the cells therefore been protracted lengthwise; so also the chief direction of the vessels is parallel with the axis of the plant.

1064. The vegetable sap ascends in these tubes, which must be therefore viewed indeed as constituting the earth-or nutritive organ.

3. AIR-ORGAN, TRACHEAL TISSUE.

1065. The plant is not merely earth-and water-organism, but also air-organism; and there must therefore be developed in it an anatomical system, which coincides with the process of air.

1066. Besides the cells and tubes naught else is found in vegetable tissue but spiral vessels; what are called scalariform tubes, annular vessels, dotted ducts, vermiform or strangulated vessels, are no peculiar formations in themselves, but only different conditions of the spiral vessels.

1067. The spiral vessels are the air-system of the plant, and therefore rightly deserve the name of Tracheæ. They exhibit the structure of the air-tubes in insects, and contain, according to the most authentic observations, air, and not sap, except in the period of adolescence, as occurs in the animal kingdom.

1068. The spiral vessels consist of one or several filaments spirally contorted, and held together by a delicate tubular-shaped membrane.

1069. They must be regarded as elongated cells, upon whose parietes the starch-granules have been placed in serial juxtaposition with each other, so as to form spirally-twisted filaments, as is to be plainly seen in many cells and also in Confervæ. This spiral condition originates without doubt from the spiral-shaped motion of the granules in cell-sap.

1070. The ultimate cause of this spiral motion, as well as the position of the parts, appears to reside in the rotation of the sun.

1071. Upon this also depends probably the winding of the stem of plants, with the spiral-shaped position of the leaves and branches, as probably even the contortions of the snail's shell and of the hairs upon the crown of the head.

1072. The production of the spiral form originates from the antagonism of the light with the matter. The number of spiral vessels is therefore less in those parts that are beneath, than in those above, the earth, or less in the root than the stem. The more indeed an organ has been exposed to the air, by so much the more do the spiral vessels preponderate, as e. g. in the leaves.

1073. An organ must necessarily be nobler in character, the more spiral vessels it contains. The plant also that, with more spiral vessels, exhibits them particularly arranged, must take a higher rank. The lowest plants, as the mushrooms, lichens and mosses, consist therefore entirely of cellular tissue; in the ferns therefore only a single bundle of spiral vessels makes its appearance. When plants become nobler, several fasciculi of spiral vessels originate; and in tracing this feature we ascend from the ferns to the grasses and lilies, up to the lower Dicotyledons. In the higher Dicotyledons the packets of the tracheæ increase for the first time to such a degree, that they form a closed circle, the fibrous ring or zone of wood.

1074. The tracheæ extend from one end of the plant to the other; many are wont to terminate only in nodes, while these are to be regarded as arrested branches. The air can therefore penetrate through the spiral vessels from the leaves even to the apices of the roots.

II. Anatomical Systems—Sheaths.

1075. These originate by vagination and separation of the tissues in the transverse direction, and prevail throughout the whole plant. The idea of the whole vegetable structure is extremely simple. Originally the plant is a vesicle in water, or cellular tissue in the seed; root and stalk also consist in their main bulk or proportion of cellular tissue, which is called parenchyma. Therein the three planetary processes reside inclosed. Such a plant is still in the recognizable state of the primary organism. In the sequel, however, through the influence of light, the polarity between light and darkness issues forth in the parenchyma, the cellular tissue obtains a linear direction, and becomes elongated into spiral vessels. The spiral vessels form one or several fasciculi, which emerge out of the parenchyma, by which they, and each packet individually, are circularly surrounded. The cellular tissue is as it were the soil, in which the fasciculi of spiral vessels are rooted as proprietary plants and out of which they grow.

1076. The effort of the three vegetable processes, to separate their organs from, and perfect them independently of, each other, is in incessant operation, both from without inwards, as well as from above downwards, because in both cases is light there, darkness here, dryness there, humidity here. The cellular tissue, that has finally become independent in the transverse direction or from without inwards, is called bark, the self-substantial tubular tissue is called liber, the non-dependent tracheal tissue, wood.

1. TRACHEAL-SYSTEM, WOOD.

1077. With increased influence of light the tracheal fasciculi also increase, and form a circle of columns in the parenchyma around the centre of the plant. Between the column, externally and internally to the same, is the parenchyma. The more, however, the columns accumulate, by so much the more does it diminish, and whereas the columns previously stood singly in the parenchyma, the appearance is now as if narrow plates only of parenchyma traversed between the columns from without inwards. Finally the columns predominate to such an extent and approximate so closely, that the plates almost disappear. They are now called insertions of cellular tissue, or medullary rays. As the tracheæ convey air, and have thus been more exposed to the process of oxydation, they generally harden sooner than in other parts.

1078. Around the fasciculi of spiral vessels the cellular tissue also strives to elongate, and begins at the same time to harden. Such extended cells, in which the light has almost disappeared, are called fibres. Indurated tracheæ and fibres are called Wood. The wood is always in the vicinity of the spiral vessels. It is a production synchronous with the latter.

1079. Only, where spiral vessels are, can genuine wood originate; but it is not everywhere, where they are found, that woody fibres must be also present, although the cells extend around all bundles of spiral vessels. If the degree of oxydation of the cells be slight, they do not harden, but continue herbaceous in texture. The parenchyma has now been separated by a circle of fibrous columns into an external and internal, or peripheric and central. The central parenchyma becomes void of sap and spongy, because the plant imbibing its nutriment on the surface, and the air and the light operating thereupon, the processes conduct it thither. This withered parenchyma is called pith, which in accordance with its origin merits no physiological consideration, nor is worthy and susceptible of any philosophical construction.

2. TUBULAR SYSTEM, LIBER.

1080. As the plant draws in its nourishment from without, so is the main proportion of the sap necessarily present in the periphery of the spiral vessels. The elongated cells in the neighbourhood of the spiral vessels, and which principally contain sap, are called Liber.

1081. Liber is necessarily present around every packet of spiral vessels, and thus with fasciculi everywhere dispersed throughout the stem. The liber is only situated beneath the bark, when the number of the spiro-vascular fasciculi is so great, that they form a closed circle in the parenchyma; it is only beneath the bark, in so far as it accompanies the spiral vessels, but can only surround the latter from without. As it is only the woody plants that have been usually examined, the false idea has thus originated of the liber having, as it were from its very essence, to be beneath the bark.

1082. In the liber is the main seat of vegetable activity. For it is soft cellular tissue with open intercellular passages, wherein the sap can move.

1083. Now as every fascicle of spiral fibres is surrounded by liber, such a fascicle must be regarded as a whole plant. A plant consists accordingly of as many plants, as it has or can have tracheal fasciculi. Every plant is a trunk of infinitely numerous plants; for every one can contain infinitely numerous tracheal fascicles. One plant is a whole vegetable world. (Ed. 1st, 1810. § 1065.)

3. CELLULAR-SYSTEM, BARK.

1084. No spiral vessels lie upon the surface of the plant, for where they originate, there the liber forms around them, and this is consequently the External. The surface of plants is therefore necessarily environed by liber, notwithstanding the greater influence of the light. The cellular tissue upon the surface of plants is, however, less rich in sap than the liber around the tracheal fasciculi, because it is too rapidly evaporated and dried up by the immediate contact of the air, light and heat. The surface of the plant is too strongly oxydized by the air, and therefore the cells harden. The sap also decomposes too rapidly and becomes rigid, so that an irregular formation only can proceed from it. The external, more inactive, or irregularly wood-converted layer of cells, is the Bark.

1085. The plant has thus likewise three anatomical systems, which are nothing new, but only the repetition or rather vagination of the three tissues; alburnum and cambium are only transitional, not special formations.

III. Organs of the Vegetable Trunk.—Members.

1086. Organs are separated parts of the body, and combinations of single tissues and systems, and are consequently a Whole in Singulars. There are, however, no uniform combinations; but one or the other system asserts its preponderance and imparts the character.

1087. In conformity with the developmental progress of the whole of nature, namely, that of always separating further its chaotically mingled parts, individualizing and yet forming them with the others into a whole, vegetation cannot continue stationary with the partition into bark, liber and wood, seeing that they are always circumscribed and form a body in common; but they must also sever this body itself into as many members as it has constituent parts. This severance makes its appearance in the longitudinal axis, because in this direction the antagonisms of air and light with water and earth are more powerful.

1088. Through the separation of the vegetable trunk three members only can originate; one with the preponderance of cells or of bark, one with that of vessels or of liber, and one with that of the tracheæ or of wood. The cellular tissue has been posited as vegetable trunk in the root, the vascular tissue as a special member is stalk, the tracheal tissue leaf. In this manner the trunk of the plant divides into three great divisions; more are not possible.

1089. Now the root is the perfected water-organ, because it is always fixed in water; the leaf is the perfected air-organ, because it moves in the air; the stalk is the perfected earth-organ, because it removes the mass out of water and air. Root is a heap of cells; leaf a plane of tracheæ; stalk a bundle of vessels.

1. WATER-ORGAN—ROOT.

1090. By the two polar systems, the earth-and air-system, the cellular and tracheal system, is the development of the plant confirmed. Thereby is it in the next place a twofold organism. By the first system it has been turned towards the planet and immersed in earth and water, by the second it has been turned towards the sun and immersed in the air. The root and the fabric of the stem, or root and stem simply, have now obtained their truest significance. Each is the whole plant, each the whole organism; in the root this is only in its original purity, but in the stem it is upon a higher stage. Root is stem in water and earth; stem is root in air and light.

1091. The root has accordingly more cellular tissue, fewer tracheæ; in the stem this condition is reversed. The root resembles young plants, or such as still rank upon a lower stage, and have but few columns of tracheæ. The root has therefore no marrow or pith. It may be said that it should have no pith, because it is usually thinner than the stem and richer in sap; but it has only the latter character, from consisting for the most part merely of cellular tissue. Root is the vegetable trunk with preponderating cellular tissue. In consequence of the antagonism between root and stem, wherein even their difference consists, the one strives to produce the Chemical, the watery earth or the mucus, but the other the Electrical, the combustible air-bodies.

1092. The root, as producing mucus or infusoria, has therefore in itself the organic process of putrefaction, in so far as the origin of mucus and infusoria is a result of the putridity. It corresponds to imbibition and digestion. To this is referrible the mouldy, and as it were fetid, condition of the root. Through the process of decomposition, which it evokes in its neighbourhood, it kills its nutriment, takes possession of it, and thus originates completely, as does every first organism, out of putrefaction, out of infusoria. To the essence of the root belong therefore not merely food, but the favouring relations of decomposition, as earth and water, whereby the access of the air, as necessary to every galvanism, has not been suppressed.

1093. The earth is not merely a mechanical station for the plant, in order to give it the perpendicular direction, but it is necessary for polar excitation, whereby the decomposition is imparted. A plant placed upright in pure water, although with the roots, necessarily perishes. Darkness is at the same time the lurking-place of putrefaction, as being that which only plays its part in localities where the polarizing and dissevering influence of light is wanting.

1094. The root always passes perpendicular into the earth, on account of its greater weight due to repletion with water. In all zones therefore the root stands perpendicular to the horizon, and thus the whole plant, although this is somewhat inclined towards the sun.

1095. The developmental stages of the root pass probably parallel to the parts of the vegetable stem.

a. In respect to the tissue, there are thus cellular roots, as is probably the case in the fungi, and with the fibrils of all roots; tubular or vascular roots as in the mosses, tracheal roots in the rest.
b. In respect to the systems the bulbs are the cortical or bark-roots; the tubers the liber-roots; the fibres the woody-roots.
c. In respect to the members of the trunk, the turnip is perhaps the genuine root, the tap-root the stalk-root, the so-called aerial roots, the leaf-roots.

2. EARTH-ORGAN.—STALK.

1096. The stalk is the idea of the whole plant, posited under the import of the organ of nutrition, of the vessels. The structure of the stalk is therefore accordant with that of the root. The anatomical systems are alike in both, bark, liber and tracheæ being in the same envelope.

1097. In the stalk, however, the opposition of tissues and systems emerges more strongly, and therefore they all become individualized also to a higher degree. The spiral vessels become more freed from the cellular tissue; the bark is more distinctly divided from the liber; this again from the wood, and in its centre the cellular tissue dries up into pith. As, however, the stalk is the first product of the light-influence, the tracheal system cannot attain as yet entire freedom. The cells have nevertheless been extended, and the intercellular passages are formed into regular tubes. The Stalk is the trunk of the plant with a preponderating system of tubes.

1098. This separation of the tissues and systems, with the endeavour upon the part of each to become individually perfect and isolated from the other, is effected by the air and the light. As root imbibes the mucous or slimy water, and sustains the chemical process in the plant, so does the stalk set the water in motion, since it exposes it to the air and light, whereby the chemical earth-process becomes separated into different saps and elemental bodies.

1099. Through the influence of the air, light and heat upon the stalk, as well as by its antagonism with the root, its elongation is determined. The greater amount of energy of the aerial polarity is in the higher regions; it is thus more excited by these than by the lower, and the bud being lighter grows more rapidly in the upper parts, and obtains an elongated form. The excitation is stronger also upon the side exposed to the light. It grows also more strongly in that direction, and thus the stalk stands indeed upwards, but somewhat inclined from the perpendicular line towards the sun.

1100. The perpendicular direction of the stalk is, however, as mechanically determined by gravity as the root. Paradoxical as this assertion may appear, it is still correct; for if we think of a moist globe, which is superiorly affected by air, light and heat, the upper aqueous or mucous parts are thus lighter, and necessarily ascend upwards through the pressure of the heavy or unheated parts, just as the air-bubbles in beer ascend to the surface. The light it is true can draw them somewhat sideways, chiefly because this side is more heated, more decomposed, and also undergoes a greater amount of evaporation; but the proper cause invariably resides in the ascent of the light parts between the heavy or immoveable. The stalk therefore grows upwards also in the darkness, and then indeed quite perpendicularly, because it has not been diverted by light. Were the light merely the Dirigent, it could not be comprehended, why towards the poles trees still stand tolerably perpendicular, and do not lie completely upon the earth. Were, however, the air that which determines the direction, plants could not thus have been inclined towards the sun, nor could we comprehend why the flowers and also the leaves obey the sun's course. Finally, were neither air nor light the imparters of direction, then the plant could shoot in no other direction than quite perpendicularly upwards, as is done too for the most part by the fungi.

1101. The winding of the stalk appears to originate from the rotation of the sun. The next cause is probably the greater heating and decomposition that occurs upon one side. In accordance with this assumption the plants upon the northern hemisphere of the earth must wind spirally upwards from left to right, or from morning towards evening, if our gaze be directed towards the meridian, but inversely upon the southern hemisphere. But this is not the case. May we therefore conjecture that plants, twisted contrary to rule, have been transported from their native soil?

1102. The kinds of stalk follow also, without doubt, the direction of developmental stages of the vegetable trunk. There are therefore cellular stalks in the fungi, vascular stalks in the mosses, tracheal stalks in the ferns. A bark-stalk is the culm, a liber-stalk the scape, a wood-stalk probably the trunk of a palm. A root-stalk is the rhizoma, a perfect-stalk the stem, a leaf-stalk being probably the shrub.

Ramification.

1103. The differencing, severing character of air and light must never be lost from our thoughts, nor also that at a height this character manifests itself more powerfully than on the surface of the earth, where the stem abandons its androgynous position in relation to the root. Through the constant process of differencing the tracheal fascicles of the plant may finally become so independent that they no longer stand in need of the others, and do not merely represent a particular plant, but are perfected also as such.

1104. This dispersion of the tracheal fasciculi will not easily take place upon the earth's surface, on account of its lesser degree of aerial polarity, but at a certain height. If several stalks emerge from the root, the bush or shrub originates, but if only one, the tree. If the subdivision first commence at a certain distance from the earth, then do Branches originate.

1105. The formation of branches demonstrates in part a great store of tracheal fasciculi, and part by an easy differential capacity in the plant. Both come to the same thing.

1106. Plants devoid of branches are similar or analogous to roots.

1107. The branches again ramify from the same cause by which the stem ramifies. A branch or twig must be regarded as an entire fascicle of tracheæ, which forms superiorly a closed vesicle or bud, that raises the bark, bursts it open into scales and then opens itself. The opening of the external tracheal tubes or vesicle becomes a leaf; a twig can therefore occupy no other position than in the sheath or angle of a leaf. Every succeeding leaf upon the twig is in the same manner a ruptured vesicle of tracheæ. As many leaves therefore originate as there are tracheal zones present.

1108. Every branch is an entire plant. All the tissues and systems are found in it. Tracheæ free themselves from the stalk, pass towards the circumference, break through the bark and carry with them the liber, whose external layer again becomes bark. The branch is only an elongated bud. The stalk is the soil or the root of the branches. Branches, that have been cut off and stuck in the earth, grow. There is nothing contradictory, in the tracheal fascicles of the branch growing downwards into the stalk. A branched tree is a complete wood.

1109. From the same cause, or by the influence of water on different situations in the earth, the root has branches. But as this influence is weaker than that of air and light, so the number, density, and length of the roots is less.

1110. The polarization of the tracheal fascicles into branches takes place all around the stalk in one situation. The influence is equal upon all sides. The idea of the formation of branches is the star. All branches have a radiate position around the stalk—all form a verticillum or whorl.

1111. Every other position of branches is only an alteration of the verticillate arrangement.

1112. In most plants the arrangement of the branches admits of being reduced to the spiral line. This position is only the verticil drawn out.

1113. This drawing out takes place through the continued growth of the stalk, in which the bundles of tracheæ, doubtless by the varied operation of light, develop themselves in order, become individualized, and issue from the stalk as branches.

1114. The crucial position of the branches depends upon the same growth of the stalk, but one in which transverse polarities are present. The irregular or dispersed position is most probably the last completion of the spiriform.

1115. The spiriform arrangement stands in relation with the formation of the tracheæ.

1116. The branches of the root observe no such regularity, partly on account of weaker polarity, partly on account of the obstacles placed in their way, at one time by the impenetrable earth, at another by want of water.

1117. The more the stem is differenced, so much the higher is it developed. Thus the richer in number the branches, by so much the more perfect is the stem. The stelliform branches belong to the first development. The plants, that have this arrangement, stand lower in the scale. To this the crucial appears to succeed, as a mediate position between the former and that which follows. The spiriform ranks higher. In it the stem is manifestly differenced more multilaterally. The dispersed arrangement appears to be the highest, because in it the greatest freedom prevails, because the poles have acted on every part of the plant, because they are everywhere in the air and in light. Plants with dispersed branches are organized air; without branches they are organized water as well as earth. Those plants only which have circles of tracheæ or rings of wood, as e. g. the Dicotyledons, ramify; these alone are, properly speaking, a conjunction by growth of many plants, and one that is truly persistent, or that bears fruit several times. The Monocotyledons being devoid of woody rings, do not, or but very rarely, ramify. Most of them therefore die off after they have once produced fruit. The ramification is a multiplication of the plant, in which the buds continue to stand upon the old stem.

Formation of Nodes.

1118. The formation of nodes, as in the grasses, is an attempt at differentialization, that has not, however, attained to perfection. A node is a branch-whorl which has continued to adhere to the stalk. Therefore the tracheæ also terminate in the circumference of the node.

1119. The formation of nodes consequently ranks directly under the stelliform formation of branches. Taken in a strict sense the formation of nodes occurs only in plants with sheathing leaves or in Monocotyledons.

3. AIR-ORGAN, FOLIAGE.

1120. If in the progressive separation of the tissues the tracheæ finally obtain the preponderance, so that they issue forth free from the envelope of cellular tissue, the leaves, or Foliage, then originate.

1121. The ribs of the leaf are the fascicles of tracheæ that have become free, and are still only connected together laterally by a thin layer of cellular tissue.

1122. The leaves can be regarded as gigantic and unrolled spiral vessels, and these again as microscopic and involuted leaves.

1123. As through the root the water-process enters the plant, and through the stalk the earth-process, so does the air-process through the leaves.

1124. It is probably the stomata through which the air is conducted into the tracheæ; the connexion, however, has not yet been proved.

Gemmæ or Buds.

1125. With the formation of branches there is at the same time a diminution of the cellular tissue, and an increase of the tracheæ. Entirely new spiral vessels commence in the branches, and are not continued or prolonged into the stalk. The further indeed the extent of the ramification, by so much less is the quantity of cellular tissue, and by so much greater the number of the tracheæ. It comes at last to this, that the tracheal fascicles, which were from all sides surrounded by dense cellular tissue, are only then loosely connected by a thin layer of such substance. This ramuscule is still therefore only a hollow stalk, consisting of fascicles of tracheæ disposed in circles, and so united by a thin cellular integument or membrane, that the whole forms a vesicle.

1126. This vesicle is a Bud. A bud is at bottom none other than the end of a twig that has become hollow.

1127. Several buds are usually involved in each other, i. e. many vesicles of tracheal rings have been encased in each other. They issue gradually forth and become shoots or buds. Buds are bulbs at the end of the branches.

Leaves.

1128. If the bud or the external vesicle ruptures, while the cellular substance becomes consumed at the apex or between two or more tracheal fascicles, it is then manifested as a leaf or Leaves.

1129. Then the second vesicle grows forth, becomes petiolated, ruptures and becomes leaf or leaves. In this manner a twig is formed, surrounded in a spiral manner by leaves.

1130. The younger leaves have been originally inclosed in the older, as in their sheaths.

1131. Every perfect leaf, i. e. every leaf-vesicle, must be regarded as the terminal extremity of an entire twig, from or out of the angle of which a new twig grows forth, that again as a bud ruptures, and from which again a twig grows forth.

1132. All leaves therefore range directly opposite to each other. A branch with many leaves is a system of branches, which grow out of each other, like the articular pieces of the grass-culm or straw.

1133. A leaf is a whole plant with all its tissues and systems; with cells, ducts, tracheæ; bark, liber, wood, stalk and branches. The leaf is a tree of special form, a tree, whose branches or tracheal fascicles all lie in one plane and are held together by parenchyma. It is the bodily expression of the position of the tracheal circle in the stem, only ruptured and to the greatest degree attenuated.

1134. In the division of the ribs of the leaf the internal arrangement of the woody fasciculi in the stem has been placed before our eyes, as by the scalpel of an anatomist.

1135. From the arrangement of the ribs of the leaf the structure of the whole plant can be recognized and its character determined. The leaf is the table of contents or index of the stem.

1136. Plants, which have no tracheæ, have also no leaf-ribs—Mosses.

1137. Plants, which have only isolated or non-ramifying fascicles of tracheæ, have parallel leaf-ribs that do not ramify—Monocotyledons.

1138. Plants, which have a circle of tracheæ, or rings of wood, have leaves with ramified ribs—reticular leaves or true foliage—Dicotyledons.

1139. The stronger indeed the ramification of the leaf-ribs, by so much the higher is the perfection of the leaf. The lowest leaf is that devoid of ribs, the higher that with parallel ribs, the highest being the reticular-veined leaf.

1140. The number and forms of the leaves that proceed from a bud, depend partly upon the number of tracheal fascicles, which pass out of the ramule into the leaf, partly upon the form of the leaf-bud.

1141. If the bud be simply ruptured at its apex or only between two fascicles of tracheæ, there then originates the spathiform leaf.

1142. If the cellular substance between several tracheal fascicles be consumed by the severing action of light and by the air, then the bud divides into several leaves.

1143. The fundamental form of the leaf is the oviform, because the bud is to be thought of as being round. Through the elongation or compression of the bud, lanciform, cordiform leaves, &c. originate. Besides it appears, that the leaf-buds, at least those of the reticular-veined leaves, burst in a circinate manner, like the ferns, and unroll themselves. Therefore the leaves are unilateral and clasp the stem inferiorly; or the petiole displays its spathiform origin. In this case the leaf is not a part of a hollow globe, but the entire globe, that has ruptured in the transverse direction at the extremity of the petiole. Such is the case at least in the demispathiform leaves of the Umbelliferous plants.

1144. The primary position of the leaves like that of the branches is thus verticillate, yet always, however, under the idea of one being encased within the other. The leaves are here only the ultimate branches.

1145. If the leaves upon the ramules issue from each other, then this happens in the same manner as with the branches.

1146. The verticillate position of the leaves is therefore the lowest, next succeeds the crucial, then the spiriform, and lastly the scattered.

1147. The spathose leaves are only to be regarded as a single bud, and therefore differ from the whorl, in so far as we regard this as the development of several leaves, that have not, however, been extruded from each other.

1148. Divided leaves originate through a higher operation of light. In them the formation of ribs preponderates, and therefore they rank higher than the undivided.

1149. From this cause the pinnate leaves are the highest.

1150. From the same cause the radical must be worse developed than the ramular leaves. They are usually non-pinnated, undivided, because they have more cellular substance in their composition than the upper leaves. In the leaf-system, consequently, the whole idea of the plant has been recontained; on the earth resides the chemical character, as is evidenced by cellular, dense and misshapen leaves; above in the air, on the contrary, the leaves are more delicate and are divided—indicating an electric character.

1151. The division and pinnation of leaves can only progress according to the odd numbers, 3, 5, 7, because the midrib determines the odd leaflet.

1152. Leaflets occurring in pairs, or equally pinnated, are arrests of development.

1153. The even number or the symmetrical form is unnatural in the vegetable kingdom.

1154. The leaves are, like the young bark, and thus the whole trunk of the plant, green, because the vegetable kingdom represents the lower totality of the earth, the planet, whose synthesis is the water.

1155. From the same cause, the chief colour of the animal kingdom is red, the colour of fire. Thus, plant is to animal, as green is to red.

1156. The division of leaves passes also parallel to the stages of rank in plants. Cellular leaves are the scales of mosses and ferns; vascular leaves, the long riband-like leaves of Monocotyledons; tracheal leaves the reticular leaf of Dicotyledons. The cortical leaf is the sheath; the liber leaf is probably the fat leaf; the wood leaf the acicular leaf. The radical leaf is the undivided reticular leaf; the stalk-leaf the free or ragged reticular leaf; the perfect leaf the pinnate. The bracteal leaves repeat all forms in the thyrsus, since they are floral leaves.

1157. The accessory leaves or stipules are none other than the remnant of the sheath-formation, out of which all the leaves, and therefore the wings of the leaf-petioles or phyllodia, have issued forth.

1158. The thyrsus has also its series of leaves; the scale-like or radical leaf is involucre and bractea; the vascular or spathe-leaf is calyx; the tracheal or reticular leaf is corolla.

1159. The vegetable trunk, namely, root, stalk, and leaf, is a perfect organism, which can exercise all the functions which belong to its individual life. If it therefore produces anything, that can be nothing new, but only itself repeated. This repetition of itself is called propagation. The organs of propagation are thus none other than a repetition of the organs of the vegetable trunk. The plant thereby steps forth out of its individuality into the province of the genus.

B. Æther Organs.
THYRSUS OR FLOWER.

1160. Hitherto we have regarded the plant as simply a planetary organism, namely, as a trunk with water-, earth-, and air-organ. But the primary vesicle does not lie wholly in the dark, but is illuminated upon its apex by the sun. Every operation, however, produces its like; thus in the plant a light-organ must also be developed. As the light evokes heat in the æther, so also does it evoke a heat-organ in the plant. As the heavenly bodies corevolve in æther by means of light and motion, or the gravitation condenses the æther into matter, so also must an organ of gravity originate also in the plant. These organs are not, however, predominant in the plant, because it is essentially planet or vegetable trunk. They can be therefore nothing else than the parts of the trunk itself with the properties of the æther or fire. They are thus a repetition of the trunk, wherein, in place of the material processes of growth, those of the light, heat, and gravity occur. The light-organ excites the heat-organ by polarization unto motion and thereby originates the organ of gravity.

1161. The process in which through polar tension the trunk has been reproduced, as the Whole in miniature, is called sexual process. The æther-organs are thus sexual organs. These sexual organs can only be a leaf-formation, because the last development of the stem is the leaf. The leaf-formations which, by means of polar tension, reproduce the trunk, are the Flower. The light-organ is the corolla. The heat-organ is the pistil. The organ of gravity the seed. The pollen upon the stigma sets the pistil in a state of tension with the trunk, whereby the sap, out of which the seed attains to perfection, ascends. Without this tension the pistil would not have had strength sufficient to perfect the seed. It would wither ere the latter had obtained sufficient nourishment.

1162. The development of the flower takes place through differentialization, individualization, or complete separation of the trunk-organs. The trunk indeed summons up all power in the leaves, to separate the three vegetable tissues and represent each as a particular organ; only in this formation it does not wholly succeed; for in a leaf the ribs or tracheal fascicles are still held together by cellular tissue. In the first place with the perfect separation of the tissues or, properly speaking, with the ex-organization of each of them unto an independent Whole, the limit of the vegetation is attained and the growth completed. This was the course of the whole of nature; in every system she has proceeded to individual consummation of the factors, to their liberation from chaos; and the developments of the systems were concluded, so soon as all factors became independent, or so soon as every factor had itself become an entire nature. Such was the case in the genesis of elements, and such in the metamorphosis of the earth-element into earths, salts, Inflammables, metals.

1163. This complete severance and individualization can no longer be effected by the air, but must be achieved by the light. The air is itself not the wholly differencing element, but derives its potency only from the light. All ultimate separation and individualization is reserved for the light.

1164. Root and stem are the water-and earth-plant, the leaf is the air-plant, the flower is the light- or rather fire-plant.

1165. In the flower the problem has been solved, of producing an entire plant simply by the light without earth, water, and air, or as it were in a merely spiritual manner.

1166. The plant is a flower that has been posited under three ideas, under the idea of earth, water, and air. As in æther or fire all the elements are dissolved; so are all the elements of the plant in the flower.

1167. The flower is truly, not merely in idea, the whole plant with all systems and formations, posited under a single idea, under that of the æther, namely, of the gravity, light and heat, or the fire.

1168. The flower as the æther-organ of the plant is not so independent as an animal, but subordinated to the planetary systems, being only a separation of the parts of the trunk, not a new formation, as in the animal kingdom.

1169. The flower as the highest formative consummation of the plant, or as the highest vegetable organ, is the extremity of the ramules. (Ed. 1st. 1810. § 1176.)

1170. Although the flower is a repetition of the whole plant, namely, of the root, stalk, and leaf; it can still only be a direct transmutation of the leaf. For every Superior proceeds from that which is immediately subjacent. The light-organ can only be developed out of the air-organ, not out of the water-organ. The transition from the water-organ into the light-organ is necessarily indicated through the aerial form. Preparations are necessary, gradual dismemberments must precede, before the isolated consummation can result. The air purifies the organs, in order that they may become participant of light.

1171. The flower is the totality of the leaves of a ramule placed upon the extremity of the latter. For a flower is the whole plant, and is the Ultimate of the plant.

1172. The flower is a terminal leaf-bud. A leaf-bud, after whose rupture the ramule cannot grow any more.

1173. The flower necessarily stands in a whorl-shaped manner, because it is the end of the branch; it is the terminal whorl of the plant.

1174. With the flower the ramule or the plant dies off, partly because it is the extremity, partly because wholly separated tissues cannot live. In the flower therefore the plant reverts to its origin. It is a ramule, whose buds have continued stationary.

1175. The idea of the vesicle lies at the basis of the flower. It is an entire vegetable vesicle, a leaf-bud that has not been drawn asunder. The flower is the last vesicle into which the stalk swells out.

1176. The flower-vesicle agrees with the leaf-vesicle. The form of the flower must pass parallel to the form of the leaves. This has reference principally to the position and number of parts. Division.

1177. The flower is the synthesis of the entire plant with complete analysis of the organs. Flower, pistil, and seed are the leaves, stalk, and root separated, and yet all combined to form a common organ. This flower regarded in its analysis is the flower proper; in its dissolution it is called fruit.

1178. The flower-vesicle is according to its essence a threefold vesicle. In it the leaf-system or the air-plant has been represented, but in like manner and of necessity also the earth-and water-plant, or the vesicles in which stalk and root have been taken up into the kingdom of light. Thus there is the leaf-, stalk-, and root-flower.

1179. The leaf-flower is in the periphery, the stalk- and root-flower in the centre of the vesicle. For the former is the metatype or copy of the leaves, the latter of the stalk and the root.

1180. The leaf-flower is the highest and the very first to be developed. It is that, which chiefly corresponds to light; the trunk-flower is, however, the lowest, the last developed, because it is only the trunk that has been prolonged with difficulty to form a flower. It is the child of the heat and gravity.

1181. It may be also said that the leaf-flower is the electrical, the stem-flower, however, the chemical. In the latter the chemical process must still act visibly, it must still be produced mucus; in the former, however, this must disappear and resolve itself into purely electrical bodies.

1182. The flower consists of three leaf-buds. The leaf-bud is the corolla or blossom. The stalk-bud the pistil. The root-bud the seed.

1183. The corolla is the external whorl of leaves, is first developed, has the form of a leaf, is a

vesicle, secretes in itself electrical, inflammable bodies, and is directed towards the sun.
1184. The difference between corolla and pistil is that of the two principal tissues, the tracheal and cellular tissue. By the light the tracheal fasciculi become finally separated from the cellular substance, evolved to a higher degree as the child or product of light, and planted outwardly. The corolla is the tracheal circle, which has forcibly gained its freedom.
1185. The pistil is the vascular substance that has become freely evolved, yet to the highest stage; in a similar relation likewise does the seed stand to the cellular substance. In the fruit therefore the flower again reverts to the primary condition of the plant.
1186. The flower and pistil are therefore those very organs which have been most antagonized in the plant. They are in the highest state of polar tension, and stand opposite each other like electrism and chemism, or as light and matter. This antagonism in the Organic is called sex.
1. FLORAL ENVELOPES.
1187. The blossom is the leaf-formation, in which the separation of the principal tissues is completely attained, where the tracheal fascicles entirely separate from the cellular substance, and become a leaf with free ribs. The cellular substance becomes the corolla-petal, the rib the stamen.
1188. The blossom passes through the three stages of leaf-varieties before it attains its completion, and divides therefore into three whorls, which correspond to the root-leaves, stalk-, and ramular or perfect leaves. The radical or squamose leaf appears in its repetition as involucrum or spatha. The stalk or vascular leaf as calyx. The perfect or tracheal leaf as corolla.
a. Involucrum.
1189. As the root puts forth numerous branches, so the involucrum or spatha frequently incloses many blossoms that constitute the thyrsus or synthetic form of inflorescence.
1190. The thyrsus or dense panicle is the whole ramage or branch-fabric repeated in the involucrum; it is therefore just as manifold as the former.
1191. The inflorescence is still as complex as the arrangement of the branches, because with every flower the ramule dies off or ceases to grow, whereby very numerous and strange relations are brought to light.
1192. The involucrum corresponds to the scale-leaves or bracts, and is therefore as a rule polypetalous. The involucral leaves stand upon a lower stage of development, are for the most part only squamose or spathiform, rarely divided or pinnate.
1193. The involucral leaves necessarily stand, as being the radical leaves of the flower, at the bottom of the peduncles or flower-stalks; each flower-stalk therefore has usually an involucral or bracteal leaf also.
b. Calyx.
1194. The repetition of the cauline or spathe-leaf in the flower is the Calyx. It does not stand therefore at the base of the peduncle, but towards its summit, and is the external bud of the blossom; it is mostly spathiform, rarely or but slightly cloven, and very rarely polysepalous; it is generally green like the leaves. Although the calyx is no essential organ in itself, still it is rarely wanting and is often itself the supporter of the corolla and stamina.
1195. As the calyx is more incomplete than the corolla, so it has usually but three lobes, and if it has five, then they are frequently placed irregularly.
1196. The developmental stages of the calyx are also three. Corresponding to the scale-like leaf or bract, it is only squamiform, as in the catkins of the hazel and fir-cones; corresponding to the spathe-leaf, it is tubular or unisepalous; corresponding to the reticular leaf, it is multisepalous, and mostly deciduous.

1197. The tubular calyx exhibits likewise three stages of development. At first it is scale-like, confluent with the ovary or epigynous—ovarial corollæ. It is next spathe-like, simply connected with the corolla or perigynous—calycine corollæ. Lastly, it resembles the reticular-veined leaf, being free from both the above organs, or hypogynous—pedunculate corollæ.

c. Corolla.

1198. From the character of the corolla or crown as being a ramular leaf all its properties admit of being deduced. It is the upper leaf-bud, inasmuch as the ramular leaves stand above those of the stalk. It is, on this account too, the internal, inasmuch as the lower leaves admit within them the upper.

1199. The corolla is homologous with the whole mass of the petiolated or ramular leaves. Thus their arrangement, form, and number taken together are not worth more than the characters of the corolla, nor are they worth less, because they are relations of deeper organs.

1200. The corolla stands also in a whorl, because it is the totality of leaves.

1201. The laws of the leaf-formation are also the laws of the corolla-formation. The corolla will therefore represent at one time a greater, at another a less, ramified leaf-system.

1202. The corolla is the last verticil in the series of leaf-whorls; for it represents the last leaf-form, and must fade, because the tissues have completely separated from each other, namely, the tracheæ as stamina, from the cells that constitute corolla leaflets or petals. No part can carry on life by or for itself.

1203. The parts of the corolla range in an alternating manner with those of the calyx, because they are the next supra-consecutive bud.

1204. In accordance with the three stages of the leaf-formation the corolla is also resolvable into three forms. The scale-or root-leaf returns in the several squamiform micropetals of what have been called the Apetalous plants; it corresponds to the involucrum—squamoid corolla. The spathe-or stalk-leaf appears in the tubular-shaped or Monopetalous corolla. It is a leaf-bud, which has only ruptured at the apex, like most calyces, unto which it corresponds—spathoid corolla. Lastly, in the polypetalous corolla the reticular or petiolated leaf appears upon its highest stage—reticular or phylloid corolla. The rank of the leaves consequently determines also that of corollæ.

1205. Although we regard the corolla as a whorl of leaves, namely, as several leaf-buds approximated, a yet clearer insight into its numbers and positional relations is attained, by viewing it only as the lobes of a single leaf.

1206. If we regard the petals of a polypetalous corolla in the import or light of pinnate leaflets, the legitimate regularity of their numerical relations, and the so-called irregularity of their arrangement is easily explained. The papilionaceous corollæ have been manifestly constructed according to the plan of the pinnate leaf; the vexillum or vane corresponds to the odd leaflet, the alæ or wings to the two anterior, the carina or keel to the two posterior pinnate leaflets; and hence their consecutive ratio of decrease in respect to size.

1207. All irregular corollæ admit of being referred to the papilionaceous form. One petal is always found, which separates from the others, is either larger or smaller, or entirely ablated, and consequently corresponds to the odd pinnate leaflet or vane.

1208. The irregular monopetalous or tubular corollæ admit also of being referred to the papilionaceous form. They need only be regarded as a confluence of petals. The labiate corollæ are one of the same character; but, on the contrary, the lower lip is trifid, the upper, and thus the keel, bifid. Here the papilionaceous corolla stands, properly speaking, reversed.

NUMERICAL LAW.

1209. The number of petals in the corolla ranges parallel with the mode of development of the pinnate leaves. But at first the odd terminal leaf necessarily originates, because a leaf must have several pairs of ribs, before it splits in a ptiloidal or pinnate manner. The primary number of petals is therefore one. This number is found in those plants called apetalous, where the small lateral petals are abortive, and frequently the terminal petal itself, as in the catkin, where only the calyx has been left remaining.

1210. The normal ratio, in which the numbers of the petals progress, is the odd or uneven. For a leaf is to be regarded as a single fascicle of fibres with cellular substance. This packet of fibres grows straight out. It is solicited by the light to give off bundles of fibres; there is thus no existing cause, why it should only give off such upon one, and not upon the other, side of the bundle. At the first division therefore it must be an odd leaflet, which is the principal bundle of fibres, while two even pinnate leaflets take their origin upon its sides.

1211. The second number of petals is therefore three; for this is the first number in which a pinnate leaf can appear. This number is much more frequently met with than the former, because all division of a terminal bud necessarily strives after the representation of the whorl. Under this head come most Monocotyledons, as grasses, rushes, lilies.

1212. Of the three petals of a corolla all three are not of equal import, two only are alike as being lateral pinnate leaflets; the third, however, is present as the odd or uneven leaflet.

1213. Upon this inequality of import depends the irregularity of many ternary corollæ, e. g. of the Orchideæ, Apiaceæ, and even many Irides and lilies.

1214. If the trinity or ternary division depend upon the pinnated leaf, so must also the next number depend thereupon.

1215. The third prevalent number in the vegetable world is the quinary. The quinary division originates, when to the two pinnate leaflets two others are superadded. The pentapetalous corolla is also an odd pinnate leaf in a whorl-shaped position with four pinnate leaflets.

1216. The pentapetalous is naturally higher in rank than the tripetalous, and this than the monopetalous, corolla.

1217. In the former four petals are of equal rank, but the fifth or odd petal differs from them. This difference is shown in its position, size, form, design, and colouring. The odd leaf is usually larger, of a rounded form, having more ribs and other spots or markings.

1218. It appears as if, with this second liberation of fibrous fascicles from the main bundle, the differentialization were closed. For most corollæ are only quinarily divided, or admit at least of being referred back to that number.

1219. It is readily comprehended why there are only three breaks present in the number of petals, that these are denoted by one, three, and five, and that they rarely ascend to seven, nine, and so on. For the posterior pinnate leaflets are wont also to abort or be arrested in the leaves.

1220. There can be no plant with originally two corolla-lobes or petals; for the fibrous bundle does not thus divide itself; one main fasciculus is always left. The cause why the principal bundle does not divide into two equal fasciculi resides in the very nature or essence of the stalk. The idea of the stalk operates throughout the whole plant. The odd or uneven leaflet is only the last expansion of the stalk. The even leaflets are its branches.

1221. For the same reason there can be no originally quaternary corolla.

1222. There can be none that is originally sexanary.

1223. But an originally septenary corolla is conceivable, if forsooth the energy of light is still in a condition to separate two bundles of fibres. How rarely this occurs is well known, and thus it still continues doubtful whether this form is not to be explained by an arrest of development.

1224. Nature can produce no originally octo-, deka-, or dodeka-petalous, &c. corolla.
1225. Corollæ with originally nine, eleven, and more numbers are not impossible. The last seem only to exist.
1226. All corollas having even numbered petals, originate through arrest of the odd leaflet.
1227. The number two originates most usually and in the simplest manner from the arrest of the number three. If it originates from the quinary quantity, two pinnate leaflets are then coarrested.
1228. The aberration is recognized either from the position of the lobes or petals that have been left remaining, or from comparison of the number in other parts, in the calyx and capsule.
1229. The quaternary corolla is a quinary without the uneven leaflet.
1230. The hexapetalous is a doubling of the tripetalous, corolla. There are two whorls of petals, provided the calyx has not become corolla-like in character. Both cases are demonstrable through the alternating arrangement of the parts.
1231. The number eight is a doubled four.
1232. Nine is indeed in most instances a product of three multiplied by three.
1233. The number ten is a doubled five.
1234. In every number of petals the law of unequal development consequently prevails.
1235. The original arrangement of the parts of the corolla is bilateral, and therefore symmetrical. In the papilionaceous corollæ this originally symmetrical arrangement is most perfectly maintained. They repeat the position of their pinnate leaves.
1236. This symmetrical arrangement is shown even in many tubular, as in the labiate and personate, corollæ. The trifid lower lip is the standard and wings; the bifid upper lip, on the contrary, the keel.
1237. The small liguliform petal of the lettuce's corolla is a tubular corolla entirely slit up; it is therefore mostly quinque-dentate.

1238. Corollæ which have only a single petal (the tubular-shaped corollæ should not be styled monopetalous, but those which actually have a single petal to the corolla) are indeed to be regarded for the most part as an odd leaflet; yet still much variety appears to take place in these developmental arrests. Thus here no division of the fibrous bundle was attained, or the lateral leaflets have wholly disappeared.
1239. In many this one leaflet also is arrested, and the corolla is wholly wanting. Such a corolla is to be viewed as a stem with radical, but without ramular leaves.
1240. It is not a matter of indifference whether the single involucre that is left remaining be called calyx or corolla; the distinction between both is philosophically correct, though at the same time also it may be frequently difficult to determine. Colour and relation to the stamina and fruit determine much; but respect must be also paid to the whole idea of the plant, whether it has radical leaves or not, whether the leaf-ribs do or do not ramify. Alternating stamina afford evidence of its being the corolla.
COLORATION.
1241. As the colour ranges parallel with the import or quality of the matter, or since the matter and colour are of one and the same kind, so also must this hold good of the colour of the light-flower.
1242. As the corolla only, and not the calyx, is the proper light-organ, so also will it only obey the light in the coloration.
1243. The corolla can no longer be coloured green, for it is no longer a leaf. Now that which

obtains another signification, which passes over into another element, must, with the function, lay aside also the old colour. The corolla is besides the perishing, fading leaf; as this begins to turn yellow or red in the autumn, so does the corolla immediately at its origin. It is a born autumnal leaf.

1244. The whole plant must be regarded as a green synthetic colour; the corolla as the analysis of the Green.
1245. The first division of the Green is Yellow and Blue. These two colours are the first which make their appearance in the corolla.
1246. Yellow is the earth-colour, corresponds to the root, and consequently indicates the lowest colour. Yellow corollæ are less developed than those which are otherwise coloured. The corollæ of spring flowers are therefore yellow; so likewise is the middle of the corolla, as is specially exemplified in the disk of the Syngenesious plants.
1247. Blue is the second colour of corollæ in the scale of dignity, or rank. Blue is displayed on the better developed corolla, and is frequently the colour in the rays of the Syngenesiæ. Blue belongs to the temperate zones.
1248. If Yellow and Blue be the divided Green of the leaves, the complementary colour to that of the corollæ must thus remain in the trunk. The trunks of plants having blue corollæ should therefore be yellow; those with yellow corollæ should furnish blue colouring matters, like the Woad.
1249. Red is the third corolla-colour, the true light-colour, in the which properly all corollæ have been immersed; and if they exhibit any other colours, these should be viewed only as instances of aberration from Red. The Reds are the splendid colours, which develop themselves in the middle of summer; in flaming red mantles are the flowers of the torrid zone veiled.
1250. Finally, the form triumphs over the colour. The light has in Red done everything which it could do for colour, having allured as it were all the colours out of the plant; it now, on the contrary, bestows its attention upon the form and delicacy of the substance. The white colour makes its appearance in antithesis to the red, and is mostly associated with very delicate structure.

1251. The cells of the red corollæ are replete with starch-granules, but those of the white are quite empty. The yellow and blue corollæ range in the middle. Red is a superabundance, White a deficiency, of nutriment. The most noble and beautiful corollæ, as well as the lowest also, may therefore be white. White and red are the general colours for all families of plants, but yellow and blue are special colours. In general the trunk is green, the corolla white, the seed black. The mediate stages are red, yellow, and blue.
STAMEN-FILAMENTS.
1252. At length we come to the last work achieved by the light in the corolla, or to complete separation of the systems or tissues. If ever bundles of fibres may entirely separate from the cellular substance, this is possible only in the corolla, as being the final light-organ. Separation must, however, be attained: for thus far do the claims of light extend. But no development remains stationary before it has corresponded with the operations of the developing agent.
1253. In the corolla, as being the highest kind of leaf, the ribs, as the fibrous fascicles, must finally separate themselves from the leaf-substance as cellular tissue. The corolla is a double organ.
1254. In conformity with the whole structure of the plant, the ribs are placed internally, the

membranes externally.

1255. The leaf-ribs, isolated and perfected as a particular organ, are the Stamen-filaments.

1256. The leaf-membranes, or probably the phyllodia, isolated and evolved into a particular organ, are the corolla-petals. These compose in the strongest sense the corolla. The filaments consist for the greatest part of spiral fibres, and the corolla-petals of the finest cellular tissue, which may be almost designated as granular. This then would be the rational import of the corolla and its stamen-filaments. Both are of similar production; they exhibit like substance, colour, and delicacy, with cotemporaneous development and cotemporaneous death.

1257. Not only are the ribs of the corolla, but those also of the calyx, liberated to become filaments. There are calyx-and corolla-filaments.

1258. As ribs, the filaments must stand in the middle of their petals, i. e. opposite to them.

1259. Filaments, which alternate with the parts of the corolla, are consequently calyx-filaments; such as alternate with the lobes of the calyx, or stand opposite to the petals of the corolla, are corolla-filaments.

1260. Most filaments, and consequently the calycine, are alternating in their arrangement. Most corollæ therefore have no longer strength sufficient to produce filaments.

1261. Flowers provided with opposite and alternating filaments have consequently two circles of them, as is the case in many Pinks. With the determination of the number of filaments the race has been therefore specified.

1262. The number of filaments stands in relation to the parts of the corolla; therefore three and five are the prevailing numbers.

1263. There is no absolute number in the filaments, but only one of relation. Corollæ with three petals have invariably also three filaments, and those with five of the former, the same number too of the latter.

1264. The number of filaments is always the simple or multiple of the parts of the corolla. Three calyx-or corolla-parts have 3×1 or $3 \times n$, filaments. 6 is not 6, but 3×2; 9 is 3×3; 10 is 5×2; 20 is 5×4; or $5 \times 3 + 5 \times 1$, and so on.

1265. The filaments do not simply follow the number, but also the conjunction, position and arrest, of the corolla. They are epigynous, peri-or hypogynous.

1266. In irregular corollæ the filaments are usually abortive; as in those of the orchideous, labiate and papilionaceous plants.

1267. The arrest of the filaments usually stands in inverse relation with the corolla. In the case of a larger-sized petal the filament is small, and, on the contrary, larger in opposite parts of the corolla.

ANTHERS.

1268. The corolla obtains its last function in the production of the highest electrical bodies, which it exhales as sweet odours. Ætherial oils ascend out of the corolla into the air.

1269. The filament, as a leaf-rib that has become free, is a moribund ramuscular extremity, which still strives, according to the law of pinnation, to produce three buds, whereof, however, the terminal one is in general arrested, and the two lateral scarce attain unto apertion.

1270. The two lateral buds of the filaments are the Anthers. They mostly open in a spathose manner, because they have not strength enough to develop themselves as perfect gemmæ or buds.

1271. The anthers are to be regarded as follicles, which mostly rupture upon the dorsal or external aspect.

1272. The starch-flour, which forms the precipitate termed albumen in the seed, here obtains in the light-organ electrical properties, and is called pollen.

1273. The pollen has a light-function in the plant.

1274. The function of the pollen must be differencing, thus vivifying and secernent.

1275. The principal antagonism of the pollen is with the pistil, upon which it must therefore act in a properly differencing manner. The pollen does not hang, like the seed, by a stalk or pedicle to the wall of the anther, but is exudated from it like chemical bodies. It is nevertheless a vesicle, like all organic parts. This vesicle consists of two membranes, and contains yet smaller vesicles, which are called fovilla or pollen-viscus. When the pollen comes into contact with the moist surface of the stigma, its external membrane ruptures, and the internal with its contained fovilla protruding in the form of a tube, penetrates the style in many instances down to the seed, whereby the germ first becomes developed or self-subsistent.

2. PISTIL OR OVARY.

1276. If the corolla be the light-flower, the stalk-flower is heat-flower.

1277. The stalk-flower, as being a repetition of the stem, must be developed later than the leaf-flower. It consequently stands superiorly upon, and so far within, this. The corolla is related to the stalk-flower like circumference to centre.

1278. The stalk repeated in the flower is the Ovarium, germen or pistillum. It is frequently converted into wood in the nut and becomes hardened into stone.

1279. Nevertheless, the pistil like the corolla is a leaf-formation, because everything that originates subsequently to, can be none other than, the leaf. It is a leaf-bud under the idea of the stalk. The pistil is thus a leaf-whorl like the corolla, and one which is subject to the same fatalities, only with this difference, that its leaves are wont to open first after it has withered, and consequently through physical forces.

1280. Every leaf that has been closed in a vesicular or tubular form is a follicle or carpel. There are therefore uni-, bi-, and tri-locular ovaria, &c. The loculi or cells of the ovarium are none other than closed carpels. As many therefore as there are of the former, so many are there of the latter, and vice versâ. The septa or partition-walls are none other than the involuted edges of the closed and confluent carpels.

1281. Uni-locular ovaria consist therefore only of one leaf. The legume is only a compressed carpel.

1282. Every carpel or every cell has its raphé or suture directed inwards or along the axis of the flower. For the leaves are always so conjoined that the two halves of the upper or inner side stand counter to each other.

1283. All other sutures are adventitious and, by their mode of dehiscence, determine those parts of the ovarium which are called valves. These sutures are either dorsal or on the back of the carpel, e. g. capsula loculicida; conjunctive, where two carpels abut against each other, e. g. capsula septicida; or finally, between both by the side of the dorsal suture, so that the valve springing up resembles a shutter, as in many siliquæ or pods.

1284. The columella of the ovary is none other than the internal edge of the carpels from which the leaf-wall has been freed.

1285. Each carpel-leaf is to be regarded as the common petiole of a pinnate leaf, upon whose lateral petioles the seeds depend. The seeds always hang therefore upon the inner angle of the cells.

1286. As the parts of the corolla alternate with the calyx, so do the carpels or cells of the ovary with the corolla; they stand therefore opposite the parts of the calyx or are situated in front of them.

1287. The parts of the ovary follow also the uneven series of numbers, one, three, five. The number two is usually found in irregular, e. g. the labiate, corollæ.

1288. If few cells be present as parts of the flower, the carpels are then to be regarded as arrested. In the personate corollæ three are arrested, but in the papilionaceous, four. The legumen is only a fifth part of the ovarium.

1289. The development of the carpels stands usually in an inverse relation to the size of the parts of the corolla. Thus the legume is situated between the two insignificant petals of the keel, opposite to the large vexillum; in the Personatæ a carpel is situated in the fissure of the upper lip; upon the lower lip, consisting of three lobes, only one carpel is situated, which consequently supplies the place of four, and is therefore also larger.

1290. The stages of leaf-formation are also displayed in the matured state of the ovarium. The radical or scale-leaf is repeated in the cariopsis; as in the grasses, orachs, nettles, and such like plants.

1291. The spathe-leaf becomes a siliqua, or hollow capsule, in which forsooth the septa are arrested, and the seeds stand upon the conjoined edges of the carpels, upon their walls, or also upon a mediate replum and placenta, as in the proper Siliquose plants with the poppies, Resedæ, Primulæ and pinks.

1292. The reticular-leaf is perfected into a capsule, where the carpels are so confluent with each other, that they form septa and bear the seeds upon the internal angle or upon the axis, as in the Rues. If these carpels separate, then polycarpal plants, as the Ranunculaceæ, Malvaceæ, Magnoliaceæ, originate. In these the median columella is the elongated floral peduncle. If they separate, without leaving a median columella between them; they are then simply called follicles, as in the larkspur and celandine. If this capsule be compressed flat; it is called legumen, as in the beans.

1293. The cariopsis generally contains one and that a large seed; the siliqua or hollow capsule is many-or small-seeded; the follicle few-or moderately-seeded; the capsule many-and also few-seeded.

1294. In the cariopsis the seed is attached to the base or apex; in the follicle in a row upon the inner suture; in the siliqua upon the carpellar edges, or their walls, and on a meso-replum or frame; in the capsule upon the inner angle or on a middle axis or column. Seed-bearing alæ to the columella are only the carpellar edges prolonged into the loculi or cells.

STYLE.

1295. What the stamen-filament is for the petal of the corolla, that is the style for the ovarial petal or the carpel, viz. the rib that has become free. As, however, the leaf-formation in the ovary is generally imperfect, so also is the separation of the tissues or systems. The style is not therefore freed at once from its root, but projects only above the leaf-substance.

1296. But as the singular circumstance occurs in the ovary that the midrib is arrested while only the marginal ribs shoot out, so is the style the elongation and coalescence of the two marginal ribs. Every stigma is therefore biacuminate.

1297. There must always be as many styles as the ovary has carpels or cells. If only one style appear, in this case it is then made up of several mid-ribs. In most instances the number of styles is recognized in the number of the stigmata.

1298. As being the rib of the ovary the style is the last ramuscular extremity of the stalk, which is resolved into mucus upon the stigma.

1299. Stamen-filament is related to style, as leaf to stalk, thus as air to earth, as Differencing to

Differencizable, as electrism to chemism or rather nutrition. This is the lower comparison; in a true sense they are related as light to heat.

1300. The light is the Active, the heat the Passive; light the Moving, heat the Moveable; light the Vitalizing, heat the Inactive, or that which becomes vitalized; light the spirit, heat the matter—male and female principle. Thus are corolla and pistil related to each other.

3. SEED.

1301. The root repeats itself in the interior of the ovary under the æther-form. The root ascends out of the earth to become an organ of gravity.

1302. After the leaves have been made independent in the corolla, and the stalk in the pistil, the root also separates and appears as a free organ, as Seed.

1303. The seeds are necessarily in the interior of the ovary; for the cellular organ can first appear, after the leaf-and stalk-buds have opened as corolla and ovarium. The blossom is a bulb, the external testa or covering of which is the leaf-, the mediate the stalk-, and lastly, the internal the root-vesicle. The stalk is adherent to the leaves, the root to the stalk; so also the seeds to the ovary, and this to the corolla.

1304. The seeds are developed in the ovary under the same relations in which the root is developed in the earth, namely, in the dark.

1305. The darkness does not allow the chemical body to attain unto difference; therefore the sap within the capsule must, instead of separating itself into spiral vessels and leaf-substance, continue to remain there undivided and without form, i. e. in the condition of simple granules, or germs of future cells.

1306. The seeds, like the root, are a mass of cells; like it they contain an accumulation of mucus, but of course more highly-formed, being separated into flour, starch, acid matter, oil, and such like substances.

1307. These seminal substances are deposited upon the alkaline, in opposition to the acid, side of the ovary; just as the root also represents the alkaline factor in reference to the stalk, in which appears the formation of acids.

1308. The seeds are the pinnate leaflets of the ovarian leaves, which continue in the condition of buds. They stand therefore as unclosed vesicles upon both edges of the carpel, as is particularly distinct in the legumens.

1309. As both edges are similar to each other, so there can be no ovary that has fewer than two seeds. In all one-seeded ovaries therefore one seed has been arrested, a fact that admits too of being demonstrated in the majority of cases.

1310. Every seed is placed at the extremity of a lateral rib of the carpel. These lateral ribs are called seed-bearers or placentæ. If such lateral ribs terminate before reaching the edge of the carpel, then the seeds stand upon the juncture or wall of the carpel. This does not, however, occur frequently, but only in the Siliquosæ, poppies, and some others. The elongated lateral rib, whereupon the seed hangs, is called umbilical cord. It is no peculiar organ, but only the seed-petiole.

1311. The direction of the seeds is possible in five ways, either upright and horizontal, transverse, or rising obliquely upwards and downwards in relation to the axis of the ovary.

1312. Every perfect seed (of Dicotyledons) is none other than a pentifoliar, involuted, pinnate leaf. The shell of the seed is the leaf-spathe or phyllodium, the two seed-lobes are the two posterior pinnate leaflets, while the germinal leaflets, or plumula, are the two anterior pinnate leaflets, together with the odd leaflet. The seed-rib or vascular cord (raphe) is continued into the seed-rootlet or radicle, and this into the petiole of the cotyledons.

1313. Every seed-coat must consist of three integuments; for every leaf consists of the lower and upper membrane, and of the interjacent parenchyma, in which the vessels are dispersed. The external leaf-membrane forms the most hard and coloured covering of the seed (testa), the internal the brown seed-tunic or pellicula; between the two lies the brown fibrous tissue, or desiccated parenchyma with the vessels.

1314. The hilum is the basis of the bud or of the seed-leaf; the seed-hole or micropyle is on the apex of the involuted bud, or rather of the phyllodium, in which the germ lies rolled up.

1315. Umbilicus and micropyle are united with each other by means of seed or leaf-rib (raphe). Both rarely stand opposite to each other, so as that the one should be below, the other above; but the apex of the leaf is usually so involuted that it again reaches the bottom of the leaf, whereby umbilicus and micropyle come into close and mutual approximation, as in the beans. The seed-petiole elongates itself into the seed-rib; this is continued upon the back of the phyllodium or testa, bends round, and returns again to the umbilicus, so as to describe a complete circle. The shell of the seed has consequently the form of the young fern or fern-capsule.

1316. The radicle is the continuation of the seed-rib, which is, however, dismembered itself, moves off, and thereby causes the micropyle to be, or, properly speaking, only renders it, free. The testa or seed-shell is consequently a phyllodium thrown over the germ, but the micropyle is the upper opening of the bud.

1317. The germ of the seed or embryo, namely, radicle, cotyledons and plumule, is therefore only the quinary pinnate leaf without the spathe or testa.

1318. Thus the whole seed, shell and embryo, completely resembles a pinnate leaf with a phyllodium, as we see it in the umbellate plants, only it must be thought of as one so involuted, that the fine leaves adhere reversed in the phyllodium. Seeds may consequently change into leaves. A seed has therefore been formed also in all its parts like a papilionaceous corolla. This resemblance speaks moreover retrogressively, for the petals of the papilionaceous corolla being viewed only as a single leaf-bud. Seeds may therefore change also into corollæ. All parts of the seed are thus an unity, a single pinnate leaf, and it is consequently impossible for them to have been patched up out of what has been called the seed-ovum, namely, the testa and embryo, which would come from some other source or out of the pollen.

1319. The seed is the whole plant in miniature: the root being portrayed in the umbilical cord, and radical leaf in the phyllodium; stalk in the radicle; caudal leaves in the seed-lobes; ramuscule in the cotyledonal petiole; ramular leaves in the cotyledons. Seeds may thus change into an entire plant. The seed is consequently nothing new in the plant, but the repetition of the same under the relations and forms of the root.

1320. It is plain that the seeds must always change into the same plant; for they are indeed nothing else but this. The identity ensuing upon propagation is accordingly nothing singular and incomprehensible; it would be so were it otherwise. With the seed the plant has but reverted again to its primary condition, to the galvanic, mucous vesicle, out of which in a secondary manner the young plant is developed like as is the first plant out of the primary vesicle.

1321. The radicle is not therefore root itself, but only emits rootlets.

1322. The germ or the radicle must observe different positions towards the umbilicus, according as the seed-leaf or the testa has been more or less involuted, and according as the germ is removed from the micropyle.

1323. The albumen or perisperm is no particular organ, but only the deposit from the sap, which the inner wall of the testa secretes. The albumen stands in no organic connexion with the parts of the seed. That therefore which has become connate with the kernel or nucleus cannot be albumen.

1324. The arillus can be none other than bud-seeds of the testa, because it is placed under the phyllodium. It corresponds to the floral involucrum or bracteal scales.

1325. As the seeds are none other than leaves that have remained stationary in the condition of root, so must they pass through the three leaf-stages. There can be therefore only three principal differences in the seed-formation.

1326. The seeds of plants with reticular leaves consist of several leaves arranged symmetrically or in pairs. They have necessarily two seed-lobes—Dicotyledones.

1327. The seeds of plants with spathose leaves consist too only of the latter, i. e. the seed-leaves remain encased in each other. They have consequently only one seed-lobe, which also incloses only one plumule—Monocotyledones.

1328. This seed-lobe is a phylloidal leaf, whose parenchyma has been superabundantly filled with farinaceous matter or flour.

1329. That which is named vitellus can be none other than the succeeding counter-leaf.

1330. What in the Monocotyledons, at least in most of them, and in the grasses, is called albumen, is not so, but only the flour of the seed-lobe.

1331. The germination of these seeds is nothing else than an elongating of the spathiform seed-lobe into a culm, from the bottom of which radicles spring forth, as out of a bulb. A monocotyledonous seed is in its structure none other than a small bulb with undivided coverings.

1332. Lastly, the third form of seeds makes its appearance in those plants which have only scale-like leaves. The seed-lobe is wanting in them, and they elongate themselves directly into the plumule or little stalk—Acotyledones.

1333. They are devoid of the distinction of testa and embryo, because, on account of the deficiency of genuine leaf-formation, they are none other than the first. For the embryo is only a small leaf. They therefore include only albumen or germinal powder, as it is correctly named.

1334. Here belong not merely fungi, fuci, lichens and mosses, but also the ferns, as being those which have only cells or squamoid leaves. For the frond with the spiral vessels does not rank in the category of leaf, but of stalk.

FRUIT.

1335. The fruit is the coalescence or blending of the three parts of the flower, or the seed, ovarium and corolla. In the flower the individual perfection of every part of the trunk was completely attained; the leaves being separated quite freely from the stalk became corolla; the stalk separated from the leaves and root became pistil; lastly, the root separated from all, became the seed. In this manner indeed each organ attained its ratio of perfection; the perfection, however, of the Whole does not alone consist in the perfection of the several parts for themselves, but in the union or combination of these individual perfections. The vegetable trunk, as being a Partial, has been represented in the parts of the flower, but as a Whole, in the fruit.

1336. The fruit is therefore the last and most perfect consummation of the plant, or the whole vegetable trunk repeated as unity.

1337. In the fruit not merely the sum of all vegetable forms is united, but also that of all vegetable matters. It is the whole vegetable body repeated mathematically, physically and chemically.

1338. The fruit therefore is also that part of the plant in which all vegetable bodies have been concentrated into flesh. Now, as the highest vegetable bodies pass over into the next, and consequently into the animal kingdom, and are therefore palatable as food, so is the fruit also in essence the sarcocarp. For this is the directly edible part of the plant, that e. g. which does not need being cooked.

1339. The nutritious substance of the fruit can be none other than a highly elevated and analysed kind of mucus, such as starch and gum, sugar and acids. Flour is that which resembles the root, sugar the stalk, acids the foliage; therefore flour is in the seed, sugar in the pistil, and acid in the calyx.

1340. There can be only three kinds of fruit, that constantly accord with the preponderance of the three parts of the flower, as seed-fruit, pistil-and corolla-fruit.

1341. The fruit that has the preponderance of the seed, or where the edible substance resides in the seed, and the ovary itself has become seed-shaped, is the Nut. The nut is the cariopsis that has become sarcocarp; it is therefore one-seeded—farinaceous sarcocarp.

1342. In the ovarian fruit the ovarium has become demi nut-like, half corolla-like or fleshy, as in the Plum. It is the carpel that has become sarcocarp—acid sarcocarp.

1343. The fruit, in which the whole ovarium together with the calyx is edible, is the corolla-fruit, the Berry. Only those perfectly soft fruits are true berries which are inclosed by the calyx, as being forsooth a part of the blossom. The berry is the siliqua or hollow capsule that has become sarcocarp. Therefore it has numerous and small seeds—saccharine sarcocarp.

1344. Finally, these fruits combine to form a common fruit, which represents the proper synthesis of all parts of the blossom, or in which seed, ovarium and flower, along with the calyx, have become sarcocarp. This is the Apple, a syncarpus. The apple is the calyx become sarcocarp, and as it usually incloses several carpels, it is therefore polycarpal and contains few seeds. It consists of seed, and ovarium and calyx, which have become flesh. The apple, as being an unopen calyx-fruit, may probably be regarded as the fruit of the trunk. It furnishes properly drink and food, is the fruit against thirst and hunger—the universal, alimentary sarcocarp. The apple contains all the bodies that have been named, viz. flour, acids and sugar. It is thus chemically also the synthetic fruit, which may be converted into the whole animal flesh, and thus become a true medium of nutrition. The nut is only garbage, the plum and berry, cherry and grape, only drinks or delicacies.

1345. Other vegetable substances, which range lower in chemical development, as mucus, bitter and colouring matters, with resins, are associated for the greatest part in root, stalk and leaves.

FRUIT OF THE FLOWERLESS PLANTS.

1346. The flowerless or asexual plants can have no genuine seed or no embryo. For the genuine seed is the repetition of the blossom under the idea of root. (Ed. 1st, 1810. § 1564.)

1347. That which has been called germinal powder is no seed or germ, but only albumen or perisperm. It has no seed-petiole, has only been exuded out of what has been called the wall of the capsule, and exhibits in its composition no seed-lobes. (Ed. 1st. § 1586.)

1348. What is termed capsule in the Acotyledones is none other than the seed-shell, whereupon it follows of itself, that the so-called seeds can have no umbilical cords or seed-petioles. (Ed. 1st; 1810. § 1573.)

1349. The capsules of ferns are involuted like most dicotyledonous seeds. The ring corresponds to the seed-rib or raphe, the fissure to the seed-aperture or micropyle. The involuted fern-capsule is a repetition of the involuted fern-frond. The little heaps of capsules or sori are consequently not pollen, but a nest of seeds surrounded by an indusium or veil, which, probably, corresponds to the ovary.

1350. The capsule of mosses is an antetype of monocotyledonous seeds; it is a spathe-leaf with the lateral suture; it springs up in a tubular manner similar to grass-leaves, that free themselves from the nodes of the culm.

1351. The hollow columella, which likewise contains germinal powder, is an internal spathe, which corresponds to the germinal leaf of grasses.

1352. The oral teeth are the dissevered parallel strips of vessels in the culm and leaf of the Monocotyledons.

1353. The urn-supporting pedicel is the seed-petiole or umbilical cord.

1354. The calyptra probably corresponds to the arillus and thus to the bud-scales; or possibly to the indusium of the ferns, and thus to the ovary.

1355. The leaf-roses would consequently be the involucral leaves of the moss-stalk; the moss-stalk itself a peduncle or flower-stalk; so that in the upper involucral leaves rudiments might indeed appear of stamina.

1356. In the lichens and fuci the whole trunk is none other than seed-shell.

1357. In the fungi, the antetype of the Acotyledons, it may be almost said that the whole stem is nothing else but albumen, the external layers of which only cling together in a membranaceous manner, and represent a kind of seed-shell. The fungus is an albumen-body, which has coagulated out of vegetable juices. In the fungus, seed, seed-vessel, ovary, blossom, foliage and trunk have become blended into one.

1358. In a perfect blossom the albumen is therefore the repetition of a fungus; the acotyledonous seed that of lichen; the monocotyledonous seed-vessel that of a moss; the dicotyledonous, however, is the repeated fern. It may be also said that the albumen were fungus; the germ, lichen; the seed-vessel, moss; the ovary perhaps, a fern, namely, its indusium.

II.—PHYTO-PHYSIOLOGY.

1359. The life of the plant consists in the co-operation of its functions. The representation of these functions is the vegetable physiology or the theory of vegetation.

1360. Vegetation depends first of all upon the two principal antagonisms of the plant, or those between the tracheal and cellular systems, or between the stem-and rootsystems, sun and planet, air and water with earth, light and matter, electrism and chemism.

1361. The functions divide into those of the æther-organs—blossom, and the planetary organs—stem.

I.—FUNCTIONS OF THE TRUNK.

1362. The functions of the trunk are those of the tissues, systems and members; thus first of the cells, vessels or ducts and tracheæ; secondly, of the bark, liber and wood; lastly, of the root, stalk and foliage.

1. Facts.

1363. The phenomena to be regarded in plants bear relation to their constituent parts, and the changes or preliminary incidents which they undergo.

A. Constituent Parts.

The chemical constituent parts of vegetables are inorganic and organic.

a. INORGANIC BODIES.

Elements.

1364. The plant contains all the primitive bodies; carbon, oxygen, hydrogen and nitrogen. The carbon forms the principal mass, and almost alone constitutes its solid parts. The nitrogen is only present in small quantity, being, as it were, a trace only of the future animal kingdom.

1365. In the plant all the elements are also active, as the æther, which, through the gravitation of the root, strives towards the middle point or centre of the earth. The light, which imparts the general polarity and decomposition, as well as produces colours. The latter appear to reside in

the starch-flour. The heat, that sustains the indifference, promotes the evaporation and the course of the sap, as also protects the plant from being killed by frost or cold. The air that penetrates through the spiral vessels to all parts, and is also met with occasionally in the hollow stalks, the interstices and cells of the pith and cuticle. It imparts the process of oxydation. The water is the proper mother of the plant, being the medium by which the nutrition is imparted. It contains in a state of absorption some hydrogen and nitrogen, a larger amount of oxygen, and abundance of carbonic acid; besides different salts, mucus, sugar, and acids. The earth, as element, bestows upon the plant a firm station, so that the water-and air-organs continue separate from each other.
Minerals or Earths.
1367. The plant is also a totality in reference to the earths. It contains all the mineral classes, and from each of these indeed the principal or fundamental minerals. It can therefore only thrive in a soil which represents the whole mineral kingdom. Among the earths the siliceous earth is very frequently found in plants, and especially in the graminaceous kinds. This, having been dissolved in the earth by potash and the rich supply even of carbonic acid, appears to be absorbed or imbibed by the plant. The argillaceous earth is scarcely met with in the plant itself; but from imbibing and storing up water for the consumption of the plant, it is without doubt its best and most necessary soil. The talcose earth is rarely found contained in plants; it, however, keeps the soil slacker, by dividing into laminæ, and being present for the most part as mica in sand. The calcareous earth is a more essential constituent part of plants, and is found therein in tolerable quantity, usually combined with phosphoric or carbonic acid. Of the salts, all plants contain a fair proportion of common salt and potash, combined too with carbonic acid; soda with saccharic or oxalic acid; probably also ammonia. Of the acids, carbonic acid appears to be alone contained in a free state in vegetable sap; the other elemental and mineral acids are united to alkalies, talcose and calcareous earths. As regards the Inflammables, almost the whole plant consists of carbon, but contains also some sulphur. The metals are represented by iron, which occurs in all plants.
b. ORGANIC VEGETABLE BODIES.
1368. These must be regarded as the repetition of the inorganic bodies. The alcohol, which does not indeed occur ready formed in the plant, but is developed out of the sugar, certainly corresponds to the æther. The ætherial or volatile oils, and the balsams and resins that are thence formed, correspond with the air. The mucus, gelatine, albumen, and sugar correspond to water; the wood, gum, starch and vegetable mould, to the earth. Of the organic salts, plants contain tannin, with azetic, benzoic, mucic, gelatic, saccharic, tartaric, citric, malic, oxalic, tannic, oleic, isatic, and hydrocyanic acids. The alkaline bodies are pungent, bitter, stupefying, and saponaceous; the fixed or greasy oils, the wax and the vegetable butters, are to be regarded as organic Inflammables; the colouring matters, as the organic ores.
1369. These bodies intermixed form the compound vegetable matters. What has been called vegetable sap is for plants, what the blood is for animals. It consists for the greatest part of water and mucus, starch, sugar, acids, and salts. It passes over into vinous and then into acetous fermentation. The starch-granules appear to form in the cells.
1370. To the secreted saps belong the coloured milky juices present in particular vessels, and consisting for the most part of water with resins, as in the celandine and spurge. The particular saps, especially those of the fruits, are very composite, consisting for the most part either of mucus, sugar and acids, or occasionally of gelatine and albumen. Solid compound matters are almost universally made up of flour, which consists principally of starch and gum; or they are furthermore mucus in the roots and seeds. The excreted or separated matters, which no longer interpose in the vegetable process, are the etherial oils, resins, fixed oils, colouring matters,

poisonous substances, gum, tannin, nectar-juices, and even water.

B. Preliminary Events.
a. WROUGHT BY EXTERNAL INFLUENCE.
1371. The influence of the elements produces different phenomena in the plant. I have felt constantly more inclined to consider, that not merely the descent of the root, but even the ascent of the stalk was simply to be viewed as a mechanical event, or one forsooth effected by gravity. The roots obey under all circumstances the gravity and would grow as far as the centre of the earth, were they to meet with no impediment; and there they would follow the revolution of the earth, and consequently become spirally convoluted upon themselves. It is almost beyond doubt that the water, which sinks downwards and, as it were like that in stalactites, invariably rigidifies or hardens at the radical capillaries, is heavier in the root. The cause of this greater weight depends upon the mucus not being decomposed.
1372. The straight ascent of the stalk also depends upon nothing else than gravity. The upper drops of mucus become lighter by means of greater heat and by decomposition in light and air, and they are therefore compressed by the heavier in the upward direction. It is always such a small drop upon the summit, which hardens into its uppermost cell. The stalk, therefore, grows upwards through the same forces and in the same manner, as the air-bubbles ascend in a glass of beer. The cause of their becoming lighter resides certainly in the vital process, which nevertheless effects in this respect nothing else than the extension or increase; but yet the cause of the ascent is naught else than the gravity.
1373. The light likewise acts upon the direction of plants and especially that of their leaves; not simply from its promoting growth by elevation of temperature and by decomposition, but obviously in a mechanical manner also; for not only do the branches of plants in a green-house grow towards the window, but most leaves turn themselves the whole day in obedience to the course of the sun. This turning must, nevertheless, have one and the same kind of cause with the growing towards light; it also is only a conatus or effort unto growth. The upper leaf-cells, being illuminated by the sun, become lighter, and are therefore directed at once, like the apices of the branches, toward the influence of light. The cells that stand perpendicularly upon the surface of the leaf are to be regarded as branches conjoined by growth.
1374. The sleep of plants depends also upon the same influence of the light. The upper leaf-cells sleep during the night, while the lower cells, especially those of the petiole, fill and consequently bend the latter upwards. The sleep of the flower must have the same cause. As likewise the alternating motion of many leaves, as in the Mimosæ.
1375. The motion also performed by the staminal filaments towards the pistil must finally depend upon this unequal replenishment of the external and internal cells.
1376. The coloration of the parts of plants is a result of the decomposition of the starch-granules in the cells by the agency of light.
1377. The operation of the heat is more intelligible than any other. That which is to move and separate itself, must have a certain degree of extension, or must be fluid, namely, aquiform. In a cold temperature the upper saps, not becoming warmer than the lower, are consequently not lighter, and on that account also do not ascend upwards. The mortal freezing of trees descends from above downwards. In other respects plants have, like animals, a self-inherent, though very feeble, process of heat. Germination proves this, in cases where many seeds lie upon each other.
1378. The air acts also mechanically and physically upon plants, by causing motion of the solid parts and by promoting evaporation. Electricity is without doubt active in the spring of the year,

and evokes the antagonism between the fabric of the stem and root.

1379. The physical operation of water consists indeed, for the greatest part, in its preserving the solid parts in a supple or pliant state. Its principal office is, however, to convey nourishment to the plant.

1380. The earths act beneficially only upon plants, if they have been all mixed with each other. Mineral salts occurring in moderate quantity in the soil promote growth; alkalies and acids are injurious thereunto. The same holds good of Inflammables and metallic limes.

b. BY INTERNAL ACTIVITY.

1381. That the plant imbibes water, and this indeed in great quantity, by its whole surface is a well-ascertained fact; but it has been by no means equally determined whether it obtains its nourishment simply through the water, or directly also from the air, e. g. the carbon, as well as the nitrogen, from the carbonic acid. The principal imbibition, however, takes place through the root; but experiments that have been made upon this subject leave it doubtful, whether in this case it is simply mucus, extract from the humus or vegetable mould, or simply carbonic acid that has been absorbed.

1382. It is moreover a fact that the green parts of plants exposed to direct sun-light consume or take in carbonic acid, and develop or give out oxygen; on the contrary, during the night, and even in cloudy or gloomy days, they absorb oxygen and exhale or develop carbonic acid. Now, as there are far more gloomy or at least cloudy than clear days, it thus becomes evident that far more oxygen has been taken up, than separated, from the air. During germination oxygen gas is consumed and, on the other hand, carbonic acid developed.

1383. The saps ascend upwards, and chiefly indeed, in the liber; on its passage different substances forming from it, which appear especially in the fruit in greatest proportion and variety.

2. Processes.

1384. The tissues of plants form three formations, which must be similar in their functions, and can only exhibit subordinate differences. The cell-formation is displayed in the cellular tissue, in the bark and root. The vascular formation in the vascular tissue, in the liber and stalk. The tracheal formation in the tracheal tissue, the wood and the leaves. There can accordingly be only three principal functions in the vegetable stem, and of these each will display minor differences.

A. Cellular Processes.

a. ROOT-PROCESS—ABSORPTION.

1385. As the root is the cellular organ proper, so in it principally resides the water-process or the commencement of chemical elaboration and analysis. Now the chemism in an organic body is called digestion.

1386. The root is the mouth or pharynx of the plant, and is therefore principally concerned with absorption. Its process is therefore the formation of mucus, or as it were of salivation. The root cannot, however, create mucus, as it was created at the conclusion of the earth-metamorphosis in the sea; it can absorb it or in the highest degree compound it out of the constituent parts.

1387. The process of the formation of mucus is a process of putrefaction; the function of the root consists accordingly in supporting a constant process of putrefaction. The soil in which the root stands must contain substances susceptible of, and the conditions necessary to, putrefaction. These substances are organic matters and water; the conditions heat and access of air. Such a soil is called humus or mould. In a pure, dry earth, no root can thrive.

1388. Carbon, from its being the earthy body, is the principal one in the formation of mucus, and the basis also of the vegetable bodies. A root can develop itself, if it stand only in a soil such as the calcareous, which contains carbon and water. The calcareous soil is as it were an original mould. It is probable that the calcareous earth is constantly decomposed by the root and its carbon absorbed. The calcareous earth is again neutralized by the carbonic acid of the water and air.

1389. There can be no doubt that the root also abstracts carbon from these elements, and converts it into mucus, or probably separates it from carbonic acid. The mucus approximates the animal nature, so that the root in its constituent parts, in its smell, and even in its structure, exhibits animal properties; animal substances therefore are also the best nutritive media of plants.

1390. That which putrefies most easily is the best manure.

1391. Through the process of putrefaction many kinds of antagonisms and attractions, by which the absorption takes place through the root-filaments, are aroused.

1392. The root has not merely one orifice for absorbing, but it imbibes upon the whole surface, from its being still immersed in the chemical menstruum. The integument of animals does the same.

b. BARK-PROCESS—EVAPORATION.

1393. The bark, as an organ of cellular tissue, which is placed wholly in the outward direction, must principally exercise the process of absorption and evaporation. Now, as there are two kinds of bark, a root-and a stalk-bark, or a water-and air-bark, so upon the former will the business of absorption chiefly devolve, on the latter that of evaporation.

1394. As the bark of the stalk possesses stomata, which are wanting in that of the root, so is this a probable reason for these apertures being organs of evaporation. This opinion is corroborated also by aquatic leaves being without stomata, while they occur in the leaves exposed to air.

1395. Meanwhile the stalk is of a twofold character; it is only the root that has ascended into the air. As an aerial root it absorbs. Without doubt the stalk absorbs the same as the root, namely, moisture from the air and carbonic acid. Experiments prove it.

c. CELL-PROCESS—DIGESTION.

1396. The cells are the crystallized drops of mucus, the fundamental mass of the vegetable and consequently the water, which converts itself into the Earthy, or wherein the Solid has been elaborated and precipitated. They construct the Solid that has been absorbed into new cells. But the Solid can only assume other forms by means of water. The solution, however, with mixture of bodies and formation into globules is digestion. The cells are thus the stomachs of which the plant has millions like mouths.

1397. The bodies absorbed must move in the cells; for chemical solution and mixture, being itself nothing else than separation and union of atoms, is consequently motion. In a single cell the motion must be upon all sides, because the atoms are attracted and repelled from all points of the cell-wall. In cells, however, which are united with others and therefore subjected to longitudinal polarity, this motion must be performed in accordance with the axis of the cells.

1398. This motion proceeds to and fro, because the extremities of the cells have different polarities, and therefore repel the same atoms, which they have before attracted. In the cells the mucus appears to be converted into starch-granules.

B. Vascular Processes.

a. Vessel-process—Conveyance of Sap.

1399. The vessels or intercellular passages conduct the sap, or the water of the plant. Their function is therefore the continued conveyance of the sap that has been absorbed from the root, and rendered solid or consistent by the evaporation going on in the bark and elaborated by the cells.

1400. The vessels of plants are notwithstanding to be compared with the lymphatic vessels of animals, in so far also as these are distributed throughout the whole body, and convey the sap simply in one direction not in a circle.

1401. As the passages between all the cells are in all directions, so the vegetable saps or fluids flow in all directions, and not to one centre as in the animal. Plants have no heart. The sap pursues a tolerably rapid course in the vessels. A fading or drooping cabbage, two feet in length, can gradually become erect in a few minutes after being put to soak in water. In other respects the course of the sap in the vessels may be seen in many plants under the microscope.

b. Liber-process—Mixture of Sap.

1402. In the liber, as being the mass of intercellular passages, the sap contained in the vessels principally accumulates, as in the thoracic duct of animals; in it the matters have not been simply conveyed and dissolved, but also mixed and converted into true vegetable sap, into blood.

1403. The tubes of the liber are those by which the chemical life is sustained.

c. Stalk-process—Secretion.

1404. The stalk is the root planted in air, and consequently its process is the differenced process of putrefaction, in which the mucus becomes further evolved.

1405. The analysis chiefly occurs in the stalk; the mucus, or rather the starch, becoming converted into sugar and acids.

1406. Sugar is the mucus of the stalk, and is found in every vegetable sap, especially that of such plants as are characterized by the systems of the stalk, and have not yet attained the formation of the reticular leaf, as the Monocotyledones, e. g. the grasses.

1407. The sugar originates from a process of fermentation; the process of the stalk must consequently be regarded as a vital process of fermentation.

1408. The process of fermentation is that of putrefaction carried on in the air, or the polar process of fermentation. Both processes consequently observe a polar relation towards each other.

1409. The sugar-process passes over finally into acidification.

1410. The Inflammables, as the ætherial oils, balsams, and resins, are formed in the antagonism of the sugar or of the acids. Here also belong most of the peculiar vegetable matters, as the milky saps, colouring matters, medicinally active bodies, poisons, and the alkaloids.

C. Tracheal-processes.

a. Leaf-process—Inspiration.

1411. In the foliage the woody rings have issued freely into the air, in order that they may offer their whole surface to its influence, and thus become electrified and oxydized.

1412. The leaf is the free, external organ of respiration to the plant; it is its lung. Through the leaf the air, and chiefly its oxygen, is transferred into the plant, just as it is through the lungs into the animal.

1413. The leaves take in oxygen gas; this is their essential function, and not that of exhaling it.

1414. The leaves only exhale oxygen gas when exposed to light. The development of oxygen in the plant is accordingly a light-and not an air-process. In consequence of this the leaves give out oxygen gas only during the day, but during the night and even upon gloomy days, where not the light but only the air is active, they take in oxygen and give out carbonic acid.

1415. The light develops the oxygen gas out of the plant in a perfectly inorganic manner, like as from every water, that can be set in a process of tension. Rumford has developed by simple glass tubes oxygen gas out of water. The oxygen gas of plants is therefore a result only of the decomposition of water in an inorganic manner by the agency of light, or virtually the separation only of the oxygen that is clinging to the water.

1416. Through the process of respiration in the plant carbonic acid has been formed and excreted. For the mucus becomes oxydized, and thereby the process also of fermentation, the product of which is carbonic acid, is promoted.

1417. The respiratory process of the leaves is the perfected process of fermentation in the stalk, in which finally, namely, in the fruit-saps, the separation of both the products of fermentation, the vinous and acetic, is prepared.

1418. Just as acids and sugar originate in the stalk, so in the foliage does their electrical antagonism, or the ætherial oils and perfumes. Sweet scents or perfumes are properties of the air, and therefore originate also with the aerial process. This is retrospectively a proof that the leaf-process is the respiratory process.

1419. Through the leaves, with which the whole surface of the earth is covered, the planet respires, and thereby the surface of the earth principally obtains its electricity.

1420. Vegetation must therefore effect an important change in the earth's electricity. The earth must be differently polarized after, to what it was before, the fall of the leaf.

1421. Thereby the northern hemisphere is differently polarized to the southern, because the latter has less soil than the former.

b. Wood-process—Nutrition.

1422. As most of the spiral vessels are collected together in the body of wood, and finally in the leaves issue forth quite free and naked into the air, so must the wood conduct for the most part air into the plant. The polarization of the other systems, of the liber and the bark, must therefore proceed from the body of wood.

1423. The greatest amount of induration must originate in the body of the spiral vessel, because in it the process of oxydation takes place in the most active manner. From the same cause the process of nutrition must also be supported by it in the most powerful manner. The wood is the chief seat of nutrition.

c. Tracheal-process—Oxydation.

1424. The structure of the spiral vessels, their resemblance to the tracheæ or air-tubes of insects, their distribution throughout the whole trunk, the air they contain which is found decidedly free in the plant, leave no doubt that the tracheæ are air-conveying organs, and consequently have, like the arteries in animals, the process of respiration directly intrusted to them.

1425. Now through the process of respiration the general polarity, and consequently the cause or fundamental principle of all life, enters the plant.

1426. The tracheæ penetrate or traverse the whole plant from the apex of the root to that of the flower. Their operation must therefore also extend through the whole plant.

1427. The tracheal system must also govern the plant by polarity, and thus in an immaterial manner.

1428. This polarity acts simply in the direction of the plant's longitude, not transversely, like the material fundamental processes.

1429. The tracheæ impart in a spiritual manner the antagonism between the root and fabric of the stem.

1430. As the tracheæ are, or constitute, the highest system of the plant, so must it be them upon which the light principally acts. The material processes of plants are kept in activity by the antagonism of light.

1431. By this only are the instantaneous changes, which follow upon the influence of light or section of the spiral vessels, to be explained. Upon this, therefore, depends the instantaneous elevation of the processes under the influence of a ray of light, and their depression, if only a cloud pass in front of and obscure the sun; hence too does the plant die, so to speak, upon the very spot, if the spiral fibres within the liber be cut through, but the latter left uninjured.

1432. The liber no longer conveys any sap to the divided tracheæ, solely because it has lost the condition to be affected by the light-polarity. On the contrary, a plant does not die so soon, if the liber be cut through, but the spiral vessels preserved. The spiral fibres conditionate consequently the motion and the excitation of the organic processes.

1433. The spiral fibres are therefore, apart even from their function of respiration, or rather because this is the highest vegetable function, that for the plant, which the nerves are for the animal.

1434. The tracheæ of plants do not ramify like the nerves of animals; but if they divide, they separate only as fascicles, which have been liberated from their origin. The tracheæ commence too directly in the mass of cells, wherever that may happen to be, and thus become what governs an organ, exactly like the animal nerves. Their analogy is greatest with the sympathetic nerves. The tracheæ, just as in the animal kingdom, are the mediators, not the founders, of vegetable life.

1435. The principle of motion must reside in the tracheæ, provided that higher, and not merely chemical movements, occur in the plant.

1436. These movements must and can only exist in those organs which consist almost entirely of spiral vessels, and thus only in the highest organs.

1437. Such are the leaves and the corollæ. Is it wished to compare the corolla, apart from its sexual relation, with an organ in animals, it can only be contrasted with the highest nervous organ. The corolla is the brain of plants, that which corresponds to light, but which here remains stationary upon the sexual stage. It may be said that what is sex in the plant, becomes brain in the animal, or the brain is only the animal sex.

1438. The most general function of the brain is, however, feeling or touch combined with motion. If the corolla could attain to a sensorial function, it would be to that of touch.

1439. It is conducted thereto; but at the instant, when it is indulged in feeling the mental capacity of the animal, it sinks down exhausted and dies. It is punished for the risk that has been run in wishing to attain unto self-cognition.

1440. Motion and touch are revealed only in the highest organs of the plant, or in the stamina. The filament moves upon the pistil and touches it with the pollen, which at that instant, however, is scattered in small particles, and leaves behind the filament in a withered state.

1441. The motion performed by the filaments appears to be a simple operation of the irritability in the tracheæ that have become soft, without undergoing chemical decomposition, but probably by sudden influx of sap, induced by the tension of air in the spiral vessels.

1442. In the highest, or the pinnate, leaves, movements, which are probably a result of the tracheal irritability, also occur, but are devoid the object of coming into contact with, or of touching, anything. The sensitive plants, as Hedysarum gyrans, move their leaves, not from any intrinsic determination upon their part, but in accordance with an antecedent stimulus, and thus not voluntarily, but probably through the influence merely of polar tension. The movements of leaves are convulsions of plants, although too an afflux of sap be caused or induced by the stimulus.

SAP-MOTION.
Galvanic Process.
1443. The motion of the sap is imparted through the antagonism of the respiratory and digestive processes. For these two processes are the combination of the Chemical with the Electric, which is the galvanism.
1444. The galvanic poles attract and repel the fluidity; thus the vegetable sap is attracted by the root and by the stalk. But the differencing or the oxygen pole is the stronger of the two. The determining principle of the movement of the sap resides consequently in the stalk, and the chief direction of the sap-motion tends upwards.
1445. At times, when the air-polarity is elevated, the sap also ascends more rapidly. As in summer, upon clear warm days. It ascends slowly upon gloomy and chill days. That in this also light and heat are playing their parts, is self-intelligible. Thereby the upper particles of sap become lighter and ascend, being pressed upwards by the lower and colder particles. As they are nevertheless by no means changed, this is a proof that, during the time so employed, polar forces also act upon them.
1446. But the root has also the endeavour to attract the sap; but as its pole is feebler in character, the stalk draws the sap from the ultimate extremities of the root into itself. If accordingly the polarity of the air becomes weaker, while the plant is losing its leaves or the organs of polarization; so is it easy to imagine why the motion of the sap becomes slower. As, however, the aerial polarity is always stronger than that of the earth, the sap must thus in winter also take the same, or upward, direction.
1447. A fall or descent of the sap can therefore never take place abstractedly forsooth from the root, in which it sinks by its own gravity. How a part of a plant, e. g. a twig, could continue alive, were the sap to have fallen or receded from it, is not to be conceived. It does not follow from what has been just stated, that movements of sap should not take place in all directions, and consequently too downwards; they must indeed occur rather than otherwise, and that indeed upon all sides; only the principal track or course of the sap must always pass in the direction upwards.
1448. The movement of the sap consists simply in an ascent and impulsion of its particles upon all sides, but without any circulation. A circulation would only be possible if the plant were an organism disengaged from the elements; but as the earth and air belong to its organization; it thus necessarily oscillates between both, and its movements also can only be oscillations of a similar intervening character.
1449. There are consequently no arteries and veins, and still less a heart in plants, as some have striven to make out.
1450. The vessels of plants are most properly to be compared with the lymphatic vessels of animals, the fluid or sap of which also tends from all parts towards one summit, namely, the lungs, while still at times retrogressive movements also seem to occur.

1451. The vegetable sap does not move in a straight line upwards, but in all directions, to the right and left, in a zigzag manner, and so on. This is proved by two incisions being made in a branch opposite to each other. The motion of the sap in the plant is more an impulsion of the sap toward all sides, with a predominance in the direction upwards, than a rapid current as in the blood. If we reflect that the motion of the sap when seen under the microscope and then magnified several hundred times, still only resembles a gentle rippling of small drops, it thus becomes clear that the true current or flow occurs only in a very tardy manner. Wherever therefore in the plant the process of differencialization may be brought into play, there the sap is impelled.

1452. Through the polarization of the sap the cells also become polar towards each other, and then even the cell-walls, whereby the cellular sap with its mucous granules is kept in constant motion. The theory of the motion of sap has not consequently been based upon the theory of capillary tubules; nor is heat alone the cause of its ascent; nor the empty space, which originates superiorly by evaporation; nor electricity in an inorganic sense.

II. FUNCTIONS OF THE FLORAL-ORGANS.

1453. These functions correspond to those of the light, heat, and gravity in the corolla, pistil and seed. The corolla irradiates, the pistil gives out heat, the seed sinks like the earth towards the centre.

1. Function of the Corolla.

FERTILIZATION.

1454. As in the vegetable trunk the principle function has been the antagonism between the aerial and the terrestriaqueous plant, so must the same function be repeated in the corresponding organs of the blossom. It oscillates in the principal antagonism between the corolla and pistil, which is the antagonism of leaf and stem, that of electrism and chemism, of light and body, of spirit and matter.

1455. The pollen electrifies, animates, or inspirits the ovarium, by which means it becomes stimulated to the development of seeds. Without this animating influence the seed had not been developed.

1456. This relation, whereby through the balance of an antagonism, a whole organism has been evoked into life, is called the sexual relation.

1457. The sex is consequently the antagonism between spirit and matter, light and mass, æther and the terrestrial elements, sun and planet, between electrism and chemism, that has been represented in an organism as totality. In sex consequently the primary antagonism of the world, or that of spirit and matter, centre and periphery, has been organically represented.

1458. The sex has hence from the beginning been established and prophesied; manifests itself also under diverse forms in the Organic, but becomes first individualized in an organic body. This is the lofty sense or signification of the sexual relation, that in it the Spiritual and Material pair together, and thereby sprout forth or germinate into a whole world. In sex the mystery of creation lies concealed.

1459. What is producing the fruit is called the female, that which awakes production, the male.

1460. Masculatity is the spirit of the world, feminality the matter which becomes animated by the former; masculatity is the light of the word, which illuminates the feminality, and it is pregnant; masculatity is the electricity of the world, which arouses the female chemism unto galvanic circulation. By the male the female becomes animated; before this it is dead and devoid the differentialization, which is necessary to every action.

1461. Impregnation is a simple act of light upon the matter, an irradiation, as it was termed, with such an exalted appreciation of its significance, by the ancients. The male imparts nothing in impregnation but the solar ray, or fluid nervous mass, in its semen, which awakes, animates, and inspirits the quiescent female. The female supplies or furnishes all the Material or, as in the plant, the fruit. In other respects it is not to be understood, as if no material whatever had been imparted by the male, but only that it is not the matter as such, which the male gives the female, that becomes fruit; but that the tension which resides in male semen, evokes at the same time, as

by a process of contagion or fermentation, a similar tension in the female.
1462. The process of tension resides originally in the male, because he is related or akin to the light; but the female first obtains the light through the male.
1463. Impregnation is an excitation by the electrical, of the slumbering chemical, process. The pregnancy is consequently an uninterrupted chemical process.
1464. The female is the first and lies deeper in the developmental history of the planet (but not in the creation), just as the digestive process is prior or antecedent to the respiratory.
1465. In the truest sense is feminality co-ordinated with the digestive, the mascularity with the respiratory, system. The female is (organically considered) abdomen, the male, thorax.
Pregnancy is a sexual process of digestion, impregnation a sexual process of respiration. In impregnation the female respires the male, whereby it receives into itself a thoracic function, becomes itself male, i. e. is then capable of producing something out of itself. Now, the female produces a fruit, which is synonymous with both principles.
1466. The semen is the fruit of the male. The male is always pregnant, and that indeed by virtue of his own power. This power is deficient in the female, which does not possess the light in itself, but only the elemental bodies that are ready and susceptible of form.
1467. The anthers are the male organs, the pollen is the semen. The pistils are the female organs, the seed-granules are properly speaking, the germ.
1468. The pollen is a most highly differenced, electrical product; the seed-granule a wholly indifferent, and tranquil mucous mass. The pollen falls upon the stigma of the pistil, and irradiation has taken place; the material fruit-capsule gains thereby so much polarity, that saps enough ascend, in order to develop the germless seed-vesicles.
1469. It is quite unnecessary for the pollen, with its sap or gas, to be materially conveyed through the style to the seed. It is only requisite for the style to be excited, dualized, electrified, and then it has life enough of its own. But it does not follow, because it is unnecessary for the sap of the floral dust or pollen to reach the seed-granules, that it cannot or ought not to reach thereunto. In many plants the pollen-tube does actually reach there and penetrate through the micropyle. In many styles it is still held as impossible, for the pollen-tube to penetrate through them to the seeds. The pollen-sap indeed simply evokes on the apex of the seed (upon the summit of the lorical rib, through whose liberation the micropyle originates) the vital process, which, without this stimulus, would perish. Thereby a new cell is secreted, from whence the germ is developed.
IRRITABILITY—MOTION.
1470. In impregnation the heaven is married to the earth; for then the spirit descends, and does not esteem itself too highly to become flesh. Impregnation is the highest immaterial action of the plant.

1471. If, therefore, the irritability of the plant at any time, or but once only, makes its appearance independently, it must be in the sexual organs, and in the moments of impregnation.
Impregnation ensues, when the two mundane principles of the plant, light and matter, have attained, as corolla and fruit, to the highest pitch of perfection; then the tension of the spiral vessels ranks so high, that they exercise their function independently of what is terrestrial in the plant, move themselves in the male filaments, touch the female organ, and die in this their highest effort.
1472. Thus has it only been conceded to the plant to be, in the instant of impregnation, an animal and enjoy animal passion.
2. Function of the Ovarium.

1473. The ovary, by its own power, is in a condition to draw towards it the chemical saps from out the stem, and as it were by its own heat to thrust new buds from its leaf-ribs, namely, the seed-pellicles or testæ. It has not, however, strength sufficient to put forth also the leaf-work, namely, the embryo, upon the apex of the seed-shell. It requires for this purpose the stimulus of the floral pollen. If the plant is very rich in sap, the ovary is so likewise, and converts itself into fruit or sarcocarp. As a rule, therefore, trees only bear a crop of fruit. If also the impregnation is less perfect, the force of the sap continues to remain inherent in the ovarian leaves; they become rich in sap, fleshy, and likewise fruity in character; trees, therefore, with imperfect or separated blossoms, as the Amentaceæ, Urticaceæ, Euphorbiaceæ, Papilionaceæ, Terebinthaceæ, and Rosaceæ, usually bring forth a crop of fruit.

1474. A stronger degree of refinement appears in these fruit-saps than in the saps of the stem, because corolla and seeds range closer to each other. They are therefore more varied and richer in substance. The fruit-substances range usually upon the side of the water or the salts, while those of the seed range upon the side of the earth or the Inflammables. The substances of the seed are flour and oil, those of the fruit sugar and acids; the former supplies food, the latter drink.

1475. Seed and ovary stand therefore in antagonism, like earth and water.

3. Function of the Seed.

GERMINATION.

1476. The seed is the plant contracted upon its centre, it is the heavy mineral mass, which can only undergo changes by the operation of the other elements, like as it arrived only at completion by the operation of the floral pollen. This acts upon it when within the dry ovary, like the water and oxygen in the dry earth. These changes are its development or germination.

1477. All the planetary elements belong to germination, and to growth the Cosmical also with all its actions. To germination belongs earth, water, and air; to growth, light, heat, and gravity; with all the four mineral classes also, such as earth, salt, Inflammable, and metal. The plant contains silicious and calcareous earth, salts, coal with sulphur, and lastly, iron.

1478. Germination is the disjunctive emergence wrought by means of moisture, heat and oxydation in the processes of decomposition and fermentation. No seed germinates in irrespirable kinds of air.

1479. The cotyledons or seed-lobes are the synthesis of the two processes; they are at once root and leaf, therefore resolvable into mucus, and may yet become green.

1480. In germination the elemental bodies of the root-and stalk-polarity directly emerge; the mucus or the flour separates into alcaline gum, that seeks the darkness, and into acid sugar, which elevates itself into the illuminated air.

GROWTH.

1481. Growth is none other than continued germination. The sap being polarized by the air becomes of necessity decomposed. One part evaporates as carbonic acid and water, the other coagulates into oxydized mucus or into cell-walls.

1482. Growth proceeds directly from the process of digestion and respiration, while its polar organs constantly remove further from each other.

1483. Properly speaking the digestive and respiratory processes are none other than growth, since both separate from each other. That, which originates between them, is the process of nutrition, the vascular system.

1484. Growth oscillates between the process of decomposition and that of fermentation; it is an uninterrupted fermentation.

FALL OF THE LEAF.

1485. If every pole of the plant has been perfected in an isolated manner, it has thus become identical with the air, and the aerial process ceases.
1486. With the cessation of the aerial process, the respiratory organ must also die off or perish.
1487. The decadence, or falling off, of the leaves is the result of the tension having been abrogated between them and the stem; it is a death by suffocation.
1488. The fall of the leaf therefore occurs in the autumn, or after the fruit is matured.

DURATION OF LIFE.

1489. The age of a plant is included between the the limits of the sap's impulse, and that which has been called its fall or descent.
1490. The actual fall of the sap is the death of the plant.
1491. If with the cessation of the influence of light, the polarity ceases entirely in the plant; it is then one year old or an annual. Every part of it dies off.
1492. In biennial plants the aerial polarity indeed disappears, but the polarity of the root remains. Flower, leaf, and stalk die.
1493. Perennial plants, also, do not entirely lose the stem-polarity, but only while they develop a new plant about the old. Flower and leaf only perish, while the water-and earth-organs remain alive.
1494. The old liber dies with every maturation of the fruit, because there the difference attains solution. But a new life develops itself in the parenchyma of the plant, and forms new liber or, properly speaking, a new plant about the old.
1495. Persistent plants consist of numerous plants, which gradually grow round about each other.
1496. In accordance with the idea of the plant, each one perishes with the maturation of the fruit.
1497. On account of the addition of the new plant about the old, the plant has also been confined to no definite magnitude and to no definite number in its mode of ramification.
1498. Indefiniteness in form, size and number, is the character of the plant, although a law lies at the basis of all this. The animal has a definite size, because several animals do not grow around each other.

III.—PHYTOLOGY.

1499. Hitherto the organs of the plant have been considered in a general point of view or as to their idea in time; to this now follows the development of the plant in a special sense, or its representation in space.
1500. The vegetable tissues, systems, and organs have only by degrees been disengaged from each other and independently perfected. The independent or self-substantial development of the organs constitutes definite or individual plants.
1501. A plant, in which all the organs are present, separately or self-substantially developed and yet combined, is without doubt the highest in point of rank.
1502. Before it attains to this separation, nature can only produce lower forms, in which fewer organs have attained to independence. These forms constitute the diversity of plants and their plurality, for nature establishes every principal form as a finished organization.

1503. There are as many plants different from each other as there are organs, namely, tissues, anatomical systems and members.
1504. The sum of all plants is called the vegetable kingdom; this is the self-substantial representation of all vegetable organs. (Ed. 1st, 1810, p. 123.)
1505. The vegetable kingdom is consequently the expression of the vegetable idea, or of the perfect plant represented in the multiplicity of individuals; it is the plant disintegrated, or

anatomised, by nature herself.

1506. Were we therefore acquainted with all vegetable organs, we should know their rank and developmental series; and thus also recognize the character, rank, and developmental series of the plants themselves, or their divisions. There can be no doubt that the lowest organs, e. g. the tissues, have been first developed and independently perfected as plants; later on they separate into anatomical systems and finally into members, whereby perfect plants must originate. The division or classification of the vegetable kingdom is consequently that of the vegetable organs. The Systematic of plants is a copy of that of their organs, or a plastic representation of the philosophical vegetable anatomy. With this every thing has been granted, which is requisite for the building or erection of the vegetable system. All principles, together with the methods, rest in the proposition that has been expressed.

1507. The artificial systems of plants are related to the vegetable kingdom, as the lexicon or dictionary is to language. Those systems which have hitherto been termed natural, but which should properly be called methodical, are related to the vegetable kingdom, as the ordinary grammar is to a language. The vegetable system must, however, be related to the vegetable kingdom, as the philosophical or genetic grammar is to language. This only agrees with the essence of the language, or is natural. The vegetable system is necessarily a philosophical or genetic one, that alone being truly or legitimately natural. (This system was first propounded by Oken, in the Ed. 1st of the Naturphilosophie, 1810; further developed in Dietrich's Garten Journal, 1813; carried out in his Naturgeschichte für Schülen, 1821, and in his Lehrbuch der Nat. Gesch. Botanik. Weimar, 1825.) The artificial system collects the materials for the edifice, but leaves them to lie without order and in confusion; the methodical or what has been called the natural system separates these materials and arranges them in homogeneous groups; the genetic, philosophical or truly natural, system, again mixes them amongst each other, but thereby actually erects the edifice. All three systems are therefore necessary and good, and no one of them merits being despised by the other; it is only when one of them imagines that it is the other, or can render the others unnecessary, that it trespasses from out its circle, and deserves reproach. Thus for Floras, whose ultimate object is to find out rapidly the names of plants upon botanical excursions, as also for the labelling of specimens in botanic gardens, the artificial system is the best; for the description, however, of foreign plants the methodical; but for insight into the whole vegetable world the philosophical or natural system. Would we compare Floras with each other, the latter system must certainly come into play; but then the matter to be dealt with is not about an excursion-book.

VEGETABLE SYSTEM.

1508. Taken in a strict sense all the diversity of vegetable structure of vegetables has reference first of all to the difference in the tissues; these being either unseparated, or separated, into special systems and members. At first the tissues lie confusedly, or without order, amongst themselves. They then separate in a concentric and tubular form into systems, that are encased within each other, like the bark, liber, and wood, which form the shaft. Furthermore they separate into members, and appear one above the other, as root, stalk, and foliage, which collectively may be called the stem; these are repeated as seed, pistil, and corolla, which together are called flower, and combined, fruit, namely, nut, plum, berry and apple. I designate by the term stock or trunk all the parts as far as the blossom; and this together with the fruit I name thyrsus. The vegetable stock, whose tissues have not yet separated into members, I style, from want of a better word, thallus. As we divide political kingdoms into provinces and circles, so also may these titles be suitably applied here. It is evident, that the plants which simply consist of tissues and have as

yet no sheaths and members, are the Acotyledones; those, however, provided with sheaths, but devoid of true roots, stalk, and foliage, Monocotyledones; those with true foliage or reticular-veined leaves are, on the contrary, Dicotyledones. The natural system of vegetables stands accordingly in the following manner.
A.—STOCK-PLANTS.
Province I. Histophyta, or Tissue-plants—Acotyledones.
Class 1. Cell-plants.
2. Duct-
3. Trachea-
II. Thecophyta, or Sheath-plants—Monocotyledones.
4. Bark-plants.
5. Liber-
6. Wood-
III. Arthrophyta, or Member-plants—Dicotyledones.
Circle 1. Axis-plants—Tubulifloræ.
7. Root-plants.
8. Stalk-
9. Leaf-
B.—BLOSSOM-PLANTS.
Circle 2. Flower-plants—Thalamopetalæ.
10. Seed-plants.
11. Ovarium-
12. Corolla-
Circle 3. Fruit-plants—Calycopetalæ.
13. Nut-plants.
14. Plum-
15. Berry-
16. Apple-
1509. A slight glance at the above table shows us the procedure of Nature. The higher she ascends, the more and more she separates, and thereby increases, the organs. There may therefore be plants which have only a single organ or tissue, as well as others, which possess all.
1510. There cannot, however, be any plant which could simply possess the higher without the lower organs. Higher organized plants are not such therefore, by virtue of their having some one organ more perfectly developed, or separated into several parts; but through this, that they actually possess several different organs. The higher grade of organization depends accordingly not upon the perfection of the Singular, but the number of the Different. The Perfected consists in the multiplicity combined to constitute unity, but by no means in the simply homogeneous multitude of the parts. Numerous stamina may render a corolla higher, but not on that account the whole plant; many digits may make a hand nobler, but not on that account, the animal. But with many digits also that hand is nobler, in which the digits are dissimilar.
First Province.
HISTOPHYTA—ACOTYLEDONES.
Devoid of, or without true spiral-vessels, leaves, corollæ, and pistil.
The vegetable kingdom ascends, in accordance with the five main positions of the organs, by five stages; these are again separable into larger groups, which may be called asexual and sexual plants.

1511. The tissues are a something internal, being, as it were, the viscera of plants or their parenchyma, which does not meet the light, and can therefore have no light-organs, which are developed only out of the foliage or leaves. The anatomical systems and organs are tissues that have become external, have attained to air and light, and are hence developed into air-and light-organs. Now, the light-organs are sexual organs. The Tissue-plants can therefore have no sexual organs; and plants divide accordingly into asexual and sexual plants. The asexual are female plants, and are consequently the first or lowest. Thus there can be thus no sexual or male plants, without the female being found that belong to them.

1512. Male or androgynous plants are only possible, if spiral vessels or tracheæ be present. They first, however, originate when the tracheæ become external, or form a circle in the stalk, i. e. are accessible by light; as in the Mono-and Dicotyledones.

1513. The asexual plants are not cryptogamic, but agamic. They do not perform self-impregnation clandestinely, but not at all; for they do not attain light-difference, and consequently not male organs. Analogues of stamina may make their appearance in the mosses, but they invariably fail to attain the development of pollen. What have been called male parts in other cryptogamia, do not merit consideration. Such projections or prefigurations are besides to be found everywhere.

1514. The asexual plants are simply formations of the tissues, of the galvanic vesicle, and are thus of a female nature. They are nothing more than a great utricle full of small vesicles, which by desiccation subdivide into germinal dust or sporules, each granule whereof attracts other mucous vesicles out of the moisture, in order to form again a large utricle.

1515. The asexuals cease in the process of vegetation, where the other plants begin. With the rupture of the gemmal-or bud-vesicle in the higher plants a new world for the first time emerges into view, such as stem, leaves, blossom, and then the ultimate bud ruptures for the first time as the pericarp, and scatters its higher organized germinal powder as true seeds.

1516. An asexual plant is one, which, without all the intermediate organs of the stem, at once represents the capsule or ovary. It consists only of the beginning and end of the plant.

1517. The higher plants differ from the lower by the interposition of new organs between the two terminal organs, namely, the primary vesicle and the true seed. It may be said, that the asexual plant is naught but seed, and that the seed of the higher plants is a fungus upon a leafy peduncle, a fungus more highly organized by light.

1518. The asexual plants have no true root, stalk, and leaf; they have not even a true bark, liber, and wood, in so far as these first make their appearance through separation. Tracheæ are first exhibited in the higher ferns, and then only as constituting a single string, which occupies the middle of the plant, and consequently forms no circle or zone.

1519. As again the true seed is a leaf-formation, and possesses therefore cotyledons or seed-lobes, such seeds must be wanting in the asexual plants; they are therefore Acotyledones. From the same cause, however, the germinal leaves or plumula must be also wanting; they are therefore germless, or anembryonic.

1520. The farinaceous or granular matter, lying next to the germ when within the shell of the true seeds, is called the albumen or perisperm; the seeds of the asexual plants are therefore nothing else but albumen. They are therefore devoid of the funiculus or, what has been called, umbilical cord.

1521. The involucre, wherein, in true seeds, the germ and albumen are found, is the seed-coat or testa; consequently what has been called the capsule of the asexual plants (of mosses and ferns) corresponds simply to this spermoderm or seed-covering, and is no true ovarium. The capsules of mosses and ferns are therefore seeds full of albuminous dust.

1522. If any of these be regarded as a capsule, it can be the calyptra of the mosses. This, too, is probably nothing else than the external testa; the proper capsule being its internal coat.

1523. The indusium of ferns incloses several capsules as they have been called, or properly seeds, and might therefore, if considered alone, be compared with an ovarium, but it is probably none other than the covering corresponding to the perichœtium that surrounds the base of the setæ in mosses. The sorus is an accumulation of seeds with pulverulent albumen contained in a membranaceous covering, the indusium.

1524. The life of the asexual plants consists simply in the galvanic process. They are the primary organisms, planted in air.

1525. As being simply galvanic process, they would require but little light and air; they therefore seek the darkness, like the roots, and thrive also in a corrupt atmosphere, in caves, mines, cellars, and such like situations. They can from the same cause thrive only in moisture, in water, upon marshy meadow lands, after rain, copious dew, and so on.

1526. They are devoid the process of fermentation, as being that which is imparted by the oxydation of air, and they therefore yield neither sugar nor acids. They are simply the organized process of putrefaction; their ultimate product is therefore germinal powder, infusorial matter. Their remaining secretions are alkaline bodies; to which belong the pungent, fetid, and nauseous excretions, as the hydrogen gas and ammonia of the fungi, the mucus of the fuci, the carbonate of lime in the lichens, the cell-threads of the mosses, the fetid principle of ferns.

1527. Very few of these plants require the course of a summer in order to perfect or complete the vital course; a single ray of light of one day's, aye, of one hour's duration, is sufficient with most of them to evoke the feeble difference, to rouse the swell of the sap, and precipitate the infusorial powder.

1528. Automatic movements, as in the leaves and stamina of the higher plants, scarcely occur in them, or at most in the ferns, from their possessing spiral vessels. They divide, according to the tissues, into three classes, into Cell-, Vessel-, and Trachea-plants.

CLASS I.
Cell-plants—Fungi.
Here belong those plants which consist simply of cellular tissue, having no sap-tubes and tracheæ. Such plants, too, possess no regular or hexagonal cellular tissue.

1529. The cellular tissue, in which there is only a single active process, cannot essentially alter its primary form. It is therefore an accumulation of round or cylindrical mucus-vesicles.

1530. Mucus-vesicles, in which the air-process is not as yet active, cannot be coloured green; but must have the colour of the earth.

1531. Plants, composed of amorphous and earth-coloured cellular tissue, are Fungi. The fungi are simply clusters of mucus-vesicles joined together in a more or less regular manner, their union being effected in dark, hollow and wet situations.

1532. They may therefore originate wherever mucous juices are evolved by the potential agency of a higher organization, and thus by putrefaction. The fungi originate by æquivocal generation. They are the anal-organizations of the higher plants and animals; the corrupted and luxuriating juices.

1533. Nevertheless the fungus is propagated by division of its vesicles, which again, in accordance with their peculiar laws of polarity, attract mucus-vesicles, and thus obtain the form of the earlier or parent fungus. This is only a more regulated kind of æquivocal generation.

1534. The origin of the fungi may therefore happen in a twofold manner, namely, by formation from other juices, and by that of their own, which is called propagation. Still at bottom both are one in kind.

1535. Their granules or vesicles are seeds, properly sporules, which are self-developed without male polarization.

DIVISION.

1536. The fungi pass moreover through stages of development, which range parallel to the vegetable classes; since it is impossible for any other organs to originate in them but such as belong to the idea of the plant. The lowest fungus can therefore change only, by endeavouring to develop in itself ducts, tracheæ, roots, and such like parts.

1537. There are accordingly as many developmental stages of the fungi as there are vegetable classes. These divisions are called families.

1538. The vegetable families range parallel to the classes. This law must hold good of all the classes. There are therefore in each class 16 families. An association of families upon each stage may be called an order.

1539. At first the fungus is none other than a mucus-vesicle or a small cluster of vesicles, e. g. an uredo or mildew. Such a vesicle next becomes longitudinally extended, and includes within itself other vesicles or granules, e. g. mould. These mould-filaments or threads unite again so as to form a common mass, which is surrounded by an external membrane, and is then called puff-ball. The pulverulent granules, which were irregularly accumulated in the puff-balls, unite at length in a regular manner to constitute a trunk of varied form, as in the ascomycetes, e. g. sphæriæ. Finally, the mould-filaments with their sporules are regularly collected together in an investing membrane, which, like a puff-ball, is supported upon a stipes or stem, e. g. the sarcomycetes or agarics. There are therefore 5 developmental stages of the fungi, which correspond to those of the classes; viz. the parenchyma, shaft, stem, flower, and fruit; and constitute orders.

1540. Each order is resolvable again into three divisions or families, which correspond to the organs. Thus there are in each class 16 tribes or families, which obviously range parallel to the vegetable organs or classes. (Vid. Tab. B.)

1541. A tribe or family is consequently the representation of a vegetable organ within a class.

1542. The genera obey the same law; for essential differences are only conceivable through the presence of different organs.

1543. Species is in the animal kingdom that which copulates without necessity and compulsion. The same definition is applicable to plants. The species range, without doubt, according to the diversities in the individual organs themselves, which admit of a great multitude of combinations, the number of which is not as yet to be determined.

1544. The component parts of the fungi are either perfectly indifferent, mucous or gelatinoid matter; or they are of an alcaline nature, being acrid, poisonous, and such like. Their odour is usually dead, disagreeable, and loathsome, or analogous to their essential process of decomposition.

CLASS II.

Vessel-or Duct-plants—Mosses.

1545. The intercellular passages or succigerent vessels of plants make their first appearance in a state of perfection, when, the cells being extended lengthways, have become hexagonal and are placed in regular juxtaposition. In these plants therefore we meet with regular cellular tissue, but still without spiral vessels or tracheæ.

1546. As the vessels or ducts constitute the fundamental tissue of the liber, while this is the principal system of the stalk; so now does the stem begin to be manifested and separated from the fruit. The seeds are no longer therefore distributed in the present class throughout the whole trunk, but developed in a special involucrum or theca, which corresponds to the puff-ball, or to the pileus of the higher organized fungi.

1547. Plants with vessels, and consequently a cauliform formation, have at once also the commencement of a bark, and next the green colour. The vascular are the first green plants, and differ chiefly through that character from the fungi. They are the Fucacæ or Sea-wracks.

1548. They have the colour of the water, because the course of the sap corresponds to the aqueous process; they are aquatic, just as the brown fungi are terrestrial, plants. Their component parts are aqueous, indifferent, mucous, and filose. Their habitation is the water itself or bogs. If they occupy dry situations, they live only when it rains.

1549. They likewise pass through the five stages of vegetation, and form therefore five orders.

1550. Order 1. The lowest or Tissue-mosses, corresponding to the Uredines; are again naught but cells or mucous pellicles, but, from growing in water, and being consequently exposed to light and a stronger oxydation, they are green—Tremellini.

1551. They multiply by subdivision, since new vesicles or granules are developed in their interior, which become separated, and subsist or continue to grow for themselves. They therefore originate also by æquivocal generation, but by such an one as constantly occurs in water and light.

1552. The second order, or the Vascular mosses corresponds to the sheaths, or to the Hyphomycetes. They are long filaments replete with granules, growing in water, and therefore green—Confervaceæ. These plants begin to ramify, and either increase by this means or by effusion of granular matter.

1553. The third order, that of the Tracheal mosses, corresponds to the stem, or to the Gasteromycetes. A membranous trunk originates in water, which in certain places secretes the seeds in special vesicles or cysts—Fucales or connate Confervæ. The fuci have at once the form of a stalk with root and leaves, because they correspond to these three organs of the axis.

1554. The fourth order, the Floral mosses, endeavours to obtain the blossom, and therefore elevates itself out of the water, but loses on that account the trunk-like character, and exhibits for the most part only membranous expansions, upon which seeds are secreted, which being usually of a beautiful colour, thus assume the appearance of corollæ—Lichenales. The lichens are fuci in dry situations. They correspond to the Pyrenomycetes or Sphæriaceæ.

1555. As the variegated colours appear in the blossom, so do they also in the lichens; but here they are for the first time chemically developed, distributed throughout the whole substance, and concealed. Most lichens yield colouring matters.

1556. As the stem is, in accordance with their signification, wanting to the lichens, they thus require a foreign trunk for their nutrition. They are therefore developed for the most part upon other plants, and principally upon the bark.

1557. Lastly, the fifth order, Fruit-mosses, originates through the development of a self-substantial fruit upon a cauliform stem—the Mosses proper.

1558. As these are the highest plants of this class, and those that directly precede the tracheal formation, so the bark already resolves itself into individual leaves, which are, however, still destitute of spiral vessels.

1559. What have been called the seeds or sporules are accumulated in a capsule-like fruit upon the summit of the stalk. This fruit corresponds to the pileated fungus, and therefore springs up in an opercular manner like the latter.

1560. But this capsule is only a spermoderm, which incloses albuminous granules that have no proper germ or seed-lobes; they are plants with seed-vessels or pyxidia, upon an open-leaved stalk.

1561. They divide likewise into sixteen families. (Vid. Tab. B.)

CLASS III.

Trachea-plants—Ferns.

1562. At first a fascicle only of spiral vessels can originate, which is necessarily surrounded by cellular tissue, and therefore lies in the middle of the plant. Such plants are the Filices or ferns.

1563. As the spiral vessels are the antetype of the leaves, so does the trunk here obtain the form of the leaf, without itself producing true leaves. For, in the ferns, the fruits lie upon the back of the apparent leaf, which can only be the trunk.

1564. The fruits, being further removed from the fungi, no longer spring up in an opercular, but in a valvular, manner like the higher capsules.

1565. Green plants with imperfect spiral vessels and blossoms, and also with naked seeds devoid of true capsules, belong to the class of Ferns.

1566. I therefore place in this class the Coniferæ or trees with acicular foliage, because they have no ovarium, but naked seeds; and besides these, some other plants, though doubtfully, on account of their very abortive blossoms, as the Naiadaceæ. There are therefore Tracheal plants without and with stamina. The first portray the stock or trunk, the second, the thyrsus or blossom; they live mostly in dry situations, and produce resins or fetid matters.

1567. First order. Parenchymatous ferns—Aquatic ferns. I here place the aquatic ferns, because, as water-plants, they occupy a lower situation, because they support the fruits upon a radical trunk, and finally, because these fruit-vesicles have two kinds of contents, all of which seems to remind us of the fuci and lichens; they correspond to the Tremellini.

1568. Second order. Sheath-ferns—Club-ferns. Here commence the land ferns, and those kinds indeed whose so-called capsules open in a valvular manner, just as in the liverworts; or almost after the fashion of a pyxidium by an orifice, somewhat as in the mosses; the trunk is provided with squamose leaves or lobes, e. g. Lycopodiaceæ and Osmundaceæ; they correspond to the Confervaceæ.

1569. Third order. Stem-ferns—Annular or Ring-ferns. Here we meet with phylloidal involuted capsules or seeds upon the back of a stem that is likewise leaf-like; e. g. the typical or true ferns.

1570. They have rudiments of roots and a stem, together with foliage, because they are the prototypes of these three organs.

1571. The ring of their capsules corresponds to the midrib of the leaf. In the preceding order the capsule was only an upsprung stalk; but it is here an upsprung and unfolded leaf, the prototype of the bud-development.

1572. The fern-capsules, namely, the true seeds, are an accumulation of leaf-buds at the extremity of the fascicle of spiral vessels. The indusium is the upraised epidermis, which opens in a spathose manner; it consequently stands in the signification perhaps of the floral spathe or involucrum. They correspond to the Fuci.

1573. Fourth order. Floral ferns—Fluviales. If the tracheal plants be exalted unto the flower, male organs cannot fail in at once beginning to develop. I place therefore in this order the Naiadaceæ with very arrested blossoms, and simply stamina without calyx and corolla. The spiral vessels are rather doubtful.

1574. Fifth order. Fruit ferns—Coniferæ. Tree with imperfectly-formed spiral vessels, stamina without corollæ, seeds without ovarium; thus agreeing with the Cryptogamia even to the stamina; they form likewise sixteen families. (Vid. Tab. B.)

SEXUAL PLANTS.

1575. So soon as the three tissues separate completely from each other into bark, liber, and wood, while the tracheæ are arranged circularly into several clusters or groups, does the antagonism of these organs also make its appearance, and exhibit itself as sex in the floral organs.

1576. This separation can only be attained through the influence of the air and light, whereby the sexual organs are conditionated. These plants have therefore the several organs of the trunk and blossom.

1577. Anthers can be wanting to none of the following plants. Now, the anthers are leaf-buds; the leaf-formation must be therefore developed also in their antagonism, or in the seed. The leaves of the seed, however, are called seed-lobes; consequently all seeds of sexual plants have seed-lobes or cotyledons.

1578. Now the perfect seed is the whole plant in miniature with root, stalk, and leaf. This formation is, however, only possible where there is a sex, or where the vegetable tissues have emerged self-substantially from each other.

1579. But the anthers are buds upon a floral rib; consequently all sexual plants must have a part of the blossom, which ranks in the signification of leaf, and thus either the calyx or, with this also, the corolla.

1580. All sexual plants must have a shaft or scape, in which its three parts, bark, liber, and wood, are to be distinguished; even so must the three parts of the stem, the root, stalk, and leaf, have the parts of the blossom which correspond to them, viz. seed, ovarium, and calyx or corolla.

1581. The sexual plants next divide into Stock-and Blossom-plants, the former of these parting into Shaft-and Stem-plants.

Second Province.

SHAFT-PLANTS—MONOCOTYLEDONES.

1582. In the Shaft-plants the invaginate character is predominant, the wood being surrounded by liber, and this by bark; they are therefore tubular in form—tubular plants.

1583. The cellular tissue preponderates in them, and the tracheal fascicles form therefore no closed circle, but are dispersed; scattered fasciculi cannot, however, ramify but only pursue a straight course and form streaks—striated plants.

1584. Ramification is wanting unto these plants, partly on account of their tubular form, partly on account of the smaller quantity of tracheal fascicles—aramose plants. A vegetable stem without branches is called shaft or scape; thus they are shaft-plants.

1585. The blossoms do not stand upon ramules, but upon the stem itself; thus, if ever ramification originates, it can first occur in the floral peduncles.

1586. As they are deficient in branches, so also are they in buds—agemmal or budless plants.

1587. The tendency to produce branches forms nodes, which are only imperfect ramal zones—nodose plants.

1588. Where the formation of branches is wanting, the minor ramification of ribs must be also wanting in the leaves; they are parallel-ribbed.

1589. The leaf is only a ruptured and expanded tube or spathe, which surrounds the shaft—spathose plants.

1590. Where branches are wanting, there can be also no petiolated leaves; they have therefore only radical leaves, from the midst of which the shaft or scape sprouts forth, only as a floral-peduncle—radical-leaved plants.

1591. If all these leaves be still sheathed or encased within each other, they are called bulbs—bulbose plants.

1592. Since the corolla is the repetition of the leaves, so also must its structure resemble theirs—spathose corollæ.

1593. But as there are here only radical leaves, so also is the corolla only a radical leaf-corolla. In the blossom, however, the radical leaf has become spathe, the cauline leaf, calyx, the ramuscular leaf, corolla. Such blossoms consist therefore either of a spathe only, as in the Aroideæ, or of a calyx—calycine plants.

1594. The stamina originate from the calyx. In calycine flowers therefore all the stamina must range opposite to the corolla-petals. These plants have only opposite, no alternate stamina. It has been generally assumed that the Monocotyledones have no true corollæ, but only coloured calyces. Philosophy agrees with this opinion; but adds thereunto that the corollæ of the Monocotyledones may be also merely spathes or involucra.

1595. The separation of the bud-formation here takes place for the first time in the blossom, namely, in the stamina. The number of the floral parts must be limited to three; for this is the first number, in which the leaf-ribs may divide. Corollæ, stamina, and capsules are ternary—trinity-plants.

1596. As the seed is only an undeveloped leaf-bud, so in it only can the number of the leaf be found. Since, however, the leaf does not ramify in the present group of plants, but is only a single spathose leaf; so also does the seed consist only of one such leaf or of one seed-lobe—Monocotyledones. The wheat-seed or grain is none other than a grass-leaf with a short spathe and very dense, highly farinaceous, lamellæ. In germinating a new leaf emerges from the short spathe; it is the germinal leaf.

1597. Thus plants having a tubular stem, and such kind of foliage, corollæ and seeds, are Monocotyledones.

1598. The chemical bodies are more diversified in these than in the remaining members of the vegetable kingdom. In the roots, as in the tubers of the Orchideæ we meet with distinct mucilage; in the bulbs with alkalies or acrid matter; and with sugar, as being a feeble conversion of the starch in the stalk. The mucilage of the root becomes, when repeated in the seed, flour. Oily matters or acids seldom occur, and fleshy fruits scarcely ever.

1599. The ovary is almost throughout this region of plants either a single spathose leaf or caryopsis; or three spathe-leaves are united together, which thus, as follicles or carpels, usually burst open upon the internal edge.

DIVISION.

1600. If we now proceed to survey the Monocotyledons, in the order just set before us, we shall recognize among them three typical groups, to which the others are allied; they are the Gramineæ, Liliaceæ, and Palmaceæ.

1601. The Grasses obviously rank the lowest as well in respect of their root, stalk, and foliage, as also of their stunted blossoms, ovarium, and seeds.

1602. To them succeed the Lilies, which have a well-marked root, a more perfect, though still always herbaceous, stalk, and a few spathoid, more numerously ribbed leaves; lastly, their corollæ, ovaria, and seeds are perfect; but still they invariably have no genuine fruits.

1603. Finally, the Palms are elevated above all by their stem being rich in tracheæ and wooded, as also by the perfection of their fruit. Scientifically, the scapose or shaft-plants must also resolve

themselves into three classes; into Bark-, Liber-, and Wood-plants.
CLASS IV.
Bark-plants—Gramineæ.
1604. In these plants the whole stalk must have assumed the form of the cortex or bark, and consequently be hollow—tubular, or culmaceous plants.
1605. A mere bark cannot ramify. But the tendency unto ramification is manifested as nodes—nodose plants.
1606. The leaf which still represents the bark, is only imperfectly slit up, and therefore still forms a tube. Such leaves are called tubular or spathe-leaves proper.
1607. Such tubular leaves being only half slit up can only shoot forth gradually from each other, and that indeed in such a manner that they stand actually by twos, encased in, or opposite to, each other.
1608. Since the blossom is, as it were, an impression, or copy of the leaves, so will it here also consist only of spathiform, involucral, or calycine leaves, and only of two, one of which, though opposite to, is surrounded by the other. Such floral parts are called glumes—glumaceous plants. If four glumes are present, then the external pair of them corresponds to the involucrum or spatha, the internal to the calyx.
1609. The corolla-petals are of necessity arrested in plants such as these, where no true leaf is as yet developed; frequently two only are left persistent as pellicles or lodiculæ.
1610. This is still more the case with the ovary and seed; in each only one leaf attains development, and the seed has entirely coalesced with the ovarium. This kind of fruit is called caryopsis.

1611. Nodose plants with hollow scape or shaft, tubular leaves, glumose blossoms, and cariopsidal fruits are Grasses.
DIVISION.
1612. The Bark-plants pass moreover, together with their subdivisions, through the five stages of vegetable organs, and they will therefore produce also a more perfect stalk, leaves, and flowers. The whole calyx will, however, never be coloured or corolla-like in its character. Cortical plants are thus herbaceous plants with hollow stalk, and with an arrested or green calyx, without sarcose or fleshy fruit.
1613. Those, which have simply glumose flowers, are without doubt the lowest in rank, as corresponding to the tissues and stock, but not yet to the flower.
1614. These again divide into two great groups, whereof the one includes plants with simply uni-seminal cariopsides, the other, on the contrary, capsules containing a free seed—Grasses and Reed-grasses. The grasses which have a cariopsis or grain-fruit reascend by two stages; the most inferior in rank do not attain to a ramification, but the flowers stand crowded together in spikes; the others, on the contrary, are pedunculated and ramify in panicles.
1615. Among the higher kinds instead of glumes there are regular flowers, of which, however, the calyx is still glumose or at least green. The cariopsis is converted into a multilocular capsule, as in the Restiaceæ, Commelineæ, &c.
1616. The first order, or the Parenchymatous grasses, have glumose flowers with a cariopsis for fruit, borne upon culms or nodose straws, and do not attain to any ramification—spicate grasses. In their seeds we meet with the greatest amount of starch developed, though doubtless at the cost of the trunk.
1617. The second order of Spathose grasses are similar, and support ramified flowers—

paniculate grasses. In this division occur grasses of a dendroidal character, and having occasionally fruit-like ovaria, as in the bamboos.

1618. The third order consists of the Cauline grasses. Here the leaves at once separate completely from the shaft, which is therefore free from nodes. Nut-like capsules, though still inclosed in glumes, also make their appearance—reed-grasses.

1619. The fourth order, Floral grasses. The leaves are still only radical; the shaft is anodal; the flowers are separated into green calyx and coloured corolla, with three or six stamina, and mostly with a trilocular capsule—Junceæ, to which the Commelinaceæ are allied.

1620. The fifth order, Carpal grasses. Hollow anodal shafts, with scarcely spathiform, mostly broad leaves, the ribs of which begin to ramify; with similar calyces and corollæ, and numerous capsules—Seeroseæ, to which the Alismaceæ and Hydrocharideæ are related.

1621. The highest kind of fruit attained by this class is nut-like, never fleshy in character. The stalk is nowhere woody as in some grasses. The grasses divide, as do all other orders of plants, into sixteen families. (Vid. Tab. B.)

CLASS V.
Liber-plants—Liliaceæ.

1622. The substance of the stalk is soft and succulent; its structure devoid of nodes; the leaves are tolerably free and ribbed; calyx and corolla coloured, both being perfectly formed and invariably tripartite, as is likewise the capsule, which has many seeds upon its inner angle. These plants are the Lilies. In the present class all parts have been developed in conformity to the liber, are rich in sap, and have become dense and fleshy.

1623. The roots are mostly tubers or bulbs, containing a superior kind of mucilage, or aromatic principles. The shaft is not hollow but, though herbaceous, filled up; the leaves are elevated upon the stalk. One division of these plants has irregular corollæ with stunted stamina and capsules, the latter containing mostly dust-like seeds, as in the Orchideæ and aromatic plants.

1624. The other division has regular 2×3ary corollæ, with, perfect glume-capsules and middling-sized seeds, as the Irideæ and Liliaceæ proper.

1625. The first order, Parenchymatose lilies. The corollæ are irregular, bilabiate, and stand upon the calyx and a membranous sexlocular capsule with very small seeds; are divided according to the pollen—e. g. pulverulent Orchideæ.

1626. The second order, Sheath-lilies. Characters similar to those of the preceding order, but the pollen is agglomerated into waxy granules—granular orchideæ.

1627. The third order, Stem-lilies. The corollæ are likewise bilabiate and situated above the calyx, but the ovarium contains few seeds, and the capsule is mostly nut-like—aromatic plants, such as the Scitamineæ and Musaceæ.

1628. The fourth order Floral-lilies, have regular blossoms, separated into calyx and corolla, placed above the capsule, and mostly furnished with three stamina. Here belong the Hypoxideæ, Hæmodoraceæ and Irideæ.

1629. The fifth order, Fruit-lilies. The leaves have not yet completely separated from each other, but still form bulbs; the corollæ are regular, have six stamina, and are placed beneath the capsule, as in the true lilies. To this order belong the Colchicaceæ, Aloinæ and Liliaceæ. Their bulbs contain mostly acrid matters. They divide into sixteen families. (Vid. Tab. B.)

CLASS VI.
Wood-plants—Palmaceæ.
Plants with woody shaft and with fruits, mostly enclosed in spadices.

1630. The desiccation of the cells and fibres is promoted by the increased process of oxydation. Where therefore the tracheæ attain the preponderance, there the conversion into wood originates.
1631. The stalks of these plants are not hollow, but have a dense interior, because the fasciculi of tracheæ lie within the liber, and there increase.

1632. The main bulk of the stalk will consist of tracheæ.

1633. As the tracheæ are longitudinal organs, and the other tissues also extend lengthways, so does the stalk or stem in these plants predominate over the other parts.

1634. In this class the most perfect leaves as regards their present stage are developed; for they are only expansions of the tracheæ, which are here present in superabundance. As regards the form also, these leaves must rank higher than those of the preceding class; the spathe is shorter, the leaf itself usually broad, many-ribbed, and frequently pinnate. The leaves are also perfect as to position, being no longer mere radical leaves, but situated upon the stalk, even on its apical extremity.

1635. The ramification gradually emerges into view, where forsooth its occurrence is possible in the shaft-plants, or in the inflorescence. It is always multiple in character, mostly spadici-, muscari-, and paniculiform.

1636. As regards the blossom the ovarium is most perfectly evolved, because it is developed out of the stalk; it is ternary and becomes elevated into a fruit with few seeds.

1637. The corollæ are frequently stunted, yet regular and 2×3ary, though insignificant on account of the preponderance in size of the fruit.

1638. In this class we meet with the first true or genuine fruits; because in it for the first time the three anatomical systems are completely separated.

1639. Plants having a ligneous stalk, free and many-ribbed leaves, ramified inflorescence and ternary fleshy fruits, are the Palmaceæ. The Palmaceæ have a woody, very hollow stem, with many-ribbed, divided, and often pinnate leaves; a muscariform inflorescence lodged in spathes, sexanary corollæ with nuts, berries, or drupes. With the palms are associated the Typhaceæ, Aroideæ, Piperales, Pandanales, Dioscoreæ, Smilaceæ, Asparagi, Convallariæ, and Bromeliæ; for their stalks are mostly woody, the leaves broad, and placed upon the stalk, the corollæ stunted, while, on the contrary, the ovarium is carpoidal or fruit-like. The five orders may be disposed as follows:

1640. Order I. Palmaceæ parenchymatosæ. Cynomoridæ, Typhaceæ, Aroideæ.

1641. Order II. P. thecales. Saurureæ, Piperaceæ, Pandanaceæ.

1642. Order III. P. axonales. Dioscoreæ, Smilaceæ, Parideæ.

1643. Order IV. P. florales. Asparagoideæ, Convallariæ, Bromeliæ.

1644. Order V. P. carpales. Palms.

1645. The plants of the first order are very imperfect herbs with spadices. Those of the second have mostly a ligneous, nodose stalk, with one-seeded fruits in spadices, without corollæ. The third have separated corollæ disposed in an open form of inflorescence. The fourth have perfect sexanary corollæ, with a frequently woody stalk and ternary, many-seeded berries. The fifth order consists of trees having large leaves, muscariform spadices, and perfect fruits, nuts, plums, and berries, ternary and one-seeded. They divide into sixteen families. (Vid. Tab. B.)

Third Province.

RETICULAR-LEAVED PLANTS—DICOTYLEDONES.

1646. With the separation of the stock or trunk into root, stalk and leaf, the latter organ attains its perfection; it becomes a reticular leaf—the plants possessing it being called reticular.

1647. The retinerved or reticular leaf is, however, only the result of a modified organization in the stalk, and indicates a ramification or foliiform arrangement of the tracheæ in the stem. The foliiform arrangement of the tracheæ in the stalk is their circular disposition. These plants have woody zones. Through this zone or ring of wood first originates the perfect separation into wood, liber and bark, whereof each formerly occupied the whole stalk or stem.

1648. The stalk is no longer a shaft or scape, but it divides into branches and twigs—ramular plants.

1649. The reticular-veined are ramular leaves, and are no longer therefore spathiform but petiolated—petiolated leaves. It is only at the root that spathose leaves may occur, and this only in the plants of the inferior classes.

1650. With the disappearance of the spathose leaves, and the appearance of the ramules, the nodes and bulbs also disappear.

1651. The blossoms stand no longer upon a radical peduncle or stipes, but upon ramules; in other words, upon a plant, which again stands upon another plant, namely, the stalk.

1652. As all the higher separations of leaves here occur, so also does the flower obtain its higher amount of separation; it becomes quinary—pentaschematose plants. The ovarium passes through all numerical conditions, being 1, 2, 3, 4, 5, and polycarpellar. In like manner all the forms of ovaria and fruits occur in the present class, such as caryopsis, follicle, legumen, siliqua, capsule, nut, plum, berry and apple.

1653. As the seed is a leaf-formation, so must it resemble the reticular leaf. But reticular leaves are not spathes or simple tubes, but ramified or separated ribs. The seed has therefore several leaves, and two indeed for the first time, which are called seed-lobes. These plants are therefore styled Dicotyledones.

DIVISION.

1654. The Dicotyledones are, in the first place, empirically divisible into apetalous, monopetalous and polypetalous, or into plants with calycine, tubular and petalous corollæ.

1655. It might be believed that the Apetalæ were, without further trouble, the lowest in point of rank; but, when closely considered, they appear as Polypetalæ with stunted corolla-petals, and are obviously allied to the Rosaceæ. Moreover, they all bear nuts, a fact which occurs in no other class, and they must be therefore placed among the fruit-bearing plants. Since, however, they are epigynous and perigynous, so must the other polypetalous Perigynæ enter into proximity with them, and in like manner forsooth come among the Fruit-plants.

1656. Thus the Dicotyledones separate into the Monopetalæ, hypogynous Polypetalæ, and perigynous Polypetalæ, along with the Apetalæ. Viewed in a scientific light, they separate according to the principal members of the plant into three districts, Axis-, Flower-, and Fruit-plants.

FIRST CIRCLE.

Stem-plants—Monopetalæ.

1657. The Monopetalæ or Tubulifloræ are the lowest in rank, and must therefore take the place here assigned them. They are still spathose corollæ. Among them are found for the most part only cariopsides and membranous capsules, rarely fruits. They are usually, too, merely herbs, rarely bushes, and still more rarely trees.

1658. They divide into Epigynes, Peri-and Hypogynes, of which the former are the lowest, the latter the highest; for in that which is left similar the coalition is an inferior sign.

1659. Their essential or typical character does not, however, reside in the blossom, but in the trunk, and that indeed in the root, stalk, and foliage. The question therefore may now be asked, whether in the Epigynes the root, in the Perigynes the stalk, and in the Hypogynes the foliage, be the principal organ.

CLASS VII.

Root-plants.

1660. That the Tubuliflorae with superior blossoms and fleshy root are radical plants, admits of being easily demonstrated. The preponderance of the root is evident from its size, its quantity of contained sap, or special chemical ingredients. A root that is rich in sap, and much denser than the stalk, is called a turnip. These plants are thus turnip-plants.

1661. Among the Monopetalae, however, there are napiform roots only in the Epigynes, namely, the Syngenesia or salad-plants, and among some Perigynes, namely, the Campanuleae, as well as the Cucurbitaceae. The Syngenesia are consequently the radical plants. The roots of the Scorzonerae, Pastinaceae, Cichoraceae, and Tussilago or coltsfoot, &c., belong to this class.

1662. That the Scabiosae and Valerianeae are directly related to the Syngenesia is likewise indicated by their roots. Unto these succeed in point of structure the Campanulaceae and the Cucurbitaceae, which have frequently, too, napiform or turnip-shaped roots.

1663. The number of the Syngenesia is so great that they fill up all the orders of the trunk. In accordance with their whole structure they are obviously the lowest, the stalk being for the most part herbaceous and placed within a circle of radical leaves, but being itself provided with few, imperfect, and scarcely ever pinnate leaves; they have moreover numerous stunted blossoms that are connate with the single or solitary seed, and crowded together, like spadices, grass-spikes, or the fungal pilei, upon a carpoclinium or receptacle.

1664. They are a repetition of the fungi and grasses; of the former in their fleshy root and inflorescence, of the latter likewise in the inflorescence and in the spathoidal root-leaves; above all, in the single large seed, confluent with the ovarium and calyx. The principle of their division must, where it is possible, and for obvious reasons, be drawn from the organs of the trunk.

1665. First order. Radicariae parenchymatosae.—Syngenesia, having radical leaves and uniform florets, tubular or wholly stunted ligular florets—Cichoraceae and Thistles.

1666. Second order. P. vaginatae.—Syngenesia, with opposite leaves and different kinds of florets, such as radiated, partly lingual and partly tubular corollae—Sunflowers, Silphieae.

1667. Third order. R. axonales.—Syngenesia, with alternate leaves and diversified florets—Anthemideae, Senecionidae, Astereae.

1668. Fourth order. R. florales.—Here the ovarium is no longer dense and confluent with the calyx, and it begins to become trilocular—Scabiosae, Valerianeae, and Campanulae.

1669. Fifth order. R. fructuariae.—Here a perfect fruit is developed, which is connate with the calyx—Asarideae, Passifloreae, and Cucurbitaceae. They have apple-like 3-5 ry fruits, and many of them have napiform roots, e. g. the Gichtrüben. They divide into sixteen families. (Vid. Tab. B.)

CLASS VIII.

Stalk-plants.

1670. Plants with a predominating development of stalk, leaves narrow, mostly opposite, quaternary corollae upon the calyx, ovarium multilocular, and containing few seeds.

1671. Here everything, both root and leaf, must be stalk-like in character; the stalk is therefore woody, the root fibrous, the foliage twig-like or narrow, like needles.

1672. This structure is chiefly found in the Heaths and Stellatae. The stalk is mostly woody; the foliage either aciculiform or leathery in texture, and never pinnated. The leaves are either arranged in whorls or opposite, a position which indicates a lower grade of development.

Moreover, they are related to the plants of the preceding class; they are either Epigynous or Perigynous. The corolla and ovarium follow the opposite position of the leaves; the former being quadripartite, the latter bi-and quaternary. Most of them grow in hot countries upon dry ground, and their virtues reside in the stalk, e. g. the Peruvian bark.

1673. The Stellatæ or Rubiaceæ are without doubt the lowest, because they are epigynous and have a quadri-petalous corolla, with a binary and frequently only follicular ovarium.
1674. First order. Cauliariæ parenchymatosæ. The Stellatæ proper along with the Coffeaceæ, all of them being two-seeded.
1675. Second order. C. vaginatæ. The Rubiaceæ with bilocular many-seeded capsules—Rondeletiæ and Cinchoneæ.
1676. Third order. C. axonales. Rubiaceæ with fruits—Guettardidæ, Hameliæ, Gardeniæ.
1677. Fourth order. C. florales. Quadripetalar Perigynæ, with similar capsules or berries—Epacrideæ, Vacciniaceæ, Ericaceæ.
1678. Fifth order, C. fructuariæ. Quaternary Perigynes with fruits—Myrobalaneæ, Olacineæ, Diospyroideæ and Sapoteæ. (For the arrangement of their families vid. Tab. B.)
CLASS IX.
Leaf-plants.
1679. Herbs having broad leaves, quinary hypogynous corollæ, and bilocular capsule. Here the whole stem has become leaf; all the parts are soft; they are herbs in the propermost sense of the word.
1680. Here belong the hypogynous Monopetalæ: Primulaceæ, Personatæ, Solaneæ, Gentianeæ, Asclepiadæ, Cariceæ, Asperifoliæ, Sambuceæ. The roots are fibrous; the stalk herbaceous, being wholly covered, and that indeed with large leaves; calyx and corolla quinquepartite, frequently bilabiate; the germen a bilocular membranous capsule, which seldom becomes fleshy, and contains few seeds. It is these plants which serve chiefly as food for cattle, and whose whole trunk is officinally known under the name of herbage; relations which express the leafy character.

1681. First order, Foliariæ parenchymatosæ. Herbs with bilocular capsules, in which there are many seeds placed upon a median cone. The capsule dehisces, while both carpels separate from each other—Primulaceæ, Scrophulariæ, Solaneæ.
1682. Second order, F. vaginatæ. For the most part herbaceous, their many-seeded carpels springing open at the dorsal suture—Orobanchaceæ, Rhinanthaceæ, and Bignoniaceæ.
1683. Third order, F. axonales. Regular quinary corollæ with seeds borne upon the margins of the two carpellar valves—Gentianaceæ, Asclepiadaceæ, Jasmineæ.
1684. Fourth order, F. florales. Few seeds in one capsule; the pistil becomes nut-like or trilocular—Labiatæ, Polemoniaceæ, Convolvulaceæ.
1685. Fifth order, F. fructuariæ. Herbs and shrubs with fruits; as nuts, plums and berries. For their sixteen families (vid. Tab. B.)
B. BLOSSOM-PLANTS.
1686. Flowers polypetalous.
SECOND CIRCLE.
Flower-plants.
1687. Calyx, corolla, stamina and ovarium perfectly separated from each other—Pedunculate corollæ, Stielblumen or Hypogynes. The blossom must be here developed in the most perfect

manner; i. e. all its parts must be complete and separated from each other. This is the case only in the hypogynous Polypetalæ.

1688. The lowest organized kinds must, from being a repetition of them, remind us of the grasses and Syngenesia. They are therefore polycarpellar or multi-ovarial.—Ranunculaceæ, Malvaceæ, Magnoliaceæ.

1689. Unto these are allied those plants whose ovaria consist of several carpels, but that are connate with each other, and mutually separate for the first time with their maturation or decay, as in the Rutaceæ, Polygaleæ, Malvaceæ, Aurantiacæ, Platanaceæ, Malpighieæ, Sapindeæ.

1690. The highest are characterized by coalition of the carpels into a single ovarium with stunted dissepiments, and by corollæ that are well developed and distinguished for colour, delicacy, and magnitude—Carnations, Violets, Cistaceæ, Siliquosæ, Poppies, Gamboge-trees.

CLASS X.
Seed-plants.

1691. Plants having a preponderance of seed, that draws after it all the floral parts.

1692. The ovaria have become seed-like, have separated from each other, and inclose for the most part only a single seed.

1693. As in the grasses and Syngenesia many flowers are collected into a spike or upon a receptacle, so here are many carpels in a single corolla—Ranunculaceæ, Geraniaceæ, Tiliaceæ, Malvaceæ, Magnoliaceæ.

1694. The stamina are usually of indefinite number, and mostly connate.

1695. All forms of stalks are here met with; such as herbs, bushes, shrubs, and trees. All forms also of leaves; spathose leaves, petioled leaves, simple and divided, yet rarely pinnated.

1696. The component parts are usually mucilage, as in the roots of the Syngenesia.

1697. They divide into two great groups, into the quinary and sexanary. Since among the quinary, herbaceous stalks with nodes and spathe-leaves, but capsules only, occur; they must be arranged in the lowest rank. The sexanary bear fruits.

1698. First order, Seminariæ parenchymatosæ. Herbs with nodes and spathose leaves, together with numerous, mostly one-seeded, carpels, attached in an irregular manner to a median columella—Ranunculaceæ and Geraniaceæ.

1699. Second order, S. vaginatæ. Trees having many-seeded carpels, coalesced like the style, Theaceæ, Tiliaceæ, Elæocarpaceæ.

1700. Third order, S. axonales. Bushes and shrubs with free, mostly simple, leaves, ovaria, mostly one-seeded, and arranged in a circle around the mediate axis or columella; anthers bilocular—Hermanniaceæ, Dombeyaceæ, Sterculiaceæ, and Büttneriaceæ.

1701. Fourth order, S. florales. For the most part trees, having frequently divided leaves and similar ovaria, yet mostly many-seeded and connate—Malvaceæ and Bombaceæ.

1702. Fifth order, S. fructuariæ. Corollæ mostly sexanary, ovaria in a circle without a columella—Magnoliaceæ, Menispermaceæ, Dilleniaceæ, Anonaceæ. They divide into sixteen families. (Vid. Tab. B.)

CLASS XI.
Ovarium-plants.

1703. Hypogynous Polypetalæ with perfect multilocular ovarium—Polygaleæ, Melieæ, Aurantiaceæ, Platanaceæ, Malpighieæ, Sapindeæ.

1704. While in the preceding class the number of the carpels was usually indeterminate; it is here limited to three and five. In the one, they usually stood around a middle columella or axis, in the other, they form a true capsule with perfect partition-walls or septa, and a single style. The number of the seeds is moderate, i. e. there are more than one, but they are easily counted. They are therefore of mediate size, having no kernels, as is the case in the nuts, but also no granules, as in the berries or poppy capsules. In the preceding class the fruits were rare; here they occur frequently in the upper orders. The number of the floral parts is, throughout the present class, five, that of the stamina five or ten, and they are seldom coalesced; the stalk also passes through all the stages of development from that of the herb through the shrub into the tree. The leaves are seldom spathe-like, but frequently coriaceous and aciculate, as in their predecessors the Heaths; many are pinnated.

1705. First order, Capsulariæ parenchymatosæ—Rutaceæ. Herbs and shrubs frequently provided with aciculate and coriaceous leaves; corollæ regular with ten stamina; ovarium consisting of five carpels, which separate when ripe, and contain few seeds—Rutaceæ, Diosmaceæ.

1706. Second order, C. vaginatæ. Shrubs and trees with similar corollæ and ovaria, which are nevertheless frequently separated and fleshy—Quassiaceæ, Ochnaceæ.

1707. Third order, C. axonales. Mostly shrubs and trees with irregular corollæ and bilocular ovarium—Polygalaceæ, Vochysieæ, Pittosporeæ.

1708. Fourth order, C. florales. Trees with a woody or berry-like ovarium, having several cells—Cedreleæ, Melieæ, Aurantiaceæ.

1709. Fifth order, C. fructuariæ. Trees; flowers quinary, ovarium mostly ternary, becoming a winged or fleshy fruit. (Their sixteen families probably pursue the order indicated at Tab. B.)

CLASS XII.
Corolla-plants.

1710. Corollæ stipaceous, having free stamina; ovaria with stunted septa, and numerous marginal seeds—Carnations, Violets, Cistaceæ, Siliquosæ, Papaveraceæ, Guttiferæ.

1711. The stalk passes through all the stages of formation, from the nodose herb unto the shrub and tree. The leaves occur likewise under all forms, modes, division, and arrangement. The plants of this class are found in all climates, and yield etherial or volatile and fatty oil with resins. They divide first of all into quinary and quaternary; the former being mostly herbs with a hollow capsule; the latter herbs, shrubs, and trees, with siliquæ or berries.

1712. Their strength or virtue resides in the flower, which, is therefore of large size, beautifully coloured, sweet-scented, is frequently appreciated and adopted as an ornament. On the contrary, the ovarium and seed are stunted. The former is a siliqua or hollow capsule, which therefore supports the numerous smaller seeds on the septum.

1713. First order, Corollariæ parenchymatosæ. Nodose herbs with spathose leaves, quinary corollæ, and ten stamina; many seeds upon a columellar or median placenta situated within a hollow capsule—Portulaceæ, Carnations.

1714. Second order, C. vaginatæ. Herbs, shrubs, and trees, with similar but mostly multi-staminal corollæ, and seeds on the capsular septa—Droseraceæ, Hypericineæ, Violaceæ, Cistaceæ, and Bixeæ.

1715. Third order, C. axonales. Herbs with quaternary corollæ and a siliqua—Siliquosæ.

1716. Fourth order, C. florales. Bushes and shrubs with quaternary corollæ and numerous stamina; ovarium a siliqua or multi-valvular hollow capsule—Capparidæ, Berberideæ, Papaveraceæ.

1717. Fifth order, C. fructuariæ. Trees with quaternary and quinary corollæ, numerous stamina, and a fruit—Guttiferæ. They divide into the usual sixteen families. (Vid. Tab. B.)

THIRD CIRCLE.
Fruit-plants—Apetalæ, Perigynes.

1718. Stunted calycine corollæ, with nuts, plums, berries, or apples.
1719. They are perigynous Polypetalæ, and include the Apetalæ and Diclines.
1720. The nut consists of a large seed, connate with the woody ovarium, and frequently with the calyx.
1721. The plum is a legumen, between the coats of which fleshy matter has accumulated, and whose internal tunic or endocarp has become woody. The berry is a many-seeded hollow capsule, which, as well as the calyx, has become soft and succulent.
1722. The apple is an ovarium surrounded by a fleshy calyx.

CLASS XIII.
Nut-plants—Apetalæ, Diclines.

1723. Ovarium woody, and inclosing only one seed. Here belong the Apetalous and Diclinous Exogens.
1724. These plants repeat the fungi, grasses and Syngenesia, and have therefore imperfect corollæ, the calyx of which has alone remained, and usually closely surrounds the nut.
1725. The stalk is indeed usually woody; yet is still found to be also herbaceous and nodose with spathiform leaves. The leaves are simple, frequently needle-shaped or else arrested. The principal ingredients are starch, as in the Gramineæ and Syngenesia.

1726. The inflorescence is mostly amentaceous, as in the Agarics, Grasses and Syngenesia.
1727. They divide into androgynous and diœcious plants.
1728. First order, Nucariæ parenchymatosæ. Hermaphrodite herbs with nodes and spathose leaves, calyx green, superior, and quinquepartite with five opposite stamina; nut mostly triangular and utricular.
1729. Second order, N. vaginatæ. Form of vegetation pretty nearly as in preceding order, but the calyx is corolla-like, and the stamina mostly alternate—Phytolacceæ, Illecebreæ.
1730. Third order, N. axonales. Hermaphrodite, calyx corolla-like, superior and wholly quaternary; herbs and shrubs bearing nuts and plums—Nyctagineæ, Daphneæ, and Santalaceæ.
1731. Fourth order, N. florales. Trees, calyx corolla-like but inferior; capsules, plums and berries—Proteaceæ.

1732. Fifth order, N. fructuariæ. Diclines; herbs, shrubs and trees without corollæ, but with nuts or plums. (Their sixteen families stand in the order exhibited at Tab. B.)

CLASS XIV.
Plum-plants—Papilionaceæ.

1733. Polypetalous calycine corollæ, with a drupe or its fundamental form, the legumen. Here belong the papilionaceous plants, Rhamnaceæ and Terebinthaceæ.
1734. The stalk is frequently herbaceous with nodes; but mostly fruticose and arborescent.
1735. The leaves here attain their highest development, and are mostly pinnated, sometimes endowed with the power of independent motion.
1736. The corollæ are mostly irregular, quinary, arranged like pinnate leaves, with ten, rarely more, connate and free stamina.
1737. The ovarium is a single carpel, owing to the four others being arrested; it is usually compressed and bivalved, with few seeds; it is a legumen, frequently converted into a fleshy fruit.
1738. The Papilionaceæ are so rich in number that they include all the orders of the trunk, and even transcend or exceed its limits; their allied families are the Rhamnaceæ and Terebinthaceæ

with fleshy fruits.

1739. First order, Drupariæ parenchymatosæ. Papilionaceæ with herbaceous, nodose stalk and pinnated leaves; corolla-petals and single stamen free; seed-lobes thin—Hedysareæ, Astragaleæ, Glycineæ.

1740. Second order, D. vaginatæ. Herbs and shrubs with ternary or tendrilless leaves; the corolla-petals or stamina frequently confluent—Trifoliaceæ, Genisteæ, Galegeæ.

1741. Third order, D. axonales. Bushes, shrubs, or trees, frequently training with pinnate leaves and tendrils; calyx large, seed-lobes thick—Vicieæ, Beans, Dalbergiæ.

1742. Fourth order, D. florales. Shrubs and trees, with tolerably regular flowers and separated stamina; legumens with frequently transverse septa, embryo straight—Geoffroyæ, Swartzieæ, Detarium, Mimosæ, and Cassiæ.

1743. Fifth order, D. fructuariæ. Regular flowers with separate stamina and multilocular plums—Stackhouseæ, Empetreæ, Celastrineæ, Rhamneæ, and Terebintheæ. (For their sixteen families vid. Tab. B.)

CLASS XV.

Berry-plants—Umbelliferæ, &c.

1744. Ovarial and calycine corolla with a single perfectly soft fruit or berry, on which are five corolla-petals, with a moderate number of stamina. This fruit is wholly edible, and has only one or two styles; stalk and leaves pass through all the stages of formation.

1745. They divide first of all into two groups, having few or many stamina; of these the fruit of one is dry, of the other fleshy. The dry fruits are also perfectly edible, as the Caraway-seeds.

1746. First order, Baccariæ parenchymatosæ. Epigynous; nodose herbs with two seeds in the calyx; only five stamina—Umbellatæ.

1747. Second order, B. vaginatæ. Mostly shrubs with quinary corollæ and bi-or quinquelocular berries—Mistletoes, Elders, Araliaceæ, Vines.

1748. Third order, B. axonales. Bushes and shrubs with quaternary corollæ, only one style and one multilocular many-seeded capsule—Epilobeæ, Salicariæ.

1749. Fourth order, B. florales. Mostly shrubs with quinary corollæ and manifold stamina; capsule or berry multilocular—Melastomaceæ.

1750. Fifth order, B. fructuariæ. Trees with many, frequently fasciculated, stamina; fruit multilocular and many-seeded—Myrtaceæ. These plants likewise divide into sixteen families. (Vid. Tab. B.)

CLASS XVI.

Apple-plants—Rosaceæ.

1751. The fruit is an apple, i. e. several carpels, containing but few seeds, adhere in one calyx, upon which there are five corolla-petals, with four to six times as many stamina. Herbs, shrubs, and trees with different kinds of leaves are here met with; the corollæ are mostly small and perigynous, the style separated, and thus polycarpal. Perigynous Polycarpæ. They grow, dispersed over the whole earth, in dry situations; several of them yield edible fruits, and are pretty generally cultivated.

1752. The apples are without doubt the most perfect fruit, as well in reference to their structure as their chemical ingredients. The apple consists of all the parts of the blossom: seed, capsule, and fleshy calyx, and is besides polycarpellar, i. e. composed of separate carpels. Its edible substance or flesh is not simply a sweet-meat, but a true aliment, which admits of being eaten after it has been kept fresh for a year, of being dried, exported, or cooked as a kitchen vegetable; in cases of exigency too it occurs in such abundance that the whole human race might live upon it, which cannot be said of any other fruit. The apple quenches at the same time the thirst, and thus supplies also the place of drink. All the other fruits are either a dainty relish only against thirst, or a simple amylaceous medium of nutrition. However, most of the plants belonging to this class produce only dry capsules and calyces.

1753. They divide into two groups, having few or many stamina.

First order, Pomariæ parenchymatosæ. Herbs with few stamina and five or more carpels, as the Crassulaceæ and Mesembryanthemaceæ.

Second order, P. vaginatæ. Shrubs with few stamina, only two to three carpels and few seeds.—Tamaricaceæ, Bruniaceæ.

VEGETABLE SYSTEM Table B
FIRST PROVINCE.
HISTOPHYTA—ACOTYLEDONES.

	CLASS I.	CLASS II.	CLASS III.	
	Cellulares.	Vasculares.	Tracheales	
ORGANS.	Fungi.	Mosses.	Ferns.	
Ord. I. Parenchyma.	Roste.	Schleim-Moose.	Wasser-Farren.	
F. 1.	Cells.	Entophyta.	Diatomaceæ.	Marsileaceæ.
2.	Ducts.	Epiphyta.	Nostochineæ.	Pilulariæ.
3.	Tracheæ.	Tuberculariæ.	Batrachospermeæ.	Equisetaceæ.
O. II.	Sheaths.	Schimmel.	Wasserfäden.	Kugelfarren.
F. 4.	Bark.	Mucorini.	Confervaceæ.	Lycopodiaceæ.
5.	Liber.	Mucedines.	Ulvaceæ.	Ophiogloss eæ.
6.	Wood.	Byssaceæ.	Leptomiteæ.	Danæaceæ.
O. III.	Axis.	Balgpilze.	Fuci.	Ringfarren.
F. 7.	Root.	Trichodermaceæ.	Cerameæ.	Parkereæ.
8.	Stalk.	Trichiaceæ.	Floridæ.	Cyatheæ.
9.	Leaves.	Lycopodineæ.	Fucoideæ.	Polypodiaceæ.
O. IV.	Blossom.	Kernpilze.	Lichens.	Fluvialen.
F. 10.	Seed.	Cytisporeæ.	Coniothalameæ.	Naiadeæ.
11.	Ovarium.	Phacidiaceæ.	Gasterothalameæ.	Podostemeæ.
12.	Corolla.	Sphæriaceæ.	Hymenothalameæ.	Ceratophylleæ.
O. V.	Fruit.	Fleischpilze.	Mosses.	Nadelholz.
F. 13.	Nut.	Tremellini.	Marchantiaceæ.	Abieteæ.
14.	Plum.	Cupulati.	Ricciaceæ.	Taxaceæ.
15.	Berry.	Clavati.	Jungermanniaceæ.	Cupresseæ.
16.	Apple.	Pileati.	Bryaceæ.	Cycadeæ.

SECOND PROVINCE.
THECOPHYTA—MONOCOTYLEDONES.

	CLASS IV.	CLASS V.	CLASS VI.
ORGANS.	Corticales	Liber-plants.	Wood-plants.
	Gramineæ.	Liliaceæ.	Palmaceæ.
Ord. I. Parenchyma.	Spiked Grasses.	Staub-Orchiden.	Rohrkolben.

F. 1.	Cells.	Secale.	Neottiæ.	Cynomoria.
2.	Ducts.	Phleum.	Arethuseæ.	Typhaceæ.
3.	Tracheæ.	Festuca.	Ophrydeæ.	Aroideæ.
O. II.	Sheaths.	Panicled Grasses.	Granular Orchideæ.	Piperaceæ.
F. 4.	Bark.	Bulrushes.	Malaxides.	Saurureæ.
5.	Liber.	Milium.	Epidendra.	Piperaceæ.
6.	Wood.	Arundinaceæ.	Vanillæ.	Pandanac.
O. III.	Axis.	Reed-Grasses.	Aromatic plants.	Sassaparillæ.
F. 7.	Root.	Sedges.	Amomeæ.	Dioscoreæ.
8.	Stalk.	Junceæ.	Canneæ.	Smilaceæ.
9.	Leaves.	Ruscus.	Musaceæ.	Parideæ.
O. IV.	Blossom.	Juncineæ.	Schwerdel.	Asparagoidæ.
F. 10.	Seed.	Restiaceæ.	Hæmodoraceæ.	Asparagoidæ.
11.	Ovarium.	Serpeæ.	Irideæ.	Convallariæ.
12.	Corolla.	Commelyneæ.	Narcissi.	Bromeliæ.
O. V.	Fruit.	Seeroseæ.	Liliaceæ.	Palms.
F. 13.	Nut.	Alismaceæ.	Colchicaceæ.	Calamariæ.
14.	Plum.	Hydrocharideæ.	Asphodeleæ.	Cocoeæ.
15.	Berry.	Hydropeltideæ.	Alliaceæ.	Phœniceæ.
16.	Apple.	Nymphæaceæ.	Tulipaceæ.	Borasseæ.

THIRD PROVINCE. ARTHROPHYTA—DICOTYLEDONES.
Circle I. Stem-plants—Monopetalæ.

CLASS VII.	CLASS VIII.	CLASS IX.
Root-plants.	Stalk-plants.	Leaf-plants.
Epigynæ.	Perigynæ.	Hypogynæ.
Root-leaved.	Stellatæ.	Personatæ.
Cichoraceæ.	Galiaceæ.	Primulaceæ.
Carduraceæ.	Spermacocidæ.	Scrophulariæ.
Mutisiæ.	Coffeaceæ.	Solanaceæ.
Opposite-leaved.	Cinchonaceæ.	Bignoniaceæ.
Eupatorium?	Hedyotidæ.	Orobancheæ.
Helianthus.	Rondeletiæ.	Rhinantheæ.
Silphieæ.	Cinchoneæ.	Bignoniæ.
Alternifoliar.	Hameliaceæ.	Contortæ.
Senecionidæ.	Guettardidæ.	Gentianeæ.
Astereæ.	Hameliæ.	Asclepiadeæ.
Vernoniæ.	Gardeniæ.	Jasmineæ.
Aggregatæ.	Ericaceæ.	Tetraspermeæ.
Scabiosæ.	Epacrideæ.	Labiatæ.
Lobeliæ.	Myrtilli.	Polemoniaceæ.
Campanulæ.	Heaths.	Convolvulaceæ.
Sicuonoidæ.	Diospyraceæ.	Pyrenaceæ.
Asaridæ.	Myrobalaneæ.	Asperifoliæ.
Loaseæ.	Olacineæ.	Verbenaceæ.
Passifloreæ.	Diospyros.	Sambuceæ.
Cucurbitaceæ.	Sapotæ.	Myrsineæ.

THIRD PROVINCE. ARTHROPHYTA—DICOTYLEDONES.
Circle II. Flower-plants—Hypogynous Polypetalæ.
CLASS X.　　　CLASS XI.　　　CLASS XII.
Seed-plants.　Ovarium-plants.　Corolla-plants.
Polycarpæ.　　Capsuliferæ.　　Siliquosæ.
Ranunculaceæ.　Rutaceæ.　　Caryophyllaceæ.
　Ranunculi.　Ruteæ.　　　Portulaceæ.
　Helleboreæ.　Diosmeæ.　　Spergula.
　Geraniaceæ.　Zygophylleæ.　Carnations.
Tiliaceæ.　　　Ochnaceæ.　　Violaceæ.
　Theaceæ.　　Xanthoxyleæ.　Droseraceæ.
　Lime-trees.　Ochneæ.　　　Violets.
　Elæocarpeæ.　Quassieæ.　　Cistaceæ.
Sterculiaceæ.　Polygalaceæ.　Cruciferæ.
　Hermanneæ.　Polygaleæ.　　Radishes.
　Dombeyaceæ.　　Vochysieæ.　Cresses.
　Sterculeæ.　　Pittosporeæ.　Cabbages.
Malvaceæ.　　Meliaceæ.　　Papaveraceæ.
　Malveæ.　　Cedreleæ.　　Capers.
　Hibisceæ.　　Melieæ.　　　Berberideæ.
　Bombaceæ.　Aurantiaceæ.　　Poppies.
Magnoliaceæ.　Malpighiaceæ. Guttiferæ.
　Magnolieæ.　Platanaceæ.　Dipterocarpeæ.
　Menispermeæ.　　Malpighieæ.　Calophylleæ.
　Dilleneæ.　　Hippocrateæ.　　Marcgraviaceæ.
　Anoneæ.　　Sapindeæ.　　Garcinieæ.
THIRD PROVINCE. ARTHROPHYTA—DICOTYLEDONES.
Circle III. Fruit-plants—Apetalæ and perigynous Polypetalæ
CLASS XIII　CLASS XIV　CLASS XV　　CLASS XVI
Nut-plants　Plum-plants　Berry-plants　Apple-plants
Nucariæ.　　Drupaceæ.　　Baccariæ.　　Pomaceæ.
Oleraceæ.　　Astragalaceæ. Umbellatæ.　Sempervivæ.
　Sclerantheæ.　Hedysareæ.　Saniculidæ.　Galacineæ.
　Atriplices.　Astragaleæ.　Carrots.　　Crassulaceæ.
　Amaranths.　Glycineæ.　　Caraways.　Mesembryanthemeæ.
Polygonaceæ.　Trifoliaceæ.　Caprifoliaceæ. Tamariscineæ.
　Plantains.　　Trifolieæ.　　Mistletoes.　Tamarisks.
　Phytolacceæ.　　Genisteæ.　Sambuceæ.　Buniaceæ.
　Illecebreæ.　Galegeæ.　　Vines.　　　Hamamelideæ.
Thymelaceæ.　Phaseolaceæ.　Lythraceæ.　Saxifrageæ.
　Nyctagineæ.　Vetches.　　Trapeæ.　　Saxifrages.
　Daphnoideæ.　Dalbergieæ.　Epilobeæ.　Cunoniaceæ.
　Santalaceæ.　Sophoreæ.　　Salicariæ.　Lilacs.
Lauraceæ.　　Mimosaceæ.　Melastomaceæ.　Rosaceæ.
　Proteaceæ.　Detarieæ.　　Rhexieæ.　　Potentillæ.
　Aquilarineæ.　Mimoseæ.　　Melastomeæ.　　Neuradeæ.

Laurels.	Cassieæ.	Grossularieæ.	Sanguisorbeæ.
Diclinisten.	Terebinthaceæ.	Myrtaceæ.	Fruit-trees.
Amentaceæ.	Empetreæ.	Lecythideæ.	Pomegranates.
Nettles.	Celastrineæ.	Barringtonieæ.	Plums.
Artocarpeæ.	Rhamneæ.	Leptospermeæ.	Medlars.
Euphorbicæ.	Terebiatheæ.	Myrteæ.	Apples.

Third order, P. axonales. Herbs, shrubs, and trees, with similar blossoms, but many seeds.—Saxifrageæ, Cunoniaceæ, Hortensiæ and Lilacs.

Fourth order, P. florales. Herbs and shrubs, with many stamina and follicles.—Rosaceæ.

Fifth order, P. fructuariæ. Trees with multistaminal-corollæ and fleshy fruits.—Monimieæ, Pomegranates, Plums, Medlars, Apples. (For their sixteen families vid. Tab. B.)

1754. In order to prove that each class of vegetables commences from below, and that all ascend in a parallel series beside each other, it is requisite only to place them in a tabular form. Further details upon this subject are to be found in my earlier works upon Natural History. The adjoining Table may here suffice. I place the families according to my "Lehrbuch der Naturgeschichte," although I am very well aware that all do not stand in the right place. That indeed no one will expect.

THIRD KINGDOM. ANIMAL KINGDOM.

1755. The animal kingdom is the individual development of all four elements.
ZOOSOPHY,
1756. Is the development of the animal kingdom in consciousness. The repetition of the animal creation is spiritually divisible into Anatomy (Zoogeny), Physiology (Zoonomy), and Zoology.
I.—ZOOGENY.
1757. Zoogeny represents the idea of the animal or the developmental history of the individual animal.
IRRITABILITY OF THE BLOSSOM.
1758. The Highest to which the vegetable kingdom could attain was the blossom; and of this the sexual parts are the completion. At the instant when the sex originated, the vegetable functions became of a nobler character; for the sexual organs are only the inferior organs refined or purified by light. The electrical and chemical process of the vegetable body are again represented, but in a spiritual manner, in the blossom. The functions of the fruit were none other than those of the elevated chemism; they were only the nobler processes of digestion and nutrition. As their purest expression of life, and that which has been produced simply by their co-operation, is the motion which resides in cellular tissue, so also was it this only, which obtained a preponderance in the fruit, and that indeed at the cost of the material processes. The germen obtained a kind of motion; yet this appears to be still imparted by material processes. In the corolla, however, this expression of life attained to completion. It is no longer simple nutrition or accumulation of sap that moves the stamina upon the female stigma, but a purely polar act; the Immaterial, the Spiritual produces the phenomena of life. These movements of copulation are by no means a coalescence, are not a nutritive act, nor the result of mechanical decissation, as is the case in many capsules; but true elevated vital actions. The parts reassume after the motion their first or original position, a feat which is performed by no capsule upon its dehiscence. Of these

movements, those of the leaves in the sensitive plants, e. g. Hedysarum gyrans, are the antetypes or prefigurations. There consequently originates with the highest development of the light-organs in the plant, a motion independent of the material processes, and of the terrestrial elements.

1759. A motion, that is liberated from the terrestrial elements, is free from their mechanism; it simply obeys the nature of the æther, which is of a spiritual, or voluntary, kind.

1760. The essence of volition or free-will does not reside, in a physical sense, in the consciousness of the action, but in the autocrasy; or in the ability to perform an action, without external or terrestrial influence. The æther-actions have originated from special polarity. Independent movements must therefore be such as have been produced simply by polarity, apart from material intrusion.

1761. The ability or power upon the part of organic bodies to apprehend polar excitation, to move themselves simply by its means, and again to restore themselves to their former state, without regard being paid to a material process, I call Irritability. That organ is irritable which can move itself without any other object than that of automacy or self-motion.

1762. Irritability belongs to the plant, but is only of that kind where the perception cannot express itself otherwise than by direct motion. In the sexual parts and probably in the highest leaf-formation, the plant is elevated to irritability; unto motion through simple perception, unto motion without design, to motion from mere lust or desire. The highest spiritual operation, of which the plant is capable, is irritability. But as everything which has attained its Highest, stands at the end of its development, so also has the plant terminated, when it has once exercised its power of irritability through copulation.

SEXUAL MOTION.

1763. All the irritable motion of the vegetable may be confined or reduced to the movements of the stamen-filaments, the other movements being merely precursors of these. What, therefore, the stamina would achieve by their motion, that the irritability does in a general point of view. The motion of the stamina is directed merely upon the stigma, in order to impart the male pollen to the female body; and thus, simply to evoke the spiritual tension, which resides originally in the male semen, as in light-æther from the dead mass, but that originally dwells in the female seed as in the dark mass of earth.

1764. Now, since the stigma bears simply relation to the contents of the ovary, and conveys everything to this, and thus, to a female utricle, which is the middle of the plant, or its body proper; so in the motion of the male organs, the conatus or effort upon their part to introduce an elemental matter, or rather its spirit into this utricle, or this body, is rendered manifest or revealed. The highest Spiritual of the plant is accordingly not a mere motion in the general sense; but one that is definite and wholly special—a motion of Ingestion. The direction or design of the first independent motion is therefore Ingestion; this again not being of a general, but a wholly definite import, namely, an ingestion of the male organ into the female.

1765. The Male is, however, characterized by its self-substantial polarity, and by its self-intrinsic life; the Female by the want of polarity, by a heterodependent life. The act of ingestion thus depends upon polarization, upon evocation and maintenance of an independent life. The fruit is vivified by impregnation, ay, for the first time then obtains life; the female becomes self-active through the vital spirit received from the male; the body is kept alive by, and only by, ingestion. The act of ingestion is the act that tends unto the self-substantial, independent life.

1766. The blossom dies, so soon as, by ingestion, it has attained this independent life. If we assume that it could not die, but retain the life that was momentarily won for several instants; then this would happen only by repetition of the first act, whereby it had obtained in an instant a self-substantial life; and thus by repetition of the ingestion. By ceaseless ingestion only can the blossom gain a lasting, self-substantial life of motion.

1767. But such a blossom, when self-dependently subsisting, could not continue in further union with the vegetable stock or trunk, for this is no longer requisite for its life's support; through the first act of vivification, through the once sprinkled pollen, it virtually becomes detached and falls as fruit to the ground; as a fruit certainly, or as a female body, unto which is wanting the continued excitation produced by the male coition. A fruit thus detached or fallen off, and retaining the male filaments, which ceaselessly exercise the function of ingestion, will be of necessity engaged in constant motion; will be a blossom, that incessantly practises copulation.

1768. As in this blossom the motion of ingestion is that alone which sustains it, and nothing more can flow to it from a stem; so also will this blossom be occupied in constant motion; and it consequently comes to pass that the action, which broke forth at the last and instantaneously in the plant, being thus the highest or most individualized, is here the first, inferior, and most general action, or one that lies at the foundation of all other processes. The free blossom is naught but movement of ingestion.

1769. The blossom, however, concentrates in itself all the lower vegetable processes, is itself naught but the aggregate of such processes repeated in the body of the light; thus the blossom, having been set free, is a vesicle of ingestion endowed with all terrestrial functions.

ANIMAL FORMATION.

1770. The vegetable blossom loses its definition as plant, so soon as it has acquired self-substantial life; it loses its definition, because as blossom, it simply lives in light, while the plant must half dwell in darkness; it loses it, because the copulating motion can be frequently self-repeated.

1771. The self-moveable or automatic blossom has consequently passed over into a new kingdom, or into one whose very definition is the self-substantial motion.

1772. A blossom, which, when separated from the stem, maintains by its own motion the galvanic process or the life within itself, and gets its process of polarization, not from a body lying external to or coherent with it (as is the vegetable stem); but only from itself—such blossom is an Animal. An animal is blossom without stem or a flower, which of itself produces its stem, this being the reverse of what takes place in the plant. The essence of the animal consists in the maintenance through its own motion of the galvanic vital process. It has been already shown above, in speaking "quoad" the distinction of the organic essences, that the exclusive or only valid difference between plant and animal, was motion arising in the latter apart from any external stimulus. We have now been brought by quite another route to the same result.

1773. If the animal is the floral vesicle living from or by itself, it can no longer lie fettered like the plant between two elements; it must be nominally free from the chains of darkness, and thus from the earth. No animal is so joined by growth with the earth, that like as in a plant this should be a co-operating pole in its processes. No animal must co-exist in two elements like the plant, but it has all elements in itself, like the blossom includes all vegetable parts. It may be said that the plant has been immersed in the earth, water, and air; these three elements having, on the contrary, been immersed in the animal. The animal is in respect to them the continent, the planet; they, however, are the continent in respect to the plant. Thus the relations to the world are completely reversed in both.

1774. An animal is a floral vesicle freely separated from the earth, and living alone through its own motion in the water and air. And here it is not locomotion that is treated of, since this by no means belongs to the animal's essence. Yet on this account the poor oyster was formerly adduced as an argument against this animal character, but unjustly; for would a man frozen up in an ice-block lose his character as an animal? The oyster opens its shell and shuts it too, as well as the crocodile opens and shuts its jaws.

ANIMAL SIGNIFICATION.

We will now proceed to knit or associate with this genetic or physiological, the physio-philosophical mode of development.

1775. Every Organic originates from a mucus-point. If this mucus-point occur in the darkness, it thus becomes a terrestrial organism, a plant; if it enter into the light, which is only possible in the water and in air, it thus becomes a solar organism, independent of the planet, self-moving around itself like the sun, an animal.

1776. An animal is a light-mucus-vesicle, a plant, so far as the root is concerned, a darkness-mucus-vesicle; but it works its way unto air and light, and becomes a light-mucus-vesicle in the blossom.

1777. A free blossom is consequently to be philosophically regarded as equivalent to the primary mucus-vesicle, which has at once developed itself in the water. Now, such an aqueo-mucus-vesicle is directly that which the blossom can first become through a series of developments, and by divestures of the Dark.

1778. The plant is an animal retarded by the darkness; the animal is a plant blooming directly through the light, and devoid of root.

1779. The animal has been posited as a light-or æther-Total upon the planet; the vegetable as a planetary Total in light.

1780. The animal is a whole solar system, the plant only a planet. The animal is therefore a whole universe, the plant only its half; the former is microcosm, the latter microplanet.

ANIMAL PLACE.

1781. No animal can become developed beneath the earth, or where it is absolutely dark and dry. None simply or solely in air. Water is the origin or source of all animals.

1782. They have originated upon the sea-shore, but not in the midst of the sea, nor in that of the land. The deluge cast up the first men. They were littoral inhabitants, and without doubt carnivorous, as savages still are. For whence could they have obtained also fruits, cabbage, and turnips?

SENSATION.

1783. In so far as the animal vesicle is a whole solar system, to it, the characters that transcend the plant, such as motion, belong.

1784. But motion is not that alone which is displayed in the floral cyst when it has become solar; but with it there is yet something higher bestowed.

1785. Like light or sun the vesicle has the principle of its polarization or determination in itself; and it is certainly itself which moves its organs in conformity with this self-determination; but it is withal in antagonism towards the elements, like the sun is towards the planets. Through this antagonism the sun is destined unto the development of light. Although the light is its own product, still it perceives the object toward which the æther-polarity is directed. This perception of the direction, whither motion should act by means of the central polarity, is in the animal called Sensation.

1786. Sensation is the relation of the Central to the Peripheric, of the sun to the planet; motion is the relation of the periphery to the centre, of the planet to the sun. The Animal emerges into being from the alternating play of the supremest antagonism in the heavenly bodies, the

Vegetable from that of the terrestrial antagonisms.
1787. Emission of light is discharging or unloading of the sun achieved by the influence of the planets; sensation is unloading of the animal by objects, by the world, and by its own organs.
1788. Sensation is therefore a positing of want in the animal. Through sensation nothing comes into, but rather something passes out of us.
1789. The animal vesicle is a sentient or feeling blossom.
SEXUAL ANIMAL.
1790. As the essence of the blossom consists in the sex, so indeed is the blossom none other than the sexual system; thus it may be said that the animal vesicle is none other than a sentient Sexual cyst. This discovery is of the highest importance for the whole of zoosophy.
1791. Two fundamental properties are originally combined in the animal, at the very instant when an animal can exist. There is never one without the other, never simply sensation, but also motion; but the latter is not simply this, but is at once also a movement of copulation. The animal is a sentient genital.
1792. Both fundamental properties are however subordinated to each other. The basis of the animal organism is the sexual system; with this the animal commences; what is further developed in addition thereunto, is only higher completion. But what it, as sexual system, does, it does only through sensation.
1793. It is natural that the animal has not been concluded with the sentient sexual system; but that the terrestrial processes also are developed like as in the plant, and perfected indeed more individually than in the latter. The main distinction is, however, this; that these processes have preceded in the plant, and the sexual system grown out of them; in the animal, on the other hand, the sexual system is the foundation, the root, out of which these processes grow forth. The first and simplest animal vesicle is a sexual cyst, a matrix or womb.
SENSITIVE ANIMAL.
1794. The animal is a twofold representation of the organization, being at one time the planetary, and at one time the solar. There is a planetary animal and a solar animal in the higher organism. The planetary animal is the plant in the animal, the galvanic animal; the solar animal is the sentient, the light-or æther-animal.
1795. The highest completion of the vegetable animal is in the blossom. There is thus a sexual or Sex's animal and a sensitive or Sensation's animal.
1796. The completed animal consists of two animals, because it is at once planet and sun, plant and animal.
1797. The vegetable and the sensitive animal have been formed parallel to each other, yet in such a manner that the former being the lowest, contains only the dispositions unto the highest. There are consequently vegetative and animal organs, which take a parallel range. The animal grows upon a vegetable body. It may be aptly said, that the root of the plant becomes the mouth or head of the animal, the stem the trunk or visceral body, the blossom the sexual parts. The three parts therefore of the animal body, and the antagonism therefore of the head with the sexual parts, unite through the medium of the vegetable stem, or the visceral body.
ANIMAL ANATOMY.
1798. The parts of the animal body divide, as in the plant, into tissues, anatomical systems, and into organs proper or members. The tissues are the constituent parts of the systems, these of the organs, and all collectively, or associated, of the body.

I. Animal Tissues.

1799. These are the mathematical primary forms, whereof the animal body consists, and divide into the animal and vegetative fundamental forms.

A. ANIMAL FUNDAMENTAL FORMS.

1800. As animality is the representation of the three conditions of æther, and thus of the gravity, with light and heat or motion, so are three tissues to be met with, which correspond to these three forms. The light emerges from the centre; the gravity occupies the whole mass; the motion oscillates between the two. The organic light-mass will therefore, as sun, occupy the centre, the gravity-mass, like the planets, the periphery, the motion-mass, like the heat, the radii between the two. The primary form is, however, the primary vesicle. If, therefore, new tissues appear in the animal, they can thus be only metamorphoses of the vesicle. The vesicle, can only resolve itself into three forms; either its contents become self-substantial—Point; or in like manner the envelope—Line; or, finally, both become an uniform mass—Globe.

1. Point-tissue.

1801. We take up or commence the study of the animal substance, with the condition under which it has originated, that of a vesicle or sensitive blossom. The vegetable texture attained a form which was prescribed by the light. Now, as the animal vesicle is, in the first place, nothing else but a Sentient, so must the texture of the original animal vesicle be commensurate with this property.

1802. The highest perfection of the blossom was, however, resolution of the texture into the original form of vesicles or granules, a retroconduction of the organic mass to the primary condition, yet under the signification of light. The Highest of the blossom was an organized and designedly-prepared granular texture—pollen.

1803. Now a substance which absolves itself from the terrestrial forms, and would assume the form of the æther, and thus of that which is most discrete, can represent no other form than that of the point. The whole æther is an infinity of non-coherent atoms. This atomic formation, when metatyped or copied in a terrestrial mass, can be none other than a granular substance.

1804. The fundamental substance of the animal is point-substance; but, since the essence of the animal consists in its being a sensitive substance, so must it belong to the latter's essence, that it be atomic or punctiform. The point-texture is equivalent to the sensitive mass.

1805. It might be believed, that as the animal is a floral vesicle, the cystic or cell-form must also lie at its foundation; only, there is another relation beyond that which occurs in the plant. This animal cyst is an already organized cyst, an organ, no longer a component part of an anatomical system; this cystic form cannot, therefore, enter into the texture of the animal mass; yet meanwhile, as is natural, the sensitive mass admits of being reduced, but only as organic in a general sense, to the vesicular form. The lowest animals, such as the Infusoria, Polyps, Acalephæ, or sea-nettles, in short, all Protozoa or mucous animals consist of this point-substance, and are wholly sensitive mass.

Nervous Mass.

1806. The sensitive mass is called in higher animals, Nervous mass. The nervous texture is a conjoined series of mucus-granules, which have become albuminous in character. The nervous mass is the least organized; it has selected the primary forms, which have been preconstructed in æther, or the densely fluid solar matter. The Dominant of the terrestrial organs can have also no other form, than such as agrees with the Dominant or governing primary mass of the planetary system; or it can have none other, because at the instant when it exists, it is sentient. At the first instant of the origin of organic matter, it can, however, originate only as infinitely numerous points; at the termination of the plant, this mass would be displayed as a light-organ; being thus engendered as such, it must forsake the vegetable forms, and assume the universal form-susceptive primary form, which is the form of the point. The floral mass, the delicate petals of the corolla, the stamina and the pollen are to be deemed the first onset or advance that is made to nervous mass. The cellular tissue becomes delicate and gradually resolves itself into granules.

1807. Granular or point-mass is, however, an accumulation of centres. The nervous mass is therefore, in accordance with the conception of the Organic, a repeated, multiplied centre. The nervous mass has therefore a light-function, i. e. the gentlest polarization in the organism. Nervous mass is light-mass.

1808. The animal substance has commenced with the nervous mass; thus with that which is the highest, and which physiologists have deemed to be the ultimate mass. The origin of the animal is from the nerves, and all anatomical systems are only free evolutions or separations from the nervous mass. The animal is naught but nerve. What it is further or in addition, is obtained elsewhere, or is a metamorphosis of nerves. The mucus of the Infusoria, Polyps, and Acalephæ is nervous substance upon the lowest stage or degree, where the other substances that are therein involved and merged, have not as yet been perfected in an isolated manner.

1809. The nervous mass indicates the absolute Indifferent in an animal, and consequently that which is polarizable by the gentlest aura or breath.

Division of the Nervous Mass.

1810. The nervous mass behaves itself also in its production like the solar mass. As from this the planets have through antagonism on the periphery, emancipated themselves, so from the nervous mass have the anatomical systems, which are subservient unto lower purposes.

1811. The development of the animal organs is a constant division of the nervous mass, whereby it becomes more and more divested of their coarse coverings, and traverses the same like radiating æther, illuminant, heating, and moving æther. It is a positing of the centre in the periphery.

1812. When also the other systems have been formed out of the identical nervous mass, still the whole animal body is naught but nervous mass, only, in a crude or inert condition. There is, consequently, no point upon the body, on which the nervous phenomena are absolutely wanting, or where they may not appear under certain relations.

1813. What remains behind of the nervous mass, has now the form of filaments or rays, which are projected from one centre, the brain, to all parts of the periphery.

1814. The nerves being individualized, and withdrawn from the coarser mass, stand in need of no actio in distans, or no nervous atmosphere (although for other reasons such a one may exist), in order that every part of the body should have sensibility or feel; for every substance is verily but an aberrant nervous substance, in which the original spirit is still inherent or abides.

1815. Each part of the body has consequently irritability, and each one has the capacity for sensation; and that, indeed, through and in itself, or not borrowed from, what have been called, nerves; as it is indeed only the coarse nervous envelope of the more delicate nerves.

1816. Yet, meanwhile it is certain, that nothing feels but the nervous mass, because every thing, which feels, does, and hath the power to do so, only in so far as it has been nervous mass. Precisely as all metals are only magnetic in so far as they are metamorphosed iron, which is the primary metal.

1817. Certain formations or textures must, on that account, have different sensations, because they have deviated more or less remotely from the primary texture; without in this any regard being paid to the number of nerves running to such systems.

1818. The transformation of the nervous mass upon the periphery is chiefly imparted by

oxydation, because here the oxygen of the water exerts a direct influence. The transformation does not, however, simply occur upon the periphery, but also internally, and that indeed in a beamy or radial direction. The external parts will be harder, the internal or radial continue soft, but be more solid than the nervous mass itself.
2. Globe-tissue.
1819. The nervous tissue alone cannot constitute the animal substance concerned in all functions, but it must with further development pass over into another. As the æther-mass cannot concentrate itself into the sun without, from the antagonism with the refraction of light, condensing into planets, so also a central mass cannot subsist in an animal, without converting itself on the periphery into one that is planetary or terrestrial.
1820. The antagonistic mass, originating in a peripheral relation in the nervous mass, will surround the residue of the latter like a bladder or cyst, just like the planetary masses, or the colours, have primordially surrounded the sun as great hollow globes. The aggregate and purer nervous mass becomes thus directly the central mass of the animal—the brain.
1821. The limitary mass originates through oxygenation. Thus the colours originate; they are an oxygenated light. Thus has every terrestrial mass originated through combustion. The planets are suns that have undergone combustion; the limitary mass is nervous mass similarly treated and deoxydized.
1822. As having been already subjected to combustion, it therefore becomes polarizable, and consequently susceptible of sensation in the least degree. The limitary mass must be rigid or fixed; for it has indeed originated through fixation of the poles, or through the strongest oxydation. The limitary mass is the most rigid in the whole animal; for it is the primary antagonism with the nervous mass, the ultimate planetary matter, which is characterized by immoveability of the atoms.

1823. The limitary mass must be typical of the earth-element, this being the most rigid, and the end of the oxydation. The limitary mass is the animal earth-mass, just as the median mass is the animal æther- or fire-mass.
1824. The texture of the animal earth-mass must be that of a crystal, but of a round globular crystal; for it is organic mass, and can consequently have been only deposited as vesicle; it is, however, earthy mass, so that the whole vesicle must be, with all its contents, rigidified. Now, the rigidified substance of a vesicle is a globe—the texture of the mass which is opposed to that of the nerves is consequently the globular form.
Osseous Mass.
1825. The rigidified limitary mass, which exhibits histologically dense globes, consists of earthy substance, and surrounds the nervous mass, is osseous mass. The osseous or bony texture is a solid globe or rigidified vesicle, being thus ambitus or boundary, as well as complexus or contents.
1826. Bone can only originate through oxydation of the animal mucous or nervous mass, whereby it is converted into a vesicular form. These vesicles are, however, by virtue of the highest oxydation, which must necessarily enter into antagonism with the highest central organ, converted wholly and thoroughly into rigid substance or earth, which is the maximum of the oxydation or fixation of æther.
1827. The osseous mass, as the organic earth-mass, corresponds to the gravity. It is the materiality in a general point of view in the Organic, and consequently the Inert.
1828. Osseous or limitary organs will become more rigid in the air than in water. The air-

breathing animals must have more perfect bones or harder limitary organs.

1829. Bones are therefore either wanting completely in the aquatic animals, or they are mucus scarcely oxydized, in other words, cartilage; or finally, almost entirely rough carbonated earth, in the corals and shells.

1830. This theory is most beautifully proved in the corals. Internally they consist of granular substance, like the polyps, or of sentient nervous mass; externally they are simply earth or globular form, which is the rudest antagonism presented to the likewise rude central mass.

1831. Bone essentially surrounds the nervous mass. The skull environs and incloses the brain, the vertebræ the spinal cord, the ribs, the visceral nerves the snail's shell all the soft parts of that animal, the coral stem its polyp-tube, the horny coat the insect.

1832. The purest and highest antagonisms in an animal are nerve and bone, and as such they are demonstrated on every occasion. The nerve is that which is soft, powerless, changeable, sentient, governing, and motion-imparting; the bone, what is hard, strong, unchangeable, non-sentient, governed, and becoming moved; the one properly speaking the spiritually vitalizing, the other the spiritually dead, or self-subsistent simply in a mineral point of view. The bone is the obedient planet of the nerve.

1833. Point-and globe-form are consequently, as regards the tissue of the substance, the first two forms of the animal body.

1834. What develops itself apart from nerve and bone in the animal, must either range between or below both; it must participate of both forms, or be only their incompletion.

3. Fibrous Tissue.

1835. The nervous and osseous substance could not range opposite to each other, without a transition or a something interposing; as little as could æther and Terrestrial, or sun and planet, between which the moved æther or the heat oscillates, and conditionates the planetary motion.

1836. Between the soft point-form of the nerve and the hard globe-form of the bone, a semi-oxydation stands midway, just as the air stands between the æther and the earth. As this is the medium element, wherein the light is refracted into colours, and thereby warms and moves the planet, so must this median animal formation be the element, through which the nerve imparts its motion to the bones.

1837. This organ stands, like the air, upon the middle rate of oxydation; the oxygen becomes alternately united with it and set free; which is neither possible in the point-form, as being that which incessantly liberates oxygen, nor in the globe-form, as being that which always holds or retains the oxygen in union with it.

1838. This tissue must consist of firm or solid nervous granules, which have been serially co-arranged in lines or radii. Such organic lines are called Fibres.

1839. The fibrous is the third original tissue, which appears in the animal organization.

1840. The nerve acts upon the fibres as upon the bone, or as a Central upon a Peripheric, as the light upon the air.

1841. Thereby the soft fibre is polarized; the poles are mutually attracted and repelled, and motion of the fibres originates, their extremities approximating or withdrawing by virtue of the polarity. Contractile fibres are called sarcose or fleshy fibres.

Flesh.

1842. The flesh is the median formation between nerve and bone. It is half nervous mass, therefore sentient, half bone, therefore moving.

1843. The essence of the motion resides in the muscle, not in the nerve. Such is the cause of motion, the muscle being the self-moving, the bone that which is moved.

1844. The flesh must surround the bone, as the air or water surrounds the earth.

1845. The flesh is a terrestrial substance, just as the bone is; the nerve is a cosmical substance, and on that account the mediator of everything.

B. VEGETABLE TISSUES.

4. Cellular Tissue.

1846. There are three, and only three constituent forms essential to the animal substance, the point, the globe, and the line; equivalent to centre, periphery, and radius.

1847. Out of these three all other forms, of whatever kind, are developed, through degradation unto vegetable structure. This form can be none other than the Cell-form. In the animal there are therefore four fundamental forms, while in the vegetable only one occurs.

1848. The cellular form may be also called the water in the animal, the globe-form the earth, the fibrous the air, the point-form the fire. Thus is the animal even in its tissues a whole universe, for it cannot otherwise be thought of.

1849. The cellular substance is the last division of the point-substance, because the nervous granule becomes hollow. A true cellular tissue first makes its appearance therefore in the higher animals.

1850. Bone, flesh and nerve are the highest organs of the animal; the viscera, which mostly consist of cellular tissue, will indicate the Vegetative in the animal. Proper animal organs can only present the above-mentioned triplicity. What is not bone, flesh or nerve, is not animal, but vegetable.

1851. Nerve, flesh and bone are mutually excited, and are independent of the cellular tissue. They are moreover the animal in the animal, the thoroughly Free and Voluntary.

1852. These three substances have therefore nothing to do with the three terrestrial processes; they do not digest, respire, nourish nor transport about the galvanic sap, but live for themselves or to their own satisfaction.

1853. The origination of the three inferior substances out of the nervous mass, is perfectly similar to the original creative process of the three terrestrial elements from the æther. The animal organism is a second world-creation, for in the organic æther an organic air, earth and water, have been produced, or it has itself become these through fixation of the poles. This capacity for resemblance between the organic and inorganic elements is marvellous; but yet there would be more to wonder at, ay, it would be thoroughly incomprehensible, if the organic elements had been created according to some other type.

1854. When once the nervous mass has become separated from the three other masses, each then commences to be self-substantially perfected, and become a particular organ, though it is still under the supremacy of the primary mass.

1855. The principal mass which constitutes the animal body, after complete separation of the chaotic nervous mass, is without doubt the fundamental mass of everything organic; being the mucus or cellular mass, in which the other elements have only been included like veins of ore. It is the cellular mass therefore which we will first consider in its process of formative evolution.

1856. As cellular mass it must be the seat of the galvanic process, and thus of the life proper. In the cellular mass consequently the three vegetative processes, or the three terrestrial elementary processes, must be firmly established; as there digestion, respiration and nutrition are.

1857. With these three processes, the three superior elementary forms, which are peculiar to the animal, as nerves, bones, and muscles, will have nothing to do, excepting in so far as they govern them. As in the plant, so also in an animal, the terrestrial processes are only the appurtenance of the cellular tissue.

1858. In cellular tissue is therefore the seat of life. But the vegetative mass simply lives that it may live, while the animal lives, in order to combine the universe with the life. The animal elements live only in order to feel and move, in order to act freely like the world; the vegetative only, that they may subsist as planet. The latter are an image of the planet, the former of the world; the one deal with matter, the other with spirit.

Integument.

1859. The cellular tissue does not continue a mere parenchyma in the animal as in the plant, but it obtains a definite anatomical form.

1860. The animal cellular tissue has issued forth from its highest formation in the plant, or out of the blossom, which is a great bladder or cyst composed of primary vesicles. It is the secondary cystic form, wherein the animal cellular tissue appears, when it becomes an anatomical system.

1861. The animal cellular tissue forms therefore everywhere large bladders or cysts, whose walls consist of primary vesicles, or of the vegetable cellular tissue. Cyst-walls are teguments.

1862. The cellular system in the animal is Tegumentary system.

1863. The idea of the integument is the wall of a cyst. There is no tegument apart from the meaning of circumscription, inclusion, and limitation. There is no flat integument, or one that could be designed after the idea of the plane. Every tegument is periphery, just as there is nowhere a surface in the universe, which could have been produced according to the level or plane.

1864. All terrestrial processes, as digestion, respiration, and nutrition, are consequently tegumentary processes. All these organs must be tegumentary organs. Intestine, vessel, lung, in a word all viscera are naught but tegument.

1865. The vegetable tissue becomes in the animal, tegumentary tissue. The tegumentary formation is the vegetable in an animal—the parenchyma, the Visceral.

1866. The lowest distinction between animal and plant resides accordingly in this, that the vegetable tissue consists of actual vesicles, which form everywhere closely compressed masses; the animal cellular tissue on the contrary of granules, which inclose a hollow space. The animal body is a hollow globe of vesicles, the vegetable body one full of vesicles. (Ed. 1st, 1810, § 1870.)

1867. Every animal cyst is necessarily composed of the element of the vesicles, and is then for the first time an organ. The vegetable bladders are, however, single vesicles, and as such are already an organ. In the plant therefore the cellular tissue is upon the lowest stage, being only an aggregate devoid of secondary form; in the animal along with its aggregation a secondary form has been imparted. In this the higher character of the animal is at once demonstrated.

1868. The above is certainly a distinction between the two organisms, but it is not the essential one; for with it, what is animal has been by no means expressed; this being first imparted in the three cosmic elemental forms, which manifest themselves through sensation and motion, and then admit of being recognized as an animal. The corolla is also a cyst, but without being an animal; because, to this animal-like tegumentary formation, the proper animal elements are still wanting.

1869. Now as the tegument is none other than the form, under which the cellular tissue exists in the animal, we must regard it as an elemental form, which has stepped into the place of the cellular. The tegumentary form constitutes the fourth form, and is none other than the primo-vesicular form elevated to a higher rank, by being composed of cell-granules, which have been formerly nerve-granules.

1870. The animal body must consist of nerve, muscle, bone, and tegument, and of no other fundamental form; in other words, of point, line, globe, and cyst.

ii. Anatomical Systems.

1871. All anatomical systems are developments and separations of the four tissues, which are prolonged as sheaths through the whole body, like as in the plant are the bark, liber and wood.

1872. They are divisible next of all into two great parties, into the terrestrial and cosmical, or vegetative and animal.

A. VEGETATIVE SYSTEMS.

1873. The vegetable systems can only be different developments of the tegument. They accord with the galvanic factors. Any further derivation of them is unnecessary. The tegumental development must be represented as the systems of digestion, respiration, and nutrition, since from it these have been sufficiently borrowed or derived.

1874. Except these three systems there can be no other tegumental system; and, if such appear to be present, they must be subordinate to these. For there cannot be subsequently any more than there was fundamentally or at bottom.

1875. In the animal, however, the galvanic processes do not remain entangled in one mass, as in the plant. But they are even characterized as animal by their individual liberation from the whole mass. In the plant digestion or absorption, and nutrition or the course of the sap, were in one kind of mass or one kind of cellular tissue, all three processes (together with respiration) being in a tolerably confused condition.

1876. The animal appears in its dignity by separation of these processes, and by the perfection of each individually, or "per se."

1877. As all life consists only in the constant conversion of what is inorganic into the Organic, so is the process of digestion or absorption necessarily the first in animals.

1. INTESTINAL SYSTEM.

1878. The chemical process of the galvanism is conversion of the Inorganic into mucus, and thus an assumption of this matter into the organic body. Now as every limit of the body is tegument or cellular tissue, so can this assumption take place everywhere. Adoption of what is external into an organic body is absorption.

1879. Absorption originates from the antagonism of the body with the earth, which is organizable, and thus with the mucus.

1880. We call this slime or mucus, nutritive matter. Wherever such matter can operate upon the body, a corresponding organ of absorption, and thus a cell or integument, will be formed.

1881. The whole body is surrounded by integument; it was originally nothing but integument.

1882. The essence of the integument consists in absorption, or in the intervention of the chemical process.

1883. The integument is the root of the animal.

1884. The animal cellular mass is, however, in conformity with its origin, a bladder or cyst that has been opened by light and air. The integument is a large cyst not closed all round, but open at one end. It is the open floral cyst, which has just become an animal. The original integument is thus Intestine. The intestine is the water-organ.

1885. The integument presents therefore to the external world, or to the nutritive matter, two parietes or walls, an external and an internal.

1886. Both walls are opposed to each other like light and darkness, like air and water. The external is the light-and air-wall, the internal the darkness-and water-wall.

1887. It is consequently only the internal wall that stands in the same relation as the root. The internal is "par excellence" root, and is thus a main organ of absorption.
1888. The external wall comes under the idea of the stem-bark, and in so far only as this has a root-nature in itself, is it likewise absorbent.
1889. As upon the external animal wall the light and air constantly operate—for without light indeed no animal originates—so is this wall more and more removed from the idea of the root, and becomes, by virtue of the influence of light and air, instead of an absorbent organ, an organ rather of decomposition—an evaporation's organ.
1890. As a cyst the internal wall incloses the nutritive matter, which originates from the mucus, and thus from the organic water. The internal wall is therefore constantly immersed in the water, and is consequently in every respect a root.
1891. As the animal is only developed in light, so must the function of the root languish in the external wall, and decay, because it is devoid of the earth, which protects it from or against the light. This deficiency is compensated for in another way, or by the formation of a cavity, into which the media of nutrition enter, and which is dark like the earth.
1892. Internal and external wall stand also opposite to each other, like water and air. The one is the water-, the other the air-wall.
1893. The nutritive matters are not decomposed upon the internal wall by extraneous influences, but they remain identical; ay, they become indifferent, because they enter into darker and warmer water.
1894. On the other hand such nutritive matters are decomposed upon the external wall; and there here therefore gradually originates, instead of the chemicalizing root-process, the polarizing process of air.
1895. In a perfect light-animal it is only the internal wall that is chemicalizing; the external has become oxydizing. The internal is a mucus-wall, but the external, on account of the decomposition of mucus, an oxygen-wall.
Division.
1896. The more an animal has been exposed to the air and light, by so much the greater is the antagonism between its internal and external wall. In aquatic animals the antagonism is at its minimum, because externally and internally there is water; both parietes are therefore mucus-walls. The external wall of the fishes secretes an abundance of mucus, as does that also of the worms, snails, and molluscs.
1897. But an internal wall is still the more mucous of the two, because it is darker and warmer.
1898. At first the animal is content with the antagonism of the walls; especially so long as it remains occluded in dark and deep water, or within other animals. Many intestinal worms, polyps, and even Acalephous animals, are but simple sacs.
1899. When the animal organization, however, ranks upon a higher stage, light, or even air, operates more upon its external wall, but upon its internal, water; thus the antagonism of the two walls is carried out to the utmost degree.
1900. Through the different, ay, opposed processes, the two walls finally adopt another structure. The external becomes denser and harder, on account of the decomposition by light and dessication by air; the internal, however, retains its original structure and consistence. Soft, aqueous, indifferent, and constantly absorbent, it is only a viscous mucus.
1901. In place of an integument of similar tissue throughout, one will originate, whose external tissue is dense and oxydized, but whose internal is loose or spongy, and indifferent. The previously uniform integument will now separate or fall into two distinct layers; into a soft

muco-cellular layer, and into a tough coriaceo-cellular layer.
1902. With the last attainable antagonism the layers finally separate; two bladders or cysts, disunited from each other, originate; of these the internal is the mucous, the external the coriaceous cyst.
1903. Now the internal cyst alone is the intestine, the external the cutis or skin.
1904. Intestine and cutis belong to one formation, or to the integument. They pass directly into each other at the mouth and anus. Their structure also is wholly similar.

1905. They are merely distinguished by darkness and light, but more closely by water and air. The intestine is the water-tegument, the skin the air-tegument.
1906. The functions of both are therefore co-related like dissolution is to combustion, along with which the evaporation has been bestowed.
1907. Intestine and skin stand in antagonism with each other.
1908. The first animal, as being sentient integument, is a sac; the first skin is also a sac; an animal around the animal.
2. DERMAL SYSTEM.
a. Branchiæ or Gills.
1909. The external wall, being constantly exposed to the air that is in the water, can adopt no other than the aerial character, and is thus like the leaf of the plant. The skin is the organ of evaporation, and with this of oxydation also.
1910. A self-oxydizing integument is called a Branchia or gill.
1911. The skin is essentially nothing else than a gill; and, if it subsequently appears as anything else, this happens only through a higher state of perfection being attained by its branchial function.
1912. The lowest animals, such as most of the worms, molluscs, and snails, breathe through the external integument; even the gills of fishes are none other than a piece of skin.
1913. Thus gills and intestine would be the first two organs, which are developed out of the tegumentary system by the antagonism of air and water. Through the gills, air, and through the intestine, water enters the body. The gill is the atmosphere of the animal, the intestine is its sea.
b. Tracheæ, or Air-tubes.
1914. As the intestine, and in general every water-tegument is prolonged into mucous tubes or absorbent vessels; so also, with a more vigorous formation, the branchial membrane is drawn out into tubes, in order to conduct the air or oxygen towards the intestinal vessels, just as the intestine conveys through its absorbent ducts the water to the vessels of the skin.
1915. This saccular inversion of the skin constitutes the tegumentary lymphatic vessels, whose original function has been to transport the oxygen, combined with the water, to the intestine. They are the original respiratory vessels, which in the higher animals become, through the pure influence of air, true Air-tubes, like the spiral vessels.
1916. The air-vessels traverse the thickness of the body toward the intestinal membrane, like the mucous vessels do toward the branchial membrane.
1917. Thus an infinite number of air-vessels will and must originate.
1918. The air-tubes are consequently the formation which is properly opposed to the lymphatic vessels. They are for the air or for the skin, what these are for the water or the intestine. Air-vessels are first displayed in insects, then in fishes, reptiles, birds, and Thricozoa or Mammalia.
1919. If the infinitely numerous air-vessels concur to form one stem, they are then called lungs, as in the higher animals.

1920. The pulmonary vesicles are nothing but ramified air-tubes, such as the insect has.
1921. The formation of air-tubes is one of a higher character than that of the gills. For in it, indeed, the function has been separated from all other functions. It is simply destined to convey the air without water.
1922. As they pass into the dark, light does not operate upon them; and they will therefore less promote evaporation than mediate unto combustion.
1923. The anatomical idea of the air-vessels, or of the lung, is a saccular inversion of the skin. The skin is prolonged into, and ramifies towards, the body. The intestine is prolonged through the absorbent vessels, as being also small inversions, toward the lung, and becomes a stem—thoracic duct. Then the thoracic duct finally unites with the lungs through the medium of the heart, which is a new formation.
1924. Everything becomes stem which attains a higher grade, and which approximates the air and light. The stem strives to be a centre, but the ramification devolves upon the periphery; the former upon the Solar or Animal; the latter upon the Planetary or Vegetative.
1925. The nobler therefore a formation, by so much the more single and stemmy is it: such is the case with the trachea of the lungs, such with the lymphatic duct.

3. VASCULAR SYSTEM.

1926. The earth or the nutritive mass acts also upon its formation, and destines the tegument to a peculiar structure. The result of the electric and chemical process, or of the oxydation of the mucus, is precipitation, mass-and earth-formation; it is thus a process of nutrition, since through it the Solid of the body, with even the branchial and intestinal membranes, originates.
1927. The earth-system can only be developed where the two last systems coincide, or where the intestinal and branchial processes enter into mutual contact; in short, where the mucus is oxydized, and thereby separated into what is aerial and rigid.
1928. This spot is only in the middle between the two. The process of precipitation and the formation of elemental matter occurs, consequently, between the two layers of integument, or at present between the two teguments that have become self-substantial, or between intestine and skin.
1929. A new formation must be evolved, whereby the two shall be held together; a formation, whereby the antagonism shall be conducted from one to the other; and thus whereby the mucus shall be conveyed to the skin, but the air to the intestine.

1930. Were intestine and skin entirely separated from each other, each would perish; the former would no longer be oxydized, the latter no longer nourished.
1931. With their separation they must continue to be attached to each other at certain points, and thus undergo eversion at certain points. Thereby tubes originate in the intestine, which pursue their course toward the skin; in this again tubes, which pass to the intestine.
1932. A tube, which receives mucus from the intestine, air from the skin, and includes both within itself, is a long cyst. A cyst, which conducts mucus to the skin and air to the intestine, is a Vessel, a vein.

a. Unclosed Vascular System.

1933. The vessel has, in virtue of its essence, two extremities, an air-extremity, which is polar, and a water-extremity, which is indifferent. Every vessel has been rooted in two systems, in the intestine and skin, and is subservient to both.
1934. The vessel is no longer a something single, like the last-mentioned or preceding cysts, but double. Every vessel hath two poles.

1935. The organization necessarily produces two kinds of vessels. A vessel, which conducts the mucus to the skin, cannot also convey the air to the intestine. There is consequently a mucus-vessel and an air-vessel, or a water-and an air-vessel, an indifferent and a different.
1936. The mucus-vessel is called absorbent, the air-vessel respiratory duct or trachea.
1937. Air-and absorbent vessel stand in antagonism like skin and intestine, like water and air. The air-vessel is the skin or the branchia, which passes to the intestine, the absorbent vessel is the intestine, which passes to the air; the one the intestinal branchia, the other the branchial intestine.
1938. So long as intestine and skin were one in kind, this vascular process was in every situation. With their separation therefore the vascular structure has of necessity originated between two opposed situations. There is no point in the skin and none in the intestine, where there might not be an air-and a water-vessel, a respiratory and absorbent duct.
1939. There are therefore numerous vessels, and consequently a Vascular system.
1940. Air-and water-vessel must abut against each other; because they are polar, because the one leads to this place and the other to that.
1941. The system of water-and air-ducts can form no closed vascular system; for they only grow towards each other, as did formerly intestine and skin.
1942. They would not have originated if both cysts had not separated from each other. These vessels are not therefore to be met with in animals that have no intestine. The transition of the water-into the air-ducts takes place in the higher animals through the union of the thoracic duct with the subclavian vein, which conveys the blood directly to the lungs.
1943. The vascular system is properly the primo-cellular tissue, which occupies the middle, and at whose extremities the two cysts remain approximated, in order that they may continue to live.
1944. The vascular is the original system, since its two extremities already carry in themselves the air-and mucus-process, so that branchia and intestine are only to be viewed as peculiarly perfected conditions of these extremities. I would even have developed the vascular system first of all, had not another course been required to render the subject intelligible.
1945. The first animal cyst is a vessel with two kinds of extremities, whereof, like the plant, one draws its supply from the water, the other from the air; but in the animal this is effected by its own motion.
1946. The branchia is nothing but a vascular tissue in the air, the intestine none other than a vascular tissue in the water. Thus, do I wish that these organs should be understood, and not as mere cystic walls.
1947. Every point therefore in the intestine and skin absorbs, and thus every part is perforated with infinitely numerous holes or pores. Here they absorb air, there water.
1948. Every absorbent point of integument is drawn out as a tube towards the respiratory system, in order to let what it has absorbed become oxydized. These tegumentary prolongations into tubes are the lymphatic vessels.
1949. In all teguments there are necessarily lymphatic vessels, but more of them in the water-than the air-teguments. On that account the lymphatic vessels are much more numerous in the intestine than the skin. They are there called chyliferous, or lacteal vessels.
1950. The lymphatic vessels are the first of all vessels. Many animals, as perhaps the Acalephæ and Distomata, appear to have only this kind of vessel.
1951. Lymphatic vessels are present in the skin, only in so far as it has resigned the respiratory function to special organs.

Meaning of the Unclosed Vascular System.

1952. The action of this unclosed vascular system is wholly similar to the motion of sap and air in the plant, there being only an ascent of the first, and a fall or descent of the last. In the absorbents the sap ascends out of the root (intestine) into the leaves (branchiæ); in the respiratory vessels the air descends from the foliage (skin) to the intestine and the whole body of cells.

1953. This vascular system is therefore the pure remnant of the plant, and has as yet assumed no properly animal character, except that both its sets of vessels or tubes are self-substantial, and ramify, while in the plant they are only intercellular passages or non-ramified spiral vessels.

1954. In insects this system has been most perfectly evolved; there the air-tubes, being well parcelled out, are in great number, and run directly to the intestine and dorsal vessel, which is, as it were, only the trunk of the lymphatic vessels, or the chyliferous duct.

1955. Now, such a vascular system, merely oscillating, as it were, between intestine and skin, can be but persistent in animals, which express only the vegetable type of organization. If other significations be introduced, so also will this vascular system be otherwise evolved.

1956. The unclosed vascular system will be present along with an energetic antagonism between skin and intestine; with an antagonism that is almost suppressed, there will be none. There are then only cells, or there is only point-substance, as in the Infusoria, polyps, and Acalephæ or sea-nettles.

1957. The dermal vessels exist only in the air-insects, because in them nothing but air and water are engaged in conflict. Externally there is desiccated horn, internally mucous water.

1958. A perfectly unclosed vascular system appears to be developed only in animals which respire air. At least there are only genuine air-tubes and lymphatic vessels in such as breathe air; e. g., in the Mammalia, birds, reptiles and fishes, the latter set of vessels being probably not present in insects.

1959. Through the predominance of the air-process, as in insects, the mucus that is conveyed to it becomes so rapidly decomposed, that no more remains behind, for which a new vessel would be necessary.

1960. The galvanic process is at every instant annulled, and only renovated by a new afflux or supply. Here the galvanism does not subsist in itself as a peculiar and independent system.

b. Closed Vascular System.

1961. The unclosed vascular system is not yet self-substantial, because it is a cæcal eversion of the intestine or inversion of the skin, being itself only a ramified intestine and skin. Every system, however, attains its perfection, by being rendered independent of its origin. Thus the leaf is the spiral vessel that has become free, the root the cellular tissue in a like condition, the blossom the liberated vegetable trunk. The vascular system will therefore aspire also to the achievement of its blossom.

1962. If the decomposition that is effected by an inferior amount of polarization, does not happen rapidly enough, for what has flowed thither, to disappear, during its afflux, by evaporation or precipitation; the rest of the mucus which has obtained the air-polarity, is now repelled by the respiratory vessels, because both have become synonymous.

1963. The oxydized superfluous mucus is at once, however, attracted by the intestine, because they are not synonymous. There consequently originates a vessel in the mass of mucus or parenchyma, that has been secreted between the intestine and skin, which begins in the respiratory membrane and terminates in the intestine.

1964. This vessel will commence at the end of the absorbents, or at a point, where it devolves upon the respiratory vessels, to take up their contents, namely the oxygen, together with the nutritive substance, and convey them to the intestine. The mucus, which previously stagnated and moved but slowly from one spot to another, is now again carried back by another vessel without interruption to the intestine.

1965. The vessel which conveys oxydized mucus from the respiratory to the intestinal system, is called artery.

1966. On the intestine, however, this polar mucus is again reduced to ordinary mucus. It has now become synonymous with the intestine, is repelled from it and attracted by the branchial membrane.

1967. One and the same fluid or sap is consequently brought back from the branchia to the intestine, and from thence again to the branchia. This last vessel is called vein.

a. Arteries.

1968. The artery is, according to its signification, an air-vessel, which is prolonged or elongated as far as the intestine. Essentially, the artery conveys nothing but air, though this is effected by means of a medium or vehicle, which is the undecomposed mucus (blood). It is an air-tube, that has been self-substantially dismembered from the skin, in order to become a special independent air-tube, or such as is commensurate with animal nature.

1969. In the artery the external integument has been repeated in the nutritive system, and hence the galvanism has become of a continuous character.

1970. The artery must therefore be the highest vessel, the most total of all terrestrial processes. For it is the air-duct united with the intestinal vessel. It contains mucus, which carries the properties of both poles of the body in itself; there is oxydized mucus.

1971. The artery contains the whole body, lung and intestine, in short, the whole animal, (whereas it was previously dispersed in two vessels,) in a fluid state within itself. From the artery therefore nutrition will directly take place; from it the animal will be formed.

1972. The air or respiratory vessels may be viewed as arteries carried to the very extreme. In the lung the arterial system has attained its highest purity, the oxygen only, without the indifferent substance, being contained therein. The trachea is the roughest artery—Arteria aspera.

1973. The arterial system, in accordance with its signification, makes its first appearance in water, because the aquatic mode of respiration is less energetic, and thus the mucus is more feebly decomposed. The Mollusca, snails, and many worms have a perfect arterial system. The branchiæ do not, like respiratory tubes, pass into the body, as in insects; but there are vessels which take up the oxygen and convey it into the body.

1974. As lung and absorbent belong to the pure air or the pure water, so do artery and vein to the water combined with the air. The two former are therefore present only where aerial respiration occurs, and the two latter where water merely is respired.

1975. The last system is only present in animals, in so far as they are aquatic in their habits.

1976. Insects, as being purely aerial animals, have therefore arteries and veins only so long as they are in the larva or worm-like condition, and as flies or perfect insects may continue to live without them. On the contrary, the purely aquatic animals appear enabled to live without true respiratory and absorbent vessels. It in fact appears, that lymphatic as well as respiratory vessels are wanting in the molluscs, snails, and worms, since the water directly bathes or washes the arteries.

1977. Animals with both systems of vessels, the unclosed and closed, must be more perfect in structure, and must at once combine worm and insect in themselves. They are insects from having absorbent and respiratory vessels, but worms as having arteries and veins.

b. Veins.

1978. The veins are developed as mucus-vessels at the intestinal extremities of the arteries, which absorb the arterial mucus (blood), after it has deposited its air on the tegumentary substance, just as the lymphatic vessels absorb their fluid from the intestine or any other part of the body.
1979. As the artery is a respiratory vessel that has become self-substantial, so is the vein a similarly conditioned, and dismembered lymphatic vessel. In the one it is the lung, in the other the intestine, that has become the free vascular system. But in the proper vascular system both lung and intestine are repeated, the former as artery, the latter as vein.
1980. These arterio-lymphatic vessels (veins) necessarily convey their arterial mucus or blood into the stem of the original lymphatic system, or the thoracic duct. For every Indifferent must be brought toward the respiratory organ.
1981. The tegumentary lymphatic vessels (absorbents) consequently unite with the arterio-lymphatic vessels (veins), before arriving at the respiratory organ, and pursue their course thither in common, where they pass over into the air-vessels. The usual notion or idea is, that the lymphatic vessels, from conveying their fluid into the veins, should be subordinated to the latter. But the true philosophical view, is of reverse import, although the veins are larger than the thoracic duct. In the investigation of such relations, recourse can by no means be had to quantity or size, but to the importance of the quality or contents.
1982. The veins are, properly speaking, subordinated to the lymphatic vessels, just as the arteries are to the tracheæ, or air-tubes, and the former therefore pass over into the lymphatic vessels.

c. Circulation.
1983. Through the veins, as arterio-lymphatic ducts, the vascular system has become a closed system in itself, because, on account of the polarity, the vein unites at both extremities directly with the artery. It is a continuation of the artery, like the air-duct is of the skin, and the lymphatic vessel of the intestine. Thereby a Circulation of the arteriose mucus or blood originates.
1984. In its essence the circulation is a combination of the intestinal with the branchial system into one anatomical system.
1985. The circulation is therefore a higher formation, since through it the vascular system repeats in itself the totality of the vegetable organism.
1986. On that account the circulation is the vital process proper.
1987. But for that reason also, the circulation is impossible in the plant, since it is devoid of arteries and veins.
1988. It makes its appearance in the aquatic animals, for they have, generally speaking, vessels. Molluscs, snails, worms, and crabs already possess a circulation; it is wanting, on the contrary, in those animals that are without intestine, and ceases in insects, when, or in whom, the air-vessels obtain the preponderance.
1989. In circulation the galvanism is restricted. In the skin, intestine, air-and lymphatic vessel there is also galvanism, but distributed upon organs that are remote and subservient to different purposes.

d. Blood.
1990. The blood signifies the earth in the animal, combined with water and air.
1991. The blood is the proper nutritive matter for the animal.
1992. The blood is earth, which carries all terrestrial elements in itself, such as the air through the medium of the gills, the water through that of the intestine, and is consequently a complete planet.

1993. The blood is a fluid planet.
1994. The blood is the fluid body.
1995. The body is the fixed or rigid blood. Blood and body are wholly equivalent, have the same elements in themselves; only here the latter are stationary, there they course along. Both consist of mucus or gelatine, albumen and fibrine.
1996. The blood is half combusted mucus, the body mucus, that has been wholly subjected to combustion.
1997. After the vascular system has attained its own circulation, or to the closed galvanism, no higher development of the tegumentary formation is any longer conceivable; as little as after the blossom anything more could originate.
1998. After all three elements are united into one point, into one system, as is the case in the circulation, where the venous blood represents the water, the arterial blood the earth and air, no new system can further originate in the vegetative body.

4. SEXUAL SYSTEM.

1999. In so far as the animal adopts into itself, or is rather based upon, the whole plant, the blossom or Sex is also developed in it. The vegetable sex consists of seed, capsule, and corolla.
2000. The seed is the first part of the blossom which is put forth in the plant. The most inferior or asexual plants have only seeds devoid of capsule and corolla, and produced without the concurrence of female and male parts. The blossom in the animal is therefore in the beginning also nothing but seeds or ova. Thus these animals are asexual. Such as the Infusoria, whose body directly divides into new animals, like the fungi.
2001. The second floral organ is the capsule, which contains on its borders or dissepiments the seeds, and on its apex the stigma, or the opening of the cyst.
2002. So also in the succeeding forms of animal life the animal capsule or uterus originates. The orifice corresponding to the stigma is the mouth of the womb; the seeds upon the septum become ovary.
2003. The lowest animals, as the Polyps, are fundamentally none other than such an uterine system. The polyp's mouth is the os uteri; the sac formed by the polyp's body is the uterus, in which ova-cysts or ovaries develop, that open into the margins of the month.
2004. In such animals the uterus and intestine, as likewise the mouth and uterine orifice are fundamentally one organ; nutritive matter and ova are also one in kind. External tegument as branchial organ is at the same time also a tegument of the uterus.
2005. The higher animals are distinguished from the lower by separation of all these intricate and, as it were, coalesced organs.
2006. Digestion, respiration and nutrition, growth and propagation, are originally of one kind. But with further development come the male parts also, which belong to the category of the organs.

Parallelism of the Animal and Vegetable Body.

All vegetative systems of the animal body being now developed, the attempt to co-ordinate them with those of the plant admits of being made.
2007. What the sexual parts correspond to in plants, needs no exposition.
2008. And just as little that the lung is the parallel organ to the leaves or foliage. The relation of the other organs is, on the contrary, difficult; the striking resemblance, however, of the sexual parts and the lungs to the same systems in plants, is ground sufficient for assuming also the parallelism of the other organs.
2009. If the root be compared with the intestine, then the stalk, as being the medium of the

vascular system, must be regarded as heart.

2010. The bark will correspond to the skin, the liber to the veins, the wood as a tracheal body to the arteries.

2011. The vegetable tissue will be transformed into the lowest organs of the animal; the cells into the mucous tissue, the intercellular passages into the lacteal vessels, the tracheæ, or spiral vessels, into the lowest kind of respiratory tubes or the cutaneo-lymphatic vessels. We have accordingly the following parallel series:

A. Tissues.
1. Cells — Mucous tissue.
2. Ducts — Absorbents.
3. Tracheæ — Cutaneo-absorbents.
B. Systems.
4. Bark — Skin.
5. Liber — Veins.
6. Wood — Arteries.
C. Members.
7. Root — Intestine.
8. Stalk — Heart.
9. Foliage — Lung.
D. Sexual parts.
10. Seed — Ovum.
11. Ovary — Uterus.
12. Corolla — Testes.

B. ANIMAL SYSTEMS.

2012. As in the blossom the light allows the whole vegetable trunk to be once more developed, though with coloured signs or marks of distinction, so also is the animal body taken up into the sense of light, and the vegetative systems are elevated unto light-or rather æther-systems.

2013. A new animal originates upon the old animal and is equivalent to it. Every perfect animal is twofold in its nature, being a planetary and solar animal, a vegetative and animal being.

2014. Three animal systems must be developed out of the three vegetative, an earth-, water-, and air-system, each being refined or purified through the agency of light, and forming the bones, muscles, and nerves, which correspond to the gravity or materiality, the heat or motion, and the light or tension.

2015. Everything that is higher in its nature can only be developed out of that which is its direct antecedent, like the blossom from the leaf. Now the artery is the ultimate Vegetative. In this therefore, must the elements of the osseous, muscular and nervous systems reside. Now, the artery consists, of four parts, of the cellular, fibrous, dense serous coat, and of the blood. The cellular coat or tunic is the remnant of the whole tegumentary formation. The fibrous tunic is the embryo of the muscular system. The serous tunic is the embryo of the osseous system; for in old age it attains to ossification, bony lamellæ being deposited around it. The blood is the embryo of the nervous system. It is only requisite for it to coagulate, and it is then nervous mass. Nervous globules are blood-globules quiescent, or in a state of rest. Accordingly, in the artery, the whole body has been actually prefigurated or typified.

2016. Every animal system necessarily exists under a duplex character, being once in the service of the vegetative systems and once for itself; or it exists as trunk and as blossom. There is, therefore, a vegetative and animal nervous system, and, under such binary conditions, osseous and muscular systems likewise.

2017. All three systems are offsets from the arteries, and therefore their constant companions. The animal systems are called flesh.

1. NERVOUS SYSTEM.

2018. The nervous mass is coagulated blood. The nervous system is thus a higher arterial system. The highest arteries, however, are the air-tubes. The nerves therefore run parallel with the tracheal systems.

2019. What the tracheæ are for the vegetative body, namely, the vitalizing and motor principle, such are the nerves for both the animal and vegetative body.

2020. The nerves are distributed like the tracheæ of plants, in individual threads or filaments, which run near to each other arranged in a fascicular manner, and then mutually separate.

2021. The nervous filaments everywhere accompany the arteries, and, as far as their ultimate ramifications, like the tracheæ of insects pass to all parts of the body.

2022. The nervous mass has been separated into an arteriose and venous mass, the former being the cineritious or gray substance, the latter the medullary or white. The nervous system is therefore an entire blood-system, with two poles; it is therefore alive and active for itself, or independent of other systems.

2023. Both nervous masses are in a state of constant tension against each other, and consequently in constant tension with the whole body.

2024. The nerves are filaments that, upon the separation of the parts of the body, have remained behind, just like the arteries are tubes individualized in the general parenchyma, throughout which the sap was previously dispersed as in the vegetable trunk. The first individualization necessarily takes place at the oral extremity. The first nerve is a ring surrounding the pharynx or gullet. Thus, if the body consists of several consecutive rings or cysts, as in the Worms, each ring will or can have its nervous ring, which inferiorly gives off nerves to the vegetative organs, and superiorly to the animal, if these be present. Such a point for giving off nerves is called a ganglion; since also every ganglion sends off nerves to join other ganglia; so both below and above a nervous cord or string will originate in the longitudinal direction. The nerves, which have been left behind in the vegetative parts, form the vegetative nervous system; those that remain in the animal parts, the animal system.

a. Vegetative Nervous System.

2025. The vegetative nervous system is the nervous mass that has remained behind, after the greatest part of it has become converted into tegumentary formations. Now, as these tegumentary formations, being surrounded in the higher animals by flesh, were thus viscera, so may the vegetative nerves be called also Visceral or Splanchnic nerves.

2026. These visceral nerves govern the vessels, the intestine, and lung; with the sexual parts also, yet in their case in combination with the animal nerves, because the sexual parts are at one and the same time organs of vegetative and animal life.

2027. These visceral nerves everywhere accompany the vessels, and are therefore like these distributed in a cystic manner between intestine and skin. They form a large cyst which concentrically surrounds the intestine.

2028. They do not, however, like the intestine, form any closed cyst, but only a cystimorphous net, like the vessels.

2029. The two nervous masses are in them separated from each other, like as the branchiæ have been distributed along the whole body and separated from the intestine. The gray or branchoid

substance has separated itself from the white medullary substance into individual ganglia, or as it were into individual nervous branchiæ. The medullary substance has also retained its connexion though only in a ramular manner, and not uninterruptedly like an integument. It is called plexus.
2030. The ganglia and the plexuses stand in mutual opposition, like the branchiæ and intestine, like artery and vein, like blood and lymphatic vessels. The ganglia oxydize, polarize; they are the active. The plexuses suffer, digest, and are the recipient.

2031. The visceral nerves, like the viscera, act for themselves, being unconcerned for the animal systems.
2032. The visceral nerves have a vegetable sensation for themselves, a sense of touch, such as the blossom might have in the instant of the pollenization.
2033. As all formations have become symmetrical through and in accordance with the vascular system, so does the vegetative nervous system separate into two parallel stems which accompany the arterial trunk. They are called intercostal nerves. These ramify and form plexuses that accord with the visceral organs, which they govern.
b. Animal Nervous System.
2034. The animal nervous system is the repetition of the vegetative, and is combined therewith to constitute an unity. The nervous cyst that was previously dissevered in a reticular manner, becomes a closed tube, which is placed upon the light-exposed side of the other animal systems, and thus upon the vertebral column. This closed nervous tube is called the myelon, or spinal cord.
2035. The spinal cord is worth as much as all the visceral nerves taken together; it is the felted system of intercostal nerves; and is, properly speaking, none other than the posterior double cord of ganglia. It consists therefore also of ganglia and plexuses, but both have coalesced, on account of the increase of the mass and its endeavour towards the attainment of union.
2036. The ganglia form a tube, which is inclosed by the tubes of the plexuses. The ganglion-tube is the gray, the plexus-tube the medullary substance.
2037. If the gray substance appear to reside within the medulla, this is the result only of involution or a folding in. The parietes of the spinal marrow stand therefore in everlasting tension with each other, like ganglia and plexuses, and like arteries and veins.
2038. The spinal marrow is the content of the bones and muscles, like the intercostal nerves or the visceral cord are the contents of the abdominal cavity, and as the blood is that of the internal and fibrous coat of the arteries. Bones and muscles are the animal parietes of the nerves, as the two vascular membranes are the vegetative walls of the blood. The skin, as branchia, incloses all, like the cellular membrane does the vessels.
2039. As the intercostal ganglia give off plexus-forming ramules; so does the spinal marrow; there are the spinal nerves. The spinal marrow is, first of all therefore, the coalescence of the two intercostal nerves.
2040. These spinal nerves are, however, animal plexuses, which partly encroach upon the visceral nerves, and partly pass to the animal systems.
2041. There are therefore two kinds of spinal nerves, vegetative and animal, and as many of them as there are divisions in the viscera and in the animal systems.
2042. The nerves pass off symmetrically from the spinal cord, because the nervous mass belongs to the symmetrical osseous system. They form therefore rings both anteriorly and posteriorly.
2043. The nervous system does not consist of individual cysts, like its two animal teguments, or bone and flesh. It is at one time the type or image of the vascular trunk and its ramifications; at

another the indifferent æther-mass, which does not crystallize; it is lastly the organic primary mass that has remained persistent, and must thus be coherent in texture. It is the blood continually streaming from the animal divisions of the heart.

2044. The whole animal nervous system is a tegumentary cyst with tubes passing off from it symmetrically in the form of rings.

2045. The spinal cord cannot be the highest. It has only the lowest signification, in so far as it stands in the service of the viscera and the sense of touch, and thus follows the position and arrangement of the bones. Thus the spinal cord is first of all an osseo-nervous mass.

2046. The nerves, as running for the most part forwards, are musculo-nervous mass; those running backwards or outwards are tegumentary or sensitive nerves. This signification furnishes us, also with the physiological function of these two divisions of nerves. The nerves are homologous with the flexors, the spinal cord with the extensors; the nerves with the air, the medulla with the earth; the former with the arteries, the latter with the veins; the nerves are thus what is more active, the medulla that which is more inert.

2047. On that account the nerves only are in intercourse with the world, while the medulla broods within itself. Consequently, both these nerve-formations are not as yet the pure self-substantial nervous blossom, which no longer imitates flesh and bone, but only itself.

Brain.

2048. The highest point attained by the lower systems are the orifices of the viscera, the mouth and the nose. The mouth is the first animal sign, which the plant gives off in the blossom from itself. The Noblest lies therefore at the anterior extremity of the animal, or in man in the direction upwards.

2049. It consequently occupies the middle point, or one that is between the anterior flesh and the posterior bones, and at the same time the spot, from which all vital processes emerge, or the mouth.

2050. The oral nervous mass is the Brain. It forms originally the posterior ganglia of the pharyngeal ring.

2051. The situation of the brain is essentially in front of or above the body, in opposition to the sexual parts, which are the lower totality.

2052. It is, however, above and behind; for it is originally situated posteriorly. The brain can therefore originate only, when the posterior medulla inclines from above forwards, making a curve in the latter direction; the brain is a spinal cord that has been bent from above forwards.

2053. The more the spinal cord is curved forwards, so much the nobler is it. This is self-evident.

2054. The brain is a spinal cord which makes the transit from the signification of bone to that of flesh.

2055. In the brain therefore the tendency must principally reside, to give off nerves, and perfect them into self-substantial nervous organs.

2056. In man the brain with its nerves curves round like a crosier, and that more perfectly than in any other animal. The spinal cord therefore in the highest formation of the brain returns again parallel to the direction, in which it has ascended.

2057. In the brain there is of necessity the greatest quantity of nervous mass. The brain is the nervous trunk, as the liver is probably the vascular trunk.

2058. In the brain the cystic formation has been most purely represented; as e. g. in the cerebral cavities or ventricles. The brain is the stomach of the nervous system or its lungs.

2059. The brain consists essentially of two substances, of one accommodated to the flesh, and one to the bones, or of one arteriose, and the other venous. The former is the gray or cortical, the latter the white or medullary substance. The cortex is the lung of the brain, the medulla the liver or the intestine.

2060. The bark or cortex is the polarizing, active, oxydizing; the medulla the patient or suffering.

2061. This nervous pulmonary substance is continued along the spinal cord and even along the nerves, there as veritable gray substance, here as the vascular membrane of the nervous mass.

Head.

2062. The brain, as being a system that has been separated superiorly from the other systems, determines the Head. The head is only there in so far as a brain is there.

2063. Head and trunk are antagonistically disposed, as Animal and Vegetable, or still more exactly, like nerve and bony flesh are to the viscera.

2064. The head is naught but a nervous organ.

2065. The concomitants of the nervous mass follow the brain, but, instead of the medulla having been previously subordinated to, or at least co-ordinated with, these, it is they that are thus related to the brain. The bones of the brain are the brain-case or cranium, the flesh of the brain, the face or countenance. On the head, bones and flesh have been disposed in the strictest manner according to their dignity or worth. Posteriorly there is almost pure bone, in front almost pure muscle.

2066. The cranium can be none other than the vertebral column continued around the brain. It consists of three vertebræ, the face of one. This will become clear in what follows.

2067. If the bones of the head are the repetition of those of the trunk, so also must the flesh of the head be a repetition of that of the trunk. Pectoral and abdominal muscles are ennobled in the muscles of the face.

2068. The face must have been principally formed by the orifice of the intestine—the mouth, and by the opening of the lungs—the nose, and by the apex of the vascular system—the members which are repeated as jaws. The month is the stomach in the head, the nose the lung, the jaws the arms and feet.

2069. The salivary glands are the liver in the head, as the mouth is its stomach. The liver that was originally also symmetric in form has become wholly symmetric in the higher organized head and forms two glands. The salivary ducts are the hepatic or biliary ducts.

2070. The tongue is the pharynx elongated upon the anterior side, because in front there is more flesh. The tongue is the extremity of the intestine converted into muscle.

2071. The nose includes pectoral muscles, the mouth arthric muscles or those of the limbs.

2072. If pectoral and abdominal muscles are repeated in the face, so also must the anterior bones, ribs, and limbs be repeated. It will be shown, in treating of the organs, that the nose is a vertebra, the jaws members, and their muscles those of the limbs. The head is the whole trunk with all its systems. The brain is the spinal marrow, the skull the vertebral column, the mouth intestine and abdomen, the nose lung and thorax, the jaws are members.

Senses.

2073. The perfect animal again consists of two animals, the spiritual or solar, and the terrestrial or planetary. The animal nervous system does not continue to remain simply in the service of the other systems, but seeks also to gradually render itself self-substantial or independent. Now, the operation of the nervous system for itself is sensation. The parts of the nervous system having become self-substantial, will be therefore pure organs of sensation. Yet as the nervous system cannot emancipate itself from the other systems, so will its highest development be attained only in combination with the highest development of the other systems. There are therefore as many stages of the self-substantial nervous development, as there are special anatomical systems.

2074. Sensation must be modified according to the processes performed by those systems, with which the nervous system combines. These systems are, however specifically distinct from each other. Sensations that are specifically distinct are sensorial sensations. Organs of sense are accordingly the combination of the highest part of an anatomical system with the nervous system. Sensorial sensations are different processes of the anatomical systems perceived in the nervous system.

2075. The first combination of the nerves with the vascular system that has become free, or with the integument, is the sense of feeling—vascular sense. The intestinal system emancipated and combined with the nerves, is the tongue—gustatory sense—intestinal sense. The lung upon its highest evolution with the nervous system is the nose—olfactory sense—pulmonic sense. These are thus the sensorial organs of the vegetative systems—senses of vegetative life.

2076. There are indeed three animal senses; but as the osseous and muscular system form in their conjoined operation but one system or the motor system—there can be therefore only 2 animal senses. The osseo-muscular or motor sense is the ear. If the nervous system becomes wholly self-substantial, the nervous sense thus originates, or the eye, in which the brain itself has been planted outwardly, and acts independently of all other systems.

2077. The vessels form the general system, and therefore the tegumentary sense surrounds the whole body. Its brain is the spinal cord.

2078. The four remaining senses are perfections of individual systems at their perfect extremity, and thus in the proximity of the mouth and the brain. They together form the head. The jaws and the tongue obtain their nerves from the medulla oblongata, and this is therefore the brain of the gustatory sense. The brain for the nose is the gray cerebral substance, because the olfactory nerves are its elongations. The ears obtain their nerves from the cerebellum, which is consequently the auditory brain. The eyes are developments of the great brain or cerebrum—optic brain. Such is the rationale and signification of the divisions of the brain.

2. OSSEOUS SYSTEM.

2079. The nervous mass consists of indifferent, deoxydized blood-globules. If these be peroxydized, then the highest oxyd of the planet is deposited in them, namely, the earth, and that indeed which was the last remnant in the order of their production, or the calcareous earth.

2080. Vesicles or cells replete with calcareous earth are globes. The osseous texture consists therefore of globes; is only a dense cellular tissue, and thus ranks nearest to the vegetable structure. The basis of the bones is at first a cellular gelatine, which, with increased oxydation, is converted into cartilage. Finally, calcareous earth is deposited in this cartilage.

2081. In the lower organized animals, who breathe for the most part by means of branchiæ, the acid combined with the calcareous earth is an inorganic, or the carbonic acid, i. e. oxygen combined with carbon, or the earthy Inflammable; in higher animals it is an organic acid or phosphoric acid, i. e. oxygen combined with phosphorus, or the aerial Inflammable. Phosphoric acid may be regarded as peroxydized gelatine, as acid of gelatine. The bone is therefore earth, salt and Inflammable.

2082. The first appearance of the osseous mass is in the oxydizing organs. It is formed from the dense or internal coat of the arteries, since in old age bony lamellæ are deposited upon this. In the hearts also of many animals bones are formed.

2083. The first regular formation of bones is exhibited in the trachea or air-tube, which has been directly exposed to the oxydizing process of the air. These first forms of the bones are rings.

2084. The antetype of the bone is, however, the intestine, as the air-vessels are the antetype of the nerves. The bone is a tube, an ossified intestine.

2085. There are two osseous systems, a vegetative and an animal; the one surrounds the tegumentary systems, as, e. g. the scales of Fishes and Reptiles, horny rings of Insects; the other the nervous systems.
a. Vegetative Osseous System.
2086. The vegetative osseous system is divisible into dermal, tracheal, intestinal, and vascular bones.
2087. The dermal bones are tegumentary rings, which surround the whole body, and are tracheal rings in so far as the skin is originally a respiratory organ. Such are the rings of the body in Insects, the shells of the Gasteropods and Molluscs, with scales and scutes in general.
2088. The tracheal bones are the branchial arches and tracheal rings.
2089. The intestinal or splanchnic bones are tubes environing the intestine, as in the corals, or imperfect annular segments, which at one time are found in the stomach, as in the Mollusca, at another in the pharynx, as in the Worms, Snails, Sea-urchins, and Holothuriæ; constituting what are called pharyngeal maxillæ. The branchial organs are fundamentally also none other than pharyngeal rings. The lingual and palatal bones, with the intermaxillary bones, belong also to the same category.
2090. The vascular bones are displayed in the hearts of many animals. The three last divisions may be called visceral or splanchnic bones; and then we have tegumentary, splanchnic, and nerve-bones.
b. Animal Osseous System.
2091. The animal or nervo-osseous system must separate from the vegetative system of bones, and be placed upon the side exposed to the light. The side of the inferior animal that is exposed to the light, or averted from the earth, is the upper surface, dorsal region or back.
2092. The back holds the same relation to the ventral side as light does to the darkness, as sun to the earth; therefore the dorsal side is of a dark, the ventral of a faint or pale colour.
2093. Back and belly are related polarwise to each other.
2094. Through the medium of the bones the distinction between back and belly has been definitely established in the animal, and, as a consequence thereof, the distinction also of right from left. Before a formation of bone exists, the animal is for the most part a round cylinder.
2095. The osseous system can in itself be only symmetrically constructed.
2096. The osseous is the only symmetrical system in an animal. The other organs are so only in so far as they follow the arrangement of the osseous system.
2097. The animal osseous system is, from its being a repetition of the intestinal canal, a tube. This tube is surrounded, like the trachea, by rings, between which the tegumentary tube suffers constrictions.
2098. The back is a series of numerous bony rings.
2099. These bony rings are the bodies of the vertebræ.

2100. The vertebræ have originated through polar repetition, through the muscular cysts.
2101. In addition to the series of vertebræ on the back, a vertebral column is formed moreover along the ventral surface, and without doubt there only where the air-organ, the branchia or lung, is situated. This inferior vertebral column is the breast-bone or sternum.
2102. All the systems, even those that are of a subordinate character in the animal, follow the direction of the main vertebral column. The intestine, as well as the vessels, are placed in accordance with it. Thus the principal trunks of the vessels take their rise along the vertebral column, the other vessels being given off from them in this situation, like the lymphatics from

the intestine.

2103. The vascular ramules which surround the intestine and the skin, run out from a main stem, and are directed in a symmetric manner downwards and upwards (in the horizontal body of animals), or towards the belly and back.

2104. If new bony rings originate, they must also take these directions. They accompany the vessels that run in a circle, as the vertebral column accompanies the vascular trunks. These annularly-disposed bony twigs or apophyses constitute in the direction downwards the ribs, in that upwards the vertebral arches. Anterior and posterior to the vertebral column there consequently originates a long canal formed by the bony rings. In the anterior canal lie the galvanic or vegetative organs, in the posterior or upper, the organs of light must be situated "par excellence." The former of these canals is called the thoracic and abdominal cavity, the latter the vertebral canal. The vertebral canal is not the bony or medullary cavity itself, but it has been formed by several bony cysts in the same manner as the thoracic cavity. It consists of the body and the two arches. These are thus posterior (superior) ribs. The vertebral canal has the same signification as the thoracic cavity has; it is only a posterior thoracic cavity. It therefore contains, like the anterior canal, viscera dissimilar in kind to bone; the one including the spinal cord, the other vessels, intestine, and lung.

2105. The bony cysts that have originated through constriction do not all harden into calcareous matter, but they remain as alternating membranous cysts. Between the rings there are permanent cysts. The membranous cysts form the joint, or articular capsule. An articular capsule is a bone which has remained soft.

2106. This change in the ossific process takes place through the attachment of muscles, concerning which we shall treat in the sequel.

2107. The whole osseous system is consequently a symmetrical arrangement of several polar cysts and rings.

2108. The vertebra is not a single ring, but is at once a tolerably compound osseous system. The whole osseous system is nothing but a vertebra repeated.

2109. The number of the vertebræ necessarily conforms to that of the pairs of nerves or ganglia of the spinal cord; for they are indeed only the periphery or envelope of the latter. The number of nerves is, however, adapted to that of the organs, which they have to take care of.

2110. Now, the nervous organs are the senses. There are consequently as many divisions of vertebræ as there are senses. Thus there are vertebræ appertaining respectively to the senses of touch, taste, smell, hearing, and sight. Now as the four latter senses make up the head, but that of touch or feeling is distributed over the whole body, and superintended by the spinal nerves, the vertebræ thus divide into two principal sections, into vertebræ of the head and trunk. The number of cephalic vertebræ is 4; namely, nasal, ocular, lingual, and auditory vertebræ.

2111. To a perfect vertebra belong at least five pieces, namely, the body, in front the two ribs, behind the two arches or spinous processes; every vertebra of the head consists therefore of five pieces also. In those vertebræ which are removed from the respiratory organ, the ribs are smaller, as in the ventral ribs, and anchylosed with the body, as in the cervical vertebræ, where they are represented by the perforated transverse processes, and in the lumbar vertebræ they disappear entirely.

2112. The formation of the cervical vertebræ, where the ribs have been impacted or interposed between the body and spinous processes, is continued into the cranial vertebræ. These are only expanded cervical vertebræ. At the base of the skull four vertebral bodies lie in a series one behind the other; the body of the occipital bone, the two bodies of the sphenoid, and the vomer.

Upon the sides of each of these bodies are situated alar processes, which correspond to the transverse processes of the cervical vertebræ, or to the ribs; e. g. the articular heads or condyles of the occipital bone, the alæ majores and minores of the sphenoids, and the two sides or lateral surfaces of the vomer. Behind these are placed the two broad cranial bones, which correspond to the spinous processes; as the occipital ridge or crest, the parietal, frontal and nasal bones. The occipital vertebra consists of the body, the two condyles and the occipital crest. The parietal vertebra consists of the body of the posterior sphenoid, the alæ majores, and the parietal bones. The frontal vertebra consists of the body of the anterior sphenoid, the orbitar wings or alæ, and the two frontal bones. The nasal vertebra consists of the vomer, the ethmoid and the two nasal bones. The occipital vertebra is the auditory vertebra; it incloses the auditory bones and the cerebellum, which gives off the nerves of hearing. The parietal vertebra is the lingual vertebra; the maxillary and lingual nerves passing through its alæ majores. The frontal vertebra is that belonging to the eye; through the orbitar plates or wings the optic nerves pursue their course, and it environs the cerebrum, from which these nerves originate. The nasal vertebra contains the olfactory nerves.

2113. Each cranial sense has thus only one vertebra, and the skull will consequently be formed of four vertebræ, whereof three appertain to the cranium, one unto the face. (See Oken's Ueber die Bedeutung der Schädelknochen, 1807.—Isis, 1817, S. 1204.)

2114. Several vertebræ are, however, found for the sense of feeling or touch, because it includes all the organs of the trunk. There must be therefore as many vertebræ in the trunk as there are particular organs placed therein. Of these there are three, the respiratory, digestive, and sexual system, or thorax, abdomen, and pelvis. To the thorax belong the neck, the arms, and the entire set of ribs. 5 vertebræ must appertain to the arms, because they have 5 digits and 5 nerves. But the ribs, and consequently also the digits, are determined by branchial vessels, are only repeated branchial arches, whose number in almost the entire class of fishes is 5. There are therefore also 5 thoracic or pulmonic vertebræ. Since the larynx consists of the 5 original branchial arches, and lies in front of the neck; so must the 5 superior cervical vertebræ stand in the signification of branchial vertebræ. The odontoid process of the second vertebra must be regarded, from its being separated from it in the fœtus, as a particular cervical vertebra. Accordingly, in the Thricozoa there are eight cervical vertebræ. The 3 inferior cervical, and the 2 upper costal or rib-vertebræ, give off through their interspaces the nerves of the arms, and are consequently the brachial vertebræ. The 3-7th rib are thus appended to the 5 proper thoracic vertebræ, which stand in the signification of the pulmonary vertebræ. To these succeed the 5 short ribs which belong to the abdomen; their vertebræ are thus intestinal vertebræ. The succeeding vertebræ belong to the sexual system, and are indeed the 5 lumbar or pedal vertebræ, because they furnish the pedal or foot-nerves, the 5 sacral vertebræ being the proper sexual vertebræ. The coccygeal or caudal vertebræ correspond to the cervical vertebræ, and are present for the sake of the sexual branchiæ; usually one and the other is arrested. Thus there are,—

3×5 Respiratory vertebræ.
3×5 Sexual vertebræ.
1×5 Digestive vertebræ.

The number of the sensitive vertebræ is consequently $7 \times 5 = 35$, which are distributed into three groups, in accordance with the principal cavities of the body, whereof the 2 terminal groups consist of 15, but the abdominal group, which combines or unites them, only of 5. The body is accordingly not merely laterally, but also in length, a perfectly symmetrical structure, which has been parcelled out in the following manner into its five stockworks, stories or floors:

I.   Dermal vertebra.
A. Sexual vertebræ.
   a. Coccygeal vertebræ   5,
   b. Sexual vertebræ 5,
   c. Pedal vertebræ   5.
B. Abdominal vertebræ   5.
C. Thoracic vertebræ.
   a. Pulmonary vertebræ   5,
   b. Brachial vertebræ   5,
   c. Cervical vertebræ   5.
II.   Auditory vertebra   1.
III.   Lingual vertebra   1.
IV.   Optic vertebra 1.
V.   Nasal vertebra 1.

This regularity is found too only in the human skeleton. The animals are irregular men. (Vide Oken's Zahlengesetz in den Wirbeln. Isis, 1829, S. 306.)

Cavities of the Trunk.

2115. The osseous system forms the trunk or body, because it follows the vascular system; the two other galvanic systems, the dermal and intestinal systems, form the large portions of the trunk or its cavities; to them is added the sexual cavity or pelvis.

2116. There are thus three truncal cavities, a pulmonary, an intestinal, and a sexual cavity, or thoracic, abdominal, and pelvic cavity.

2117. The osseous system will develop itself most feebly around the abdominal, because it is the indifferent cavity. Therefore there are either no ribs at all, or they are so short that they do not reach to the anterior vertebral column or the sternum. The short or false ribs are, according to their physiological sense, abdominal or splanchnic ribs. The thoracic ribs must be perfectly developed, i. e. abut against both vertebral columns, be entire ribs; the entire or perfect ribs are thoracic or pulmonary ribs. The sexual ribs are arrested on the pedal and coccygeal vertebræ, but on the proper sexual vertebræ, namely, the sacral bone, they are still present as rudiments.

3. MUSCULAR SYSTEM.

2118. As the intestinal system reappears in what is animal under the condition of a vertebral column, so also does the aggregate of the vascular system ascend, and the vessels become animal.

2119. The animal vessels are the muscles, or filled-up vessels. The polar process enters the body through the vessels; thereby the tubes obtain two strong poles, and are drawn out lengthways. They are fibres, and consist of a series of strongly oxydized blood-globules.

2120. The fibre is chiefly apparent in those vessels, such as the arteries, in which the influence of air operates more forcibly. Now an artery has, in addition to the external cellular coat, two coats, like the first animal body, or one wall turned towards the mucus, and one towards the air. The internal arterial wall is enteroid, the external dermoid in character; the one being simply granular, the other fibrous membrane. The two membranes separate into two cysts or tubes, which likewise adhere within each other like intestine and skin. The external will become fibre, the internal bone.

2121. There is a vegetative and an animal fibre-or muscle-system. The one is associated with the tegumentary formations, the other with the bones and nerves.

a. Vegetative Muscles.

2122. The muscles of vegetative life are simply fibrous tunics, as in the arteries, and are found in the skin, in the intestine, and in the vessels.

2123. The tegumentary muscles lie under the skin, and are inserted into it, or into the dermo-osteous system, when such an one is present. If the fibrous membrane be strongly developed under the skin, then it receives the name of a Panniculus carnosus.

2124. The intestine has also its fibrous coat, which, upon the stomach, anus, and pharynx, is frequently developed as a panniculus carnosus.

2125. The same holds good of the vessels, especially of the arteries and trachea.

2126. The fibres are either elongated or else annular fibres. The latter obtain the preponderance at the extremities of the tubular formations, as on the pharynx, anus, lips, and eyelids.

2127. We have consequently a dermal, splanchnic, and osseo-muscular system.

b. Animal Muscles.

2128. In the oxydizing part of the vascular system the formation of fibres must be preponderant over that of cells, and thus at the root of the lungs.

2129. The vessel becomes in the lungs a fleshy cavity. The fleshy vessel is the heart.

2130. The heart is a vascular fragment, with a preponderating development of fibrous membrane.

2131. This fibrous membrane is developed at a spot where all vascular systems enter into mutual proximity, such as the respiratory ducts and the intestinal lymphatic vessel, the artery and the vein.

2132. In the union of all the highest galvanism is attained, and then the formation may launch out into what is animal.

2133. The heart is the animal in the plant.

2134. The first heart is an arterial heart. There is originally no venous heart. In the embryo, particularly in that of the bird, this is extremely distinct; the arterial heart also emerges for the first time, and per se, in the lowest animals, in the Molluscs, Snails, and even in the Fishes, although in the latter it is regarded as a venous heart.

2135. The arterial heart is the central, the venous heart the peripheric.

2136. The heart is the prototype of the muscular system. All muscles must be a metatype of the heart.

2137. The muscle is hollow. It is a cyst.

2138. The muscular system is a manifold and serial juxtaposition of fibrous cysts or of hearts. In this respect the muscle has been formed in a corresponding manner to bone. Both are rows of cysts.

2139. But the muscle, as being the external fibrous tunic, is the enveloping or external cyst.

2140. In idea, the muscle can only directly envelop the bones, and not the other parts, for it ranks upon the same grade of development with the bone; it is the arterio-fibrous wall, while the bone is the internal arterial wall.

2141. Bone and flesh stand in antagonism like air and earth. The muscle is that which is polarizing—moving, the bone what is polarized, moved. The muscle is heart, the bone the moved blood. Bone and muscle are related as that which is contained and what is containing. The muscle is the wall of the cyst, the bone, the fluid which has been secreted from it and rigidified.

2142. As therefore the muscle is an individual cyst, a heart, which cannot invest the whole body like a single large cyst, so also must the muscular contents be only a discrete cyst. The uninterrupted character of the bones depends therefore upon that of the muscles, and the latter upon the meaning of the heart.

2143. A physiological rationale of the joints is accordingly afforded in the heart.

2144. A bone is a rigidified, ossified heart; the osseous system is a series of mutually dependent, alternately ossified and non-ossified (arteriose and venous) hearts.

2145. The muscular cyst includes the soft bone, or the joint.

2146. At its two extremities there is preponderance of oxydation, whereby the soft osseous cysts are combusted into hard calcareous earth.

2147. The muscle is a cause of the alternating ossification.

RELATION TO THE OSSEOUS SYSTEM.

2148. As the formation of bone predominates upon the side exposed to the light or the nervous, so does that of muscle upon the shady or vascular side. The abdominal side of the animal is the muscular, just as the dorsal side is the osseous. Upon the thorax, abdomen, on the members which belong to the anterior region of the body, and on the face, the muscular layer is by far the most predominant. Posteriorly, however, or upon the back, it is slightly wanting, and the bones there project.

2149. The back is related to the anterior surface (when regarded in man) as bone is to muscle. What is in front is muscle, what is behind is bone. The anterior side is therefore more active, nobler, more powerful and more spiritual than the posterior. Posteriorly stands the earth inert and rigidified, while in front is the air in ceaseless capacity for, and actually in, motion. The anterior muscular layer is more active and powerful than the posterior.

2150. In every muscular cyst there are two kinds of layers, an anterior and a posterior, or stronger and weaker layer.

2151. The stronger layer is the flexing, the weaker the extending. For the limbs have been necessarily bent forwards. The direction alone of the joints at once resides in the structure, which is even determined by these relations. Such a muscular layer, which mostly consists of several bundles, is termed a muscle.

2152. A muscular cyst consists of flexor and extensor muscles. The individual muscle is therefore only a piece of a cyst, and therefore not in itself hollow. It is only an entire muscular layer of flexors and extensors which is the pattern of the heart. The flexors are the strongest and are placed in front, the extensors lie behind.

2153. In the heart the flexor layers have not yet separated from the extensor layers, because the vegetable flesh has as yet no symmetry in itself.

2154. Flexors and extensors occur in pairs; because the osseous system is in pairs.

2155. There resides in the osseous and in the muscular system no cause for a diversity in the two halves of the body. Thus, if there be a difference, it must only reside in the asymmetrical galvanic systems.

3. Organs.

2156. Organs are parts of an anatomical system, which separating, combine with a part of another system, and thereby obtain a peculiar or special function.

2157. For each system therefore there are as many organs, as there are possible combinations of systems. There are vascular, intestinal, pulmonary, sexual and tegumentary organs, with furthermore, osseous, muscular and nervous organs.

A. VEGETATIVE ORGANS.

1. INTESTINAL ORGANS.

2158. The intestinal or splanchnic system falls first of all into three great divisions, into that of

the viscera, the sex and the head; the visceral intestine separates moreover into pulmonic, venous and tegumentary intestine, and this in constant accordance with its combinations and functions.
a. Visceral Intestine.
2159. In the digestive system it is the chemical process that takes precedence. This, however, divides into three moments, into that of solution, separation and formation, or crystallization, which is here absorption. Thus the intestine separates also into a solvent, secernent and absorbent intestine, and that indeed through its combination with the pulmonic, vascular and tegumentary system. The solvent viscus is the stomach, the second or secernent, the duodenum, the third or absorbent, the small intestine (jejunum and ileum).
Pulmonic Intestine.
2160. All solution is accompanied by oxydation. The gastric juice is in its action an acid.
2161. The gastric juice obtains its oxygen from the spleen. The spleen is the stomachic lung. This is attested by its position and close approximation to the stomach; by its black, venous, and deoxydized blood, which in certain diseases has been secreted even in the stomach; by its want of any excretory duct; by its tissue, which resembles that of the oxydized placenta; furthermore by the natural character of this function when contrasted with the unnatural quality of other useless functions that have been assigned to it; and lastly too, by the fact that, without this opinion be admitted, it must remain a superfluous organ, and as regards use, unknown. After a series of years, during which this doctrine has been combated upon all sides, without, however, a single reason, save and except that it was not believed, being advanced by its opponents, I must still persist in the correctness of the above view.
Vascular Intestine.
2162. In the duodenum the analysis of the food takes place through the medium of the bile. It is consequently the biliary intestine or stomach.
2163. The biliary intestine does not stand upon a par with the other intestines, but has an equal rank with the stomach. It is therefore not confined within the mesentery, but can expand itself like the stomach; it has its vessels and nerves. In it the separation of the chyme into chyle and excrementitious matter takes place.
2164. What the spleen is for the stomach, that is the liver for the duodenum; it is the hepatic stomach, and consequently the vascular stomach.
2165. The liver is the ramification of the intestinal canal along with the whole vascular system.

2166. Now, as the analysis is the principal function in the whole process of digestion; so is the liver the chief organ of all digestive organs.
2167. The liver is the centre, the brain of the digestive system, because it is the blossom, the synthesis of the vascular system. From it everything emanates, and upon it everything which concerns digestion, ay, the whole body, retrogressively operates. If the liver suffers or undergoes functional derangement, the whole vascular and tegumentary formation then becomes a liver, as is exemplified in the disease called jaundice.
2168. The bile effects the analysis or separation through its basic or alkaline character, since it combines with the acids of the chyme, and thereby forms the excrementitious matter.
2169. That which mediates between the acid and what is alkaline is the fluid secreted by the pancreatic gland. The pancreatic gland is the ramification of the intestine along with the arterial system.
Tegumental Intestine.
2170. That in the jejunum and ileum or the small intestine proper, and in that alone, absorption

and thus the tegumentary function, but nothing else, occurs, is well known. By this absorption the chyle becomes removed from the intestine, so that the excrement alone is left.
b. Sexual Intestine.
2171. The sexual is pre-eminently that which is exsecernent or excretive, since one sex strives to reintegrate itself upon the other, becomes ingestive for the other, but egestive for or in relation to itself. It is therefore an essential property of the sexual parts, that they secrete and excrete.
2172. Every galvanic system, which has been adjoined or appended to the sexual parts, is excretive. They are, taken in a rigid sense, the only system of excretion. The co-operative processes of the sex are those of the vegetative systems, but they take place in the reverse direction. The latter convey inwards, the former outwards. The kidneys are a reversed or evacuating lung, an excrement-forming liver; the urinary cyst is an exspirant, a trachea containing refuse matter; the urethra is a glottis reversed; their diseases are therefore similar. They thus expel products of the individual respiratory system.
2173. And in this is the sexual apparatus distinguished from them, namely, that it expels the products of all systems, the products of the whole organism, or the organism itself. In the semen the whole male body with all its parts transudes, or passes over in a fluid state, into the female parts; in a child, the female together with the male body, passes over—formed or fashioned into the world.
2174. The sexual intestine must therefore be expulsive also. It is that which conducts the intestinal juice and the refuse of the aliment out of the body.
2175. The evacuating or efferent intestine is the large intestine. It is consequently the Sexual intestine.
2176. The large intestine is related to the abdominal or small intestine, exactly like the urinary cyst is to the kidney, and this to the vascular system. The small intestine passes therefore by perforation into the large, not the latter into the former. The small enters by perforating the large intestine, and therein evacuates its alimentary residue, as into a special-cyst, that has nothing to do with the intestinal system. The large intestine is the excrement-cyst, just as that which has been called bladder is the urinary cyst.
2177. The large intestine no longer digests, but only receives what is left from the process of digestion and expels or casts it out.
2178. The large intestine begins with a cæcal extremity, or with an obtuse cyst, and opens into the anus, exactly like the original animal cyst, the polyp. The blind extremity is called cæcum. With this the small intestine communicates by perforating it at an acute angle, and that indeed in a direction which runs towards the obtuse cystiform extremity; so that both intestines lie in mutual juxtaposition like a fork, whereof the pharynx and the rectum are the two prongs, while the cæcum is the handle.
2179. The two intestines are consequently by no means correlated. In the perfect animal there are two intestinal systems thoroughly distinct from each other; two intestines, which belong to two different animals, the sexual and encephalic animal, or the plant and the animal. The genesis of the large intestine and all its relations, which are principally seen in the cæcum and rectum, plead for this philosophical derivation of the two intestines.
2180. The rectum appertains wholly and entirely to the sexual system, and especially to the uterus. It is devoid of mesentery, and supplied by special vessels; it stands in most obvious sympathy with the uterus, with its diseases, and with menstruation. Even hemorrhoids are a sexual disease, a disease of the sexual intestine.
2181. The anus is thus the intestinal mouth of the sexual animal. In the lowest organized animals,

the oviducts seminal and urinary ducts together run into it, as into a proper mouth. The anus is a true oral cavity in the fishes, reptiles, and even in the whale.

2182. The pharynx opens into the mouth, and so does the rectum into the anus; the trachea opens into the month, and so does the urethra into the anus; the salivary ducts open into the mouth, and so into the anus do the oviducts, with the spermatic canals in the lower animals. The rectum moreover lies behind the urinary cyst, like the gullet behind the trachea.

2183. The sphincter or occlusor muscles of the anus have been formed like the sphincter muscles of the pharynx. The anus is a mouth without a head, and therefore a mouth without lips, or an pharynx.

Cephalic Intestine.

2184. The union or combination of the intestine with the animal systems, such as the nervous, muscular, and osseous, occurs in the head.

2185. In so far as it combines with bones and muscles, it becomes an organ of motion, and, when with the nerves, an organ of sensation. The animal or sarcal intestine is pharynx and mouth.

2186. The motor organ is an organ of prehension. The organs of prehension are endowed with independent motion, and therefore move towards the food. The first general organs of motion are the members of the body. In the higher animals the thoracic members are at once the prehensible organs. Instead of the aliment being obliged to flow, as unto the plant, in a state of suspension in water, the animal moves itself toward its food.

2187. The members are the first organs of prehension. But these are repeated in the head as jaws and teeth. The teeth are the second organs of prehension, but the first which belong to the cephalic intestine; they are called organs of seizure or prehension.

2188. The organ of digestion is, however, one of a chemical character. It must be therefore repeated as such in the head. This is shown in the salivary glands. The saliva is the animal gastric juice, and is therefore directly solvent in its properties. It is poison.

2189. After and during the operation of the saliva, the aliments in the mouth are subjected to the action of the molar teeth and ground or chewed. These comminuting or crushing organs are only a repetition of the act of prehension, and consequently belong to the prehensile or biting organs.

2190. The mouth is the stomach repeated in the head.

2191. The union of the intestine with the nervous system is the tongue.

2192. The intestine, when repeated in the head's muscular system, is the organ of deglutition, as is seen in the pharynx and oesophagus.

2193. The prehensile and manducatory, venomous, gustatory, and deglutitory organs, are the forms into which the intestinal system divides, when it is repeated in the after-animal, or in the sphere of animal life. The gustatory organ is the nerve-intestine; the prehensile organ the bone-intestine; the organ of deglutition the muscular intestine; while the poison-organ is the proper cephalic intestine, or the stomach.

2. VASCULAR ORGANS.

2194. The vascular system regarded "en masse," or in a general sense, has to participate in the nutrition of the body; so far it takes the place of the cellular tissue, and cannot therefore be developed for itself into any special organs. If, however, certain vessels separate from the general system, and combine with other systems unto the performance of special offices or functions, organs then originate, which, nevertheless, do not, in a rigid point of view, belong to these systems.

2195. There are therefore as many vascular organs as there are possible combinations; thus with the tegument, the lung, the intestine, the sexual parts, and with the systems of animal life.

a. CUTANEO-VASCULAR ORGANS—BRANCHIÆ.

2196. The development of the vessels unto a special organ in the integument are Respiratory organs; or, more properly speaking, the development of the integument to constitute a special organ in combination with the vessels, is a respiratory organ.

2197. In the commencement the branchiæ are only a vascular network upon, and therefore subordinate to, the integument. They pass, however, through all possible stages of development, until they have subjected and converted also the integument into a vascular system; a point which is attained in the formation of the lungs.

2198. The branchial membrane already commences in the earthworm to concentrate itself and dilate, so as to form what have been called the sacculi or pouches; in the leech it is folded inwards, so as to form lateral vesicles or cysts, that prognosticate the air-passages or stigmata of insects; in the Nereids the vessels project upon the back from above the integument as free branchial ramules, a formation, which is again found in the nudibranchiate Gasteropods.

2199. These branchial ramules form at first two rows extending along the whole dorsal region; by degrees, however, the posterior set disappear, and the cervical branchiæ only are left, as antetypes of the gills of fishes.

2200. In the Mollusca the branchial vessels unite to form laminæ upon the ventral sides, and are already surrounded by a kind of thoracic cavity, the mantle, as is exemplified in the tectibranchiate Gasteropods. Here again the branchiæ are either a simple vascular net enveloped in a fold of the mantle, or elongated into filaments disposed like the teeth of a comb, or into laminæ, &c.

2201. The lateral branchiæ of the Nereids usually emit filaments resembling feet, upon the root of which the true branchiæ are then placed. In the Crustacea these filaments harden into actual feet provided with joints. The feet are therefore nothing else than branchial filaments, which have lost their vegetative function.

2202. In many worms the same branchial filaments are converted into hairs or bristles, which are therefore none other than desiccated branchial filaments.

2203. Even the hairs of Mammalia and the feathers of Birds have been left as remnants of the original branchial formation.

2204. Where the branchiæ have assumed the laminated form, they are surrounded by a similarly formed covering or operculum. The shells of the Mollusca are, as regards their signification, none other than branchial opercula, or gill-covers; the same holds good also of the gasteropodal and crustacean shells, and at bottom of every calcareous and horny covering of the body.

2205. To the same category belong also the opercula of Fishes, and even their scales. Fundamentally, the whole epidermis is only a product of respiration or oxydation.

2206. The first saccular or cæcal inversion of the integument, as in the Leeches, Molluscs, and Snails, is at once a predominance of the tegumentary, obtained through the medium of the branchial formation, whereby the skin commences to become a self-substantial organ of respiration.

2207. In the Scorpions the branchiæ are introverted sacs or cysts, into which, however, instead of water, air already enters.

2208. This insaccation is converted in the Spiders into more distinct air-cysts, which finally ramify in the higher insects and become true tracheæ.

2209. Lastly, the respiratory system obtains the upper hand to such an extent, that, together with internal air-tubes, external branchial laminæ are also developed, as in the Molluscs; but in them the tracheæ obtain the preponderance over the blood-vessels, so that these laminæ dry up and become wings or fins.

2210. The wings of Insects are branchial laminæ, converted into air-organs.

2211. The wing-coverings or elytra are branchial opercula, and correspond to the shells of bivalve Mollusca.

2212. Every insect therefore must properly possess four wings and two wing-coverings, of which last, however, rudiments only appear to be left in the nocturnal Lepidoptera.

2213. In the higher organized animals those gills only which are nearer the head are persistent, the posterior or lateral branchiæ being gradually arrested.

2214. These lateral branchiæ remain in Fishes as lateral mucous openings, which constitute the lateral line.

2215. The cervical branchiæ are limited to the number five, which has already begun to be established in the Crustacea; namely, at the origins or roots of the five anterior pairs of feet.

2216. The number five probably derives its origin from the vegetable kingdom, and that indeed from the genesis of the pinnate leaves, so that one kind of numerical law appears to prevail with respect to this organ in both organic kingdoms. It is probably also an imitation of the five organs of sense.

2217. The vessels of the branchiæ of Fishes are accompanied by bony rings, which correspond to the feet of Crustacea.

2218. All Fishes have, with few exceptions, five branchial arches.

2219. When in Fishes the sarcous system begins to produce viscera, then the five branchial foramina pass inwards, and only a single respiratory aperture is left for them in the fleshy body; namely, the external branchial foramen.

2220. In the lower animals water or air passes in and out through the same respiratory aperture, but in Fishes these two courses are preserved distinct; for, except in the Lampreys, the water enters through the mouth, and passes out through the branchial aperture.

2221. Here the attempt is manifested in a still greater degree, to bring the process of respiration wholly under the control of what is animal, this being in the next place, and for the first time attained, when respiratory apertures only are left upon the head.

2222. The respiratory apertures of the head are the nostrils, which are suddenly manifested in Fishes, but are in them simply subservient to the sense of smell, and not as yet to the respiratory act.

2223. All the higher animals have, like fishes, branchial apertures in the neck, only they coalesce and become obliterated at an early period, or so soon as that respiratory process appears, when air passes through the nostrils. In the Salamanders and Frogs these branchial foramina are persistent for a longer time, frequently through the whole period of life; but in Birds and Mammalia they disappear, while they are in the embryo state.

2224. When the branchial apertures close, the vessels separate from the arches, and betake themselves to a glandular body lying in front of them. The thyroid gland is the remnant of the former branchial formation, and is therefore found only in Reptiles, Birds, and Mammalia.

Lungs.

2225. In Fishes the internal organ of respiration is already indicated by a cæcal eversion of the pharynx, which is surrounded by the branchial arches. This membranous cæcal eversion is called the swimming-bladder, which in the higher animals becomes double on account of their symmetry, and is then called lung.

2226. In the Fish the aquatic and aerial process of respiration are present together, the former being external, the latter internal.

2227. The branchial arches having coalesced, are converted in the higher animals into tracheal rings, into the larynx and the posterior cornua of the hyoid bone, if these are present, as in the Reptiles. The larynx is therefore no special organ, but only a remnant of the branchial respiratory apparatus.

2228. The laryngeal vessels are, like the thyroid glands, branchial vessels, and in Fishes, therefore, the branchial vessels do not correspond to the pulmonary vessels, but to those of the trachea. The pulmonic vessels of Fishes are the blood-vessels of the swimming-cyst, which convey blood directly into the heart, whereby this organ obtains the signification of the left or arterial heart.

2229. When the branchial apertures have coalesced, the nose then opens into the mouth or into the trachea, and in this way the nostrils assume the complete character of aerial foramina.

2230. The nose is therefore originally an organ of smell, and then a part of the respiratory system. It is the animal lung.

2231. As the secretion of bone is a product of the more powerful process of oxydation, so do the bony rings multiply beneath the branchial arches or the larynx, and are called tracheal rings. In the feebly respiring Reptiles the trachea is therefore still in a membranous condition, but in Birds and Mammalia, it is surrounded by many rings, or is a repetition of the larynx.

2232. In Birds even a kind of inferior larynx originates, which is provided with muscles and can produce tones.

2233. The ramification of the trachea into two bronchi or branches is constantly progressing, and at length these tubes divide into a great number of vesicles, which together form the lungs. The lung, which was in the commencement a simple saccular inversion of the integument, has now become a self-substantial organ, to which the respiratory vessels have been subordinated. The lung also divides upon each side into five lobes.

b. VASCULAR ORGANS OF THE INTESTINE.

Liver.

2234. The self-substantial development of the vascular, and its separation from the general, system, is most perfectly attained in the Liver.

2235. In the liver, as being the vascular system, which combines with the intestinal canal, the venous system has become independent. The portal vein arises from the intestinal canal, collects into one trunk, and again ramifies in order to unite with the biliary ducts, which are only a ramified saccular eversion of the intestine. This conjunction is represented by the liver.

2236. The liver as being a venous organ stands therefore in opposition to the lung, and produces, instead of oxydes, a basic body, or the bile. The liver, as being the venous system which has become free, is to be regarded as the highest development or blossom of the vascular system.

2237. It is for the vegetative body, what the brain is for the animal; and hence, therefore, the similarity of structure and the sympathy between both organs.

Spleen.

2238. As opposed to the liver, the arterial system also develops itself upon the intestine as a respiratory or branchial organ. These intestinal branchiæ are found in several of the lower organized animals, especially in the Holothuriæ and Aphroditæ.

2239. In the higher animals they are aggregated into a special organ, through which the gastric juice obtains oxygen; this is the spleen. The spleen is the branchia of the stomach; it has therefore no excretory duct and requires none.

2240. Lastly, the salivary glands also in the mouth as well as in the duodenum, the abdomino-salivary glands, and even the odoriferous glands on the rectum, such as the castor-and civet-pouches, are complications of vessels with intestinal ramifications.

C. VASCULAR ORGANS OF THE SEX.
Kidneys.
2241. The vascular organ of the sexual system is the kidneys.
2242. As the urine is chiefly characterized by the urea and thus by a basic principle; it thus corresponds to the bile, and consequently the kidneys directly to the liver or the lung reversed.
2243. But there is also a sexual branchia in those inferior animals which breathe through the anus, such as many aquatic larvæ.
2244. The remnant of this in the higher animals appears to be the allantois, and in the body itself probably what have been called the primordial kidneys.
2245. The combination of the vascular system with the animal systems is the sense of touch.

3. RESPIRATORY ORGANS.
2246. The respiratory organ is a development of the integument.

2247. The perfect organ of respiration is an air-organ or lung; combined with the tegumentary system, it is a water-organ or branchia.
2248. There are dermo-branchiæ or the branchiæ proper, as in the Worms, Molluscs, Snails, and Crabs.
2249. Intestinal branchiæ in the Holothuriæ, constituting in the higher animals a spleen.
2250. The sexual branchiæ are probably the primordial kidneys.
2251. The branchiæ when united with the osseous system are the branchial arches of Fishes, which subsequently separate into larynx and thyroid gland.
2252. The self-substantial development of the integument into the respiratory organ is a lung.
2253. When united with the vascular system or the vegetative systems generally, the integument forms the tracheal system of Insects.
2254. The tracheæ are spiral vessels as in plants.
2255. The respiratory organ, united with the motor system, is the lung proper, which is situated within the thorax, and covered by ribs.
2256. The proper lung divides also, like the intestine, into two parts, into the cystiform, pharyngeal expansion of the larynx, and into the pulmonic substance, which is equivalent to the stomach, and wherein the analysis of the air takes place.
2257. The trachea, and especially the larynx, is moreover an entire thorax, a ribbed skeleton upon a small and membranous scale. In the larynx, the animal thoracic structure, consisting of ribs and muscles, resides prognosticated. The larynx has originated from the coalescence of branchial arches. The ribs are a repetition of the branchial arches.
2258. The diaphragm is a formation, which admits, by no reasoning from its anatomy, but only genetically, of being explained. Originally, the whole body was only an abdomen, upon the external surface of which the branchiæ were appended. A remarkable instance of this is at once afforded by the Snails, and also the Fishes. As the gills were converted into lungs, a special body or the thorax originated for them, which encroached upon the abdomen. Now, the abdominal wall left between these two cavities, was just the diaphragm.
2259. The diaphragm is not a transverse wall or septum. The idea of such a cross-rail gainsays all sound physiology. It has been the abdominal wall.
2260. The union of the lungs with the nervous system is the nose.

2261. The nose is the cephalic thorax; but it has also the thoracic contents, or the very lung repeated in itself.

2262. The multicavernous ethmoid bone is the lung in the nose, the two nostrils being the stigmata or foremost openings of the trachea. The nasal muscles are homologous to the cartilages of the trachea, and especially to those of the larynx.

2263. The velum palati is the diaphragm between nose and mouth, or between the cephalic thorax and the cephalic abdomen.

Coverings.

2264. The animal coverings are desiccated respiratory organs appertaining to the integument.

Capillary Vessels.

2265. The principal function of the vessel is that of secretion, whereby nutrition is imparted. This secretion must take place in the whole body, in so far as it is opposed to the lung. The vessels then pass over into the finest canals, and are called capillary vessels.

2266. The capillary system of vessels is an organ in antagonism to the lung; what enters the latter organ, passes out through the former.

2267. The capillary system of vessels is the property of the tegumentary system. It may almost be said that wherever there is integument, there capillary vessels are present. The most perfect evolution of capillaries is the integument. It is the proper exsecernent, in opposition to the intestine, which is the absorbent organ.

2268. Evaporation is the essential tegumentary process.

2269. The product of the evaporation is mucus.

2270. In the evaporation, however, the mucus becomes analyzed by the influence of the air and light.

Epidermis.

2271. The external mucus exuding from the skin becomes oxydized, the inferior on the contrary reduced; the oxydized becomes vitreous and transparent. It is the Epidermis.

2272. With the maximum of oxydation the epidermis passes over into a vitreous transparent horn, e. g. scales.

2273. The scales, which invest the toes and digits are called claws, and finally become nails. The finger-nail is nothing but a scale, which in this situation has become particularly large and strong.

2274. That which is reduced beneath the vitrified epidermis determines the colour of the skin. With a demi-oxydation it is uncoloured, appears white. Where the integument is thin, the red colour of the blood appears through it; such integument is therefore white upon the whole, but red in individual places.

2275. With the most complete reduction effected by the highest operation of light, its substratum or under-layer becomes black. The mucus passes over into reduced carbon. Beneath the vitreous epidermis there is thus a metallic pigmentary membrane.

Hairs.

2276. Capillary vessels, which simply convey mucus, but are elongated singly above and from out the cutis, are Hairs. The idea of the hair is a capillary vessel, the contents of which are no longer blood, but reduced mucus. It is an indifferent capillary vessel. The hair is hollow and contains an oil, which determines the colour.

2277. In the hairs the nutritive system is prolonged above the body.

2278. The hairs and scales are the general terrestrial system of the body determined by the air.

2279. Thus the earth has sprung up into a plant. Scales and hairs are like true vegetable leaves, which still continue their process in the animal; they can no longer it is true make their respiratory process hold good for the animal process itself, but are at present content with only oxydizing the material of evaporation. It is properly the integument only, not the entire body, which respires through hairs and scales.

2280. The hairs are dried branchial filaments, and therefore continue to occupy in Man, those situations only where in the lower animals branchiæ or tentacula are placed. Around e. g. the mouth, upon the head, under the arms or in the axillæ, and around the sexual apertures.

2281. The feathers are dried ramified branchiæ, pinnate leaves.

2282. The hairs participate in the electric process of the whole body.

2283. What determines the colour in the plant, determines it also in the animal; only in the plant the colour is more coarsely precipitated, and nothing therefore glimmers from the interior of the body, but all is green. In an animal, however, the colouring matter is transparent, and the interior is thus rendered visible.

2284. With the external coverings, such as scales, and lastly with the hairs, all organs of the trunk, in so far as they are derived from what is vegetable, are exhausted. The vegetable animal, as trunk, is completed, and we must now therefore turn to the sexual organs.

4. SEXUAL ORGANS.

2285. The sexual organs are developments of the integument upon a higher grade, and combinations of the same with the animal systems, like as the blossom is a repetition of all vegetative systems.

2286. They range in the middle between the vegetative and the encephalic animal, and are therefore a totality for themselves, or a sexual animal.

2287. There are vegetative and animal sexual parts.

I. Vegetative Sexual Organs.

2288. Are special developments of the intestine, vessels and branchiæ.

a. SEXUAL ORGANS PROPER.

2289. The proper sexual parts are a repetition of the digestive system on its transition to what is animal, or to the organs of sense.

1. Female Organs.

2290. The female parts are a floral capsule, consisting of an ovary or sac, stigma and ova. All higher development takes place, however, through separation of the complicated organs and processes.

2291. The three parts therefore of the gestative sac or matrix separate, each part becoming self-substantially perfected. The orifice elongates into a cervix or neck, which is gradually rendered more distinct from the matrix, and in its highest and self-substantial completion is called oviduct.

2292. The matrix elongates also at its cæcal extremity or fundus, as well as at its open end. The germinal bodies also become self-substantial, separate gradually from the fundus of the matrix, and are then independent ovaria.

2293. As in the highest animal formation they also assume the animal symmetry, while at first they were only single, or manifold, as in the plicated capsules of Star-fishes, and therefore remained stationary at the number two; so is the ovisac prolonged into two long cornua, or trumpet-like tubes, which at the commencement, indeed, still include the ova, as in Insects and Fishes, but are subsequently left entirely free.

2. Male Parts.
2294. The leaf-formation is elevated to the corolla or the male parts in the plant. They are, however, only the repetition of the plant upon a higher stage. Hence in the animal also the male parts will be upon a higher grade of position to the female.
2295. As the stamina surround the ovarian capsule in the plant, so do animal stamina stand around the orifice of the oviduct; as penes.
2296. In the lowest ovisacs the penes are circularly disposed around the orifice; as is exemplified by the tentacles of a Polyp; by degrees, however, they are resolved, for the sake of symmetry, into two, and then project upon the sides of the orifice, as in the Serpents and Lizards. In the higher animals the two penes coalesce into one.
2297. This penis situated at the orifice of the oviduct is the clitoris. The vagina is separated from the clitoris; but in the male organs they both combine or blend with each other, and the vagina becomes the seminal canal or that of the penis.
2298. As the orifice begins to assume a male character, and the external parts to be more forcibly developed in the direction outwards, so, on the other hand, do the internal parts recede the more, and remain simply as ovaria or ovisacs, in which the ova, instead of being formed as such, dissolve into pollen, mucus, or male semen. These ovaria, which now secrete semen instead of vitelline vesicles, are called testes.
2299. The testes originate, while the mucous ova are being reduced to primary mucus, or to infusoria.
2300. What is male originates through an organic process of decomposition or putrefaction of the ova. The semen is an organic product of decomposition.
2301. The semen must contain infusoria. A semen which does not contain infusoria, is ovum-like, or feminine. Except at the time of heat or rut, and thus when the animals have a female character, the semen possesses no infusoria, and is in that case simply albumen.
2302. Semen, which is devoid of infusoria, is incapable of impregnation. How can one wretched female ovum impregnate or fructify another?
2303. As the trumpet-like tubes or oviducts belong to the ovaria, so are they developed with the testes, and now convey semen instead of ova. The female tubes become the vasa deferentia or spermatic ducts, the uterine cornua, the vesiculæ seminales.
2304. Between the vesiculæ seminales and the vagina or the penis the uterus shrivels up into the prostate gland, into which the vasa deferentia open, like the oviducts do into the uterus.
2305. As the matrix is the proper female organ, so will the spermatic ducts seek to combine with the penis, or at least open self-substantially upon the os uteri. The testes open either through the spermatic ducts into the vagina, as in Fishes, Reptiles, and Birds, or into the penis, as in the Gasteropods, Insects, and Mammalia.
2306. Male and female parts are therefore perfectly homologous, the former having a greater development of the external division, the latter of the internal.
2307. The female parts have undertaken the business of vegetation, or that of the viscera, the male that of the animal excitation.
2308. Since the male parts are no new formation, but the female parts themselves, only characterized by internal arrest, and external increase of development, so do male and female parts appear incapable of occurring together in any animal. Perfect hermaphrodites would, accordingly, be impossible; for where testes are, no ovaria could be, because the testes are themselves the ovaria, or the latter only changed.
2309. Hence androgynism would be possible only by one ovarium remaining as such, and the

other being converted into a testis.

2310. This development appears only possible when the two halves of the body are unequal. Asymmetrical animals only can be androgynous. In the snails one of the two molluscan shell-valves has been more largely developed than the other, and therefore one half also of the body is greater than the other. On this account many hermaphrodites are found among these animals.

2311. Thus, there ought to be no androgynous animals with at the same time two ovaria and two testes. Nevertheless, this very peculiarity occurs in many of the lower organized animals, e. g. the worms.

2312. Accordingly, the principle of androgynism is in general to be found in the want of symmetry. Symmetrical animals are as a rule diœcious. No hermaphrodite is found among the Insects, Fishes, Reptiles, Birds, and Mammalia.

2313. If such occur, they are formations that have remained persistent at the lower stage of development, or upon the transition of the embryo through the Snail-type of organization; they are thus monstrosities, or malformations.

2314. These malformations also, when occurring in the higher animals, can never possess more than one testicle and one ovarium. The uterus then ranks midway between its condition as such, and that of a prostate gland; the spermatic tube or urethra opens after a female fashion below the root of the penis.

Impregnation.

2315. Since the male sex is related to the female, as corolla to capsule, as leaf to stalk, as air to water, and as light to matter; so is it related also as integument to intestine, as lung to lymphatic vessel, as artery to vein, as nerve to flesh or muscle, as Animal to Vegetative.

2316. Copulation is therefore an irradiation.

2317. Already, in the course of the heavenly bodies, has the highest act of the animal, that of copulation, been preindicated or portrayed. The creation of the universe or world is itself nothing but an act of impregnation. The sex is prognosticated from the beginning, and pursues its course like a holy and conservative bond throughout the whole of nature. He therefore who so much as questions the sex in the organic world, comprehends not the riddle or problem of the universe.

2318. If the female parts have effected a complete transition into the male, so are the sexes necessarily separate and distinct.

2319. Since the male parts are the female that have been more highly developed, so there resides in the latter the constant conatus or effort to convert themselves into the male.

2320. This metamorphosis, or conversion, is, however, no longer possible in the female parts, when already finished or fully formed, but is only attainable in a new attempt being made upon their part, to transform the fluid mass into ova.

2321. Gestation or pregnancy is none other than the propensity of the Female to convert itself into the Male.

2322. The fœtus is the male in a female, or the fœtus is the male sexual parts in the female.

2323. In idea, every fœtus should be male. But, if masculary be attained with the first production, then the second necessarily subsides into the female. In this manner there of necessity originates an equiponderance in the number of both sexes.

2324. Moreover, if the sexual parts be regarded according to their proper signification in the animal, they are the upper intestinal system, as it has been developed in the mouth, and that indeed in such a manner, that the female parts are the vegetative form or the oral cavity, the male, the animal form, or the tongue with the salivary glands; the former represent the process of deglutition, the latter that of taste.

2325. Into the uterus or into the prostate gland the excretory ducts of the sexual glands, or the spermatic canals and oviducts, pursue a convergent course, like the salivary ducts do into the oral cavity.

2326. The testes, as also the ovaria, are forebodent of the salivary glands. The semen and oviducts are salivary ducts; they open by twos and symmetrically. Semen and ova are secreted like saliva. Semen and ova have also a salivary function, though of this the former is endowed with more than the latter.

2327. The ova correspond, as being an object of the semen, to the object of the saliva, which is the food or aliment. The saliva imparts to the bole or mouthful of food its first animal signification; it renders it for the first time capable of passing over into animal organs; it impregnates the morsel. The semen renders the ovum capable of effecting the transition into an animal; it spits upon the ovum.

2328. Impregnation is a process of smearing with saliva, conception a process of deglutition.

2329. Pregnancy is a process of digestion and formation of blood.

2330. If the internal sexual parts denote the internal visceral parts of the mouth, so must the external parts of the one correspond to the external of the other. The labia pudendi correspond to the lips, and the clitoris to the tongue, which is more perfectly represented in the penis. Both tongue and penis consist of two halves; in cases where the former is fissured or divided, the latter is so also, as in the Serpents and Lizards. In many animals, as the dogs and other Mammalia, there is likewise a bone in the penis, which corresponds to the os hyoides or linguale. The salivary ducts have combined with the penis; or, taken in a more strict sense, it may be said, that in the penis the tongue has coalesced with the oral cavity, so that both form a canal,—that of the penis, into which the salivary ducts (vesiculæ seminales) open.

2331. The sexual passion or venereal desire is a gustatory process of the sexual animal, the copulation being at one and the same time a matter both of taste and deglutition.

3. Germ or Embryo.

2332. In the embryo the whole animal already resides in miniature, as does the plant in its seed.

2333. The embryonic intestine is the vitellus or yelk.

2334. The embryonic integument is the amnion.

2335. The embryonic vascular system is the chorion.

2336. The embryonic sexual system is the allantois. The above propositions can only be perfectly developed in the physiology.

b. VASCULAR ORGANS OF THE SEX.

2337. The hæmatopoietic or blood-preparing vascular system is the lung; the blood-destroying, exsecernent vascular system is the kidneys. The kidneys are the lung reversed. The liver decomposes the venous, the kidneys the arteriose blood.

2338. The kidneys are the individualized vascular system of the sex, as the liver is that for digestion. They accord with the liver in their glandular structure, in the renal pelvis, which is like the gall-cyst, in the ureters that resemble the biliary ducts, and lastly, in the general signification of the urine as a product, wherein the whole organism, the whole blood-system has been excreted, like the bile, in which the venous blood undergoes the same process.

2339. Every disturbance of the digestive function acts in a striking and very direct manner upon the urine. The jaundice is apparent in urine; and what else is diabetes than a malady analogous to diseases of the liver? In urine we recognize what the bile has done with the food; the urine is the fluid nutritive system, and consequently the fluid organism "in toto;" the sexual blood, or sexual bile.

2340. The urine is the purest mirror or reflex of the bodily condition, and hence ourology or the doctrine of urine is of the most universal importance in semeiotics or symptomatic pathology: the ureters correspond to the tracheal branches or bronchi; thus, the urinary cyst to the trachea; the urethra to the larynx.

2341. In many animals the ureters open directly into the cloaca, as in many Fishes, Reptiles.

2342. By degrees the cloaca is retracted towards the ureters, and then originates a cloaca, which is both urethra and urinary cyst, as in Birds.

2343. In the higher animals, where a perfect ourocyst or urinary cyst has been evolved, the urethra opens into the anterior wall of the vagina, in front of which the urinary cyst then lies, in a similar manner to the trachea in front of the pharynx.

2344. In many Fishes the urinary cyst is absent, their pulmonic sac being also but feebly developed and only remaining as an asymmetrical swimming-cyst—the ureters too open directly into the cloaca, just as the swimming-cyst opens into the pharynx. The pharynx of Fishes, as being surrounded by the branchial arches, is at once pharynx and larynx, like as in many animals cloaca and urinary cyst are of one and the same kind.

2345. In many Reptiles (such as the Tortoises and Frogs) the urinary cyst gives off two cæca or blind sacs, as is the case in the larynx of many apes.

2346. In the Bird the two cul-de-sacs of the urinary cyst are much more developed and have assumed the form of two cæcal intestines, so that they have been actually viewed as such, and their number two been assigned as characteristic of the bird, the other animals meanwhile having only one. The cæca intestinalia of Birds are the lateral and upper extremities of the urinary cyst. The true cæcum of the bird is the vitelline canal or duct, just as in the fishes and all higher organized animals; this being distinctly retained in the aquatic birds.

2347. In the Bird the rectum properly opens into the urinary cyst between the two blind sacs or cæca, and that indeed with a regular vulva, which is a sphincter muscle.

2348. The cloaca of the Bird is an urinary cyst, into which the anus opens.

2349. The orifice of the cloaca is properly that of the urethra; ova and fæces are moistened with urine. In a bird both these egesta are combined.

2350. Like the urethra, so is the trachea membranous in Fishes, and in most Reptiles also.

2351. Urinary cyst and urethra stand in sympathetic relation with the trachea and larynx, and have also similar diseases, such as catarrh, inflammation, &c.

2352. The proper proof, however, of the urinary cyst belonging to the respiratory system resides in its genesis. It proceeds from the allantois, which in Birds is a decided organ of respiration, a branchia.

2353. Out of this urinary cyst issue in the embryo what have been called primordial kidneys, which shrink or dwindle down at a later period, but have entirely the structure of branchiæ.

2354. Here then is a sexuo-respiratory process, which corresponds to the anal respiration, upon a higher stage, of many worms and aquatic larvæ, e. g. Holothuriæ and Libellulæ. This anal respiration is in its signification a sexual respiration.

2355. Even in the acephalous bivalve Mollusca and Snails the respiratory openings are almost always situated in the proximity of the anus; their respiration is still a sexual respiration.

2356. For the first time in Insects it becomes a respiration of the trunk; and for the first time in the higher animals a truly animal, namely, a cephalic respiration.

2357. The urinary system is a double system; it conjoins in itself the two highest galvanic processes, that of secretion and excretion.

2358. Secretion is an hepatic character, excretion a pulmonic character. Secretion belongs to nutrition, excretion to respiration. Excretion is an exspiration, secretion an influx or infusion. Secretion is related to excretion as water is to air, as liver to lung, as basis to oxygen.

2359. Secretion takes place, e. g. of bile and saliva, in so far as the processes of the body, especially those of the digestion, are promoted. Excretion is only effected in so far as that the organs, into which what is secreted enters, may obtain a tracheal signification, i. e. evaporation. All excretory orifices are in a certain sense larynges or tracheal orifices. Thus, this relation between secretion and excretion would have been discovered without recourse to conjecture being had upon our part.

2360. The urine is "par excellence" a double product of this kind. It is secreted in the kidneys for simply one object, like the bile. It is excreted, because it belongs to the sexual system, which is essentially exsecernent.

2361. The object or purpose of the urine has not been subverted in all animals. In Birds, where the urinary cyst and intestine are confluent, the urine enters like the bile into the intestine, at least at a spot where there is intestinal matter, which it renders fluid.

2362. In Insects and Snails it appears to invest as a mucus the ova, and serve for their attachment. The same appears to hold good of the spider's web. What has been called purple juice and ink (in the Cephalopods) ranks probably in the signification of urine.

c. SEXUAL INTESTINE.

2363. The sexual intestine is the colon or large intestine, which, in this respect, belongs to the sexual system, as has been indicated at § 2171.

II. Animal Sexual Organs.

2364. The bones of the sex are the feet with their appurtenances, such as the pelvis with the lumbar, sacral and coccygeal vertebræ. The muscles, as well as the nerves, are self-understood. But of these we will speak in the sequel.

B. ANIMAL ORGANS.

2365. All organs, which are purely animal, are penetrated or traversed by the nervous system, just as the lower systems are by the tegumentary formation. No higher organ is the wholly pure evolution of one system, but the systems ever combine with each other; and this combination, when individually represented, constitutes the organ.

2366. Organ is distinguished from system in not pursuing its course through the whole body, nor consisting simply of one and the same mass, but by its occupying a definite part of the body and being composed of several systems.

2367. Every organ has therefore a special and specific function also.

2368. The systems of animal life divide only into two kinds of organs, into those of sensation and those of motion, into the solar and planetary, or central and peripheric.

a. MOTOR ORGANS.

2369. Bone and muscle are not societies, but only poles of one system. There is therefore no mere bony organ, and no mere muscular organ. Meanwhile we will here regard them in particular or "per se."

1. Osseous Organs.

2370. The first bones were branchial arches or tracheal rings. When the lungs were developed from the branchiæ, the branchial arches were repeated as ribs or pulmonary arches. Lastly, should bones be formed, which are to be wholly in the service of the animal or the nervous system, so also must they be wholly liberated from the vegetative organs, and become self-substantial, i. e. have nothing else to do, but move. Free motor organs can be none other than ribs that have become free.

2371. These free ribs must inclose the respiratory organ or the integument, which has become an organ of animal life; they are the members or limbs. If we think of ribs, whose office is no longer to inclose lungs, which must no longer be subservient to the uninterrupted vital motion of respiration, and which are no longer united by pleura into a closed cyst or sac—will not such simply retain the self-substantial voluntary motion in themselves? will they not abandon the inferior cystic form, and represent the same, though but ideally and voluntarily? will not such a thorax open in front, like the intestine has opened at its nobler extremity?—will not such ribs be members, arms, digits? The members or limbs are the members of the trunk, or ribs that have opened in front; they are the thorax that has opened in front; and hence are nothing new, but only something emancipated or set free. Such ribs can be none other than motor organs; for they were previously nothing else. Then, however, they performed motion in the service of the viscera; at present, where they are absolved from this service, they execute it only in accordance to the will of the head, simply according to its will, for they are verily nothing more than motor ribs. Where, however, or in what region of the body will the ribs attain unto such freedom? Without doubt, in the neighbourhood of the head, and thus at the very spot where the lungs derive one of their extremities. The limbs are therefore cervical ribs.

2372. The arms are a thorax, consisting quite purely of bone and muscle, and represented as isolated or detached from the viscus, or the lung; on this depends their nobility, or, in other words, upon what is vegetative having been wholly left behind.

2373. The arms, when clasped together by the fingers, are a thorax without viscera, without heart and lungs; they are destined to inclose a whole body in the embrace.

2374. By an embrace that which is embraced has been made our viscus; it has been adopted as our animal heart, and as our animal vital organ—or lung. The embrace has an exalted physiological signification, and precisely that which it unconsciously possesses in the state of pure love. Nature always thinks more nobly than we do. We follow blindfold her beautiful regulations, and she rejoices in the sport.

2375. As the fundamental number of the branchiæ is five, so also must the limbs represent five ribs; they split into five digits or fingers. The feet of the Crustacea and Insects generally do not correspond to our feet; but to our fingers. The lower organized animals have only toes, no feet. The five thoracic feet of the crab correspond to our fingers; its five abdominal feet to our toes.

2376. There are three limbs in accordance with the three totalities of the body, truncal, sexual, and cephalic members, or arms, feet, and jaws.

2377. The members of the body or trunk belong to the thorax, because it is the respiratory system. The abdomen has no members; what have been so-called, are in their signification sexual limbs.

2378. Had the animal no sex, it would have no posterior limbs.

2379. As the three lower cervical and the two upper dorsal vertebræ belong to the arms, so also they appear to commence with five ribs, but then to become arrested, and again emerge to perfection in the digits.

2380. The shoulder appears to consist of five ribs, but this does not as yet admit of being clearly pointed out. Meanwhile, it is certain that the scapula, acromion, and coracoid are particular bones, to which may be added the clavicle.

2381. The middle finger is the elongated radius, and is therefore the longest or radial digit. It is that which is persistent, if only one finger has been left, as in the horse. The ring-finger is the ulnar finger. It is that which, along with the former, appears in the bi-ungulate animals; the spurious or dew-claws are the auricular and index-finger; the thumb is the last ramification, is therefore always arrested, and frequently present only as a wart-like excrescence or papilla.

2382. All animals, which have true digits, are furnished with five of them, more or less completely developed. If what has been called the metacarpal bone of the thumb be numbered, which it must, as a digital articulation or phalanx, every finger has thus one carpal bone, and each bone of the fore arm also one.

2383. The sexual members or feet correspond in all their pieces to the arms: the pelvis is the shoulder repeated; and certainly, the iliac bone is equivalent to the scapula, the ischium to the coracoid process, the os pubis to the acromion, the marsupial bone to the clavicle.

Cephalic Members.

2384. Both pairs of limbs are repeated in the head, because in it the whole trunk is repeated; the upper jaw corresponds to the arms, the lower jaw to the feet. Each jaw consists of two members, which are ankylosed in the higher animals at their point of meeting or in front, but in Fishes are partly, and in Insects completely, separated.

2385. Each jaw consists of the same bony divisions as the limbs of the trunk, of scapula, humerus, and fore-arm; or of pelvis, femur, and tibia. This is easily to be demonstrated in Birds, Reptiles, and Fishes.

2386. The digits are repeated in the teeth. The teeth are claws.

2387. There are therefore five kinds of teeth, which correspond to the five digits. The thumb becomes the canine tooth; the index-finger the false molars; the middle finger the laniary molar, the ring-finger the second, and the little finger the third true molar tooth.

2388. The intermaxillary with its incisor teeth belongs, as well as the palatal bones, to the pharynx; and is a visceral or intestinal maxilla.

2389. The lower animals therefore, as the Fishes, have almost nothing but intermaxillary and palatal teeth. They act principally upon the lingual teeth. The reptiles have still palatal teeth, which higher up in the vertebrate series disappear.

Symmetry.

2390. As the cervical ribs have nothing more to inclose, have no longer to respire, but only to move; their symmetrical development is thus undelayed. The symmetry is at first wholly attained by the act of opening or apertion.

2391. The limbs are the most symmetrical organs. They are symmetrical in each smallest part, and these parts again assume a symmetrical arrangement in regard to each other. They are the ideal of symmetry.

2392. They are, however, the free living symmetry. They can create by their motions symmetrical forms. The symmetry consists chiefly in, and has been produced only by, the motion.

2393. The symmetry of the motion is the most exalted, for it is that which is endowed with life. The symmetry of the form is that which is dead.

2394. The symmetry of the form appertains to inorganic nature, the symmetry of the motion is the property of animals.

2395. Dancing and acting are the highest organic symmetrical movements, and also the highest symmetries. They are the symmetry of the motor members, wrought by motion.

2396. Music is a much higher symmetry of motion.

2397. Speech is the highest spiritual symmetry; it is the dance and histrionism of the spirit or mind.

2. Muscular Organs.

2398. The muscles are everywhere attached to the bones, and help to form with them the same set of organs.

2399. The muscles of the larynx are therefore the antetypes of the costal muscles, these of the member-muscles, and the dorsal muscles of the scapular and pelvic muscles.

2400. The muscles of the limbs are found in a state of treble repetition. That the muscles of the arm and leg are of one kind, admits of being demonstrated with tolerable facility. But it is necessary, that in doing this regard be paid to the ligaments.

2401. The ligaments are only stunted muscles. Without bringing them into account, the muscular system does not admit of being explained and understood.

2402. The crural muscles are again found in the lower jaw.

2403. The brachial muscles, on the upper jaw or face.

2404. The movements of the facial muscles correspond to those of the limb-muscles. Upon this depends the interpretation of dumb-show, or the art of physiognomy.

3. Nervous Organs.

2405. Nervous organs are liberations of individual parts of the nervous system, and their endowment with a peculiar function or sensation.

2406. The liberations of the nerves are combinations with the other anatomical systems at the spot where they have attained their highest evolution.

2407. Each system, however, has its peculiar process. Through the reception or adoption of the systems into the nervous system, a peculiar sensation must therefore originate.

2408. Peculiar sensations are sensations of sense.

2409. The combinations of the anatomical systems with the nervous system, whereby the former have been subordinated to the latter, are consequently sensorial organs, or organs of sense.

2410. In the sensorial organs the processes of the several systems attain unto sensation. They are brains of the anatomical systems.

2411. There are as many senses as there are different anatomical systems, and consequently senses belonging to vegetative and animal life.

2412. The number of vegetative systems is 3; the vascular, intestinal, and pulmonary system.

2413. The most complete combination of the vascular with the nervous system is the integument—constituting the tegumentary sense, or sense of feeling.

2414. The most complete combination of the intestinal system with the nerves is the tongue—intestinal or gustatory sense.

2415. The most complete combination of the lungs with the nerves is the nose—pulmonary or olfactory sense.

2416. Among the three animal systems, bones and muscles produce, from their association, only one action—the motion. The most complete and perfect combination of the motor system with the nerves, is in the ear—osseo-muscular or auditory sense.

2417. The nervous system has become a self-substantial organ in the eye—the nervous, or visual sense.

2418. There are therefore only 5 senses; they are none other than repetitions of the anatomical systems in the sensation; they are the highest developments which are possible in the lower systems, being the blossoms or heads of such systems.

2419. These systems are, however, processes of the universe taken up into the organization. In its organs of sense, the processes of the universe are thus felt or perceived. The senses are world-organs, and are therefore placed in contact with the world, or occupy an outward position.

2420. The vascular system is the nutritive system. In it the blood coagulates into the solid parts of the body. The feeling-sense is therefore sensible of the nutrition or rigidifying process of the body. Now, the Solid of the planet is the earth. The sense of feeling perceives therefore opposition—it is an earth-sense.

2421. The function of the intestinal canal is digestion. In taste, the process of digestion is felt. But digestion is a solvent, a hydrapoietic process; in taste the water is therefore felt—it is the water-sense.

2422. Respiration is a process of oxydation. In smell the respiratory process is felt. Oxydation is, however, an air-process—it is air-sense. Thus do the three vegetative senses feel the elements of the planet—are planetary senses.

2423. The animal systems are symbols of the æther, of the gravity with the heat or motion, and of the light.

2424. The motion is only moved matter, and thus a combination of the muscular and osseous system. The ear therefore perceives the motion of the primary matter, or the atomic motion—it is gravity-sense, æther-sense.

2425. The light is the tension-process of the æther; to see, is therefore in an organism to emit light or shine—light-sense.

2426. The signification of the senses is twofold; they are anatomical systems which have become nerves, and on that account also elements that have attained unto sensation.
1. Sense of feeling—vascular, tegumentary, nutritive sense, earth-sense.
2. Gustatory sense—intestinal, digestive sense, water-sense.
3. Olfactory sense—pulmonic, respiratory sense, air-sense.
4. Auditory sense—osseo-muscular, motor sense, æther-sense.
Matter-Sense.
5. Visual sense—nervous, tension-sense, light-sense.

2427. The sensorial organs are not simply combinations of the anatomical systems with the nerves, but also with the bones and muscles. These have been entirely taken up into the signification of the animal body.

2428. Each sense has its own nervous, osseous, and muscular system.

2429. The sense of feeling has its bones and muscles in the limbs.

2430. The sense of taste, its bones in the lingual bone, its muscles in and upon the tongue.

2431. The sense of smell, its bones in the nasal bones, its muscles being frequently very much developed in the snout or proboscis.

2432. The sense of hearing, its bones in the auditory ossicles, its muscles in the auditory conch.

2433. The sense of sight, its bones in a ring surrounding the sclerotic coat and in the eyelids, its muscles in the ocular muscles.

2434. Besides the proper sensorial nerves, each organ of sense is provided with nerves for the motor system and for the vegetative systems, especially those of secretion.

2435. The integument has, besides the nerves of the tactile papillæ, vascular and motor nerves.

2436. The tongue is supplied with nerves of motion and digestion, and has therefore three pairs of nerves.

2437. The nose receives motor and respiratory nerves from the fifth pair.

2438. The ear has likewise three kinds of nerves; the auditory nerves, facial nerves, and a branch from the fifth pair, not to mention those that are distributed to the auditory conch.

2439. The eye has, in addition to its wholly special motor nerves, a number of others, which superintend its vegetative systems, such as the iris and the secretions of the humours.

a. VEGETATIVE SENSES.

1. Vascular Senses.

2440. All senses are only conditionated by the peripheric nervous mass because they are combinations of the nervous mass with the blossoms of the inferior systems.

2441. The most general system of the animal is the vascular system, which is represented externally as integument. The animal was in the commencement nothing but integument, and this again naught but vascular and nervous mass, so that the whole integument was thus an organ of sensation.

2442. Through the integument the animal becomes an individual, or a something distinct from the aggregate of nature. Now as the integument is principally the organ of sensation, so is the primary sensation that act, by which the animal is distinguished from nature. The tegumentary sense is the sense of distinction, of limitation.

2443. Through the act of discriminating, a something foreign or extraneous is granted us. The immediate perception of what is extraneous, is called feeling. Tegumentary sense is sense of feeling.

2444. The feeling-sense is the first in the animal.

2445. It is that which is general in the animal.

2446. The whole animal is naught but a sense of feeling.

2447. Out of the feeling-sense all other senses must be developed, just as all other systems are developed out of the tegumentary formation.

Organs of Touch.

2448. Where, however, the skin has attained to a higher grade of formation, or where it has combined with higher systems, there also will the sense of feeling be supplied by the former.

2449. The combination of the integument with the osseous and muscular system, and with a nervous system of its own, takes place in the limbs. Since the motor members are only a liberated thorax, so no other sense can belong to them but that of feeling, which the thorax previously possessed.

2450. But these sensitive organs are moved, and therefore voluntary organs, digits, or former branchiæ. Moveable or voluntary organs of feeling are called tactile organs. The feeling of the motor members is touch.

2451. The highest feeling necessarily consists in touching, because it has in that become active, while before it was only passive.

2452. In the situation of the sensitive papillæ, the origin of the digits from respiratory organs admits of being still recognized. They are arranged in spiral lines upon the points of the fingers.

2453. The sexual organs belong as tegumentary developments to the sense of feeling. There is no special sexual sense.

2. Intestinal Sense.

2454. Opposite to the general feeling or the integument, the function of the intestine is evolved. In the trunk it is simply busied with its own processes; when it first mounts or ascends into the head, it becomes subordinated to the nervous action.

2455. The combination of the intestine with bones, muscles and peculiar nerves, is presented in the tongue. This is the sensitive organ of the intestine.

2456. The tongue is a feeling-sense in water, just as the skin was that in the air. For it is the blossom of the digestive process. To the tongue therefore belongs the digestive or water-organ of the mouth, which is constituted by the salivary glands.

2457. The sensation of what is fluid in its chemical relations is called taste. Gustation is not a peculiar process, but obviously only the nervous commencement of the digestive process. On that account also the gustatory sense still lies concealed in a cavity. The whole buccal or oral cavity still belongs to the sense of taste.

2458. As in the sense of feeling the motor system still predominates, so does it also in the tongue, as being the second sense, which has been liberated from the plant. The nervous mass is in this sense not preponderant over the muscular and bony mass.

2459. The tongue is still to be regarded as an organ of touch, though one in which the flesh or muscle has gained a mastery over the bones, while in the true tactile organ the bones determine the principal forms and functions. The tongue is a nervous organ in the muscle, the hand is such in the bone.

2460. The hyoid or lingual bone is none other than the first branchial arch, and consists pretty nearly of the same pieces as the arm.

2461. Compound lingual bones, such as occur in many Reptiles, have originated from coalescence of several branchial arches.

2462. Like the limbs, so is the tongue originally a double organ. In most Reptiles it is longitudinally fissured. Such animals have usually also a double penis. In all animals the tongue is divided into two moieties, which are only connate by means of suture. The penis also consists of two connate penes.

2463. As in the tegumentary sense, the nerves could not be peculiar or special in kind, but took their origin from all parts, and particularly from the spinal cord, so also is this the case with the intestinal sense, which is still only an internal tegumentary sense. The lingual nerves proceed from several situations, and that too from the upper part of the spinal cord.

2464. The oral cavity also consists, properly speaking, of mere tactile organs, which have been repeated in the head. Thus, there are tactile organs which are subservient to the gustatory sense, in biting, chewing, and swallowing.

2465. The lips are tactile organs upon the threshold or brink, as it were, of the gustatory sense.

2466. In the oral cavity, however, the glands of the intestinal canal are repeated. The salivary glands secrete fluid, like the pancreatic glands.

2467. The sense of feeling is present in all animals. They are only animals by virtue of it; but the sense of taste appears to be first formed at a later period, after the intestine has separated from the integument; in the animals that have no intestine, its existence is problematical, and even in Fishes and Birds it is but poorly developed.

3. Pulmonic or Lung-sense.

2468. When the respiratory organ is hoisted up into the head, and there becomes an organ of sensation, it passes over into a sense.

2469. That the nose is the thorax, together with its viscera, repeated in the head, has been already remarked. The many convolutions of the turbinated or olfactory bones correspond to the ramifications of the trachea; the nasal cartilages to the tracheal or laryngeal rings; the olfactory membrane to the pulmonic vesicles.

2470. The process of the lungs repeated in the head becomes smell, like that of the intestine became taste. The olfactory sense is the highest blossom of the arteriose vascular system or the branchial net. On this account the olfactory membrane is the most delicate, and the densest tissue of arteries and veins.

2471. The nose is related to the mouth, as the thoracic is to the abdominal cavity; the olfactory membrane to the tongue, as the lung is to the stomach. It is a cephalo-thorax. The nose is not therefore so completely closed as the mouth, but is opened through the two most anterior air-holes or spiracula. The nasal apertures or nostrils are the last persistent remnants of the spiracula, after all those upon the sides of the body have been closed up.

2472. It is the last organ of sense, which has been evolved from the trunk. It is therefore nobler than the two others, and has also a nobler object, the air.

2473. The nerves of the olfactory organ are peculiar to it, and are encephalic nerves. As the sense of smell is the pulmonic or arteriose sense; so also does the arteriose substance of the brain combine with this organ. The olfactory nerves consist of cineritious or gray substance, and are only prolongations thereof.

2474. This is the only phenomenon of the kind met with among all nerves, but it is commensurate with the character or signification of this organ. A sensible pulmonary organ can only have arteriose nerves. As the liver is throughout venous, so is the nose throughout arteriose in quality.

b. ANIMAL SENSES.

2475. It now only remains for us to consider the motor and the sensitive system proper upon their highest stage of development. The motor system, when represented in the nervous system, is a peculiar organ of sensation, as is likewise the nervous system itself upon attaining its highest state of development.

4. Osseo-Muscular Sense.

2476. The lowest condition of the motor system is the limbs, which represent no peculiar sense, but only the sense of feeling refined or set in motion. This motor system ascends into the head, and there no longer exercises its motive powers in prehension, progression, &c., but solely unto sensation. Now, a system, which converts its function into that of sensation, is a sense.

2477. The sensorial organ, which simply through motion, or a resistance presented to that of the atoms, produces sensation, or wherein the motion as such is felt, is the Ear.

2478. The ear is none other than the ultimate development of the bone and muscle, when they are brought under the direct dominion of the nerves.

2479. The auditory ossicles are the limbs subtilized or refined in character. They have joints, are provided with muscles, and move exactly like the limbs. It may be said; that the stapes is the scapula, the incus the humerus, the malleus the fore arm, and the concha with its cartilages, the hand with its digits.

2480. The ear, like the bulk of the limbs, has originated from branchiæ. In Fishes, the auditory ossicles have degenerated into the branchial opercula.

2481. The ear-trumpet (or Eustachian tube), which opens into the mouth, is the internal branchial aperture.

2482. The motor system, however, belongs to the trunk, whose viscera are repeated also in the ear, or in a cavity which has been called its labyrinth. The three semicircular canals appear to correspond to the intestine, the cochlea to the trachea.

2483. The ear has not only a nerve, but likewise a brain, of its own.

2484. The cerebellum is the auditory brain, and from it the acoustic nerves take their rise. For, since the ear is the sense of the whole motor system, and consequently of half the animal, it inevitably follows, that an appropriate or special nervous mass must be developed for it, just as the myelon or spinal chord has been for the trunk. An organ, which is so constantly active, must necessarily have a large nervous mass. The cerebellum is consequently not a brain in the general sense, but one which is wholly individualized. It participates in the motion, which, as sound, is transmitted to the animal.

2485. By its signification as well as by its own brain, the ear gives us to recognize its elevated rank above the other senses.

2486. The ear must stand in relation with the body's limbs.

2487. The ears first make their appearance in animals, associated with a tolerable development of the limbs. With the exception of a few instances, the ears first become apparent in Fishes, being in them at least furnished for the first time with true ossicles and semicircular canals. The ear, like the limbs, with which it constantly maintains a parallel course, is very slowly perfected. In the Fishes which have only fins (as instruments of locomotion) it is wholly concealed within the cranial bones; in the Reptiles it emerges or is more exposed to view; but for the first time in Birds and Mammalia where, generally speaking, the limbs also are first perfected, it attains its completion; in them only is the cochlea fully developed, and an auditory meatus which opens externally.

5. Nervous Sense.

2488. In all lower organs, and even in the senses which have hitherto been commented upon, the nervous system was not the chief, but only the co-ordinate agent. It has only by its conjunction with, assisted in elevating the character of, other systems, so that their material might be converted into spiritual processes. But the nervous is also a self-substantial system, and must therefore attain likewise a free development.

2489. With the highest organs of the nervous system, the relation which has been hitherto maintained, must be reversed. The inferior systems will now be the co-ordinate agents.

2490. The highest nervous organ can only possess that function, which is originally peculiar to the nervous system, i. e. the most delicate polarization, the light-function. There is the Light-sense.

2491. The eye is nothing but a nervous system represented in a state of purest organization, just as the ear was the purest system of motion.

2492. In the eye it is the brain itself, which expands, in order to turn unto or face the light.

2493. As the ear has its own brain, so also has the eye; the cerebrum is the optic brain. (Ed. 1st, 1811. §. 2317.) This is the rationale of our having two brains.

2494. Now the cortical or gray substance has been already diverted from the cerebrum to supply the olfactory sense. Its medullary or white substance remains for the eye. The medullary brain is the optic brain. The medullary is consequently the nobler part of the nervous system.

2495. The brain's medulla is homologous to the light; its exterior or cortical portion is related to the material light, the air.

2496. The eye is only a medullary brain, posited peripherically and in a nervous manner. The brain itself has elongated and become integument.

2497. The visual membrane (or retina) is the cerebral substance cystically expanded. It must be regarded as being originally a closed bladder or cyst. (Ed. 1st, 1811. §. 2321.)

2498. The optic nerve is itself hollow, and unites the cerebral with the orbitar cavity.

2499. The vitreous body, which fills out the cyst of the retina, is the cerebral medulla which has become transparent, or a semi-fluid albuminous mass.

2500. The sclerotic coat of the eye is the continuation of the brain's dura mater.
2501. The vascular or choroid coat of the eye is the continuation of the encephalic pia mater. All parts of the brain have been consequently continued into the eye.
2502. Now, what the brain is for the earthly body, that also must it be for the eye. The eye is not simply brain, but a representation also of the whole body. The brain can, forsooth, be nowhere without its body; if therefore it is elevated into the eye, so also must it take up and elevate the body along with itself.
2503. The eye is an entire body, a whole animal. Again, in the next place, the animal systems, such as limbs, thorax and abdomen, have been most distinctly represented in it. The light is seized, respired, digested, and hence felt by the eye.
2504. As the light represents chaotically the whole of nature, but this material nature enters completely into the animal through the processes of the trunk, so does the light enter it through the eye. The eye is the chaotic representation of all material processes of a body.
2505. The limbs or members of the eye are repeated in the ocular muscles and the sclerotic or bony ring; in many Fishes the eye stands upon a flexible pedicle, as in the Crabs. The ocular muscles move the eye in different directions like a hand.
2506. The sclerotic corresponds to the corium, the cornea to the digital unguis, or finger-nail.
2507. The choroid or vascular coat, is the respiratory system or lung in the eye. The iris corresponds to the larynx, the pupil to the glottis; its expansion and contraction is a respiratory movement.
2508. The choroid coat incloses also an osseous mass, the lens—a vertebral body. The morbid states of the lens are osseous diseases, such as gout.
2509. In the chambers of the eye, water, as being a product of digestion, is constantly secreted.
2510. The orbitar cavity is a mouth with salivary glands—giving vent to tears.
2511. The lachrymal canal is a branchial duct, which opens into the nose, like the Eustachian tubes did from the ear into the mouth.
2512. The eyelids consequently correspond to the lips, and are in like manner fringed with hairs.
2513. As the body has everywhere two halves, and laterally also presents two entire organisms, so also is the formation of the nervous system a double one. Each eye is an entire body.
2514. In the two eyes the halves of the body have completely separated as entire bodies, and each has attained self-substantiality. Each eye is a free animal in the animal body. Each eye is therefore circumscribed by its own integument—is a free animal. It is endowed, like the hand, with omnilateral motion; it has cavities, i. e. its bodily cavities and humours, or inclosed masses—viscera.
2515. An organ, which again repeats in its miniature the whole animal itself, of which it is only a part, must necessarily be the highest point unto which an organism can attain. With the eye the organization, and consequently nature, has been concluded.
2516. The eye is a parasitic animal, of the same kind with the animal upon which it exists.
2517. In a certain sense all sensorial organs are parasitic animals in the animal; only they are not all of the same kind with it. No one of the other senses has e. g. repeated all lower systems in itself, and it is therefore to be regarded only as a subordinate or half-animal, which lives upon that which is more perfect.

Senses of the Sexual Animal.

2518. In essaying to speak concerning the sensorial organs of the sexual animal, we shall only encounter in the latter the emotions of the vegetative senses, and these indeed disposed according to their rank.

2519. The sense of feeling is most perfectly developed in the legs, whereof the pelvis represents the scapula.

2520. The external sexual parts are the analogues of the gustatory sense; the female parts being indeed those of the mouth; the male, which are frequently furnished with bone, of the tongue. The jaws are not repeated in a sexual animal, except in Insects, namely, as the pharyngeal maxillæ.

2521. The analogue of the nose is wholly stunted, and is only left persistent as a trachea or air-tube in the urethra.

2522. In other respects the cavity of the sexual parts is a trunk-cavity "per se," like the thoracic and abdominal; the pelvic cavity contains the viscera of a whole animal.

II.-PHYSIOLOGY.

2523. Physiology is the doctrine or science which treats of the functions of the animal. Like the organology was, so also must the science of the functions be developed. There will be functions of the whole animal, and of the tissues, systems and organs.

A. FUNCTIONS OF THE ANIMAL IN GENERAL.

2524. The first act of the animal is an assimilation to the universe, whereby it also receives or takes up into itself the primary function of the universe. This is the perception of the circumscription and totality in itself, its self-manifestation, the feeling of self. The first action of the animal mass is that of self-sensation. Now, through the self-sensation, self-substantiality has been granted.

2525. The animal is consequently a Whole in Singulars only through the feeling of self.

2526. As the universe is only an analysis of the self-consciousness of God; so also can the development of the animal and the formation of its organs, be none other than an analysis of the self-sensation. All other functions (just as all masses are only a metamorphosed nervous mass) are but differently polarized self-sensations, or these as it were dispersed.

2527. The self-sensation of an individual body is not, however, inclosed within itself, like the primary relation of the universe; because it is not the universe, but only a fraction of it, which has sprouted forth like a bud from the great planetary body. The self-sensation is not therefore a simple feeling of self, but a feeling also of something foreign or extraneous, and thus the animal self-sensation becomes, in reference to Nature, an act of discrimination or discernment.

2528. Now, the animal is distinguished only from Nature by the act of being liberated from it. It is therefore comprehended in a constant liberation or severance. The life of the animal continues only by a constantly renewed and indefatigable severance, by a desertion or falling off from Nature.

2529. In so doing, however, it detaches itself from Nature, as being a part thereof; the severance is therefore a conversion of the nature into an animal. In this consists the reciprocal action of both, viz. that the animal is constantly seeking to assimilate the nature unto itself.

2530. The ability or power to assimilate the nature, is called excitability.

2531. Excitability is the most general phenomenon of the organic mass, and appertains both to plants and animals.

2532. But in the animal excitability, the free self-sensation, within which a free motion is necessarily inherent, is superadded or originates. This excitability unto motion I call irritability.

2533. Thus this irritability belongs only to animals.

2534. Irritability does not depend directly upon motion, but throughout upon sensation. Without sensation no irritability is possible. If the sensation ceases, so also does mobility, or the capacity for motion, cease.

2535. Since irritability originates from the antagonism of the animal with the world; so is it parallel to an antagonism of the heavenly bodies, or to that of sun and planets. The mutual operation of these two heavenly bodies is, however, an interchange of polarity, a polar excitation. The irritability is a polar process; but one which is pure and devoid of material excretions, just as the sun excites the earth without eliciting therein any material change or transition. The animal becomes polarized by the incentive agent or stimulus.

2536. Through the irritability there originates a double polarity in the animal. In the first place one between the world and the animal; in the second, one between the exterior of the animal and its interior. The world-polarity gives the feeling or sensation, the body's polarity the motion.

2537. In the sensation the animal always transcends itself; there is thus only excitability. In the motion the animal abides or remains within itself; but the self-sensation proceeds from both conditions. Accordingly, in self-sensation the world, and the animal within the animal, convene or come together. The animal is itself universe, and it at the same time comprehends the great universe. Now, in both conditions or in both functions has the animal been turned towards the world and also towards itself. In feeling it turns itself towards the world, in order to adopt this spiritually into, or repel this from, itself; in motion it turns itself towards the world, in order to materially adopt or to repel it. In both cases of assumption it turns itself towards itself.

2538. Both these properties belong to the animal body; if individual organs are deficient in them, it happens therefore, from their having superadded or adduced other properties, which are predominant, to the original ones. This is the essence of what is animal in a living body. Thus, if from an animal all the vegetable systems could be subtracted, it would do nothing but feel and move.

B. FUNCTIONS OF THE ANIMAL IN SINGULARS.

I. Functions of the Tissues.

a. FUNCTIONS OF THE ANIMAL TISSUES.

1. Functions of the Point-Tissue.

2539. The point-tissue is also spiritually represented, and this spiritual manifestation or expression is the stamp or impress of the arrangement of the matter into points.

2540. The point-matter is the matter of sensation; the sensation is, however, imparted by polarization. The act of feeling is a conduction of the polarity from point to point; this is the light-polarity. This light-like polarity transmitted from point to point is sensibility.

2541. The sensibility is not the result of a peculiar nervous fluid, or even of mechanical vibration, &c., of the nerves; but of an antagonism between animal and world, brain and tegument, sun and planet.

2542. The point-tissue is in every respect the æther which has become the mucus, and has in every respect also to act in accordance thereunto.

2543. The sensibility or the nervous power, viewed in relation to the smallness of the body, operates instantaneously upon every spot. It is not first conducted or tardily conveyed hither and thither; but everywhere, where an antagonism, a stimulus, a planet is, there also is the "vis nervosa." Where a planet is, there also is the solar tension.

2544. But abstractedly or ideally considered, the nervous force certainly requires a time to elapse for the propagation of the stimulus, just as the light-tension darts through the æther only in a given time. As light traverses at the rate of 40,000 miles per second, so must the nervous power, if it obey similar laws to those of its prototype, and the length of a man be estimated at 1-5000th of a mile, glide through the human body in 1-200,000,000th of a second, which, "quoad" the phenomenon, cannot, of course, be observed.

2545. Accordingly, it is theoretically certain that the nervous power does not operate with absolute or unconditional velocity. To the validity of this statement, observations of many kinds, especially those made in diseases, and by astronomical computations of the strokes of the chronometer, afford additional evidence.

2546. Thus, for a stimulus impinging upon the brain, to act or be transmitted from thence into the toes, there elapses 1-200,000,000th of a second. In the event of disease, a retardation in the process is conceivable, and may admit even of being observed.

2. Function of the Globe-tissue.

2547. The dense or solid form is the matter, when it has died away; the function also, or the crystallization, died therein, so soon as the crystal was represented. The office of the bones is only to preserve the counterpoise to the nerve, to supply a limit, and therefore a support to its action, by which means the achievement of the latter first becomes possible.

2548. The business of the bone, is to play an antithetic part to the nerve, and nothing else, or to serve the latter as a "terra firma," upon which it may execute its plans. The bone suffers, and in this consists its office. The other offices of the bone, such as its being the firm and solid framework of the body, its protecting the nervous mass, &c., are subordinate matters, which are at once clearly understood.

3. Function of the Fibrous-tissue.

2549. There is the active motion. The primary motion is, however, a result of the polar tension induced in the æther by the light. Just as the heat has been produced in the æther by the light-polarity, so has been the animal motion in the fibres by the polarity of the nerves.

2550. Every fibre, when it is in motion, has necessarily two polar extremities. For the fibre stands between the artery which is oxygenous, and the nerve which is basic, or between a zinc and silver pole.

2551. In the conditions of rest the two poles must be neutralized, or, generally speaking, not in existence. This is only possible by disjunction of the galvanic circuit or chain.

2552. Since the artery is constantly operating upon the fibre, but the nerve only, when it has been stimulated; so must this change in the condition of the fibre reside in the change of the nervous influence.

2553. In unirritated conditions the nerve does not act upon the fibre; the latter is not illuminated, nor does it stand in a state of tension with the nerve, which is not in itself polar, but is only, when stimulated, a sun towards the fibre, or polar within itself. If the chain be then dissevered, the fibre is in a state of indifference.

2554. Thus, in the event of tension, the nervous extremity of the fibre becomes negative, the arterial extremity positive; both ends necessarily attract each other in order to discharge or unload themselves. This attraction is an abbreviation or contraction.

2555. After the discharge the fibrous extremities are homonymous; they repel each other, and relaxation or extension ensues.

2556. The motion of the fibres is consequently a perfectly galvanic process, devoid of material interference, between nerve, fibre, and blood. The nerve is silver, the artery zinc, and the fibre moist card. All theories of motion based upon swelling of blood, elementary change, and such like causes, are not animal, though organic in character; the former are injections, the latter

chemical processes.
2557. Through the fibrous tissue being shortened and elongated, the entire animal is self-displaced.
2558. Through the fibrous tissue the animal appears as animal, through the point-tissue it is (and that in essence) an animal. God is in himself, but he first appears in the world. This elucidates or explains the relation of "noumenon" and "phenomenon."
4. Function of the Cellular Tissue.
2559. This tissue has no other function than that in the plant. Many saps or juices are effused into the cells. The latter are engaged in constant expansion and contraction, whereby they decompose and propel these juices. They are the proper seat of the nutritive process, since they coincide with the capillary vessels.
2560. But as every rigidification is combined with volatilization, so does a process of evaporation occur in them at the same time.
2561. This process is the property of the teguments or membranes, both of those lining the internal cavities, as also of such as are placed externally.
2562. But the process of evaporation is principally the function of the true skin or corium. As an organ of evaporation the corium is similar to the lung. It is consequently aeriform water, which of necessity constitutes the principal mass of evaporation.

2563. The water is, however, mucous in character, the mucus becomes oxydized in the air, and through this also carbonic acid is formed.

Heat.
2564. The chief function of the cell-formation is the process of heat. The cell-process and heat-process are one in kind. Heat is the product of the process of condensation and rarefaction, and thus of the nutrition and evaporation, both of which happen in the cells, or, if we please, in the capillary vessels.
2565. In the skin the process of temperature is individualized.
2566. All temperature depends upon the process of condensation and rarefaction. At one time it is the nerves, at another the vessels, anon, external influences, which alter or modify it. Animal heat, like the cosmic heat, has been produced by change of the fixation. This change, however, takes place principally in the nutrition and evaporation.
2567. The fat is the "residuum" or the antagonism of the process of evaporation or water-formation. It is therefore deposited everywhere on the water-organs, beneath the integument, along the intestine and the vessels, around the kidneys, and so on.
2568. It is, like the adipocire, a product of the aqueous decomposition. On this account it is subservient to the process of temperature. It is the only isolator of heat, while all other animal bodies are conductors.
II. Functions of the Systems.
2569. The functions do not concern the whole body, but only its principal parts.
a. VEGETABLE FUNCTIONS.
2570. The vegetative functions are matter-changing processes, which, consequently, correspond directly to the chemical processes, or are rather of one kind with these, but altered by organic elemental bodies.
1. Functions of the Intestinal System.
2571. The general function of the integument consists in the secretion of juices, of which there

are principally two kinds, the general and special.
2572. The general intestinal juice is mucus; it acts by rarefying.
2573. The gastric juice is of an aqueous, mucous, and acid nature. It acts as an acid and is endowed with chemically solvent properties.
2574. The bile is of a basic, inflammable, alkaline nature. It acts also chemically by analysing and precipitating.
2575. The saliva is the gastric juice of the head. It is a juice secreted under the influence of the sensibility, and is on that account indifferencing, and nullifying in its effects; it is the highest poison.
2576. As the indifferent saliva precedes the gastric juice, so does the indifferent pancreatic fluid the bile.
2577. The proper function of the intestinal system is the digestion with all its divisions. There is an animal and a vegetative, or oral and abdominal digestion.
a. Oral Digestion.
2578. Oral digestion is a mortifying or putting to death of the food.
2579. Since what is organic only serves as aliment for the animal, but nothing can be assimilated to the latter, without its having been previously reduced to the original condition of Infusoria, so also must the first act of the digestive process depend upon this, or converting the organic into primary organic bodies.
2580. This reduction to the primary condition is a putting to death of the organic individual. Organisms only which have been killed, can be converted into infusorial matter, and are then nutriment for the animal. The first act of digestion is consequently an act of putting to death.
2581. The act of killing consists of two moments, the mechanical and dynamical, or in lacerating and poisoning.

Laceration.
2582. The mechanical act of putting to death commences with the search after nutriment, and thus with the movement of the feet; to this succeeds the prehension, or seizure with the claws or hands.
2583. This motion of the limbs is then repeated in the cephalic members, or the jaws.
2584. The seizure of the food with the cephalic arms or jaws, is the infliction of a wound commensurate with the position and form of the teeth. The teeth are digits of the cephalic limbs, or being devoid of any fleshy layer, claws. A gripe with such digits is in itself the infliction of a wound. For, in order that the food be grasped with only sufficient firmness to admit of its being drawn into the mouth, the sharp digital points of the mouth, i. e. the teeth, must make an incision therein.
2585. To seize, bite, wound, and kill, is in an animal one and the same act. For, so soon as the food or prey is seized, and consequently wounded, it is drawn deeper into the mouth, and therefore a new grasp and bite is essayed.
2586. Upon this the food is twirled about in the mouth by the muscles, or is chewed, whereby the parts are separated into mechanical atoms.
Poisoning.
2587. Mechanical atoms are, however, not yet dead, because the vital polarity is not extinguished. It is requisite for a dynamic act, which suppresses all galvanism, to operate upon them.
2588. As this act of killing has directly for its object the death of the Organic and that alone, so is

it a venenation or process of poisoning. To poison is not merely a chemical act, but one having for its immediate purpose, to separate the connexion of the organic atoms.

2589. The first act performed by the animal is that of poisoning.

2590. Poisoning, as being a direct aggression upon and destruction of life, is a destruction of the galvanic process. Now, the galvanic process is destroyed by suppression, destruction, and separation of its poles. The demolition of poles is a polar equilibration. Poisoning is equilibration of poles on the galvanic organism, or their neutralization.

2591. The primary change of poles is, however, in the blood. Poisoning is a balancing of the blood-poles.

2592. Indifferent blood is no longer blood, but chyle or infusorial primary mass.

2593. The nutritive fluid or chyle is poisoned organic substance. There is no aliment except what is wrought by poison.

2594. The poisoning of the aliments must take place through the medium of their blood, or their sap. If poison therefore does not get into the blood, the death does not ensue. The digestive juice is poison only for the blood, but not for the other systems.

2595. Chemical bodies which induce death, act by destruction of the bodies themselves, not merely by their neutralization or excessive polarization. They do not operate very differently from iron when in a state of red heat. Such is the case with sulphuric acid, nitric acid, alkalies and arsenic.

2596. Now, if these bodies be called poisons, there are thus three kinds of poisons, mineral or chemical, vegetable and animal poisons.

2597. The chemical poisons destroy the mass; they convert the Organic into minerals.

2598. Vegetable poisons reduce the animal to the plant; they do not effect the destruction of the mass in a general sense, but only that which is purely animal, or the nervous system—they are nervous poisons.

2599. The animal poisons destroy what is vegetable in the animal, or the galvanic process—are blood-poisons.

2600. The saliva is the digestive poison.

2601. The saliva is not present, in order to macerate, and so prepare the dry aliments for digestion, but to poison them. Everything else is only a subordinate operation. Now, the process of poisoning takes place only in the blood. The saliva does not act without inflicting a wound.

2602. Strictly regarded, all saliva is poison. There are examples of the saliva of birds, and even of men when inflamed with passion, having acted as poison.

2603. All other animal poisons, contagious principles or miasmata, are analogous to the saliva, being partial salivæ, partial poisons. Diseases of the skin produce tegumentary poisons or salivæ; glandular diseases, glandular poisons; nervous diseases, nervous poisons; diseases of the lungs, pulmonary poisons. The saliva is the blood-poison; the hydrocyanic acid, the nervous poison.

2604. During mastication saliva imbues every atom of the food, whereby the mass is throughout neutralized, or annulled.

Docimasy or Testing.

2605. The oral digestion or that of the mouth cannot, however, occur by itself, without its being a nervous process also. The intestinal function taken up into the nervous system is the taste.

2606. Taste is the chemism resident in the head, the digestion in the nerves. Taste therefore happens according to the theory of the digestion or the chemism.

2607. The lowest element of the chemism is the water, the highest repetition whereof in the earth is—the salt. The former is therefore the object of abdominal digestion, the latter of nervous digestion. To taste is to digest salt. In order to taste the salt, every part of the food must be brought in contact with the tongue; this is only possible through comminution of the aliment by means of the teeth. The mastication dissolves the aliments mechanically into atoms, just as digestion will resolve them chemically into "infusoria."

2608. That which should be digestible, must have a saline character, must be soluble. The tongue is accordingly the test-organ of the digestive process.

Deglutition.

2609. At first the tongue tastes only with its apex or tip; but after the salts which operate upon the tip are neutralized and examined, the tongue will also taste with its root, the two extremities of the tongue and their gustatory sensations being thus mutually opposed. It therefore takes the morsel upon the root and presses it against the palate, whereby the pharyngeal muscles clasp and swallow it.

2610. In deglutition the object of the tongue is not to give over or surrender the morsel to the pharynx, but to enjoy it "per se." During this fruition, however, it is robbed of it. Deglutition is therefore a result of an undesigned mechanical contrivance. Each organ works for the other, while fancying that it works for itself, which again it actually does, while it reaps the enjoyment so derived. Has the tongue finished tasting the food, then the pharynx obtains it involuntarily.

2611. To the saliva, and thus to a morsel, the acidulous gastric juice is polarwise related; both therefore seek to unite, and with this their organs also, namely, pharynx and stomach. The stomach obtains the preponderance, because it is acidulous; the pharynx moves towards it and with it also the morsel.

Rumination.

2612. If the food, when received into the stomach, continue acid by virtue of its nature, or from not having been properly chewed and imbued with saliva and so neutralized, it is then homonymous with the gastric juice. The stomach therefore seeks to neutralize it, by restoring it again to the action of the saliva.

2613. Acid aliments cause vomiting. Grass which has not been masticated, and therefore enters acid or non-killed, and susceptible of fermentation, into the stomach, is regularly brought back into the mouth, and to the saliva, i. e. is ruminated or rechewed.

2614. Rumination is a regular act of vomiting, which has originated from the antagonism of the saliva and gastric juice and from the acid nature of the morsel.

b. Intestinal Digestion.

2615. The intestinal digestion is the perfect chemical process taken up along with all its moments into the animal.

Gastric Digestion.

2616. The stomach exercises by means of the spleen the oxydizing process of the intestine, the solvent function, and thus the action of the water. The gastric juice is related to the food, like water is to earth. The gastric digestion is liquefaction, unto which oxydation makes the preliminary step.

2617. Through the process of liquefaction, the poles are only potentially augmented, but are not dissevered, nor new substances formed. The gastric digestion creates no new bodies, but only mixes the old in the most homogeneous and intimate manner with each other.

2618. The gastric digestion is an animal process of fermentation. In deranged states of the stomach there is therefore a propensity to acid eructation, development of carbonic acid, and even formation of sugar.

2619. This digestion consequently reduces the (animal) food to the signification of the plant. Gastric digestion is a process of vegetable germination. The salivary process is a reduction to the animal death, digestion a reduction of this dead something itself to a lower kingdom.
2620. The gastric juice alone, like an acid, effects the solution of the aliment; with this, the movement of the gastric walls, which only tends to produce an easier mixture of the alimentary particles, having nothing to do.

Hunger.

2621. Through the digestive process the gastric juice is consumed by the food, and the stomach deoxydized. If there is a deficiency of dephlegmatizing and deoxydizing aliments, then the peroxydation of the stomach must produce a feeling—called appetite. If this be not appeased, the oxygenic tension in the stomach is elevated or increased, and then begins to become unpleasant; this is Hunger.
2622. Here the feeling of the stomach's peroxydation is an obstructed process of fermentation, dependent upon want of food and alkaline principles.

Thirst.

2623. The feeling of the reverse condition to the above is Thirst. It originates through a too rapid deoxydation of the stomach, through deglutition of the gastric juice on account of an excess of food. But it may also originate from a deficiency of gastric juice, or from an alkaline tendency in the latter; just as hunger resulted from a superabundance of gastric juice, or a disposition to form acids.
2624. Thirst is the feeling of too powerful a digestion, or of too rapid fermentation, whereby the product or leaven of the latter becomes, as it were, bankrupt; upon this the blood flows in greater quantity, in order to secrete the gastric juice; the arterial nature becomes elevated, and finally, an inflammatory condition, associated with a sense of dryness, originates, and is propagated as far as the mouth.
2625. Thirst and heat rank, like hunger and cold, parallel to each other. The feeling of dryness appears as heat, that of moisture, as cold. Cold therefore at once extinguishes or quenches the thirst; but produces hunger, which again is mitigated by heat.

Biliary Digestion.

2626. The stomach is the pulmonic intestine; the duodenum, through its combination with the liver, is the vascular intestine, having a predominant venous character; or the one is the arteriose, the other the venous stomach.

2627. The gastric digestion has reduced the food to the state of a vegetable; but this does not yet suffice for it to become a nutritive matter, namely, a new organism. As the primary organism originates, so also must the body originate in the process of nutrition, and thus from the element of the Organic. Now, this primitive matter or element is the mucus, the "infusorium." There is still therefore a process necessary to redissolve the vegetable and convert it into protoplasma or primary mucus.
2628. This, however, occurs only through the process of putrefaction being wrought in the food. This process is not simple solution, but decomposition, or separation of the constituent parts. The acid vegetable parts, which had been active through the process of fermentation, must be therefore decomposed, and in such wise, indeed, that what is mucous shall be separated from them.
2629. This separation or analysis takes place by means of the Bile, which combines with the acid

of the chyme.
2630. The biliary digestion is an alkalizing or saponaceous process. Through it the chyme becomes separated into nutritive juice or chyle, and into excrement.
2631. The excrement is what is vegetable, or the product of fermentation combined with the bile. The chyle is what is infusorial, or the product of decomposition. Nutrition therefore begins anew. Its process is a "generatio originaria."
2632. In digestion the processes of both organic kingdoms, namely, the fermentative and putrefactive, are repeated.
Absorption.
In the intestine, not merely lung and vascular system, have been represented, but also the integument or organ of absorption.
2633. After the animal food has been reduced to the vegetable, and this again to the infusorial state, it can be taken up by the body. It now becomes absorbed in the small intestine. The small intestine is the tegumentary system, or the root-bark.
2634. The chyliferous or lacteal vessels stand in antagonism with the lung, or the skin as being the original respiratory organ. It is only, therefore, the infusorial chyle that has been absorbed, not the excrement, because between the latter, as the product of oxydation, and the lacteal vessels, repulsion takes place. The chyle, having been absorbed, enters into the thoracic duct, and from thence into the lungs.
Evacuation.
2635. Through the absorption of what is fluid, that which is excrementitious becomes more solid, and is thus given over or transferred into the vegetable, sexual, or large intestine.
2636. The excrement is now found in another, i. e. in a lower, or vegetable animal. It therefore obtains the direction of all sexual secretions; it is thrown out or ejected, and in a reverse direction, because the anus is the sexual mouth.
2637. Digestion is thus through all predicaments, from its incipient dealings with the highest life unto the plant, and from this to the mucous globule, a thorough process of putting to death.
2638. The nutritive will be through all predicaments, from the infusorium to the plant and to the animal, a thoroughly vivifying or life-inspiring process. Digestion is descension, nutrition is ascension.
2. Functions of the Respiratory System.
2639. The branchiæ and lungs are the air-organ of the animal, the foliage. The animal, like the vegetable foliage, is oxydized from water or air, by which means the animal sap, which hitherto is only a root-sap, becomes differenced into an aerial sap.
2640. No animal can live without oxygen gas, because the air is the condition of the galvanic process.

2641. The oxygen passes over materially into the blood or the chyle. Beyond this it is an indifferent matter for physiology, whether the blood simply derives the positive tension from the air, or combines the positive oxygen materially with itself. In both cases the same heterogeneity originates. Were oxygen, however, not to enter the body through the lung, it could not then be seen, whence its ingress might be effected. In other respects, every change of matters is established with material combinations and separations.
2642. The carbonic acid of the respiratory process may originate accidentally, as when it is formed on the integument. In so far is the expiration one and the same with the evaporation.
2643. The chyle ascends directly from the small intestine into the leaf-fabric, or into the lung.

Both organs are to be compared with an entire plant, whereof the intestine is the root, the lymphatic vessels the stem, the lung the leaf-fabric or foliage.

2644. The chyle moves in the lymphatic vessels like the vegetable sap, and thus by polar tension between lung and intestine. Lung and intestine have been diametrically opposed. The conductors of the antagonism are the lymphatic vessels.

2645. The lymphatic vessels do not absorb by virtue of their own contractility, nor by open ends; but by physical pores like the cells of plants.

2646. It is the chyle, which is oxydized in the lung; the venous blood is a subordinate object.

2647. Through the oxydation it becomes coloured like the vegetable sap does in a leaf. But the latter sap obtains only the colour of the terrestrial oxydation, of the terrestrial oxyde or the water, and becomes green; while the chyle gains the colour of the cosmic oxyde, or the fire, and becomes red. The red, fire-coloured sap is the blood. The function of the lungs is consequently a formation of blood.

2648. The blood is an infusorial, a green mucous sap, which has been elevated to what is aerial. The blood consists of aerated Infusoria—blood-globules. These are the red substance of the blood. In the lower animals only, where the blood, like the body, contains but few blood-globules, is it colourless.

2649. Through the respiration an electrical difference enters the blood, whereby it becomes separated into several substances.

2650. The water in the chyle is the inorganic menstruum, wherein the mucus, as being a fundamental matter of the vegetable sap, has been dissolved. Through the digestion the mucus becomes animal—gelatine.

2651. The gelatine is the basi-constituent part of the chyle, being a product of the digestion or the root-process. Gelatine is the vegetable mucus repeated and ennobled in an animal. This ennobling took place through the digestion.

2652. "En route" too, the chyle mixes with the nervous blood, and already undergoes a degree of oxydation, whereby the gelatine becomes converted into albuminous matter.

2653. Through the respiration gelatine and albumen are elevated to aerial importance, and still more oxydized; by this means the albuminous matter becomes elevated, and passes over into fibrine. The fibrine is the last product of respiration. Fibrine is the starch-meal in an animal.

2654. The electric duplicity, induced in the blood by the process of respiration, exists accordingly between fibrine and gelatine, while the albumen constitutes or forms the indifference. The fibrine is repelled from the lung, because they are both homonymous; it is the oxydized gelatine. From its communicating its polarity to the whole mass of blood, this also is repelled from the lungs.

3. Functions of the Vascular System.

2655. The capillary vessels of the body range opposite to those of the lungs, just as the biliary does to the splenic stomach, as alkali to acid, as precipitant and secernent to what is non-separated.

2656. The capillary vessels therefore attract the pulmonary blood, separate it, secrete and form new constituent parts; and then, after it has become homonymous, they repel it back again towards the lungs.

2657. The circulation only subsists through the polarity which exists between lung and capillary vessels, between lung and body, between oxydation and reduction, combination and separation.

2658. If the intestine and body be the root, the lung the foliage, so is the lymphatic and vascular system the stem, in which the most perfect substances have been formed. In the circulation the matters for the animal systems must have been fully formed, such as for the bones, the lime; for the muscles, the iron; for the nerve, the albumen.

2659. The liver is the principal organ of the circulation. There also must one principal preparation of the blood take place. On account of its parallelism with the brain, the condition probably resides in it that conduces to formation of the albumen, namely, of the Indifferent, which is the nutritive matter of the nervous mass.

2660. In the vegetable sphere of life there are, in addition to the liver, but two mutually distinct excretory organs, namely, the intestine and the skin.

2661. In the mucous secretion of the intestine the condition, requisite to the formation of phosphate of lime for the bones, probably resides.

2662. In like manner in the formation of the carbonic acid and the water in the integument, the basis for the formation of phosphate of iron for the muscles may reside. In the circulation the intestine would thus be the lime-, the skin the iron-, and the liver the medulla-forming organ.

2663. The two extremes of the circulation, or intestine and lung, form gelatine and fibrine; the circulation itself forms the purely animal matter. Out of the integument and lung grows the muscle, from the intestine the bone and gelatine, out of the liver the nerve. Muscle is integument and air, bone is intestine and chyle, brain is liver and blood. Thus each has a function that is peculiar to it; each organ has its business to perform in the diffuse fabric of the animal body.

2664. Through this variety or change of the offices, the circulation first becomes possible.

2665. If the secernent process be therefore suppressed, the animal then dies as rapidly as if it had been suffocated. It is a suffocation of the opposed pole. Query? does not many a fit of apoplexy depend upon this?

2666. The circulation has consequently two factors, the lung as oxygen-pole, the capillary vessels of the body as hydrogen-pole, the blood as the indifferent water. The circulation is a galvanic process.

2667. In all extremities of the body the arteriose blood becomes deoxydized, decomposed; it is therefore basic and homonymous with the capillary vessels, so that it is consequently repelled, and driven back into the veins.

2668. It can, however, flow nowhere else than to the lung, because there resides its opposite pole. Being again oxydized in, it becomes homonymous to, the lung, is repelled by it, and again attracted by the capillary vessels of the body.

2669. The circulation is therefore a result of dynamic forces, not of mechanical functions. It would occur, were the vessels to be glass tubes.

2670. The pulsation of the heart is not a cause of the circulation, but inversely rather, its consequence or effect.

2671. In the circulation the whole organism, or intestine, lung and integument, is combined. It is therefore the fundamental system, which includes the whole mass of the body.

b. FUNCTIONS OF THE ANIMAL SYSTEMS.

1. Of the Osseous System.

2672. The functions of the osseous system are simply mechanical relations, such as solidity, form and motion.

2673. The motion of the joints presents to our notice interesting relations, especially in reference to the motion of the vertebræ, ribs, limbs and jaws upon each other, but these could not be here treated of in detail.

2674. The acts of swimming, creeping, standing, walking, running, leaping, climbing, and flying,

do not exhibit simply mechanical, but truly philosophical moments.
a. Bodily Motion.
2675. Swimming can take place through simple contraction of the body, without locomotive members. It is the continuation of the vesicle's first process of origination; as in the Infusoria and Polyps. In the Worms and Serpents it is effected by an undulating motion of the body, whereby the water is struck with oblique surfaces; there it is fibrous motion. In the Holothuriæ and many aquatic larvæ, it is a propulsion, effected by expelling water from the anus, and thus by squirts, there being consequently contraction of the body or arteriose straitening of the sphincter muscles. In Fishes, swimming is a rowing or remigial stroke, produced by lever-motion, together with the oblique slap made by their tail.
2676. Creeping is either a shortening of the body by fibrous motion, as in the Snails, or an undulatory motion, as in the Serpents.
b. Pedal Motion.
2677. Standing is the position of a lever "in equilibrio," the creation of the proper centre of gravity.
2678. Walking or running is an exchange or alternation of the equilibrium, a combination of standing and creeping.
2679. Leaping or hopping is a flight with the feet.
2680. Climbing is an use of the feet as hands.
c. Alary Motion.
2681. Flying is lever-swimming in the air. It takes place by means of aerial branchiæ, namely, the wings in Insects, and by the thoracic extremities in Birds, which are also none other than animal branchiæ.

2682. Soaring is creeping in the air.
2683. Hovering is standing in the air.
2684. Pouncing is hopping in the air.
2685. Diving is hopping in the water.
2. Functions of the Muscular System.
2686. The muscular system performs, in an active sense, what the osseous system does in a passive. The strength or power of the muscles, and their leverage attachment, is here especially to be regarded. The contraction of the fibres is a charging, by nerves and blood, of the two fibre-poles.
2687. The fibres are charged by the air. It is, in the most general sense, the respiratory vessels by which the muscle is charged. This is strictly the case in Insects, where the tracheæ traverse all the limbs, and directly conduct to the flesh the polarity of the air. In animals, however, with a closed circulation, the arteries undertake the conveyance of air upon the blood, and it is then the latter fluid which streams into the muscles in order to charge them.
2688. Thus if an artery be ligatured, the limb is crippled or lamed. The artery, however, imparts only the positive pole, and consequently of itself produces no shortening or contraction of the fibres. The oxydation takes place at the lower end of the muscle; here, therefore, the latter passes over into tendon.
2689. The nerve is the second condition of the muscular contraction, since it evokes in the fibre the negative pole. Thus if a nerve be ligatured, the limb is likewise motionless.
2690. If the poles be brought by contraction into close approximation, the fibres must re-extend, so soon as the influence of the blood or the nerve ceases.

2691. Now since the blood is constantly streaming in, the reason for the muscular rest must reside in the nerves. The rationale of voluntary motion is consequently the nerve. The relaxation or extension is an unloading of the fibres.

2692. The muscular motion is an electrical process, a motion of blood in the Solid.
2693. Through the polarization of the fibres the muscle is formed from the arteries. The muscle is therefore an individual biconical piece of fibre, having unequal cones. Oxydation takes place at the muscular extremity; here, therefore, originates the sinew or tendon.
2694. A fleshy cyst—or heart—which includes an osseous cyst, must subdivide into several fibrous cones or muscles. One reason of this is the "fore and aft," another is the quantity of the essential vascular branches.
2695. The muscle contracts upon application only of a stimulus.
2696. Every stimulus induces motion only as a result of polar excitation. Every stimulus polarizes; for even the gentlest contact is like the friction, and produces electrical antagonism. It therefore amounts to the same, whatever stimuli, whether mechanical, chemical, or spiritual, have been applied to the muscle. One acts like the other.
2697. If no motion supervene upon, or is even suppressed by, the contact of a body, the nature of the body must then be indifferencing.
2698. Relaxing, laming, life-destroying matters, are indifferencing, or cause a suppression of the poles.
2699. Overcharging principles, e. g. lightning or strong electric sparks may also produce relaxation. These destroy the function of the fibres, and act therefore worse than the in differencing matters.

3. Functions of the Nervous System.

2700. The function of the point-substance is also that of the nervous system, for this is only the point-substance, fashioned and arranged into stalk and branches.
2701. But even on this account the nervous tension proceeds only according to a determinate line, while before it penetrated through the whole mass.
2702. The nervous tension takes place in a nervous system only between a special organ and the nervous centre.
2703. In itself the nervous system is an Indifference, and such then are all the organs upon which it acts, when regarded in reference to this operation; they might, through other functions, be polar.
2704. The nervous system becomes differential either through its two kinds of substances, or through extraneous influence; in the first case it thinks or moves, in the second it feels.
2705. Sensation is in the nervous what motion is in the fibrous system, namely, a polar condition, in which the two ends have a tendency to come together, in order to discharge or unload.
2706. In the sensation the extremity of the nerves strives to approach that of the brain; it is therefore a contractile effort in the nerves, like motion is in the muscular fibre. Juxta-posited granules could not, however, shorten; therefore, the already calculated polarity courses over them.
2707. All external stimuli act upon the nerves and polarize them, but not the muscular fibres.
2708. The fibre becomes directly polarized only through the internal stimuli, or by the blood, galvanic tension, &c.
2709. The most general function of the nervous system consists in its assuming polarity from the world, and imparting this to other systems, apart namely, from the alternate operation of its two

substances.

Nervous and Motor System.

2710. The first system is that of motion. The tranquil presence of the nerve in the fibre produces indifference in the latter; so that it has no susceptibility for the polarization that is effected by the artery. Doubtless the arterial blood passes, for the most part, close by the fibre through other capillary vessels.

2711. But if the nerve be polarized it is then basic, negative, and enters into opposition with the arterial blood, which now streams into the capillary vessels of the fibre, and renders the two ends of the latter heterogeneous.

2712. Thus, if the nerve be cut across, it remains always indifferent, and no motion can any longer ensue. But this does ensue so soon as the nerve is galvanized. A proof that the nervous influence is homologous to the galvanic tension.

2713. If the nerve becomes involuntarily negative, then originates spasm.

2714. If the nerve continue morbidly indifferent, then paralysis originates.

Nervous and Vegetative Systems.

2715. The second great system is the tegumentary formation. In this too its ordinary functions occur, such as secretion, evaporation, nutrition, elevation or depression of heat without nervous influence; or they occur while the nerve only acts indifferently.

2716. But if it acts by polarizing, as in the muscular motion, then the tegumentary processes are at once changed. What is material becomes extinguished, and the simply Irritable is manifested in the cells, capillary vessels, and so on.

2717. The process of temperature becomes thereby instantly changed, because the decomposition is changed. The heat is increased by the rapid suppression of the evaporation; it is diminished by the rapid increase of the latter.

2718. The art and manner in which the nerves act upon what is vegetable, is consequently an elevation of this Vegetable to what is animal. It must resign its processes, and simply undertake those of motion. Exactly the same takes place in the secernent organs. They perform their offices themselves through the galvanic tension of the vessels. But if the nerve act upon them they secrete more powerfully, and exactly from the same cause whereby the muscle is self-moved. Thus, if the nerve operate feebly, whereby it is in an irritable state, or if its action be wholly withdrawn, then the secretion will become less.

2719. The more irritable the nerves are, by so much the more animal does the vegetable organism, or the splanchnic system, therefore become; so much the less is it produced on the mass.

2720. Irritable animals and men are therefore meagre or lean. The two cases are uniformly related to each other. Lean men are irritable, not perhaps because the nervous ends are not covered with fat, such as there might be over the integument, the ear, the tongue, nose, &c., but, because where the nutritive process is inactive, what is animal necessarily preponderates. There are consequently three kinds of nervous actions, or phenomena, though all nervous activity in entire systems is similar, or but one. The difference is only derived from the organ upon which the nerves act. There are, consequently, no special nerves of sensation, motion and secretion; or such, forsooth, as have only one of these offices to take care of. Thus, were a nerve of sensation to be prolonged or drawn over a muscle, it would excite motion, and in the case of the liver, secrete bile, &c.

Mesmerism.

2721. When the sensibility mounts to the highest degree, each mass-function will then almost

cease, and the organs of sense feel the weakest operation of the stimulus.
2722. Since every stimulation is a polarizing act, and each body is in polar activity towards the other at every conceivable distance, so may an extremely irritable nervous system also perceive the feeblest polarizations.
2723. The eye perceives the polarization at a remote distance from the body whence it proceeds.
2724. In hearing the vibrating body still indeed acts directly upon the ear through the vibrations of air. But a finer, i. e. more irritable ear, hears farther than one that is more dull.
2725. With an elevated sensibility the other systems also may therefore perceive the polarization of the bodies, without coming into contact with them.
2726. To perceive objects in the distance, i. e. merely their polar influence, is called Mesmerism, or animal magnetism.
2727. Now, if it is once possible for other senses besides that of sight to extend their perception into remote space, it no longer matters as to the magnitude of the latter. A feebly charged electrical machine only attracts bodies that are near, one strongly loaded, those which are more remote; such is the case too with weak and strong magnets.
2728. The integument at once perceives electrified surfaces at certain distances; now, as every surface appears electric to the integument, so must the latter, if its sensibility is very much elevated, perceive something of that sort in every proportionate distance.
2729. But homologous polarities only act upon each other, and therefore traverse thoroughly through heterologous bodies. Thus the magnet attracts the iron filings through the table-board, unhindered by the interposed wood, and without any regard being paid to this, or its being even perceived.
2730. The senses may therefore perceive their homologous polarities through other bodies, walls, and such like. By virtue of their perception, they stand in relation to them.
2731. To the very sensible nervous system the vegetable system and its impulse is a foreign object, which detaches itself from the above system, just as the objects of sense have done from the sensorial organs. The vegetable, or in general the material body, appears therefore to the mesmerized like a strange world—they see their own organs—are clair-voyants. Mesmerism therefore comprises nothing which could contradict physiology.
Vegetative Nerves.
2732. The splanchnic or visceral nerves are also distinguished from the animal by their being in a constant state of tension, and hence keeping the processes of their system in constant repair.
2733. The rationale of this resides in the two nervous substances having separated into ganglia and plexuses.
2734. This also serves to explain another phenomenon, viz. that the nerves indeed, but never the brain, attain to perfect rest; because the former is without ganglionic or cortical substance, the latter is throughout surrounded, and obviously interwoven, by it.
2735. Therein lies the reason why the viscera do not sleep.
Sleep.
2736. The condition of the nervous system, as hitherto represented, is called that of being awake. It is the interlude played by the nerves with the world, and with the animal body.
2737. When in a state of health the first interlude ceases, then the other also is over or past. The world-nerves, however, operate only upon the animal systems, upon the senses and the motion of the muscles; it could therefore be these only wherein the nervous function, in compliance with this cessation, is suppressed. Now, muscular rest originates through suppression of the tension between nerve and muscle. In sleep this rest is also derived from the same means.

2738. The tension between nerve and muscle can only cease, if the tension also between the termination of the brain and nerves be suppressed. We are now reduced simply to the consideration of the nervous system, and may, in treating of sleep, pay no attention to the muscular system.

2739. But whence comes the tension in the motor nerves? Obviously only from the cerebral tension. This can originate only in two ways; either through the special organization of the brain, one or the other substance being preponderant, or through the influence of external stimuli.

2740. The encephalic substance becomes more potent than usual, if by rest the cortical substance becomes more arteriose in character. This tension is communicated to all the nerves, sensitive as well as motor, and continues in their interlude with the world and the motor system.

2741. If this encephalic tension is not too potent, it remains only in the brain, without the ability to polarize the nerves also. It then only produces cerebral phenomena, thoughts or dreams.

2742. Dreaming is an encephalic tension excited by the organization, not by the world.

2743. Dreaming is the first step in the liberation of the animal from the vegetable system—it is the first step towards mesmerism.

2744. In a perfect or middling state of health, where the nervous is not very much separated from the tegumentary system, we do not dream.

2745. In a healthy condition an external stimulus would be, consequently, the only cause of waking, did not the long repose bestow a preponderating influence upon the cortical substance. Dreams therefore happen in the morning.

2746. Waking is the intercourse with the world, not with self. If one wakes also from intercourse with self, still the former is synchronous and coexciting.

2747. Thus, if intercourse with the world ceases, sleep originates. If the vegetable intercourse with it also ceases, then death originates. Waking is "consensus" with the world.

2748. Sleep is a death of the animal systems.

2749. Every awaking is a resurrection from death, a new sympathizing with the vegetal body, from which the animal body again originates.

2750. As the animal originally took its rise from, and only through the plant, so also is this repeated in an individual. The plant is the ever-living, ever-verdant, or green, out of which the animal daily sprouts forth as a blossom.

2751. The animal intercourse with the world is also interrupted in two ways, and there are therefore two modes of falling asleep.

2752. The first cause resides in the want of stimulus. The nerves of the senses are not polar, do not therefore excite the brain, nor does this again affect the motor system. The muscle therefore arrives at a state of non-tension; it becomes relaxed, and along with it necessarily the organs of sense, which are thrown into activity by muscular motion. The arms and fingers, whose business is to touch, sink down; the feet which move, and thereby warm and animate the body, are slackened and bent together; the body is in the recumbent posture; the eyelids drop, the light no longer plays upon the visual organs, the external and internal auditory muscles flag also, and the sound is no longer borne upon the ear. Now also does the tension of the senses with the brain cease, and with it the sensation—there is sleep.

2753. This sleep arising from want of stimulus is a faint sleep, and rendered useless by dreams. For there is actually no cause present why the encephalic tension should entirely cease. Men, who do not fall asleep through fatigue, but from want of work, sleep restlessly, awake easily, and again readily fall asleep. Their life is dreaming.

2754. The other cause of the polar suppression in the nerves is like that of the extension of the muscles, or their falling to sleep; it is thus the discharge of the too strongly excited poles. With too high a degree of fibrous tension, which also originates through too long a continuance of the tension, the fibre is placed in a state of activity, which consists in the antagonism being balanced by approximation of the ends. Were nerves, when greatly tensed, capable of being shortened, they would also discharge themselves, and come at least for one instant to rest—they would sleep.

2755. The falling asleep of the fibre is its sleep, though it also does not last long. So is the expansion of the heart its sleep, so is expiration the sleep of the thorax.

2756. In all polarizable organs there is a change or alternation of waking and sleeping, which endures a longer and shorter time. This periodicity depends upon the energy of the polar influence, and upon the size and susceptibility of the substance.

2757. Every substance has its own periods of waking and sleeping, of action and repose. The pulse sleeps shorter than the breathing; this again shorter than being hungry; this again for a briefer time than the sexual function.

2758. There are organs, or systems, which are nearly always in a state of slumber, e. g. the osseous system, because in it the polarity is extinguished. It is only in states of inflammation that it wakes up. Others scarcely ever slumber, e. g. the cellular system, because in it indeed no pole is yet fixed, and in the alternation of poles its life consists.

2759. There is a similar change of poles in a nervous system, and it indeed halts or stops for a middling time. Through the persistent influence of the external world, the nerves of the senses are thrown into such a state of tension with the brain, that blood cannot flow thither in sufficient quantity, in order to maintain the two cerebral substances in mutual antagonism. Brain and nerves become therefore indifferent; muscles and sensorial organs lose their polarizability, and their intercourse ceases with the nerves as well as the world. The brain, and everything else, has now been discharged, and a deep sleep without dreams, an animal death, ensues.

2760. It has been arbitrarily asserted, that no sleep is possible without dreams, but for this statement there are no existing grounds. Whence should the dream come if there is no tension in the brain, or if it has previously undergone sufficient self-exhaustion?

Periodicity.

2761. The sleep of the nerves ranges parallel with that of the planet. It might be said that such was the case from habit or custom, but it is, properly speaking, dependent upon a parallel process of organizing that occurs at the origination of the animal. The matter stands thus: the germ originates in the morning; until evening stimuli act upon and polarize it: in the evening they cease, and with them the tension. The muscles relax upon every movement, and rest necessarily follows in the manner above delineated. With to-morrow's morn the world again acts until evening upon the germ; it wakes up, and the same course of events happens as on the first day. Finally the substance becomes organized according to this periodicity; it becomes, forsooth, not more energetic than is necessary in order to admit or receive a charge of one day; towards evening it is exhausted, is neutral, and rejoices with the relaxation of the muscles, that the world no longer acts upon its frame.

2762. Thus, we may attribute this phenomenon both to custom and synchronous formation; nevertheless one ought not to forget that the organic formation is a metatype, or has been imitated from something that has gone before, and that consequently the law of periodicity has not been bestowed upon both at the same time, but first of all unto nature, and through this to the body, which is its image or likeness.

2763. The nervous periodicity ranges in accordance with the periodicity of the light, or is thus parallel with its archetype, and consequently with the day and night. Day is the waking, night the sleeping of nature. But the animal has originated in and by this alternating change of nature. It is spiritually, as well as corporeally, nature's likeness.

2764. The sexual function in a perfect animal, as in Man, has been adapted to the periods of the year; in other animals other natural periods exert their sway. Animals are commonly pregnant by the month, e. g. for one, two, four, five months, &c. The human species requires three quarters of a year for pregnancy, one quarter for giving suck to the babe, and then it can again conceive. Pregnancy thus lasts a year, and has been based in the sun.

2765. If the female does not become pregnant, the sexual passion, or instinct, is repeated at the expiration of a month. It sleeps about one month, and then awakes for some days. The periodicity may be divided into vegetative and animal, thus,—

   a. Animal   = World-periodicity.
1. Sleep   = Rotation of the Earth.
2. Menstruation   = Lunar revolution.
3. Pregnancy = Solar revolution.
   b. Vegetal   = Earth-periodicity.
1. Digestion   = Water, Ebb and Flow.
2. Respiration = Air, Electricity.
3. Pulse   = Earth, Magnetism.

Awaking.

2766. The awaking takes place of itself through the origination of a new polarity in the brain during the afflux of arterial blood, whereupon the dreams follow which precede the act of waking. During sleep the plant continues to act, the unloaded cortical substance becomes again oxydized and charged; tension arises between it and the medulla, and with this come the dreams. This encephalic polarity is imparted to the sensitive and motor nerves, and the organs open. If external stimuli are superadded, all this happens somewhat earlier. The act of awaking is invariably, however, a restoration effected, or brought about, by the plant, and especially by the circulation.

2767. Man would therefore wake up had there also been no world of the senses; but then he would not continue awake, but at once relapse into slumber, and sleep the eternal death.

III. Functions of the Organs.

2768. The functions of the organs are the functions of the system associated or combined, just as the organs are but ultimate evolutions of the systems existing under similar relations. The organic functions are always therefore in a state of concatenation with other systems, and there can be no organ which acts in an isolated manner.

2769. By this character a new field for sympathy has been opened. All the organs operate through sympathy. Sympathy is therefore the result of parallel systems, or also of antagonisms between the factors of a single system. Taken in a strict sense, there are no vegetative organs. The organs are therefore limited to the encephalic animal, such as are those of motion and the senses, and to the sexual animal.

1. Functions of the Encephalic Animal.

A. ORGANS OF MOTION.

2770. Just as the nerves have a function in themselves, and one directed towards the subordinate organs, so also has the motor system.

2771. The motor system is in the first place doomed to serve the whole body, since it flexes it, moves it forwards and backwards and upon all sides. It is related principally to the movements of the vertebral column, and serves in numberless animals to effect the act of crawling or creeping.

2772. Then again it will serve individual parts of the body, such as the belly in its evacuations, or the sexual animal in emitting urine, &c. It ministers unto the thorax in the act of respiration, which is a very complicated process. The thoracic muscles are to a certain degree co-ordinated with the constantly polar nervous system, and become thereby and in part involuntary. But one main reason of this appears to be the air that is constantly renewed in their cavity.

2773. In the act of respiration two orders of muscles are active, the proper-pectoral muscles, and the abdominal muscle, which has been displaced from the thorax, or the diaphragm.

2774. As originally the thorax took its rise at the expense of the abdomen, so also is every inspiration an elevation of the thorax and a displacement of the abdomen. Every breath or in-draught of air expands and produces the thorax, but narrows and arrests the abdomen. The diaphragm expresses this contest. Its contraction being a result of the respiratory tension, expresses consequently a preponderance of the thorax, and in obedience to this, narrows and diminishes the abdomen. It is supplied, in conformity with its origin, by nerves from the upper cervical vertebræ, since prior to this the abdomen extended as far as the head, and the branchiæ were appended to it as lateral strips of integument—as in Fishes.

2775. As the thoracic cavity is drawn by the diaphragm's descent towards the abdomen, so by the pectoral muscles is it raised upwards to the head. The latter movements are what is animal in the respiratory process, for they lift the thorax towards the head. Pectoral muscles and diaphragm stand opposite to each other, like limbs and trunk; the ribs are the limbs, the diaphragm the upper abdominal tunic or covering. Through this antagonism what is limbed, as belonging to the thoracic cavity, is drawn upwards, what is abdominal in its character, downwards; the result is expansion, and through this pumping in of the air.

2776. The air is in part voluntarily swallowed like the food, as e. g. by the movement of the ribs, in part involuntarily by the diaphragm. The diaphragm may be termed the heart of the abdominal cavity.

2777. In the lower organized animals, where merely branchiæ are present, the animal motion encroaches but little upon the act of respiration. In Molluscs and Snails, the oxygen is almost always on the branchiæ, as is the case also in Insects. In Fishes the water is still taken in like the food through the mouth, and driven out or expelled by the pharyngeal muscles from between the branchiæ. In them the air is in both respects swallowed. In many Reptiles the air is indeed drawn in through the nostrils, but conveyed into the lungs by a true act of deglutition. These kinds of functions are necessary, because as yet the whole trunk is abdomen, and the thoracic cavity has not as yet separated from it; hence the diaphragm is wanting.

2778. The respiratory originates therefore from the digestive organ; it is freely developed only from the abdomen, and is at length entirely liberated from the latter as a self-substantial cavity, e. g. first of all in the Mammalia. Respiration is originally but an act of deglutition, which has gradually become perfected, by the animal trunk being associated with it to a greater extent.

2779. The pumping in of air has therefore become in the highest degree a process of suction, and in this resembles the digestive function when it has become animal.

2780. Then also the air passes no longer through the mouth, but through the nose, as being the peculiar opening of the thoracic cavity in the head. Even the Fishes have as yet no nostrils opening into the mouth.

2781. The thoracic motion is a limb-motion, and would become locomotion, were the ribs not conjoined. In many of the lower animals the branchiæ are at once organs of motion, such as fins or rudders.

2782. Every inspiration is a self-manning towards the animal; every expiration a retro-depression unto the plant; the abdomen in this case regains the upper hand from the diaphragm reverting to its usual position, and the thoracic cavity narrowing. Respiration is a becoming animal.

2783. The motions in themselves, without reference to the trunk, are the motions of the limbs, as standing, walking, &c. The movements of the arms and feet are sympathetic, because their muscles are of equal signification.

B. FUNCTIONS OF THE NERVE-ORGANS.

2784. These functions have relation only to the nervous system itself, because all nerve-organs are elevated above the trunk, and live in themselves. They are simply the functions of the sensorial organs.

a. Functions of the Vegetable Sensorial Organs.

2785. These must be regarded as those that still encroach upon the inferior organs. They are not, however, the inferior processes themselves, but their ascensive formations into the nervous system. This therefore works henceforward only in and by itself, but yet in relation to the inferior processes.

1. Function of the Sense of Feeling.

2786. To constitute the sense of feeling the integument, in other words the nutritive or vascular system, has assumed a nervose character, and consequently that which is in communication with the materiality of the external world. The function of this sense will therefore have materiality only for its object.

2787. The integument is the organ by whose means the animal is absolved or liberated from the world. The sensation belonging to it, is none other than the perception of this diversity subsisting between the two.

2788. Through the tegumentary sense, the world becomes a something external in relation to the nervous function; while previously it was such through the medium of the skin for the lower organs only, viz. as an object of absorption. The discrimination of materiality is called Feeling. The sense of feeling is the earth-sense.

2789. The sense of feeling perceives materiality, like the nerves perceive all objects or all stimuli, through polar excitation. Every pressure, every contact is polar excitation.

2790. The sense of feeling is characterized by poles only being excited in it by absolute proximity or immediate contact. Just because it is the first sense, through which the animal is set free, so must that which is liberated be at once perceived in the moment of liberation, and thus in immediate contact. The sense of feeling is a polarity of contact, a polarity without distance. The stronger the contact, by so much the stronger is the polar excitation—there is increased pressure. The gravity acts simply by pressure. The perception is therefore resolvable into one of pressure or contact.

2791. Different degrees of pressure necessarily impart different amounts of feeling. Perception of the different degrees of pressure made by an object betrays its inequalities of surface. The sense of feeling is also the sense for determining inequalities, for the Soft and Hard, for the Solid, Fluid, and Gaseous; all these feelings, however, are referable to the contact.

2792. Through diseased conditions, the polarizability of the sensitive nerves may become exceedingly elevated, and they then perceive the polarity of contact prior to the contact having taken place. For the two bodies invariably excite poles that are antagonized to each other. Were other bodies not to approach them more closely, or else act upon them more energetically and so extinguish the polarity; they would remain at an infinite distance in a state of polar relation towards each other. Feeling can therefore be extended to an indefinite distance. Hysterical, mesmerized, and even healthy human beings, feel further than they grasp or touch with the hand.

2793. Homogeneous polarities, or those of the same kind, are throughout nature to be found also by means of others; e. g. electrical polarities are not disturbed by the intervention of magnetic ones. Such is the case also in feeling. That which is related to oneself is felt, although it is more remote from us than other objects, upon which we do not bestow any attention, or towards which we do not turn our poles.

2794. The sense of feeling differs according to the diversity of certain points in the integument, and is thus nobler in character, the higher the rank which these may hold. Thus it is most feebly developed in vegetable situations, where hairs, nails, claws, and scales are placed. It must attain the highest grade of perfection in the animal organs, and thus in the limbs or their parallels, the lips.

2795. In the limbs, by reason of their mobility, the feeling becomes voluntary. It is then wholly in our power to strengthen or weaken the contact, to press gently or firmly, and allow these periods of feeling to succeed each other rapidly or slowly.

2796. Feeling associated with motion is called touch; this condition of the organ the tactile sense. The tactile sense is by no means different from that of feeling; it is only the combination of feeling with motion.

2797. The fingers are the most perfect organs of feeling, because they are the most moveable parts of the body, and therefore they are organs of touch.

2798. As simple feeling perceives the asperities of bodies, so does touch the forms. The perception of forms is based upon the form that resides in the tactile organ itself.

2799. All possible forms reside in the motion of the fingers.

2800. Each hand is a semi-ellipse, in which the four fingers are the periphery, the thumb the radius. Both hands together form a perfect ellipse with two radii.

2801. Now, in the ellipse are involved all geometrical figures. The hands include in their movements the whole of geometry.

2802. We can only perceive the forms of nature, because they all reside in ourselves, because we can create them. This, and none other, is the meaning or sense of the doctrine of pre-established harmony.

2803. The sense of touch is also the sense of form. The fingers constitute so perfect an organ, that it is scarcely requisite for us to estimate all its value. In it the whole body is repeated simply in forms, in spiritual motions.

2804. Through the hand the whole planet becomes an object unto us. It is the hand which instructs us how to know this terrestrial world.

2805. The greatest perfection is attained by the greatest variety or multiplied diversity of the organs. The limbs are simply destined for motion, yet feeling also resides in them, because they are invested with integument. Could the limbs therefore divide into motor and tactile members, the conceivable sum total of perfection must be attained. This division is present alone in Man. The feet have become simply locomotive members, because they are those of the sex; but the hands have become tactile members, because they are those of the encephalic animal.

2806. It is not, as has been imagined, the hands, as hands, which confer nobility upon our species, for by their means an essential half of the animality, or the power of locomotion, is lost; but it is the preservation of all possible functions in an animal, and in such a manner that each stands upon its highest grade of perfection. The highest perfection cannot, however, be attained where two functions are attached to one organ. Should both hands and feet exercise the sense of touch, the motion is impaired; should both move the body, then the sense suffers.

2807. The four hands of the Apes are therefore an imperfection, which we have no need to envy them. They can, properly speaking, only climb and grasp with them, but not run. Each gait of theirs becomes therefore ungainly, the horizontal as well as the upright or perpendicular, and they make use of both alternately; because too the act of handling, i. e. of climbing, is their only proper movement. Now, in the act of climbing all the limbs are brought into requisition, and consequently a free voluntary touch and a free progression disappear.

2808. The feet support the body and stand in its service.

2809. The hands are, on the contrary, supported by the body, are free.

2810. The wings also support the body.

2811. Feet and hands define man. Through the two only does he become free.

Tegumental Covering.

2812. The skin, as being originally a branchia, is provided also with its branchial operculum. This is the epidermis.

2813. Scales are plications or folds of the epidermis, which accord with the situation of the branchial vessels; they are therefore arranged generally in a circular form around the body, and are more definitely varied upon the dorsal than ventral aspect of the body. Large scales, or those which may be regarded as formed by the confluence of several scales, are called scutes.

2814. It is merely the epidermis which is concerned in the production of the squamous covering; but if the former branchial vessels emerge themselves so as to project above the integument and become dried, then hairs originate.

2815. If these hairs ramify, then there are feathers.

2816. It is only therefore the Mammalia and Birds who possess a proper external covering or garment.

2817. The claws or nails are scales upon the extremity of the branchial arches, which have become animal in character, or in other words, are digital scales; they are animal branchial opercula.

2818. The nails are demi-claws, and therefore leave the points of the digits free. Free digital apices or points constitute the most perfect organ of touch, because this is then divided into two parts and because the nail increases the amount of resistance.

2819. The organs of defence are therefore an appendage of the sense of feeling, as are the bones of the motor system.

Splanchnic or Visceral Senses.

2820. These senses will not be found to disown their predecessors; and just as the latter extracted the qualities from terrestrial matter, so also do they. But while the former had to deal with the material, the present set will have to transact business with the spiritual qualities.

2. Function of the Gustatory Sense.

2821. Digestion is a chemical process, and one indeed wherein actual mixture and decomposition take place; it is therefore also, and "par excellence," an aqueous process. For matters that are actually decomposible are alone submitted to digestion, since it is of too coarse a character to perceive the proneness of such bodies to decomposition.

2822. It belongs only to a higher grade of perfection, or to a nervose condition of the digestive process, to perceive the rationale of the decomposition, or the spiritual conflict, which prevails between the matters, when they are about to separate.

2823. Now, the organ which only perceives the qualities of matters, without reference to their actual separation, is a sense. Upon the highest grade of perfection the digestive passes over into a sensorial function.

2824. Tasting is the first commencement of the digestion in the nervous system, where the aliments are felt just before the analysis into their polar quantities has taken place. The gustatory is a water-sense.

2825. For the exercise of the sense of taste the same conditions are requisite as for digestion, viz. solution and capacity for decomposition. Without the capacity for being dissolved, and without the occurrence of actual solution, nothing can be tasted, any more than digested. The saliva is the gastric juice for the tongue.

2826. If water be the basic element in the process of digestion, so in gustation must the higher water, or the salt, be the basis of taste. Salt alone is sapid, and every thing in order to be tasted, must possess saline properties.

2827. The tongue, by means of the saliva, passes over gradually into salt. Salt is the last extremity of the tongue. The salt-formation is a member of the gustatory formation. Hence tasting is only an ascension of the inorganic to the animal tongue. The salt is the gustatory sense of the earth.

2828. The general object of taste is the salt of the sea. It alone can and must be converted into a pleasing taste or relish. What is general in nature, is the antetype of that which is equivalent in an organism. Sea-salt and the tongue or saliva are one in kind.

2829. Every thing admits only of being tasted, in so far as it is salt; every thing is but savoury, in so far as it is marine salt.

2830. As the component parts of sea-salt are acid and alkali, so also are both these the extremes of tastes. Tastes are divisible accordingly into acid and alkaline savours.

2831. As salt is a product of what is inorganic, so will the sapid bodies of this kind be pleasant objects for gustation, provided they do not act chemically nor in excess. Therefore, what is saline, or acid and alkali combined, is agreeable to the taste, even if its action should last a long time.

2832. On the contrary, the proper organic savours, which admit with difficulty of being reduced to the former or inorganic kinds, if they do not prove at once nauseous or unpleasant, yet become so by lengthened operation, e. g. the Sweet, the Bitter.

2833. According to these savours has the organ of taste been regulated. It hath in itself also polar relations. The point or tip of the tongue tastes what is acid, its root more readily what is bitter, its dorsum or back what is poignant or moist.

2834. The matters are not mechanically analyzed upon the tongue; it does not therefore taste the several component parts, but only their chemical behaviour or relation in the water, in other words, their reaction.

3. Function of the Olfactory Sense.

2835. In the lungs the air is materially analyzed and deprived of its oxygen; but, when these organs have assumed a nervose character, the tension only of the air unto the analysis will be perceived. Now, the action of the air is the electrism. The nose perceives only the electrical condition of the air.

2836. The sensation of the electrical relations is called Smelling. The olfactory sense is an air-sense. We smell nothing but the electricity, and neither the contact, nor the impressions, &c. of the particles that find their way into the nose. These particles would have no effect upon the nasal organ, if they did not stand in an electrical relation unto it.

2837. Now, the electrical bodies in nature are the resins or Inflammables. What salt is for the gustatory, that is resin for the olfactory sense. The nose is an electrical, a resinous organ.

2838. In like manner the solubility of bodies in the air is as requisite for smell, as that in water is for taste. The water is the menstruum of the sapid, as the air is of odorous, bodies; and this indeed of necessity, because water and air are the antetypes of these mineral classes.

2839. In order to be an odorous body or object of smell, the resin must resolve itself into air, or become aeriform. Aeriform resin is æthereal oil. Bodies such as those which part rapidly with their electricity, substances containing hydrogen, æthereal oils and burnt spirits, are the usual objects appreciated by the olfactory sense.

2840. The hydrogenous body is therefore provided with a sweet scent. Most bodies which are evolved by fermentation, in so far as they are electrical, are odoriferous. Most blossoms smell agreeably, because they secrete aerial matters.

2841. The products of putrefaction emit a fetid odour, because they indicate the presence not of aerial, but aqueous and terrestrial matters. Nearly all animal bodies stink, besides many secretions of the sexual parts, because they belong to the vegetable nature.

2842. The objects of taste have their residence in what is inorganic, but those of smell, as being objects of a higher sense, have it in the vegetable kingdom. The succeeding or auditory sense has the animal kingdom for its object, while the eye scans the universe.

2843. The nose is in every respect an electrical organ; it is an electrophore, or rather a battery consisting of numerous plates. Of the truth of this, its numerous tortuous passages and laminated bones are striking proofs.

2844. That the nose consists of a great number of blood-vessels, as well as of arteriose olfactory nerves, is quite commensurate with its character or signification, as being a higher pulmonary organ.

2845. The objects of the three vegetative senses are the three elements of the planet, earth, water and air; in the first of these reside the relation of the gravity, rest, and crystallization; in the last the relation of the electricity; in the water that of the chemism. The sense of feeling is an earth-sense, that of taste a salt-sense, that of smell a resin-sense.

b. Functions of the Animal Senses.

2846. The objects of the animal senses are no longer matter, nor its chemical qualities, but the higher relations of the solar system, and the highest organizations, the animals themselves. Throughout the supra-planetary solar system there is naught to be conceived but motion and light in action. Wherever there is æther, it is in motion; the corresponding organs of sense must therefore perceive these two relations. Now, since the animal is also motion and light, and this alone, so by means of these senses will the innermost of animality be at once perceived. Animals become acquainted with themselves only through these senses, and by them only enter, in so far as they are animals, into communication with each other. In so far as they are mere masses of matter, they are capable of self-perception through other senses. These senses may therefore also be called cosmical, while the three former are terrestrial.

2847. They correspond thus with each other. The tactile sense is a precursor of the sense of motion, and represents the motion, gravity, and pressure, after a terrestrial manner; the two splanchnic senses are the precursors of the light-sense, since they dwell upon the qualities of the matter, while the light also is only a quality of the æther. The sense of smell, being as it were an air-sense, will in particular border most closely upon the light-sense.

2848. Through the two cosmic senses the universe is translated into the animal, like the planet is through the terrestrial senses; through the former also the animal spirit, which is a transcript of the universal spirit, passes over into other animals. They are the senses of the highest instruction, of freedom.

4. Function of the Auditory Sense.
In the æther resides the movement of the world.

2849. To the motor system that only which is its equal, and thus the movement of nature, can of necessity become an object. The motor system represented as a sense, cannot, however, perceive the borrowed or derived motion, not the planetary or massive motion, but the primary motion of the æther. The planetary motion is related to the primary motion as the oxydation is to the electrism, as chemical analysis to chemical affinity, and consequently also as respiration to smelling, as digestion to tasting, in short, as the material metatype to the spiritual antetype.

2850. The limbs are the planetary motion organized, and therefore perceive only this material motion—pressure. Touch is related to the animal sense of motion, as digestion is to the tasting.

2851. Smell and taste no longer perceive the bodies in the very act of decomposition, but their laws or their spiritual operations; so will the motor sense not perceive, like the sense of touch, the mass when in motion, but only the motor laws of the mass.

2852. These laws of motion are those of the primary motion. This is, however, a product of the light in the æther, an effect of polarity, and that indeed the first polarity which was manifested in the universe. The motor sense therefore perceives only a motion which has originated through primary polarity.

2853. Such motion is not relative in kind, i. e. it does not affect several portions of the matter in reference to some other matter; but it affects the whole matter internally, or its atoms, so that all matter may remain in its place and yet every atom of it be moved.

2854. This motion is like the motion of heat in the matter. By it heat is excited. For internal motion of the atoms, when aroused by polarity, so that every atom enters into a state of motion against the other, is a discharging of the poles, and consequently development of heat.

2855. This internal motion has, however, been produced by an external; for the external motion acts by contact, and this is a process of polarization. Now, the interior of a mass is only moved by repeated contact, or through the restlessness of polarization and by a proper amount of force, or one which is proportional to the mechanical resistance of the mass to be excited. The last of these is the stroke or blow, the first the vibration of the body. By vibration or oscillation only can a body be internally polarized; for, if it does not oscillate upon the shock being applied, it is still indeed set in motion, but "en masse," so that the internal parts remain in a state of rest.

2856. Oscillation is distinguished from continuous or progressive motion by its affecting the atoms of the body, while the latter acts upon the body itself. Through the vibration heat is engendered, because the poles are free and the matter passes over into æther.

2857. Vibration must endure the longest in solid bodies, and thus in that which belongs to the earth. Among these the hard bodies must take precedence, because the soft are of an aqueous nature. Among the hard bodies again the heaviest must vibrate most effectually, because they offer a longer resistance, and do not yield so soon as light bodies to the effort made at separation. The purest representative of the earth-element or the metal is thus the best instrument or means of vibration, and consequently the object of the motor sense.

2858. Thus as the salt of the earth-element is the object of taste, and as the resin of the earth-element is the object of smell, so would the metal be the object of this motor sense.

2859. But no sense-object is without its medium for transmission, except in the case of the sense of feeling or touch. The salt is only tasted by means of the water, the Inflammable only smelt by means of the air; the metal's primary motion could not therefore be perceived directly by the auditory sense. It must be propagated through the medium which ranks next to the heat, and whose atoms insinuate themselves most easily into those of the vibrating body—thus through the air. Man perceives the primary motion, in which things tend to resolve themselves into æther, through the air. By the metal, or by every vibrating body, the vibration of the air is communicated.

2860. This vibration is not, however, a general movement to and fro, but a dissolution of the material bands. This dissolution can only take place according to the laws of the primary motion. They are rigidified in the solid masses as crystalline forms. Every law of motion is a crystalline form which has become free or spiritually manifested. Through the vibration forms are engendered in bodies, which are commensurate with the substance and form of the mass and the degree of vibration. These forms, being as it were the ghosts or phantoms of crystals, are called sonorous figures.

2861. If the air be displaced when in a state of covibration, it is not thrown into undulatory circles or waves, like water into which a stone has been cast, but in each of its parts the sonorous figure of the rigid body is repeatedly represented. The vibration of air is a progressive motion of sonorous figures.

2862. If the sonorous figures are not incommensurable, several may be at one and the same time in a single portion of air, without interfering with each other. They harmonize, because they have originated according to concordant laws. But if they are products of different laws, they are then confused, and an indeterminate offensive vibration originates, just as savours become loathsome if they depart from their laws.

2863. These figures of the air are only perceived by the Ear. The ear is the only sense in which the motor system is represented in a pure state, devoid of any vegetal signification, and simply endowed with nervose nobility. The ear is therefore the only organ which can perceive the primary motion of the matter; for like acts only in or upon its like.

2864. The metals are the ear of nature, the salt her tongue, the resin her nose, the earth her hand.

2865. The power or capacity excited by sonorous figures of covibrating according to the same laws, constitutes Hearing; the phenomenon is called Sound. Hearing is a primary motion in the musculo-osseous system of the ear, which is communicated to the auditory nerve. The auditory sense is æther-sense, metal-sense. Hearing is magnetizing.

2866. The sonorous figures are formed in the auditory organ, and even in the auditory nerves, just as they have been represented upon an infinitely small scale in the air. The nerve becomes in hearing a sonorous figure.

2867. It is not the mere motion in the auditory organs which produces the sensation of sound; the nerve certainly perceives each movement in the ears, because none is possible apart from or without primary motion; only such a motion is no sound, but only a noise. What has been written in the tingling metal according to eternal laws, is transcribed or copied in the auditory nerve; it is only this writing, but no massive motion of the air, which is legible by the nerve.

2868. Melody is a retrogression of the matter into æther, of the formed world into the primary world; through melody is the spirit of the world revealed. The ear is the first liberation of the

animal from all terrestrial matter; through the ear the animal becomes for the first time spiritual.
2869. Melody is the voice of the universe, whereby it proclaims its scheme, or its innermost essence. Hence the wondrous, mysterious action of harmony, the secret sovereignty of music. Music is the expression of the ardent desire to revert to the primary idea. It makes man unconsciously yearn after a condition which he knoweth not; it transports him unconsciously into this condition of divine repose and godly bliss.
Speech.
2870. That which melodizes proclaims its spirit: the melody of animals displays their internal law.

2871. The musical system of all animal laws is Speech.
2872. Speech is the representation of all nature's sonorous figures in the human organ of sound.
2873. Through speech Man delineates himself in spiritual outlines or sketches, which, devoid of matter (or body), he sets down before himself. Such sketches are easily seen through, since to them every material covering is wanting, and they lie purely before the sensation, as the law, the will of nature.
2874. Through speech Man appears as a double essence. He is a corporeal essence; and the spoken word appears before him in the same outlines, but without body. When conversing, Man is a self-manifestation unto self.
2875. Previous to speech no self-consciousness originates.
2876. Without an auditory organ there is no self-consciousness.
2877. To the organ of hearing, however, belongs also the auditory nerve and the cerebellum. Without a cerebellum there is no self-consciousness.
2878. While Man appears unto himself, he appears also to change. Nature is gloomy, incomprehensible; the spirit is clear, and enlightens her.
2879. Manifestation is only possible through self-manifestation; through a doubling of itself, through its expression.
2880. Animals appear only, in so far as they are individual self-manifestations of Man.
2881. With speech Man creates unto himself his world. Without speech there is no world. For the Apes there is no world, but only tree-fruits, female and male.
2882. Through speech Man becomes acquainted with or learns to know himself; through it he becomes a self-substantial essence, which resembles God, because it creates for itself its world, and recognizes itself, i. e. speaks.
2883. Words are forms of our body mathematically laid down.

2884. A single world is dead, so also are many.
2885. Words, which are connected together according to organic laws, form an organic system, are at once alive, and have a meaning.
2886. Speech originates gradually like the organs, like Man. Speech grows like a plant; at first it is only root, next it puts forth a stem, then leaves, and finally blossoms, when it is the perfect expression of the animal body.
2887. The organ of speech is composed of the three terrestrial organs of sense, the air-, water-, and earth-sense.
2888. The air-organs are the principal medium, because they must produce the sonorous figures; the tongue imparts to them the specific modification; but the lips and jaws, as being motor members, afford the articulation or the movement proper. The lungs and nose breathe out the

tones; the tongue digests them; the lips move them, and fashion them into perfect bodies—words.

2889. A word is at once for itself a regularly inter-articulated body. The sounds are its members, its organs or fundamental formations.

2890. Speaking is a gentle respiration, carried on by the mouth, nose, and limbs or jaws.

2891. As respiration has a special thorax, so also has speaking. The speech-(or voice-) thorax is the larynx.

2892. The larynx represents the ribs and arms, which all move in order to form a sound. The tongue is, so to speak, the head upon this thorax.

2893. The nose imparts euphony to the sounds. It tests their fragrancy. The tongue gives them a special quality, their chemical character or taste; the teeth and lips furnish the cadence as a kind of joint to the sounds, or in other terms the words.

2894. Four organs of sense belong to speech, viz:
Touch in the Jaws.
Taste in the Tongue.
Smell in the Nose.
Hearing in the Ear.

2895. The ear receives the products of the three vegetative organs of sense. It is a synthetic sense.

2896. The tongue gives the vowels; the jaws the consonants.

2897. In accordance with this the vowels are the body of speech, and the consonants the limbs or members, whereby it effects its movements.

2898. Vowels express time, consonants space; the one the chemical import, the other the form.

2899. The vowel E expresses the present, A that which has just past, O that which has quite passed, U that which has passed long ago, I the future.

2900. The more consonants there are in the words, so much the richer is the language; the more vowels, the poorer it is. Such is the speech of savages or wild men.

2901. The speech of animals is a vocal or vowel-speech.

5. Function of the Optic Sense.

2902. As the primary motion of the world is manifested to the animal through the ear; so does the primitive cause of the motion, that of every activity and every phenomenon, i. e. light, appear unto the nerve-sense.

2903. The light-sense is similarly formed or modelled according to the light of nature, and kindles also the light within itself, just as the light originated in the æther; viz. through primary antagonism taking place in its own substance.

2904. Light is the binary division of the æther-mass, not an antagonism between it and some other matter; in like manner sight is a binary division of the nerve-mass in itself, without antagonism to other organs.

2905. Sight is the tension of the æther directly continued into the animal æther, just as taste was a chemical action continued into the animal chemism, and smell an electrical process continued into the animal electrism.

2906. In sight the nervous mass is completely self-antagonized, is a phenomenon unto itself; the eye is the brain placed opposite to the brain.

2907. Sight is thus a tension between optic and central brain; as illumination is tension between planetary and solar æther.

2908. Illumination and vision are of one kind, only occurring in two different sorts of worlds. The planet sees by means of the illumination, the animal illuminates or gives out light by means of seeing. Sight is a light-sense.

2909. Now, the illumination is a fixation of æther, a coloration, and thus a descent of the æther into what is terrestrial. In seeing we perceive the æther, as to how it becomes world; in hearing we perceived the world, as to how it became æther.

2910. Seeing and hearing are opposite functions; the first indicates the creation, the latter the return of the creation into chaos.

2911. Through sight we become acquainted with the universe, through hearing with the miniature universe, or man. Sight passes out of us, hearing into us; or through sight Man is posited in the world, through hearing among his fellow-men. Sight is the speech of the world, hearing that of the planet.

2912. Sight is the speech or language of the universe, hearing the language of Man. Through sight the world reveals unto us its spirit or its thoughts; but through hearing man only discloses what are his own. As words are but the representations of the disintegrated body of Man, so are the world-forms the representatives of the disintegrated body of the primary spirit. The word is a rigidified, crystallized thought of Man; a natural body is a rigidified, crystalline thought of the primary act—a word of God.

2913. Through hearing self-consciousness originates; through sight consciousness of the world, universal consciousness. By means of the former we only become acquainted with human relations—which is understanding; by the latter with those that are universal—this being reason.

2914. Without the ear there would be no understanding, without the eye no reason.

2915. Understanding is microcosm, reason macrocosm. From what is intelligent we expect human wisdom, from what is rational world-wisdom.

2916. That the light also has a medium, whereby it acts upon us, is at once evident from our existing in such a medium; but it could also act directly upon us, were it not inevitably broken up beforehand into colours in traversing the media. All terrestrial elements may be a medium for the transmission of light, such as the gaseous, fluid, and solid, i. e. if they are transparent.

2917. We perceive only coloured light, because our organ of light is only a rigidified colour, a material light. There is no pure light for us; there is none also in general.

2918. Sight is thus a terrestrial light-tension, a colour-becoming.

2919. This happens only through refraction. The eye is a refracting medium. It is distinguished from the brain by being a translucent, refractive, encephalic substance.

2920. Light does not stream into the eye like water into the sponge, but it progresses gradually into and operates upon it.

2921. The eye, in order to experience the sensation of light, is placed in a similar tension to the air, water, or crystal. This tension between it and the brain is perceived by the latter as illumination. The eye is a prism, in which the brain sees the world, in which the brain observes its own tension, or the production of colour. Sight is a deoxydation of the eye.

2922. The optic nerve is an organized ray of light, the brain an organized sun, the eye an organized chromatic sun or rainbow.

2923. Just as the sonorous figures are delineated in the ears, and as the nerve perceives these, but not any concussion of the air; so also does the optic nerve perceive not the light in general, but its terrestrial formation or the chromatic image, which has been propagated into the eye.

2924. In an eye, while seeing, the world is depicted; as in the ear, when hearing, the crystalline forms of the air are delineated.

2925. The eye does not on that account see two worlds. For the chromatic image is nothing else than that which is external to or without the eye. It is verily one and the same influence of light, which acts continuously in a straight line between the chromatic image and the object apparent or beheld.

2926. As a stick thrusts us from the side whither it comes; so does the chromatic image from the side whence the light comes. The exit and entrance of the light are not distinct from each other. The objects could not therefore appear reversed, because we do not see the image in the eye, but feel its process of deoxydation along with its direction.

2927. The objects of the eye are colours. Like as they are related in nature, so also must they be in sight; for they are only the eye extended, or this again is only the formed colour.

2928. We see nothing else but colours, no bodies. For the eye there is no material world. It directly perceives the spirit, and indeed its own spirit, or the world of light.

2929. There is no pre-established harmony, but complete conformity between the world and organ of sense. (Vid. Oken's Essays, "Ueber das Universum als fort-gesetztes System der Sinne," and "Erste Ideen zur Theorie des Lichts." Jena bei Fromann, 1808.)

II. Functions of the Sexual Animal.

A. VEGETAL SEXUAL ORGANS.

1. Of the Sexual Intestine.

2930. As the sexual animal is in every respect the encephalic animal reversed, so also is this the case in respect to its functions. The sexual intestine gives out through its mouth or anus, while the other or animal gut takes in. As being the intestine of the vegetal animal it receives the fermented product of digestion, or the excrement, and conveys it backwards towards the sexual orifice or mouth.

2931. The intestinal function of the sexual animal is an act of vomition. The evacuation is an act of vomition, seeing that the intestinal contents are moved backwards.

2932. The sexual stomach is the rectum. In it the excrementitious matter is accumulated, so that it may be ready for being broken away in the process of evacuation.

2933. The commencement of the sexual intestine is the cæcum, its extremity the anus.

2. Functions of the Sexual Lung.

2934. We may distinguish two circulations, the splanchnic or visceral, which takes place between the lung, intestine, and liver, and the great circulation, which instead of going to the viscera passes to the other organs, and is called the bodily or systemic circulation. The liver excretes the product which results from the splanchnic circulation; the systemic circulation has its organ also, but one which does not secrete a special, but general product.

2935. The general secernent organ of the whole body together with all its systems is the sexual system, which having been elevated by virtue of this general character to the rank even of an animal, is a true sexual animal. That which is a general and not a partial excretion, is imparted by the sexual animal. It is the animal reversed.

2936. Thus the secernent organ of the general circulation must belong to the sexual system, and perform in it that which the liver, or in other words the reverse of the lungs, has done in the splanchnic circulation. The kidneys are the lungs reversed.

2937. If the bile be an extract from the visceral blood, so is the urine an extract from the body's blood, and is consequently the purest reflex or pattern of the former.

2938. The urine is sexual blood, just as the excrement is a product of the sexual digestion. The urine is reversed blood.

2939. The formation of urine is a retro-formation of the blood into digestive fluid or sap. The urine is blood of the sexual animal which has become chyle. It has both properties in itself. It is discoloured blood, and consists for the greatest part of water and salts, all of which are characters belonging unto chyle. It, however, contains urea, which corresponds to the noblest parts of the blood. This substance, like fibrine, consists for the greatest part of nitrogen; it may be called dissolved or decomposed fibrine. It imparts colour to the urine; it is converted by oxydation into lithic or uric acid, and is precipitated of a red colour analogous to that of the blood-globules. In addition to this, albumen, gelatine, carbonate of lime, and phosphorus, consequently the whole blood, are present in the urine.

2940. In urea the muscle flows or runs out of the animal, in albumen the nerve, in lime and phosphorus the bone, in gelatine the tegumentary together with the visceral system, lastly, in water the menstruum of the digestion and respiration.

2941. Thus the urine, just like the blood, is the whole body rendered fluid, but only in a sexual manner, namely, as being half decomposed.

2942. So the bile, from its not representing the whole body, does not contain the latter in itself. It properly contains only the excretion of the intestinal process.

2943. The kidneys stand accordingly opposed without distinction to all the organs, in so far as all of them are affected by the circulation. Their remote sympathy, or if we please, their antagonism, is with the animal systems, or with bone, muscle, and nerve. With the osseous, as being the profoundest system, there is of necessity also a close sympathy. In diseases of bone the bones, as well as the morbid matter, flow away principally through the urine. Their most intimate sympathy must be with the organs of circulation, with the lung, liver, intestine, and skin. As the skin is also an organ of evaporation, so is the antagonism between it and the kidneys of a direct or immediate kind. The skin is the kidneys expanded into a large cyst. These are in turn, just as the lung is, the inverted skin.

2944. A lung in the reversed animal can do nothing else but expire. It only expels the evaporated matter of the sanguinary system, but takes in none, so as to alter or support the blood. The sexual animal aims at the destruction of the animal. The urinary cyst, as being the remnant of the allantois, and of the primordial kidneys or sexual branchiæ, is simply destined to purposes of expulsion. It is the larynx reversed. Micturition takes place through contraction of the cyst, as does expiration by that of the lungs in Reptilia. It is a cough.

B. FUNCTIONS OF THE ANIMAL SEXUAL ORGANS.

2945. The sexual functions proper correspond to the sensorial functions, though to these upon an inferior stage. They are sensorial functions, which are simply occupied with the materials of the senses, so that they are vegetative senses. They are a prefiguration of the sense of feeling, taste, and smell.

1. Functions of the Male Organs.

2946. The testes secrete semen in the same manner that the salivary glands do their fluid or juice.

2947. The semen is sexual saliva, and is thus sexual virus or poison. Like the saliva destroys that which is living, so does the semen. The saliva, however, destroys it in order to form a new animal out of the food; with the same motive the semen destroys it. But both differ in this, that the saliva takes care of the body to which it belongs, while the semen attends to another body, the fœtus or fruit.

2948. The saliva is only the highest condition of the digestive fluid, and is thus a totality merely of the intestinal system; the urine is the total product of the vascular system in its antagonism with the lung; but the semen is the product of the whole body. Through the semen the whole body, rendered fluid or reduced to the primary form, runs away. The semen is the chyle already prepared for all parts; but because it is in a sexual animal, it thus takes the reverse direction and passes out.

2949. A fluid, in which the whole animal has been dissolved, is parallel to the nerve-or point-mass. The semen is a fluid point-or nerve-mass, the fluid brain.

2950. Even what is spiritual directly resides in the semen; it need only assume a form and the cerebral functions commence.

2951. The penis, as being the sexual tongue, has only retained the sensibility of the sense of touch and the function of ingestion.

2. Functions of the Female Parts.

2952. The female aperture is the pharynx for the ingestion.

2953. It is by means of the female parts that the whole sexual system becomes for the first time equal to the perfected animal; through them for the first time the male tongue obtains an oral cavity.

2954. In the total representation of the sexual animal the female parts environ and include the male. This moment is called copulation.

2955. Copulation is the representation of the entire animal out of two incomplete ones. The sexual animal is only an entire beast in copulation, and is only then to be considered as equal to the encephalic animal. Copulation is a representation of the hermaphrodite.

2956. This propensity for bringing about the sexual animal's completion is sexual passion or lust.

2957. In copulation the male parts are "par excellence" the sensorial organ, the female only the recipient mouth. Both are properly organs of sense, but the one is operative or active, the other patient or passive.

2958. Previous to copulation the female parts are consequently inactive, just as digestion is before taste. As digestion first commences after taste has given up the food and excited the stomach to activity, so also the sexual function first commences in female animals, after the act of tasting is past.

2959. Through copulation what is female becomes masculine. It now secretes for the first time self-substantial semen. Through impregnation the female ovaria are first excited to secrete the saliva, which the whole animal contains in a state of solution.

2960. As the chyle becomes what it is, or is derived from the saliva and food; so from the semen and vitellus proceeds the fœtus, but in such a manner that the female substance gives it the mass, while upon this the male only bestows the polarity.

2961. Thus, were the male semen to actually solidify into the fœtus; it is still not its mass which comes into consideration in the latter, but only its polarizing strength. It supplies the place of the nervous system. This power appears to reside principally in the (seminal) Infusoria or animalcules, just as that of the blood does in the blood-globules. Meanwhile both are only signs of the maturity of their respective juices, as the Infusoria in sea-water are a proof also that the sea can produce other animals from its mucus or slime. The Infusoria are the primary mass of the Organic. Its life is only a manifestation of the seminal polarity. The Infusoria are semen which has been poured out over the earth. Propagation is only possible through reduction taking place to the infusorial primary mass.

2962. The semen and the ovum first meet or come together in the uterus.

2963. The ovum is the mean between vegetable and animal semen. As the former is distinctly formed and at once represents, upon a small scale within itself, the principal parts of the future plant, so does the ovum; but only in parts, from which the animal organs first of all grow forth, upon which having commenced, the former or vegetal parts are cast aside.

2964. The ovum is the entire animal in idea, or in design, but not yet in structure; it is the thought unto the animal; it is related to the animal as the thought is to the word.

2965. The ovum has therefore no organ of the animal preformed within itself, but only the materials requisite thereunto. But the materials are not so general in character, that like as from the infusorial mass, everything could become or derive its existence from everything else. But they are at once destined for definite organs, as the vitellus for the intestine, the albumen probably for the integument.

2966. In an ovum therefore the animal resides preformed only in a spectral or phantom-like manner. There are principal masses present in it, from which the principal organs originate.

Mammæ.

2967. In the oviparous animals the secretion of the vitellus is distinct from that of the albumen; the one takes place in the ovarium, the other in the oviduct or uterus.

2968. By degrees the albumen-secreting vessels advance further outwards upon the orifice of the sexual parts, and are then called milk-organs—Mammæ.

2969. Mammæ are only the vascular bundles of the oviduct placed in the direction outwards, or albumen-glands of the integument.

2970. Mammalia are those animals where the ovarium has completely separated into albumen- and vitellus-organ.

2971. Those mammæ which have scarcely been detached from the oviduct, and become free, are necessarily more incomplete, and are situated in the neighbourhood of the sexual parts—as udders.

2972. As the separation of the substances composing the ovum is an advancing-step towards their improvement, so also is the removal of the albumen-glands from the vitelline sac a nobler condition. They cannot, however, remove further off than to the thorax, because this is the highest post or station of the vegetative parts.

2973. Milk is a vegetable product of the animal.

2974. Numerous mammæ are a lower development.

2975. The milk is albumen, which has been secreted by tegumentary glands; it is animal albumen. The lacteal or milk-organs therefore belong to the sexual system.

2976. As the male parts are only the female otherwise developed, it is to be comprehended why the male animals also have mammæ. They are probably the embryo's principal organ of absorption.

Functions of the Uterus.

2977. Now, the uterus obtains the sexual food in a living state, and from its being such is affected by it.

2978. The uterus must be thus a world for the living germ. Two processes are, however, indispensable for the germ, namely nutrition and respiration; both of which are furnished by the uterus.

2979. The uterus is to be regarded as the water, the sea in which the germ or embryo is developed. The water decomposes into basic nutritive matter and oxygenous matter of respiration. The water of the uterus is the blood; this is separated through the antagonism of the fœtus into mucus and oxygen. The mucus penetrates into the amnion, the oxygen into the chorion, or the placenta.

2980. The fundus of the uterus is more arteriose than the os uteri, and stands therefore in

opposition with it.
3. Development of the Fœtus.
a. Anatomy.
2981. The germ may be regarded as a vesicle, full of nutritive matter or albumen, situated within the cavity of the uterus, the walls of which act upon it.
2982. As the fundus of the uterus is the arterial pole, so it oxydizes the vesicle and repels that part which lies close upon it. Through this originates a saccular inversion, as in the mesentery of the peritoneum, and the vesicle separates into three divisions. In itself it is amnion, the inverted part is the integument of the embryo, the tube which unites these the umbilical cord.
2983. The amnion is thus the root or primary cyst of the integument.
2984. Through continuous oxydation blood-vessels are developed upon the surface of the amnion, which finally withdraw to constitute a special integument, which is called chorion. Its vessels are in like manner repelled from the fundus uteri, and prolonged into the inversion of the umbilical cord and the embryo. The chorion is thus the root or primary cyst of the vascular system.
2985. These two bladders, sacs, or cysts are the only general ones which circumscribe or invest the entire embryo, because there are only two general vegetative systems, namely, the tegumentary and vascular.
2986. The embryo has not originated freely in these shut sacs, but only through their introversion; it is itself a portion of these sacs.
2987. The embryo properly lies external to its envelopes, as the intestine does in respect to the mesentery.
2988. Just as the two general vegetative systems have been developed from primary cysts, so also are there sacs for the two special vegetative systems, or the intestinal and sexual; but, on that very account, these cannot be general sacs, nor any longer envelop the embryo.
2989. At the entrance of the inversion of the umbilical cord is situated a small vesicle, which divides and is prolonged into the two intestines. It is therefore the root or primary sac of the intestinal system, and is called in Man the 'vesicula umbilicalis,' in the Mammalia the 'tunica erythroides,' and in oviparous animals the 'vitellus.'
2990. In the same situation is placed another cyst, which is prolonged into what has been called the 'urachus' and the urinary cyst, and from which the primordial kidneys, the true kidneys, and the sexual parts are developed by sacciform eversion. This sac is called the 'allantois,' 'tunica allantoides,' and is consequently the root or primary sac of the sexual system.
2991. These cysts or sacs are consequently not envelopes that serve for the protection of the fœtus, but its developmental organs, which disappear, so soon as their prolongations into the fœtus itself enable them to exercise their functions.
2992. There are thus as many developmental sacs or cysts, as there are vegetative systems present, viz.:
a. Two General Cysts.
1. The Vascular cyst—Chorion.
2.   Tegumental cyst—Amnion.
b. Two Special Cysts.
3. The Intestinal cyst—Vitellus.
4.   Sexual cyst—Allantois.
2993. It is only the vegetative systems which take root in the fœtal envelopes, but not the animal systems. There is no developmental cyst for the nervous, muscular, and osseous systems.

2994. The persistent vegetative systems are the developmental organs for the animal systems; as the intestine for the bones, the veins for the muscles, the tegument or branchial sac for the nerves.

2995. The fœtus consists of three floors or stories like a house, whereof one has been based upon, or rather developed out of the other, viz.:
a. Of the Developmental cysts.
b. " Vegetative systems.
c. " Animal systems.

2996. According to time the sacs are developed in the following series. The first sac is that of the vitellus or the intestine, which is also the first that is present in the development of the animal kingdom. Upon this vitelline membrane the blood-vessels ('vasa omphalomesenterica') are developed, are prolonged into the body with the intestine, turn again outwards and form the chorion. From this the amnion next separates into the envelopes, and the integument upon the embryo. Lastly, the allantois appears, and in its prolongation the sexual parts.

2997. Originally the whole chorion is replete all around with vessels; but as the process of oxydation occurs most powerfully upon the fundus uteri, so are the vessels developed most abundantly in that very situation, and form the placenta.

2998. The placenta is no peculiar organ, but only the more energetic part of the chorion.

2999. It must necessarily be placed around the insertion of the umbilical cord, because at this spot the inversion takes place, on account of the strong oxydation.

3000. The placenta is always situated upon the fundus uteri, because it originates only through its influence. It cannot therefore begin to suck, like the mouth of a leech, in a fortuitous or voluntary manner anywhere. If it is found occupying some other situation, it is a proof that the oxydizing process of the uterus has been displaced. This is consequently an abnormal situation.

3001. Opposite to the fœtal vascular system is first of all developed the general system of animal life, namely, the nervous system, and indeed the spinal chord, or what has been called the 'carina' or primary streak.

3002. The development of all other systems oscillates in this antagonism of blood and nerves.

3003. In the antagonism of the placenta is formed the liver, which in the embryo is one of the largest organs, and in its antagonism the brain is developed.

3004. In the antagonism of the amnion the integument is evolved, and in its antagonism the branchiæ and lungs.

3005. In obedience to the antagonism of the arteries and veins, the vitelline sac divides into small and large intestine. The one pursues a course towards the arterial extremity of the body, the brain, the other towards the venous or sexual parts; mouth and anus.

3006. Finally, from the antagonism of the allantois the sexual parts emerge at the very extremity of the body, which is placed opposite to the mouth.

3007. The osseous and muscular system first makes its appearance, when the vegetative parts are present.

3008. The intestines, the vessels, and the commencement of the sexual parts are originally situated in the umbilical cord, which is itself surrounded by the integument. The umbilical cord is therefore nothing else than the posterior extremity of the body or abdomen, through which the embryo respires and is nourished.

3009. The first respiration and deglutition is therefore a respiration and deglutition effected by the sexual parts, as in the lowest animals.

b. Functions of the Fœtus.
1. Nutrition.
3010. The juices which are contained in the developmental sacs are nutritive juices or chyle; they contain principally albumen. The chyle of the intestinal vesicle depends upon the vitellus.
3011. The fœtal water in the amnion is derived from albumen in the ovum; it is here secreted by the oviduct, but in the Mammalia by the internal wall of the uterus, and absorbed by the general envelopes.
3012. The cause of this secretion resides in the decomposition of the blood through the influence of the chorion. Thus if the maternal blood become deoxydized, it must necessarily revert to the condition of chyle. This chyle is the fœtal water.
3013. The fœtal water is absorbed by the embryo through the integument.
3014. No blood is transferred from the mother directly to the fœtus.
3015. The blood-vessels of the uterus and placenta do not open into or communicate with each other.
3016. The fœtal water corresponds to the albumen or the white of eggs, not to the vitellus. This becomes also during incubation consumed in the formation of the body of the chick, and not the vitelline mass, which is first of all destined for the intestine.

3017. Towards the termination of pregnancy, when the fœtus is endowed with muscular motion, the fœtal water is also absorbed. Nutrition is therefore in the commencement an absorption carried on by the integument, and lastly by the intestine.
2. Respiration.
3018. The respiratory organ of the fœtus is the chorion, and in particular the placenta. Its tissue is like that of the branchiæ or spleen.
3019. The arterial blood transmitted through the umbilical vein is conducted through the foramen ovale into the left ventricle of the heart, and from thence directly to the principal organ of the fœtus, the brain and spinal cord. From hence it returns, venous in quality, to the right side of the heart, and from thence through the ductus arteriosus into the inferior or descending aorta, from which, pursuing its course through the umbilical arteries, it again reaches the placenta, in which it undergoes a renewed oxydation.
3020. If therefore the umbilical cord be compressed, the fœtus dies suddenly, as happens indeed accidentally in certain cases, which completely resemble those of death produced by suffocation. The chick when in the egg dies, if the egg-shell be coated over with varnish, or if the egg be submitted to the noxious influence of gases which are devoid of oxygen.
3021. The respiration effected through means of the placenta, admits also of being proved by the sudden change which occurs in the circulation after birth. No arterial blood being then any longer brought to the heart through the umbilical vein, the left heart is no longer stimulated, and the foramen ovale collapses and is closed. Thus, all the blood enters the right side of the heart, and finding no thoroughfare into the ductus arteriosus, it is driven forcibly into the lungs, which now expand, and thereby leave an empty space between the pulmonary vessels, into which the air rushes in.
3022. The first act of respiration is therefore the result of the lungs being injected with venous blood, and thus of necessity ensues. When, on the contrary, cases arise causing pulmonary suffocation, the blood then regurgitates to the umbilical vessels, in order to reach the original branchia or placenta.
3023. Another process of respiration is found to take place in the vessels of the allantois. Its fluid

becomes oxydized and penetrates through the urinary cyst into what have been called primordial kidneys. This kind of respiration through the anus is persistent in many Worms and larvæ of Insects.

3024. Finally, there is a respiratory process on the body of the embryo itself through the branchial apertures on its neck; these in the Frogs and Salamanders still continue visible after their exclusion from the egg. This oxygen must be derived from the fœtal water.

3025. Without doubt also the vitelline vessels respire, and thus each vegetative system has its own process of respiration. The intestine respires through the vitelline vessels, the vascular system through the vessels of the chorion, the sexual system through those of the allantois, the integument through the branchial apertures. The lungs belong to, and perform respiration for the whole body.

Decay of the Developmental Organs.

3026. When all the organs are developed, the intestines are drawn together with the vitellus into the abdomen. The chick is still nourished several days after exclusion from the egg by the vitelline mass, which passes through the vitelline canal into the intestine. The vitelline membrane subsequently becomes flaccid, and finally disappears through maceration. In the Mammalia the umbilical vesicle separates at an earlier period from the intestine, and continues to lie in the umbilical cord.

3027. The secernent point of the vitelline membrane or of the umbilical vesicle is the cœcum.

3028. There are therefore two intestinal systems, which branch off in a bifurcated manner from the cœcum, namely, the sexual intestine and the small or truncal intestine.

3029. All embryos have originally umbilical herniæ, which do not originate through protrusion of the intestines out of the abdominal cavity, but through their entrance into the latter being retarded.

3030. The umbilical herniæ therefore indicate an earlier condition of the animal, which has originated through arrest of the development.

3031. At birth all the enveloping membranes die, and their point of liberation is called the navel or umbilicus.

3032. The animal has originated through the umbilicus, and through this has it respired.

3033. All animals which breathe by the anus, do so properly through the navel; animals of this kind are fundamentally umbilical animals.

Parallelism of the Fœtus with the Animal Classes.

3034. During its development the animal passes through all stages of the animal kingdom. The fœtus is a representation of all animal classes in time.

3035. At first it is a simple vesicle, stomach, or vitellus, as in the Infusoria.

3036. Then the vesicle is doubled through the albumen and shell, and obtains an intestine, as in the Corals.

3037. It obtains a vascular system in the vitelline vessels, or absorbents, like as in the Acalephæ.

3038. With the blood-system, liver, and ovarium, the embryo enters the class of bivalved Mollusca.

3039. With the muscular heart, the testicle, and the penis, into the class of Snails.

3040. With the venous and arteriose hearts, and the urinary apparatus, into the class of Cephalopods or Cuttle-fish.

3041. With the absorption of the integument, into the class of Worms.

3042. With the formation of branchial fissures, into the class Crustacea.

3043. With the germination or budding forth of limbs, into the class of Insects.
3044. With the appearance of the osseous system, into the class of Fishes.
3045. With the evolution of muscles, into the class of Reptiles.
3046. With the ingress of respiration through the lungs, into the class of Birds. The fœtus, when born, is actually like them, edentulous.
3047. After birth it is suckled or fed. The milk is the nutrition continued by means of albumen; for the mammæ are verily only the albuminous vessels of the Bird, which are placed free and external in the Mammiferous animal. After the time for sucking is past the young one obtains teeth; and thereby becomes for the first time independent of the mother, and passes over into the class Mammalia. Now, should the sketch here afforded of these parallels be not in all respects correct or justifiable, still sufficient proof remains, that a perfect parallelism is found to take place between the development of the fœtus and that of the animal kingdom.
3048. Animals are only the persistent fœtal stages or conditions of man.
3049. Malformations are only persistent fœtal conditions, or animal formations in individual animal bodies.
3050. Diseases are vital processes in animals. Pathology is the physiology of the animal kingdom. A human fœtus is a whole animal kingdom. (Vid. Oken's 'Die Zeugung,' 'Beyträge zur vergl. Anatomie,' and 'Ueber die Nabelbrüche.')
Periods of Life.
3051. If the young in the ovum or in the mother's body resembles the aquatic animals, and has passed through their organization; so after birth does it belong to the air-breathing animals and traverse their organization.

3052. One period is that of sucking; the edentulous condition of the Bird. This is called the suckling age, baby-hood or infancy.
3053. One is that of the eruption and persistence of the milk-teeth; the condition of the Rodentia, a repetition of the gelatinous animals. This is the age of childhood.
3054. One is the eruption of the permanent set of teeth until the attainment of puberty, or the development of the sexual functions; the condition of the Marsupialia; repetition of the Conchozoa or Shell-animals. This is boyhood.
3055. From the state of puberty until the development of the faculty of understanding; the condition of the Shrews and Bats, the repetition of Insects. The age of youth.
3056. The period of the understanding passes through the Ungulata, and is a repetition of Fishes, Reptiles, and Birds. This is the first stage of manhood.
3057. After the understanding the reason matures; this is the condition of the higher Mammalia up to Man; it is a repetition of the Mammalia. This is adult age or mature manhood.
3058. Then follows the decadence or dying off of the sexual functions; a retrogression through the animal classes. Gray old age.
3059. Finally, the reason and understanding die; childhood returns and terminates with the death of the vegetable in the animal.
3060. Death results through the sexual animal.
3061. Death is only a continuous growth through retrogression into the organic primary matter or Infusoria.
3062. Death is an organized decomposition.
3063. Decomposition is a forming of seeds, ova, and fœtuses.
3064. Dying is a multiplication of self.

III. ZOOLOGY.
3065. Zoology is Zoogeny divided and self-substantially represented. What in Zoogeny was the organ of a single indivisible animal here becomes the organ of a separated animal, or becomes a self-substantial animal.
3066. The self-substantial animals are only parts of the Great animal, which is the Animal Kingdom.
3067. The animal kingdom is only one animal, i. e. the representation of animality with all its organs, each of which is a whole for itself.
3068. A single animal originates, if a single organ frees itself from the general animal body, and yet exercises the essential animal functions.
3069. The animal kingdom is only a dismemberment of the highest animal, i. e. of Man.
3070. Animals become nobler in rank, the greater the number of organs which are collectively liberated or severed from the Grand animal, and which enter into combination. An animal, which e. g. lived only as intestine, would be, doubtless, inferior to one which with the intestine were to combine a skin; and that animal again must be regarded as higher than the latter, which should present, in addition to these organs, vessels, liver, branchiæ, tracheæ, and lastly bones, &c.
3071. Animals are gradually perfected, entirely like the single animal body, by adding organ unto organ. The animal kingdom is developed through the multiplication of the organs.
3072. Each animal ranks therefore above the other; two of them never stand upon an equal plane or level. Animals are distinguished by their position of stages or degrees from each other, by the number of their different organs, but not by the division of a single organ.
3073. The animal system cannot be arbitrarily disposed according to this or that organ, just as it may chance to meet the eye; but only in accordance with the rigid prescripts of the animal body's genesis.
3074. The animal body separates into two series of organs, which, corresponding with, pursue a proximal course in relation to, each other; into the Anatomical systems and the Sensorial organs, unto which the sexual parts appertain.
3075. The number of the sensorial organs is 5, and they thus stand according to their genetic development one above the other:
Tactile sense or Skin.
Gustatory sense or Tongue.
Olfactory sense or Nose.
Auditory sense or Ear.
Optical sense or Eye.
3076. In animals, which are characterized by the sense of feeling or touch, the other sensorial organs must be either still wanting, or if present but imperfectly conditioned, i. e. not constituted like those of man, who is the type, pattern, or paragon for every formation.
3077. Their sensations are limited to those of general touch or feeling, and of those derived through the medium of the other senses we meet with but feeble manifestations.
3078. Their body itself will only be a tegumentary body, with the organs subordinated to the integument, namely, the viscera. They are therefore devoid of a true tongue, of a nose, and of ears and eyes consummated after the fashion of these organs in man; they are devoid of an osseous, muscular, and myelonal (spinal chord) system, and therefore of the nose "in toto," as being the anterior extremity of the myelon.
3079. Such are what have been called the Invertebrate animals, which are consequently, in

accordance with their physiological signification, Splanchnic or Tegumentary animals.

3080. The tongue exhibits for the first time in Fishes a resemblance to the human structure, while their nose, ears, and eyes have not yet attained the latter grade of perfection. To the nose are wanting the posterior nasal foramina, to the ears the external auditory meatus, to the eyes the palpebræ and power of motion.

3081. In the Reptiles the nose opens for the first time into the mouth, and serves for the thorough passage of air. It is thus developed as in Man, while to the ears the external auditory meatus and cochlea are wanting, the eyes being barely endowed with lids and motion.

3082. In Birds, for the first time, the external auditory meatus, as well as the cochlea, is exhibited in its perfection, while the eyes are scarcely gifted with motion, and have only the inferior lid perfect; the tongue and nose, with the limbs also, have again become retrograde in character.

3083. For the first time, in the Mammalia, the eyes are moveable and covered with two perfect lids, without the other organs of sense having suffered degradation through this completion of the eyes.

3084. Thus in respect to the Senses there are only 5 animal divisions of equal value or worth. They should properly be called classes; but, as the lowest division, from comprising within itself the viscera or the vegetative systems, is very rich in contents; were we to call these divisions classes, many inequalities in rank, and hence also in the number of the orders and families, would originate—

1. Dermatozoa   Invertebrata.
2. Glossozoa   Pisces.
3. Rhinozoa   Reptilia.
4. Otozoa   Aves.
5. Ophthalmozoa Mammalia.

3085. Now, with the sense of feeling, or the integument, the sexual system is associated or conjoined, and that indeed as the first or lowest development of the tegumentary system. Nevertheless, the sexual system divides into two groups, into the sexual organs, which are impressed with a true sensorial signification, and into their product, or the sexual juices, and the ovum or fœtus. With these two divisions the development of the integument proceeds "pari passu."

In the ovum the tegument and its contents are not as yet separated. Both consist of a transparent mucous or gelatinous mass, as is exemplified by the vitellus and albumen; such is the case also in the Infusoria, Polyps, and Acalephæ or Sea-nettles.

In the sexual organs, however, both parts separate into membranous capsules or cysts, and glandular contents, as in the roe or ovary; milt or testes; kidneys, with furthermore the oviduct, penis, and urinary cyst. The latter are sentient membranes. Such is the case in the Bivalve Mollusca, and Snails.

In the next place the tegument becomes, for the first time, a self-substantial organ of sensation by appearing as an envelope of the body; the vesical form is then repeated, and by this means the annular character originates as a series of cysts arranged one behind the other; it is thus a veritable skin, from which, finally, the members sprout forth, as in Worms and Insects.

The Dermatozoa will accordingly range in three stages.
1. Blasto- or Oozoa.
2. Sexual animals.
3. Cutaneous "

An annulated tegument, cutis or true skin, appears for the first time in the Worms, with here and there lateral filaments and tentacula. True feet and antennæ appear in the Crustacea or Crabs. Lastly, feet and wings in the Insecta or Flies.

3086. The external sexual parts, especially the male, first make their appearance, and that indeed with a very striking amount of development, in the Snails, and in like manner the body of the Bivalve Mollusca or Mussels has become almost a complete mass of ova or roe. In the Cuttle-fishes the first traces or rudiments of urinary organs appear. The animals which belong to this group are accordingly the Conchozoa or Shell-animals.

3087. Animals, which are directly resolvable into sexual fluids, or that represent parts of the ovum, are the gelatinous Infusoria, Polyps, and Acalephæ. Unto this category belong the Protozoa or Mucus-animals.

3088. The complete subdivision of animals, according to the organs of sense, would consequently stand thus:
I.   Dermatozoa   Invertebrata.
 1. Oozoa   Protozoa.
 2. Glandular animals Conchozoa.
 3. Cutaneous   "   Ancyliozoa.
II.   Glossozoa   Pisces.
III.   Rhinozoa   Reptilia.
IV.   Otozoa Aves.
V.   Ophthalmozoa Thricozoa.

3089. Unto these organs of sense the anatomical, or internal parts, are subordinated, and range parallel to them in a striking manner. The following is their order of succession in accordance with that of their origin:
1. Intestinal system.
2. Vascular "
3. Respiratory "
4. Osseous   "
5. Muscular   "
6. Nervous   "

3090. That the vegetative systems are correctly arranged after this manner, is proved chiefly by their order of development in the animal series.

3091. The animals, occupying the lowest grade, are nothing but an intestine, which is in many instances scarcely separated or distinct from the tegument; they are devoid of vessels and branchiæ, and are barely provided with self-substantial ovisacs. Their body consists of one or two concentric cysts of an homogeneous and transparent substance—Intestinal animals, Protozoa.

3092. When the intestine is freed from the mass of the body, both then obtain the form and substance of tegumental cysts, whereof the external being only an intestinal tunic, thus represents the peritoneum. They are now, however, united by means of a vascular system, which is again surrounded by a cyst, and thus by a pleura. Their body consists of three concentric cysts; intestine, peritoneum, and pleura. It contains too a liver and self-substantial sexual parts—Vascular animals, Conchozoa.

If these cysts be repeated in the direction of the axis, the tegument becomes an annulated skin.

An annulose animal is a multiplied cystic animal. Associated with this character, the respiratory organs are gradually evolved to form vascular plexuses, branchiæ, feet, tracheæ, and wings, while the sexual parts are for the most part separated—Ancyliozoa, Respiratory animals.

3093. The osseous system first appears in the Fishes, along with imperfect, mostly tendonless, and only white muscles, as also a myelon that is only developed into a stunted encephalon, in which the brain-organs of the Thricozoon are in great part wanting.

3094. Typical, or true muscles, provided with tendons, and of a red colour, are first exhibited in the Reptilia.

3095. A complete nervous system, pretty similar to that of the Thricozoa or Pilose animals, having a cerebrum and cerebellum, with similarly distributed and delicate nerves, is first displayed in Birds.

3096. According to the anatomical systems there are, therefore, six divisions of animals:
A.   Splanchnozoa.
  1. Intestinal animals Protozoa.
  2. Vascular   "   Conchozoa.
  3. Respiratory   "   Ancyliozoa.
B.   Sarcozoa.
  4. Osseous animals Pisces.
  5. Muscular   "   Reptilia.
  6. Nervose   "   Aves.

3097. The Pilose animals first originate through completion and combination of all the organs of sense. They are Æsthetic or Sensorial animals.

3098. The arrangement of animals, according to the organs of sense, coincides consequently with that derived from the anatomical systems, and each animal division is therefore determined by two principal organs, by a vegetative and an animal. Every animal is at once a vegetable and animal body, the inferior kinds being partly so, but the highest or the Pilose animals entirely so, or in every respect, i. e. in them are found all the anatomical systems, and all the sexual and sensorial organs.

3099. The signification of animals is accordingly as follows:
I. Anatomical Systems.   II. Organs of Sense.
A. Vegetative Systems.   A. Tegumental Sense.
1. Intestinal animals. 1. Oozoa   Protozoa.
2. Vascular   "   2. Glandular animals Conchozoa.
3. Respiratory "   3. Cutaneous   "   Ancyliozoa.
B. Animal Systems. B. Cephalic Senses.
4. Osseous animals. 4. Glossozoa Pisces.
5. Muscular " 5. Rhinozoa   Reptilia.
6. Nervose " 6. Otozoa   Aves.
7. Sensorial " 7. Ophthalmozoa   Thricozoa.

The ovum divides into vitellus, and albumen, along with the calcareous shell, and into germ or envelopes; the intestine into pharynx or stomach, bowels and absorbent vessels, thus:
1. Gastric animals   Vitelline animals   Infusoria.
2. Intestinal "   "Albuminous "Polyps.
3. Absorbent "   Involucral   " Acalephæ.

The sexual parts divide into female, male, and urinary organs; the vessels into veins, arteries, and hearts, thus:

1. Ovarial animals   Venous animals   Mussels.
2. Orchitic   " Arterial   "   Snails.
3. Renal   " Cardiac   "   Kracken.

The annulated tegument divides into papillæ, feet, and wings; the respiratory organs into tegumental network or rete, into branchiæ and tracheæ, thus:
1. Papillary animals   Reticular animals   Worms.
2. Pedal   "Branchial   Crabs.
3. Alary   "Tracheal   Flies.

When parallelized with the organs of plants, the following remarkable relationships between them are rendered apparent:

1.	Cells	Stomach	Vitellus	Infusoria.
2.	Bark	Intestine	Albumen	Polyps.
3.	Root	Absorbents	Envelopes	Acalephæ.
4.	Ducts	Veins	Ovary	Mussels.
5.	Liber	Arteries	Testes	Snails.
6.	Stalk	Hearts	Kidneys	Kracken.
7.	Tracheæ	Retia	Papillæ	Worms.
8.	Wood	Branchiæ	Feet	Crabs.
9.	Foliage	Lungs	Wings	Flies.
10.	Seeds	Bones	Tongue	Fishes.
11.	Pistil	Muscles	Nose	Reptiles.
12.	Corolla	Nerves	Ears	Birds.
13.	Fruit	Senses	Eyes	Thricozoa.

A. DIVISION INTO PROVINCES.

3100. The animal body divides first of all into the vegetative and animal. There will therefore be animals in which the former, and others in which the latter, systems predominate. The kingdom consequently separates into a vegetative and into an animal province. The vegetative parts are all tegumental developments, and thus the creatures in whom they prevail are Splanchnic or Visceral animals, but the animal parts are developments of the flesh, and constitute the Sarcose animals.

First Province. Splanchnozoa.

3101. Unto the Splanchnic or Tegumental animals are wanting bones, muscles with the nerves belonging to them, and thus the neural axis or encephalon; they are consequently devoid of bone, muscle, and brain, being in a word asarcose or fleshless animals.

The tegument is, however, the general organ of sensation or feeling; they are thus Sensitive animals.

3102. In them, developments only of the sense of feeling can occur, such as sensitive papillæ, tentacula, feet, and wings. All the remaining organs of sense can only be exhibited as rudiments of a very feeble or stunted character. They do not possess a true tongue, nose, ears, and eyes, i. e. constructed after the type of these organs in Man. The eyes only, from their being the sense of the animal system proper, can assume a definite kind of development.

3103. But, these organs of sense are the sensorial organs of the head, or rather are the head itself; the Tegumental animals are therefore devoid of a true head. They only possess such an one, in so far as it is determined by the tegument and nervous sense, by the mouth and the eyes.

These animals are what have been called Invertebrata, a name which is, however, defective, from its indicating the absence of one part only or of a single animal system, while the word flesh comprehends bones, muscles, and nervous mass; they are asarcose animals. But even this last appellation is not correct, because it is negative. Their positive system, or that under which they actually exist, is the tegument; so that the name of Dermatozoa or Sensitive animals, is the only proper one.

As the tegument includes the viscera and forms therefore the trunk alone, they may thus also be designated Trunk-animals.

Second Province. Sarcozoa.

3104. As in the animal body, bones, muscles, and encephalic system are suddenly associated with the tegumental system; so also does a second series of animals, having these systems, suddenly originate. Now, as the first formation of the osseous system is the vertebra, so all these animals have, as is generally understood, a vertebral column, and are on that account indeed Vertebrata; but, they are much more than this, and hence the title is of too limited a character. Besides there are among them animals, in which only the chorda dorsalis is present, without the ring of the vertebral body being formed around it.

Along with the animal systems the head is for the first time developed with its organs of sense—Cephalic animals.

These animals have consequently, in addition to the general sense of feeling, a true tongue, nose, ears, and eyes—Cephalæsthetic animals.

B. ANIMAL CIRCLES.

3105. Animal Circles are representations of entire anatomical systems as self-substantial bodies.

3106. Now, the vegetative body divides into three principal systems, into the Intestinal, Vascular, and Respiratory system, with their functions, or the digestive, nutritive, and respiratory processes. There are therefore Intestinal, Vascular, and Respiratory animals.

Circle I. Intestinal Animals.

3107. The intestinal system is the first form of body, or that from which the other systems have not yet separated. The body of these animals consists therefore of the homogeneous primary mass—the animal protoplasma, mucus or slime—Mucus-animals.

Now, the protoplasma is a hollow globule. The intestinal system is therefore nothing else than the original cystic form. Thus, there are Cystic animals like the Infusoria.

Again, cysts can increase in no other way, than by dividing into their like, or engendering cysts within themselves. The first kind of increase or multiplication of species is thus fissiparous, or produced by division.

The newly-engendered cysts are to be compared with the vitellus, and when they have attained perfection, to the ovum. They are therefore Oozoa or Ovum-animals.

In these animals consequently there are no separated sexual parts, namely, in addition to the vitelli, neither testes nor renal organs, or at least only obscure indications of them. In these Cystic animals the lowest feeling only, that of the sex or general sensation, can occur.

The Oozoa or Cystic animals, when compared with the plants, are the first flower that has been set free, or in other words, a flower which no longer stands polarwise upon a stem, because it is not developed in the differencing air, but in the indifferent water. It may be said that nature, having brought matters as far as the development of the sexual parts, then quits or passes out of the vegetable world; while these parts, or even the entire plant, requiring now no longer the stem and root, become a root themselves, and in behalf of this enter the water.

Animals, which have the form of flowers, are round or radiiform. There are Radiata or Radiated animals.

3108. These Anthoidal or Flower-like animals are Infusoria, Polyps, Acalephæ; being single or double concentric cysts.

3109. We may regard the Flower-animals as the fundamental mass of the sexual parts, which has attained unto free motion. They are sex throughout, or nothing but sex; it cannot therefore be said that they possess sexual parts like the plants, but that they are sexual parts. They are sexual parts that swim.

3110. Formerly, most of these Anthoidal animals were actually taken for real plants, on account of their floral and ramular form, and even their very substance; so slightly withdrawn are they from the vegetable kingdom. The whole difference between the two is effected by the water. Could we transpose them into the air, they would then become real plants.

3111. Now, as the vegetable flower is not a mere sexual system, but is also stock or trunk; so also is the animal flower at once an organ of digestion, respiration, and nutrition. The lowest condition of these organs is, however, only one of absorption, evaporation, and rigidification; these processes will therefore be present also, though only upon the lower stage—they are Intestinal animals; for in a simple intestine the same processes could occur, only within each other, whereas they are mutually dissevered or set apart in intestine, lung, and capillary system or parenchyma.

3112. The sexual parts are themselves viscera, or the viscera themselves are sexual parts, just as the fungus is at one and the same time root and seed-capsule. The sexual parts themselves absorb, respire, and nourish.

The floral sac is not therefore a mere sexual sac, but also an absorbent sac; upon a somewhat higher grade it is even a digestive sac, the sac-wall itself being a respiratory and nutritive paries or wall.

The sexual function has at once become an ingestive function, tending unto nutrition, or the deglutition of the food is itself an act of copulation.

The sexual capsule in these animal flowers can as well be termed stomach as uterus, and its wall ovary as well as branchia.

3113. As being of a sentient, mucous nature, they are point-substance or nervous mass. The tentacula are higher organized stamina, and thus occur as cilia, surrounding the oral aperture or mouth, as in the Infusoria. These palpi or feelers are, from being organs of ingestion, both male penes as well as digits or tongues, as in the Polyps. Their structure is still wholly tubular, while their elongation appears to be for the most part effected by injection with water—they are absorbents, as in the Acalephæ. The Oozoon brings forth young in the same cavity, in which it digests and by which it respires, and impregnates itself with the same filaments, whereby it seizes, swallows, and tastes its food.

In the bottom of the cavity of the Germ-animals, granules develop, which are born or escape through the floral opening—pharynx, and again become similar beings. In others the granules also sprout forth attached to the walls of the cavity, remain there some time united with the parent animal, and thus completely represent the mode of propagation in plants by means of gemmules or buds. Among the Polyps and Acalephæ it is well known that the ova issue from apertures near the mouth; while in the Actiniæ this is stated to take place from the stomach. The ovaria are, as is well known, situated between the stomach and parietes of the body.

3114. The intestinal animals are an entire animal organism, but only in the chaotic condition. They are the fundamental tissue, the cellular system of the animal, and the higher animals are only separated cells.

3115. The propagation is in every respect similar to that of plants. Now, as the seeds are the whole plant upon a small scale, so is the granule or ovum the entire animal; it is liberated through the pharynx and continues to grow merely by increasing in size. But if the young animal protrude through the tegument, that is a true gemmiparous propagation.

3116. Those Oozoa, which like plants can develop buds, consist of several animals, and may be cut in pieces like plants; when each piece becomes again an entire animal.

3117. The Oozoa represent those products of nature which are prior or antecedent to the animal world; namely, first of all plants, and further still the inorganic kingdom also, or the earth, since they have originated in the water and can be as well developed from the stones as the Lichens. There are therefore Lithozoa or Stone-animals, and Phytozoa or Plant-animals, among the Oozoa.

3118. Would we compare these animals with the parts in plants, they then represent their cell-development, or cells, bark and root. They are themselves either vesicles, as the Infusoria, or barks, as the Corallines, or a root-like fascicle of tubes, as the Acalephæ.

3119. In the animals, however, the cells have become stomach, the bark intestine, the root absorbents. The Oozoa represent therefore the aggregate of the intestinal system, the primary mass of the animal body.

Their whole body is a digestive body, or parenchyma, traversed in many species in all directions by tubes or absorbents, as in the Acalephæ.

As yet no nervous filament, nor any muscular fibre, &c. has been separated from their mass, as in like manner a tegument has barely been freed or separated from the intestine. These animals are nerveless and sine-tegumental, precisely because they are wholly made up of nervous mass and tegument.

Circle II. Vascular Animals.

3120. In the next place the digestive separates completely from the tegumental function, and each constitutes a function for itself, but which now being separated from its fellow could no longer subsist. Between the two therefore is formed the nutritive function in the vascular system. Now, the vascular system of the self-substantial intestine is the liver. This organ will therefore for the first time make its appearance in this animal circle—Hepatic animals. Upon a higher stage, salivary glands are also developed on the intestine, and will here likewise begin for the first time to appear—Salivary animals, Snails.

3121. Through the separation of the viscera from the remaining substance, this must necessarily remain behind as a hollow cyst or tegument investing the former. The true free tegumental formation is therefore by no means accidental, but is necessarily bestowed with the viscera in the course of animal development. This tegument is peritoneum. There are animals which are invariably bicystic, but consisting of concentric cysts—abdomens.

Around the ventral integument, however, the vascular system also forms its tunic; this is the branchial membrane or operculum—pleura, mantle. There are tri-cystic animals—intestine, abdomen, and thorax disposed concentrically around each other—Mussels. Their body is therefore not articulated, but its parts are still inserted within each other.

The Vascular animals are consequently multiplied Protozoa; or Mucus-animals of the second power—Ovum2.

The sense of feeling ascends upon its second stage, when its organ, having been freed from the mass of the body, surrounds the viscera as a self-substantial tegument. The feeling is then no longer one of a merely general character, but is a definite perception of external objects, a passive power of feeling.

3122. True muscles could not as yet originate in this skin, and that for obvious reasons, although fibres are present; for the former are to be brought under the signification of arterial fibres.

3123. Cilia, when furnished with fibres by whose agency they are rendered moveable and susceptible of inversion, are called tentacula, which here occur under every variety of form.

3124. Would we compare the Vascular animals with the plants, they must then represent their ducts, liber, and stalk. The heart is the stalk, the arteries liber, the veins tubes or ducts. These animals have assumed also upon the whole the cauline or cylindrical form—Cauline animals.

3125. From their having obtained in addition to the intestine only the vascular system, they are still governed by the water, and live therefore for the greatest part in this element. They have the first mode of respiration and thus an aqueo-respiratory process—possess branchiæ but not tracheal tubes.

3126. The sexual parts, which in the Germ-animals were still coalesced for the most part with the body, here become self-substantial through the separation of the teguments, make their appearance as a repetition of the digestive system under the condition of a free or separate system, and are evolved into true ovaria and even male parts—Sexual, Glandular animals.

3127. The first step towards the evolution of male parts is, however, only half achieved. Only one testis originates, while the other remains behind as ovarium—Androgynous or Bisexual animals.

3128. These animals, which are characterized by the vascular system, and by the first self-substantial or external sexual parts, which indicate organs of sense, are the Conchozoa, such as the Mussels, Snails, and Kracken.

Circle III. Respiratory Animals.

3129. When once the intestinal and vascular system, through perfection of their individual parts, such as the liver and branchia, and through separation of the sexual parts, are completed; then the individualization of the tegument steps into view, and it becomes an independent respiratory system.

3130. Through the increased process of oxydation the tegument hardens and is converted into horn. All induration, however, only takes place in opposition to soft places. The tegument therefore separates into hard and soft rings—Annulate animals.

3131. The annulate tegument is a tracheal tube wholly converted into a body. To distinguish it from the general tegument it may be called the cutis or skin—Cutaneous animals. The annular tegument may be regarded as a series of cysts placed one behind the other. The Annulate animals are therefore multiplied Malacozoa; Mucus-animals of the third power—Ovum3.

The respiratory organs, being in their lowest condition, will not as yet be freed from the tegument; the vessels simply form a network or projecting filaments and lamellæ—reticular branchiæ; as in the Worms. The tentacular organs, from being yet soft and thus scarcely moveable, are still very imperfect. Upon the lowest stage the tegument or skin simply feels; in the next place papillæ, and finally filaments, originate, especially about the mouth—Cutaneous, Papillary and Filamentary animals.

3132. If the tegument, as being the original branchial membrane, is converted into horn; then the branchiæ cannot continue as retia, but must elongate above the tegument into filaments, ramules or lamellæ.

With this, these elongated branchiæ separate into two organs, one part of them becoming indurated in like manner with the general tegument, and supporting the other as gill. Horny branchial filaments, which contain vessels, nerves, and fibres, are called feet—Pedal animals.

3133. The limbs or members of these animals are simply hollow tegument, hollow hair, and are therefore thoroughly different from the bones or the animal system.

The tegument thus hardens around the soft parts and the viscera. A horny coat of mail originates, and thus we have horny or mailed animals, in opposition to the Malaco-or Conchozoa.

3134. Beneath the horn, however, there must still be soft skin, and this becomes fibrous by the strong oxydation which it undergoes. Fibrous fascicles are attached to the coat of mail and to the hollow limbs, and are consequently within the tubes.

3135. These fibrous bundles are not flesh, but a fibre-drawn tegument, so that there are also no true muscles. They must on that account too be numberless.

3136. The articulations or joints are external not internal; they thus consist of tegumental tubes, not of bones, abutting against each other, and are not surrounded by flesh. Hence, like all the preceding groups they also are devoid of flesh—asarcose animals like all the preceding ones.

3137. In the branchiæ, however, it is only the cauline portions which become horny, while their ramules continue to perform the respiratory function. The branchiæ therefore are appended to the extremity or roots of the feet; or rather these latter form the branchial arches.

Annulate animals, having true or hornified and likewise annulated feet, are called Crustacea or Crabs.

Thus, in these animals the sense of feeling obtains special and moveable organs; they are Tactile animals.

Tactile organs are prolongations of tegument moved by muscular fibres, which in some degree adapt themselves by pressure unto the forms of objects, or have the power of seizing and retaining them—such as feet, antennæ, maxillæ, palpi.

3138. But if the tegument be entirely converted into horn, and the respiratory vessels have in this way disappeared within it, then the internal respiratory organs are formed by inversion of that part of the tegument which is between the rings, and through these openings the air penetrates to the internal parts—stigmata, tracheæ.

3139. The tracheæ can first of all originate when the respiratory process has attained its highest development, or, in other words, when the creature breathes air.

3140. Lastly, in the air-breathing Ancyliozoa even the external branchial lamellæ harden and are converted into wings—Pterozoa, Insects or Flies.

The wings of Insects do not correspond to the wings of birds; they are not feet, but pedal appendages or branchiæ, and thus are no new or unknown organ.

Circle IV. Sarcose Animals.

3141. The second animal province may be regarded as the fourth stage in the self-substantial development of the anatomical systems, although, properly speaking, it ranks as regards its value or worth upon a level with the three former circles, and resolves itself directly into three stages, which accord with its three systems. But since these stages are at the same time also classes; they should, for uniformity's sake, retain the latter name.

3142. The animal kingdom accordingly divides into four great districts or circles.

Circle I.   Intestinal Animals,   Oozoa—Protozoa.
II.   Vascular   " Sexual animals—Conchozoa.
III.   Respiratory " Cutaneous " —Ancyliozoa.
IV.   Sarcose   " ; Cephalic " —Vertebrata.

C. ANIMAL CLASSES.

3143. The animal Classes may be designated as the self-substantial representation of a stage in the development of an anatomical system or of the inferior organ of sense, and in the Sarcose animals of these systems themselves or of the cephalic senses.

3144. There are then as many classes as there are stages of development or systems. Thus the intestinal system separates into pharynx or stomach, intestine and absorbents.

3145. The vascular system into veins, arteries, and hearts.

3146. The respiratory system into branchial membrane or skin, into branchiæ, and into lungs or tracheæ, i. e. air-tubes. Taken in a strict sense, such divisions do not constitute classes, as has been already remarked.

3147. It is only the animal systems, which do not divide into several functions, but remaining upon a level with each other, are simply repeated in the higher organs of sense.

First Province. Splanchnozoa, Dermatozoa.

FIRST CIRCLE. INTESTINAL ANIMALS, OOZOA.

3148. The intestinal animals are nothing but depressed cysts. They rank therefore upon the lowest stage of development, and consist of mucus or granular nervous mass—Protozoa, Gelatinose animals.

3149. It is an established fact that, with animals as well as plants, the first function consists in imbibition or absorption, and the body must consequently be an absorbent cyst or a pharynx, which nevertheless takes up the food in the same way that the tegumental lymphatic vessels absorb. We can therefore style these animals pharyngeal or gastric, although the name is not perfectly correct—Infusoria. Then the intestine associates with the pharynx or stomach, so that they are Intestinal animals—Polyps or Corals.

Lastly the intestine sends out absorbents, and then the animal consists of pharynx or stomach, intestine and absorbents; such may be termed an Absorbent animal—Acalephæ.

The Intestinal animals therefore divide according to the developmental stages into three Classes.

Class 1. Gastric, Vitelline Animals.

3150. The lowest animals commence with the water, which has scarcely become mucus; they are nothing but drops, vesicles, which swim about independently—Protozoa or Primitive animals.

3151. The Protozoa correspond to the vitellus or the male semen, which is nothing but vitellus in a state of solution. They are the animal semen of the planet, the animal dissolved. Animal generation cannot take a deeper commencement or origin. The stone, which is decomposed into carbon mixed with water, can become nothing less than a point. They are the animal germinal powder. The fungus is a tissue of vesicles, which dissolves directly into seeds—fungus-powder or dust. Thus, they are ovaria or testes which have dissolved into seed, fluid testes—Vitelline, Seminal animals.

3152. The vitellus or semen is point-or nerve-mass dissolved. The Vitelline animals are sentient or nerve-points, which have combined all other processes in this identical mass. The divided point-mass ranks, however, in the signification of vesicular or cellular tissue. These animals are nervose cells.

3153. Nerve-cells must originate in every water, because every water is in a state of tension with the earth and the air; thus, dissolves the former and absorbs the latter. The water itself is a digesting and respiring mucus.

3154. The nerve-cells have an internal cavity, from the surface becoming oxydized and consequently converting itself into a denser layer, or into tegument. This, however, can only take place at the expense of the internal mass, as being that alone which abuts against the external parietes and becomes rigid.

3155. As the animal life is not simply a single act of rigidification, but a repetition of the same, alternating with solution, so must the Protozoon again replace the mucus-granules which have been disposed of from its interior; it must eat.

3156. It is a matter of indifference for the philosophy, whether the reception of food is effected by one or several mouths. There are Acalephæ and even Intestinal worms (Entozoa), which absorb by several mouths, almost like plants.

3157. But in the animal the mouth or mouths is or are definite organic openings, not merely interspaces or pores as in the plants; for they rank in the signification of the blossoms, or of composite parts.

3158. It can be therefore said that every animal hath a mouth or mouths, and consequently a stomach or stomachs.

3159. Their motions consist in abbreviations or straitenings of the cyst. Indications can scarcely be present of the secretions of higher organs, such as the intestine, vessel, branchiæ, liver, and such like parts.

3160. On the contrary, developments of the tegument and nerves may occur, the former as cilia which serve also as organs of motion and as branchiæ, the latter as ocular points or rudimental eyes; for both are none other than nervose tegument. The mouth of these animals is still passive, and subordinated to the water. It is surrounded only by cilia, which by the vortices produced by their movements in the water, impel or draw the latter into the mouth and with it the food. Such animals are called Infusoria.

3161. As the Infusoria are the seed or vitellus itself, so also are they the very ovum, and require no special sexual parts for propagation. They absorb, and are so nourished, and if there is a sufficient quantity of the mass to admit again of its division into several points, they thus divide. A magnified Infusorial animalcule has become, as it were, an ovarium or a testis, which then produces seed by dissolving itself into the latter. They are a constant conflict of the organ and its product, of the Solid and Fluid, a vitelline and orchitic process.

Class 2. Intestinal, Albuminous Animals.

3162. With the separation of the cyst into an internal and external layer, or into intestine and tegument, the animal must necessarily ascend a stage higher, since it now contains two systems different from each other, and is consequently a double Infusorium.

3163. The form of the Intestinal animals passes gradually from that of the globe into the tube—Tubular animals.

3164. They are tubular nerves surrounded by a tegument.

The cilia also are perfected and lengthen into filaments, which no longer perform simply vortical movements, but now actually seize and convey the food self-substantially into the mouth. Such animals are called Polyps. Their multiplication takes place no longer by division or fissure, but by ova and shoots, or ramification.

The oviducts or egg-tubes lie between the intestine and tegument, and open upon the margin of the mouth between the tentacula. In many instances the ovisacs, or cysts, hang also freely from other parts of the body, as in the Sertulariæ.

The sprouts or offshoots detach themselves from the parent and become independent animals; but they frequently continue to stand as ramules upon the maternal animal, though they nourish themselves independently of the latter.

3165. If the process of oxydation be augmented, then, the tube's external wall indurates, becomes coriaceous, and, lastly, ceratoid or horn-like in texture. The nerve-tubes or the animal proper can now swim no longer, since one kind only of motion is left it, namely, to protrude itself out of, and then retract itself within, the tube. It consequently falls to the ground, and while the external mucus hardens, it clings to the former; such are the Sessile or fixed Polyps.

3166. Sessile Polyps having coriaceous or horny tubes are called Plant-animals, Zoophyta, Phytozoa.

3167. The adherent, dried and dead external tegument of the Polyp is called stem. Such a ramified stem completely resembles a plant.

3168. These woody or herbaceous stems are not rooted in the earth, but have the power of adhering firmly to every substance, to stone, glass, shells, and such like bodies. They do not therefore draw in nourishment through any root.

3169. The ramification is often wholly plant-like in character, resembling that of a shrub with separate ramules, which even assume too the form of leaves, and the animal tubes that of flowers.

But frequently the ramules grow together also by their extremities, giving rise to a trellis-work, the production of which in plants is impossible. The soft animalcules, which come in contact, cleave unto each other, and grow together like wounded parts in the Sarcose animals.

3170. Upon the surface of the ramules or the leaves are apertures, out of which the mucous substance protrudes the radiated mouth. But these mouths are frequently, especially in the Cystic corallines, of two different modes of formation. The one kind are cysts without filaments, and contain ova, which are developed and fall out. The others have filaments, which move and do not produce ova. The former look like seed-capsules, the latter like flowers with stamina, while the entire trunk resembles a monœcious plant.

3171. With increased oxydation calcareous earth is deposited in the external tegument or rind, and the stem is converted into stone—Lithozoa, Lithophyta, Corals.

This calcareous earth contains the most general acid, or the carbonic, and thus oxygen combined with the inorganic carbon, while the bones contain phosphoric acid, oxydized gelatine.

3172. As the calcareous earth is, properly speaking, only a granular deposit in the tegument, like it is in the cartilages of the higher animals; so is it not to be regarded as a free worm-tube, but as the body itself. Meanwhile it forms a tube open above, from which the mouth of the animal projects.

3173. As the animal ramifies, so also do the stony tubes increase, and there originates a phytoidal or plant-like stem, but one consisting of a stony mass.

3174. Thus the Coral is the earth-animal, and indicates the globular or osseous mass under its first grade of formation in the animal kingdom.

But there are also Polyps, whose stem only originates through saccular inversion of the upper portion of the animal's body; yet this is only distinct in the soft stems. In most of the species, where a separate intestine is to be found, it is probably only the upper part of the body which is so inverted. Meanwhile there are some, whose intestine forms a circle by returning upon itself, and opens into an anus.

3175. If then the Infusoria are the vitellus or seed of the animal kingdom; in like manner the Corals are its ova. The carbonate of lime is the shell surrounding the albumen, while the animal or intestine is the vitellus—Albumen-animals.

3176. Nature forms these living ova, when she takes vitellus and albumen out of the sea-slime, invests them with a shell derived from the earth, and hatches them, after they have been vivified

by sun and air.
Class 3. Absorbent, Involucral Animals.
3177. Did the former animals remain in the condition of ova, from want of a perfect vascular system; they are next developed, so soon as vessels appear and form a vascular plexus, into fœtal involucra or envelopes. These animals are vitellus with the vascular membrane.
3178. The Absorbent animals are no longer simple vesicles, but large cysts or capsules like the developmental envelopes of the fœtus, along with a choroid plexus—Involucral, Fœtal-animals. This choroid plexus does not, however, consist of arteries and veins; but is only a ramification of the intestine, so that the vessels are of a lacteal character—Absorbent animals.
3179. In these animals there is no longer any egg-shell, but everything has been taken up into the galvanic circle; the shell has itself become organic and life-imbued. Their substance is still mucous or albuminous; they are still vitellus, though converted into a vascular tissue.
3180. They therefore cling firmly nowhere; but swim about freely, like brain-masses converted into radiated cysts.
3181. Free Mucus-animals, traversed by vascular plexuses, are Acalephæ.
3182. There are Acalephæ which are simply air-sacs, like the air-sacs of ova, to which hang ramified vessels as absorbent tubes—Cystic, Tubular Acalephæ.
Others represent hemispheres with numerous absorbent tubes, which concur in the middle of the animal to form a kind of stomach, from which again other tubes pass towards the border, in order to elongate into tentacula. Thus the absorbent vessels have become motor and sensitive organs.
Besides this, many have around the mouth four large lobes, which must be viewed as the antetypes of the sentient lobes of the Bivalve Mollusca.
Lastly, others have a true mouth, which leads to a similar gastric cavity, out of which the same vessels emerge and ramify. Both kinds are called Hutquallen or Acalephæ.
There are yet others having the same structure, but oviform, with respiratory lamellæ upon the absorbent tubes, Rippenquallen or Beroes.
3183. An Acalephan is a brood-egg, which swims freely about without a shell.
3184. The vessels are quaternary in their distribution, and form a cross, like the involucral vessels of the chick.
3185. In these animals the ova first begin to be detached and agglomerated together in definite situations so as to constitute ovaria. The number here is also four.
They lie usually within four cavities surrounding the stomach, and into these wide apertures situated near to the mouth lead; they are at the same time regarded as respiratory cavities.
In other species vesicles, wherein seminal animalcules are developed, occupy the same situation. Here therefore is recognised for the first time a separated sex.
In the Röhrenquallen or Physaliæ the ovisacs mostly depend externally in the form of bells.
Besides this we find in the Acalephæ all sorts of laminæ, which are probably organs of respiration.
It is not known what is the meaning of the air-sacs and of what the air consists.
Most of the species emit light like globes of fire, just as many Infusoria do also. It is probably a phosphorescence given out by the mucus when passing over into a state of decomposition.
Very many have also the stinging property of the nettle; but, whether the cause of this is chemical or mechanical, is not yet exactly known.
SECOND CIRCLE. VASCULAR, SEXUAL ANIMALS.
3186. So far, or in the ascending scale up to the Acalephæ, the animal is only viscus with an absorbent canal, which is at the same time a canal of evacuation, without a distinct intestine

being set apart for that purpose; such is the general rule.

3187. After the Acalephæ the formation undergoes a change; the distinction between the exterior and interior is prominently displayed, and the internal wall becomes freed as a separate and perfect intestine along with a mouth and anus; the external wall appears as a free tegument. But, seeing that two concentric and separated cysts could not subsist without combination by means of the nutritive system or the vessels; a perfect vascular system is formed, divided into veins, arteries, and hearts.

The tegument, wherein the vessels become self-substantial, is the branchial membrane. There is therefore placed around the intestinal body a vascular body also, or branchial membrane, which is consequently pleura or mantle (pallium), as in the Mussels. The body of the intestine consists of intestine and peritoneum; that of the vessels of branchiæ and pleura or mantle.

These animals are hence bisystemic, being both Intestinal and Vascular animals; but, since the vascular system is a new addition, it is thus characteristic of the circle, and its members must be therefore called Vascular animals. With the vascular system, however, all its further developments have been bestowed; thus especially the complication of the vessels with the intestinal ramifications or, in other words, the liver—Hepatic animals. The salivary glands also are a similar complication, and in this series therefore they make their appearance.

Lastly, the renal glands are such a kind of vascular organ, or branchiæ of the sexual parts; they also begin to be astir in this series.

With the separation of the systems into separate teguments or membranes, the sexual parts also separate. The ovarium becomes an independent organ furnished with its own excretory ducts; the male parts are individualized to form veritable testes furnished with excretory ducts, or even with a penis. Still this is all effected only by degrees, or as yet within the confines of this circle.

These animals divide according to the viscera into Venous, Arterial, and Cardiac animals; according to the sexual parts, into Ovarial, Orchitic, and Renal animals.

Class 4. Venous, Ovarial Animals.

3188. With the protrusion into view of the vascular system, the veins are chiefly developed along with their principal organ, the liver, as being the bond of union between the circulation and the intestine.

3189. The animals, which first bring to bear in addition to the intestine a liver, are the Bivalve Mollusca.

3190. With the veins arteries also originate, but with a preponderance of venosity or the venous character. The blood is lymphatic, colourless.

3191. The cardinal venous organ or the liver evokes also into existence a corresponding organ of respiration, namely, free branchiæ with the tegumental form—branchial laminæ or leaves.

3192. In the middle between the branchial laminæ and the liver, the heart is evolved; this organ comprising always a ventricle with an auricle, but consisting of a venous, membranous tissue almost devoid of fibres.

3193. The first heart is in other respects arteriose; it receives the blood from the branchiæ and transmits it to the liver as well as to the remaining parts of the body, from which it proceeds directly to the branchiæ without entering any venous or right heart.

3194. In these animals, as is well known, four branchial leaves lie externally on the belly, which includes the intestine with a large liver, and hangs together with the branchiæ as a separate purse within the mantle.

3195. In the Mussel a structure originates for the first time, which can be compared with a

thoracic or pectoral cavity. What covers the branchiæ, must stand in the signification of the thorax or chest. The pallium or mantle of the Mussels is pleura.
3196. Their shells are branchial opercula (as in the Fishes). They are secretions from the mantle, and everywhere accompany the branchiæ.
3197. The locket or hinge of the shell-valves corresponds to the rachis or spina dorsi, as is especially distinct in the Teredines or "ship-worms."

The shells of Bivalve Mollusca are a calcareo-thoracic box, open in front, inverted behind, and moveable like ribs.
3198. The two sphincter or occlusor muscles signify shoulder and haunch.
3199. These animals begin for the first time to exhibit bilaterality or symmetry, because in them there is stirring the idea of the osseous formation. From the branchiæ being situated symmetrically upon both sides, the cardiac chambers are also symmetrical.
3200. The pectoral tunic (mantle) usually elongates at the anal extremity into two tubes, respiratory tubes, through which the water is drawn in and thrust out or expelled. Such is the case in the highest Worms or Holothuriæ, only there the respiratory tube leads into the body itself. A similar arrangement is found in the Echini. Many aquatic larvæ from all kinds of Insects or out of different classes breathe through anal tubes. All these animals consequently repeat the Mussels, and this formation admits of being followed out up to Man, where it is left as allantois and primordial kidneys. The thorax of the Mussel thus opens by the anus. But, since the thorax is that which is here preponderating and is nearly the whole animal, so does the anus open into the posterior respiratory tube.
3201. The Mussel can be regarded as an animal consisting of three cysts or sacs inclosed within each other, namely, of intestine, around this the ventral pouch, and around this again the thoracic sac or mantle. If, moreover, the shells be regarded as a cyst, the animal then consists of four sacs. The heart and branchiæ lie within the thoracic cavity; intestine, liver, and ovarium within the abdominal or ventral cavity. The Mussel is thus a doubled Acalephan.
What is termed foot in the Bivalve Mollusca is nothing else than the abdominal tegument dilated in front to form a muscular ridge.
3202. The Mussels are embryos, in whom the liver originates, and whose chorion acquires a placenta. Then again, as the embryo is nearly all liver, and hangs surrounded by a watery fluid within the widely-expanded chorion and amnion, so does the abdominal pouch hang within the pallial cavity, or in the wide, water-full chest.
3203. In the abdomen there is still only the ovarium, and that indeed very large. There are properly two ovaria, each of which, according to my observations (vid. Göttinger Gelehrte Anzeigen, 1806. Stück 148), opens laterally under the shoulder-muscle and gives exit to the ova, which then enter within or between the branchial fringes, in order to be there developed.
3204. Here the respiratory organs are still at the same time a kind of uterus. The ova may be oxydized therein, like the embryo is in the uterus. These branchiæ are probably still to be compared with, or designated as, sexual branchiæ.
3205. In the back of the shoulder these animals have a highly vascular organ with two excretory passages, which open near to the orifices of the oviducts. I formerly regarded it as a kidney, but, according to more recent observations, it should be a testis. In the anterior part of the foot there is frequently situated a gland which ejects a gelatinoid moisture, which hardens into glutinous filaments—the byssus or beard. It is probably a memento of the tentacula of the Acalephæ. There is besides also an organ in the feet of many Bivalve Mollusca, which occasionally squirts a

watery juice out to a great distance. I have occasionally found this organ in our Teichmuschel. Its signification is probably the same as that of the preceding organ.

3206. The organization of the Bivalve Mollusc can be thus most distinctly described; it is an abdomen, wherein is an intestine with mouth and anus, a liver and a double ovarium; upon the sides of this abdomen are situated the branchiæ in the form of four laminæ; around the branchiæ and the abdomen is the pleura or mantle, which is always open posteriorly.

3207. The mouth is placed directly at the anterior extremity of the abdomen, is devoid of neck and head, as also salivary glands; it is consequently not a true mouth, but only a pharyngeal aperture. Thereupon, however, are situated four sensitive lobes, which are in structure exactly like the branchiæ—cephalic or pharyngeal branchiæ. They are the further formations of the four arms of the Acalephæ.

3208. The Mussel has a perfect splanchno-neural system with ganglia and a pharyngeal ring, which probably corresponds to the nerves that sweep around it.

3209. The Mussel has no other organ of sense than that of the passive sense of feeling, the tegument. It cannot once move its sensitive lobes voluntarily; it has no lips.

3210. The abdomen only elongates in most of them to form a moveable, variously constructed foot or keel, which cannot, however, creep but only push. The progression of the Bivalve Mollusca is backwards, as in the Acalephæ. In the Snails the ventral surface first becomes a creeping sole.

3211. The Mussels repeat the Infusoria; are Infusoria with a bivalve calcareous testa or shell.

Class 5. Arterial, Orchitic Animals.

3212. In the preceding class of Bivalve Mollusca, it is in truth the abdominal viscera only, such as the intestine, liver and ovarium, which have been perfected; and then the veins and arteries with a membranous heart. The cephalic organs, eyes, maxillæ, salivary glands, and even moveable lips with tentacula, are wanting, as well as the muscular heart. Lastly, the arterial sexual system or a self-substantial testis and the penis were absent.

Now, Mollusca, which have eyes, maxillæ, a muscular heart and a ventral sole or foot, with salivary glands and a penis, are Gasteropoda or Snails.

3213. The Snails possess salivary glands, a trace of the tongue and of maxillæ, moveable lips and tentacula, with thus an approximation to the head, unto which the eyes are rarely wanting—Salivary animals.

3214. With the development of the head or rather of its inferior organs of sense, the antagonism makes its appearance also in the ovarium. A half of the ovarium is converted into the testis. The Snail is therefore a Mollusc, which is female upon one side, male upon the other.

3215. The androgynous or bisexual animal is, as a general rule, asymmetrical.

3216. The mantle also or the branchial cavity obeys this want of symmetry. The branchiæ of one side dwindle down; those of the other turn with the mantle towards the head, and the respiratory aperture occurs upon the back.

3217. With the one-sided evolution of the mantle, one shell also is only developed, while the other is stunted or placed under arrest. The Snail's shell is one of the Bivalve Mollusc's shells, its operculum is the other. This last is stony, horny, and finally is entirely wanting.

It is remarkable that the right shell has been pretty generally perfected, while the left dwindles down into the operculum; all the Snail's openings are therefore upon the right side, such as the anus, with the orifices for the escape of the ova and semen.

Male animals are situated on the right side, female on the left; or where the right side was predominating, there originated the male sex, where the left, the female.

3218. As the orifice of the mantle and the shell is properly the aperture of the branchial foramina; so can it be said, that the Snail were a Mussel, which does not simply extend the foot, but also the mouth or head to the branchial opening; it is a Mussel reversed.

3219. Every thing else in the Snails is arranged in accordance with these fundamental organs and forms.

The cephalic portion of the intestinal canal is indicated also by muscular fibres. The pharynx and the mouth can contract and expand, seize and bite off; the former frequently admits of being protruded and retracted as a fleshy proboscis with perforating maxillæ.

3220. Since the muscular fibres are only tegumental fibres, and therefore lie within cavities; they thus act like as in the feet of Insects.

The tentacula of many Snails are moved like the feet of Insects; but as they are not horny, but soft, they are turned either inside out or "vice versâ." The oviduct and seminal duct or penis follow the same mode of formation. They are likewise everted and inverted.

3221. These members of the Snail are true Insect-limbs that have remained soft, and are thereupon susceptible of inversion and eversion. Were an Insect's foot soft, every one will admit that it would then yield so as to become inverted, if the fibres pulled upon it. The limbs of Insects have thus only become stiff, and are thereby Snail's horns that resist inversion. All these members are teguments, and give the lie only unto limbs; for it belongs to the essence of a limb, that it be dense.

3222. That which suffers eversion or its converse is no limb, but only a sheath, a prepuce. Nearly the whole Snail is but a prepuce, a "membrum virile" or "verge."

3223. There is no class of animals, in which the testes and penis are found so disproportionately developed as in the Snails—Orchitic, Penes-animals.

3224. The vascular and nervous systems are related pretty nearly as in the Mussels. But the heart is fleshy and has, by reason of its unilateral or single branchia, only one auricle also.

3225. The Snails repeat the Corals in the cylindrical form of the body, the tubular-shaped shell, and the return of the intestine upon itself towards the mouth, as in many Corallines.

3226. In them also the organ, which virtually corresponds to the kidney, appears to be astir, namely, what has been called the calcareous gland, situated within the branchial cavity, and which opens not far from the anus.

Class 6. Cardiac, Nephritic Animals.

3227. Hitherto there has been found only a single "cardia," namely, the left or arterial heart, which, receiving the oxygenated blood from the branchiæ, propels it to all parts of the body—for the purposes of nutrition. But now also the right heart suddenly makes its appearance, which drives the blood into the branchiæ—for oxygenation. This must doubtless be regarded as a higher development, especially since, as I have shown, in the higher animals as well as in the embryo, the right heart is first perfected subsequently to the left.

The Conchozoa with a double heart are the Sepiæ or Cuttle-fishes. These are in consequence correctly called Cardiac animals, and must be regarded as the fundamental form of this stage. With this completion of the heart other structural changes, which probably stand in intimate connexion with it, appear. In the Snails indeed a kind of renal organ, or what has been called the calcareous gland in the branchial cavity, has been already shown to exist. Whether the shoulder-gland in the other Mollusca belongs also to the members of the present class, may be left undetermined. This kidney, like other vegetative organs, pours out its secretion quite involuntarily. In the Cuttle-fishes it is, however, combined with an organ, by means of which it can discharge its fluid, or the ink, voluntarily as in the higher animals. The kidney or ink-gland with the ink-sac is therefore characteristic likewise of the Sepiæ, and admits of our calling these creatures Nephritic or Renal animals.

At the same time the whole form of the body is altered; it becomes cylindrical, and is provided upon the ventral aspect with neither a muscular keel nor sole, whereby it might glide or shove itself along; thus herein also a resemblance is manifested to the higher animals.

This motionless body is on the contrary endowed with independent organs of locomotion, namely, with fins or arms, which are wanting in the Snails and Bivalve Mollusca. The labour of motion is consequently removed by the organs which minister unto it from the body, and thus we trace another similitude to the higher animals, which transport their body from place to place by wings or feet. These animals can be therefore called Cylindrical Snails, out of contrast to the Sole and Carinate Snails or the Bivalve Mollusca.

Closely allied now to these cylindrimorphous Conchozoa are all those Mollusca in whom the keel (carina) or sole is wanting; i. e. such snail or mussel-like animals, which either cannot transport themselves at all, or effect this by means of fins or podoidal appendages, in short all that have been called Snails, having no sole, and all supposed to be Mussels, but without a keel. Now, the Heteropoda or Pterotracheæ and the Pteropoda or Clionæ have projecting processes which act like fins.

The Brachiopoda or Terebratulæ, though unendowed with the power of locomotion, have, as their first name implies, arm-like organs.

The Cirripedia or Lepades, being likewise without the power of continuous motion, possess podoidal or foot-like appendages. In both these families, however, these appendages are nevertheless true motor organs; for by means of them they seize their food, an act which hitherto has been seen to take place neither in the Snails nor Bivalve Mollusca.

Lastly, the prehensile organs dwindle down into mere filaments or small lobes, but the body always retains its cylindrical form without keel and sole, as in the Ascidiæ or Meerscheiden. In the Salpæ only the cylindrical body is left, unto which appendages are still, and not rarely either, attached, serving them wherewith to lay hold of each other. In accordance consequently with their external form all these animals appertain to one class along with the Cuttle-fishes. I call them Kracken.

The renal organs are still not to be found in all of them; the structure alone of the heart agrees essentially, despite of its simplicity, with that of the Cuttle-fishes. In the Ascidiæ, simple as well as compound species, it is indeed but a single, though muscular pouch. Now, this pouch drives the blood alternately, at one time into the branchiæ, at another backwards into the body, and is thus in the first act a venous or right heart, in the latter an arteriose or left; consequently, according to function this single pouch is compounded or made up of the two hearts.

The branchiæ deviate entirely from those of the Mussels and Snails, and exhibit a very complicated structure; in the mussel-like families they present the form of a trellis-shaped sac, as in the Ascidiæ, or that of filiform appendages to the feet, as in the Cirripedia, or they are funiform as in the Brachiopoda. In the Pteropoda they are very varied; in the snail-like Heteropoda they are mostly pectiniform or tuft-shaped; in the Sepiæ or Cephalopoda phylliform or fin-shaped.

The external character ccommon to the whole class is the cylindrical body, unto which may be aptly conjoined the possession of special organs of motion; whether such organs be prehensile in function, media of support, or veritable instruments of progression.

The snail-like Pteropoda, Heteropoda, and Cephalopoda, have a head, but this is wanting in the mussel-like Salpæ, Ascidiæ, Cirripedia, and Brachiopoda.

THIRD CIRCLE. RESPIRATORY, CUTANEOUS ANIMALS.

3228. Respiratory animals are Dermatozoa with a predominating system of respiration.

Now, the respiratory system is the tegument, which here then must attain the highest pitch of perfection.

This again takes place through the increased process of oxydation, which produces induration of the parts.

The vessels, which surround the tegument, must nourish this portion of it to a greater degree and render it harder or more compact than any other, whereby alternating expansions and contractions thereof originate, in other words, the structure of the trachea.

3229. The whole body of the Respiratory animals becomes a trachea, a series of rings. The Respiratory are therefore the Annulate animals.

The rings are to be regarded as cysts, which are mutually juxtaposited or repeated, not being inserted within each other as in the preceding animals, but behind each other, so that the present circle consists of multiplied Cystic animals.

3230. The Annulate animals must represent the developmental stages of the tegument, while the viscera make a retrograde step—Tactile animals.

3231. The tegument has two functions: it is an organ of respiration and of general feeling or sensation. In both cases it passes through three stages of development.

It is either entirely a branchia, i. e. reticular in character; or the branchiæ become individualized in certain situations and partly horny, constituting branchiæ proper; or finally, the tegument is converted into an air-breathing organ, tracheæ or air-tubes.

3232. It either feels with its whole surface, or by papillæ; or the branchiæ change into horny feet; or lastly, into alary appendages or wings. The Ancyliozoa consequently divide into Reticular or Papillary animals; into Branchial or Pedal, and into Tracheal or Winged animals.

3233. As the arterial character preponderates in these animals, so do the venous organs, especially the liver and renal structures, retrograde.

3234. The body is now an intestinal and tegumental body with predominating respiratory and sentient organs; here therefore sensitive papillæ, antennæ, feet and wings, appear in abundance.

3235. With the retrogression of the viscera, the glandular structure also and large proportional bulk of the sexual parts disappears. They assume the form of the intestine and tegument, i. e. become tubes.

The spawn like the milt consists, as a general rule, of only two long membranous tubes, running near the intestine.

3236. The external sexual parts belong to the sense of feeling and follow the developments of the tegument; they make their appearance as antennæ, and are usually accompanied by foot-like accessory organs.

3237. The nervous system in like manner accompanies the tegument. It consists of two gangliated cords passing along the ventral surface.

3238. As the maxillæ are only repeated feet, so they are rarely wanting, and are, like the latter, disposed after the fashion of scissor-blades. The same holds good of the antennæ; they are mostly in pairs, and consist of a series of rings projecting from the head.

3239. The eyes are usually present, but only as tegumental organs, or placed upon the points of the antennæ.

Class 7. Reticular or Papillary Animals.

3240. The repetition of the Intestinal or Vascular animal, taking place under the dominion of the respiratory system, must be accompanied by the vascular system. The tegument is a vascular, a branchial membrane, a branchial network or skin.

3241. A body with predominating tegumental and intestinal system, is cylindrical. It can only become gross or lump-shaped, if the glandular viscera, such as liver, ovarium and testes, get the upper hand or prevail.

3242. A tegument which appears as a branchial membrane, is annulate or ringed.

3243. An animal, having an annulate respiratory membrane without annulated feet, is a Worm. Upon the lowest stage the vascular system is still similar to that of the Vascular animals; its blood is a colourless fluid—White Worms, Entozoa.

In order to be a Worm, it is sufficient to have an annulate tegument. Even if the intestine be wanting and the tegument supplies its place; the character is nevertheless complete. Here it is the skin, which digests, while in the Intestinal animals it is the intestine, which digests and breathes. Since it is here the tegument, which undertakes the offices of the whole body; the vessels, liver, and salivary glands are wanting unto the intestine.

The sexual parts are also frequently arrested and the ova appear to originate in the tegument.

Many undergo division without injury to themselves; yet the detached portions do not become again entire animals.

In most of them, however, the sexual parts of both sexes are present, being associated with, though separate and distinct from, each other; those of the female mostly opening upon the sides of the body.

As being Dermatozoa, they have for the most part special organs of sensation around the mouth, under the form of papillæ, filaments, spines and cups. The nervous system, wherever it is to be met with, consists of a ring surrounding the pharynx, and of a double ganglionic cord lying along the ventral surface of the body.

3244. Upon a higher stage of development the arterial system gains the preponderance, and the blood becomes mostly of a red colour—Red-blooded Worms.

3245. With the arterial the fibrous system also makes a more decisive appearance. The tegument is a fibrous tunic—it is itself an artery. All Annulate animals with a fibrous tunic, which can consequently contract, belong to this class, whether too they have red blood or not, like the Holothuriæ and Star-fishes.

3246. The Lumbricales or Earth-worms and the Hirudines or Leeches respire obviously through the whole tegument, though a special organ of respiration does begin to be evolved, in the first family in the "saddle," in the second in the lateral vesicles or cysts.

3247. In others the branchial vessels make their appearance above the level of the tegument as filaments or ramules, and are arranged in two rows, as in the common Sea-worm and in the Nereides.

3248. Lastly, they accumulate about the neck or head, as in the Amphitrites and Serpulariæ.

3249. There are also Worms which respire only through the intestine, its vascular system being bathed on all sides by water, as in the Aphroditæ. This water becomes, as it would appear, simply imbibed by the tegument in Thalassema, but through an opening at the anal extremity of the body in the Holothuriæ.

3250. These Worms have also no liver, or at the very best only traces of that organ, by reason of the preponderance of the arteries.

3251. Papillæ or sensitive filaments gradually sprout from the external branchial filaments, lie along the sides of the body, and are the prelude of the feet; as in the Nereides—Papillary, Filamentary animals.

3252. Others of these filaments become horny and appear as setæ or bristles, placed somewhat similarly to the above, as in the Earth-worm.

3253. Mouth and head are more perfectly developed than in the Entozoa. The first can readily expand and contract, frequently protrude the pharynx like a proboscis or tube, and not unfrequently has manducatory pincers, like as in Insects.

3254. On the head there are mostly annulate tentacula with muscular fibres, and frequently simple eyes.

3255. In those species which draw the water itself into the body and use it for the purposes of respiration, the formation of the mouth has risen higher, and the pharynx been provided with maxillæ, in number five or ten—Echini, Holothuriæ. These maxillæ form in themselves a peculiar skeleton around the pharynx, which is ranged circularly instead of by pairs.

3256. The nervous system is directed according to the relations of the tegument and intestine. It forms two ganglionic cords running along the ventral side of the body, and, where it makes with the nerves going to the maxillæ a ring around the pharynx, corresponds to the pharyngeal or pneumogastric nerves.

3257. In reference to the sense of feeling it may be said, that the Worms were those among the Sensitive animals, which feel by the whole tegument or body. Their body itself is a tentaculum.

3258. The sexual parts are likewise intestini-and dermiform, not glandular in figure like the ovarium and testis of Mussels and Snails, but tubular, as in the Entozoa. There are usually two oviducts and two seminal tubes to be found.

3259. So far as we are acquainted with the Red-blooded worms, they are androgynous, at least as regards the Earth-worms and Leeches, and their sexual parts are indeed tolerably symmetrical; yet they do not open posteriorly, but far forwards on the ventral side of the body, this being the case with even the male sex-organs.

Class 8. Branchial, Pedal Animals.

3260. The representation of the Cutaneous animals is not yet the entire completion of the tegument. It attains a higher grade by separating in accordance with the visceral systems, and in the conversion of the branchiæ into sensorial organs or feet.

The branchiæ endeavour to separate self-substantially, and form a body of their own. The branchial body is the thorax.

By this attempt, the abdomen originates "per se" as the body of the intestinal and sexual system. The branchiæ upon it are arrested or stunted. In like manner the body of the nerves separates itself from the thorax and becomes head.

But meanwhile this separation is only imperfect; the head and thorax, if otherwise distinguished by their size and the feet, are for the most part connate; abdomen and tail are usually fused together, and only to be recognized as distinct by the limbs or members.

With this greater division of the body each division of it obtains its articulation into rings; but they will be hardened by virtue of the separation of the branchial network, or will become horny—Horny animals.

3261. With the conversion into horny texture of the body, its branchial filaments or tentacula must undergo a similar change, and divide likewise into rings like the general or bodily tegument.

Annulate, horny branchial filaments, are feet—Pedal animals.

A part of the branchiæ remains hanging to the feet, mostly to their basal joints, and is supported by animal motor members or limbs; the respiration becomes voluntary—Branchial animals.

The Annulate animals, having branchiæ and feet, are the Crustacea or Crabs.
The branchiæ and feet are, of course, most largely developed upon the thorax.
Those upon the abdomen continue small, and those upon the tail or sexual body dwindle in size and change into other organs, such as fins, sacs, filaments, pincers.
3262. The number five of the branchiæ is also exhibited in the feet. There are mostly five large pairs of feet on the thorax, and as many small ones on the abdomen, as in the Crabs.
On the tail they appear more dwindled in size, and are frequently reduced to a smaller number.
Each part of the body has properly five rings; and these parts being the thorax, abdomen and tail, there are thus fifteen rings in all.
In the Crabs it is distinctly demonstrated that the maxillæ are nothing else but feet; there are therefore for the most part, after taking into account sundry arrests and coalescences, five pairs of maxillæ also.
The Crabs have usually two pairs of antennæ, whose signification is unknown; one pair is probably the elongation of the ear.
All Branchial animals have eyes, frequently too supported upon articulated pedicles.
They have a double nervous cord upon the ventral surface of the body.
They have, like the Mussels, a heart upon the dorsal aspect, and arteries with veins.
The intestine opens into the apex of the tail, and is surrounded by a liver.
The sexual parts still open for the most part on the thorax, and indeed by two orifices, as in the Worms.
There are no more androgynous or bisexual beings in the present class.
In some few the branchiæ already enter the body and become air-tubes, as in the Scolopendræ, Spiders and Scorpions.
The Branchial animals repeat, as being the second class of their circle, the Corals and Snails; their "quasi" coat of mail is therefore harder, often richer in calcareous ingredients, and, in addition to this, lies frequently upon the thorax or branchia as a special testa, called the scute or shield.
Class 9. Tracheal, Alary Animals.
Annulate animals, whose branchiæ have undergone a partial conversion into tracheæ and into wings, are the Insecta proper or Flies.
3263. A Worm with feet, tracheæ and wings, is an Entomon or Insect.
3264. The first separation takes place in respect to the three tegumental segments of the body, the abdomen, thorax, and head. All three are, in the Insects, more separated from each other than in the Branchial animals, and united together usually by a narrow tube; even in cases also, where they are connate with each other, they are still easily recognized by respect being had to size, form, or appendages.
Every Insect is divided into three segments. In the abdomen are the organs belonging to the Worm, such as the intestine, and a fatty body which appears to be an analogue of the liver, a dorsal vessel, tubular sexual parts and air-tubes (tracheæ), but nothing else.
The abdominal feet now disappear entirely, and even the number of thoracic feet diminishes, owing, doubtless, to the production of wings.
3265. The thorax alone is reserved for the limbs or locomotive members. It never carries more than three pairs of legs and two pairs of wings. Of the viscera, it contains nothing but the

pharynx, while in the Crabs, important intestinal organs and even the liver, are situated within its cavity. It is in Insects therefore nothing else than a medium of support to the respiratory organs which have become limbs. Hence the thorax never has more than three rings, namely, one for each pair of legs. The wings invariably stand upon its two posterior rings.

3266. Since the limbs of the Insect are only the lateral filaments of the Worms, which have become hard and consequently hollow; they are as such not to be termed true feet, but are only to be compared with branchial arches or ribs; a step by which their greater number also admits of being understood. They are not to be compared with our feet, but unto the toes, which have been separated from each other as far as the rings of the body. The Crab has properly five thoracic and five abdominal toes. All its thoracic feet taken together are only equivalent to our hand.
The feet of the higher animals are only Insect-feet connate or coalesced.

3267. In other respects they already typify or prefigure true limbs, as well from their position as by the division of their joints. A perfect beetle's foot divides exactly like the limb of Man, into femur, patella, tibia, tarsus, and phalanges. These parts of the leg must not, however, be so absurdly divided and named, as has hitherto been unfortunately the case in our systems, where the femur has been called coxa, the patella trochanter, the tibia femur, while the toes have been lumped under the name of tarsus. (Ed. 1st, 1811. § 3087.) The regular number of the toes or tarsal joints is five, so that they correspond to our digital phalanges, to the metacarpal, and the anterior carpal bone.

3268. The wings are the branchiæ of the Mollusc that have been set free; they are placed therefore upon the back and are four in number. In many Insects there is still a pair of wing-like scales in front of the four wings, as in some Lepidoptera. They perhaps correspond to the shells of Molluscs, are branchial opercula.
It is only from this point of view that the structure of Insects admits of being fully understood; and apart from this it is absolutely devoid of all analogy. Thus it is possible for only six legs to take their origin from the thorax in the direction downwards, and nevertheless for there still to be wings upon the dorsum or back. The wings of Birds are by no means homologous to the alar appendages of Insects; they are, as is well known, the anterior members themselves, and there is no longer therefore in the Bird any feet attached to the thorax below, as in Insects. Besides, if the wings did not signify arms in the Bird, then it must have four legs. Thus in the Insect the wings could not also mean feet.
Their structure also speaks in favour of our view of the Insect's wings. They are known to be completely traversed by respiratory tubes, are true, or only desiccated, branchiæ—aerial gills. (Ed. 1st, 1811. § 3088.) Wings and feet are dependent from the same ring of the body, and are thus like the branchiæ and feet of the Crabs. Let the Crab's branchiæ elongate and dry, and they will thus be wings.

3269. Since the wings are newly eliminated organs of the sense of feeling, so are they here characteristic of the animal, and are consequently of greater importance for the purposes of division and arrangement than the organs of the head, which in all the lower animals is only an apparent head, and cannot therefore serve to characterize groups, &c.

3270. That the tracheæ in Insects have been developed out of the branchiæ by saccular inversion, is a fact displayed in a particularly distinct manner by the Scorpions and Spiders, who still possess at bottom internal branchial laminæ, unto which, however, air instead of water finds its way. It may be said, that with the general conversion into horny texture, the arteries of the Mussels were transmuted into internal tracheæ, and the branchial flabellæ into external organs.

Would we interpret the body of the Insect in a strictly philosophic sense, the parts must then receive very different names to what they now bear. Properly speaking, in our own species the thorax has no limbs, but only the neck. The upper limbs are not lungs, but branchial organs, and it is the cervical vertebræ, from between which nerves are sent off to the arms, for it is just upon the neck also that the branchiæ have been left. What is termed therefore the thorax in Insects, would be properly their neck. Their abdomen would consist accordingly of thorax and abdomen, and it is this also, which carries on the principal share of the respiratory process. It therefore consists generally of ten rings, and has ten pairs of spiracula, namely, twice five, for both thorax and abdomen.

Or again, the abdomen may be regarded as an intestinal and sexual cavity, and the thorax may be left with its own name. In that case there would be five spiracles for the sex, five for the intestine, and perhaps only two for the thorax.

Would we proceed yet further; the head can then only be viewed as pharynx, and consequently as neck.

3271. In most Branchial animals that live in the water, a perfect circulation is present, because, by reason of their feeble amount of respiration, all the blood is not consumed. This is also the case in the young of air-breathing Insects, so long as they may have to grow. But when they have ceased to do this, so strong an amount of tension emerges in the circulation, owing to the increased respiration of air, that blood remains but in scanty quantity to be carried back by the veins, and the arteries now for the most part convey the air in a pure state, namely, uncombined with blood, as in the higher animals.

3272. As the air-tubes pass to all parts of the body, like the arteries whose place they now supply, so does the nutritive juice become everywhere oxydized and converted into parenchyma or tissue. The nutritive juice doubtless transudes at once through the intestine and penetrates to all parts, as in the Plants.

3273. Of the vascular system nothing at last remains but the dorsal vessel, whose ramifications appear to disappear entirely. According to its analogy with the Crabs, Scorpions, and Spiders, it is the aorta. It appears, as if in Insects the circulation dies off from the living body.

The whole Insect is an air-organ, an aero-vascular stem. All the organs respire directly, such as the intestine, the motor fibres, nerves, sexual parts, and wings. There is no part unto which tracheæ do not pass, and in this respect completely resemble the arteries of other animals.

3274. The intestine is invariably furnished with an anus, and that too situated quite posteriorly. It is usually expanded into several stomachs and has appendages, almost as in Fishes, which virtually correspond to the abdomino-salivary or pancreatic gland.

3275. The salivary ducts open into the mouth, and thus as in the Snails and Kracken, whose similar structures they repeat.

3276. Whether what have been called the biliary vessels are what the name implies, and really convey bile from the fatty body to the intestine, or whether they are lacteal vessels and discharge their fluid near the rectum into the dorsal vessel, does not yet admit of being determined. They have been thought to constitute an urinary apparatus; only certain cysts which occur in connexion with the sexual parts, appear rather to correspond to the latter.

3277. The nervous system consists, as in the Worms, of two ganglionic ventral chords.

3278. In the head, the feet again, and probably also wings, undergo repetition. This repetition is nowhere so distinct as in Insects. He who can still entertain any doubt of the maxillæ being arms, let him resolve to descend into the Insect-world, and he will become in place of a sceptic, a believer. (Ed. 1st, 1811. § 3095.) What have been called manducatory pincers move outwards as well as the legs, and seize like arms, are only arms. Their gripe takes place in the lateral or horizontal direction, and resembles that of scissors or shears.

3279. Where, moreover, there are three pairs of legs on the thorax, we find also three pairs of maxillæ, namely, an upper and lower maxilla, and a labium or inferior lip, which consists of the same parts as the maxillæ, only they are united by tegument.

3280. Where there are five or more pairs of legs on the thorax, as in the Crabs, there also are found as many pairs of maxillæ.

3281. Upon these maxillæ, palpi are likewise situated, which are probably nothing else than what has been termed tarsus, thus repeated in the head also. They are only arrested upon the upper maxillæ or mandibles. (Ed. 1st, 1811. § 3096.)

3282. The wings appear to be repeated upon the head as antennæ. Thus the head also is in the Insects a perfect trunk.

3283. Upon the head there is nothing more than the eyes. They have also become horny in texture, whereas before they were completely membranous, as in the Snails.

The eyes have been subordinated to the sense of feeling; they are nervous papillæ placed beneath a transparent tegument upon the apex of a tentaculum.

Insects therefore have a number of eyes; they stand either separate, as in the Worms, and are then called simple eyes, or they are crowded together, constituting compound eyes.

3284. Of the other sensorial organs a papilliform elevation in the pharynx is frequently exhibited as a tongue.

3285. Ears are found in the Crabs at the root or base of the antennæ, or in them there is only a tympanic cavity with an ossicle inclosed within; in the Insect there is nothing of the same kind. But meanwhile, since many species attract each other by means of sounds, they must thus possess the faculty of hearing. The antennæ therefore probably correspond to the auditory conch. The auditory conch of the Mammalia ranks also in the signification of the hand, and thus of the organ of feeling.

Nevertheless the antennæ may be transformations of the wings. The auditory ossicles are indeed members which have originated from branchiæ. Probably the antennæ are the auditory ossicles themselves which have emerged outwards, as in the Fishes and Reptiles.

There is no trace of a nose; nor can there be any, for it is the anterior orifice of the vertebral canal, and this is wanting in all lower animals. A noseless animal is an inferior or Tegumental animal.

3286. The sexes are separate, because they are Air-breathing animals and perfectly symmetrical in form. The ovi-and seminal ducts are likewise symmetric and in pairs. The ovipositors and penes are perfect, as in the Snails, because they can be shoved forwards, but not everted as in the latter.

3287. The position of the sexual parts at the posterior extremity of the body has been permanently established, with exceedingly few exceptions, as in some aquatic Insects.

That they correspond to the head, or are imperfect cephalic organs, is most distinctly shown in the Insect. They are usually environed by valves, pincers, and filaments, which resemble maxillæ and palpi.

3288. The Ovum-animals multiply by division, by granules or gemmæ, i. e. shoots, the Sexual animals by membranous ova, the Arthritic animals by horny ova. These last, or egg-shells, are hard, and frequently also so strung together or laid by the parent upon each other as to represent the annulate body of some Insect.

3289. In the ova of the Dermatozoa or Tegumental animals the vitellus only appears to exist, being without albumen, which first seems to make its appearance along with the animal systems. As the animal separates into higher and lower substances, so also does the ovum or microzoon. The simplicity of the Tegumental animals has been prophesied or foretold in that of their ova. The Insecta, as being the third class of their circle, repeat the Acalephæ and Kracken. With the last they bear much resemblance in form and in their motor organs.

3290. Would we arrange the Worms along with the spiral vessels or ducts, the Crabs with the woody rings in the stem, we must term the Tracheal animals foliage or leaves. Their wings are pinnate leaves, and among the Orthoptera many occur which, as well in the form of the body as of the wings, appear as if they had been just liberated from the Papilionaceous tree. The tales or stories, about leaves changing in the torrid zones into Insects, are not without meaning; for poetry is none other than the Ideal of natural history.

Metamorphosis.

3291. We now proceed to retrace our steps. The Insect is a Tegumental animal represented in limbs. There can be therefore present in the Insect no other development than in the Cutaneous animal, which works itself up into the Branchial and Tracheal animal.

The Insect passes through three stages prior to its attaining the adult or perfect condition. It is at first Worm, next Crab, then a perfect, volant animal with limbs, a Fly.

3292. The representative passage of the Insect through the preceding classes in the course of development, constitutes its metamorphosis. Thus the Insect's metamorphosis obtains a meaning and an explanation. Upon the whole, the history of every kind of pregnancy is none other than the passage which takes place through all the animal classes, as I have first represented in my book upon "Generation" (Von der Zeugung, 1805); but in no class of animals are the periods so dissevered or drawn apart as in Insects. It therefore comes to pass that these animals are the equivalent transcripts or copies of a system common to them and the preceding animals.

3293. The Worm is the first condition of the Insect. It is represented, by the larva, which, according to its diversity of size, is called maggot, caterpillar, grub, &c.

3294. The larva is only a Cutaneous and Intestinal animal. It knows of nothing else but eating like the Worm; it has no sexual function, no lust, nor pain; it can scarcely move; in many the feet are wanting, as in the larvæ of Flies, which thus resemble the Entozoa; many have a crop of lateral papillæ, like the caterpillars, which resemble the Nereids.

3295. The change into a nympha, chrysalis, or pupa, commences with the horny induration met with in the Crab, and in the higher organized kinds of Snail. The pupa is the embryonic Crab or its antetype, it is the Snail in its shell.

3296. In the third condition the Insect makes an advance above the branchial condition, and casts aside the Snail's or Crab's shell; and is then the perfect Insect, the Fly.

3297. The metamorphosis is accordingly the embryonic transition of the Insect, after extrusion from the egg, through the three classes of its circle. In the ovum state it only passes through that of the Intestinal and Sexual animals.

3298. This is a retrospective proof that the higher animals also pass in the ovum through the condition of the lower animals, but after birth through the classes which directly precede them. What holds good of the Insects, does so also of the preceding groups of animals, although in a less degree; the higher class of each circle still passes after birth through one or other of the inferior classes.

Thus, the Acalephæ first appear under the form of Infusoria; next change into Polyps, and then obtain for the first time the form of the perfect Acalephæ. The same phenomenon occurs in the naked snails or Slugs. They have, when freshly hatched from the egg, a small mussel-like shell, which they lose at a later period.

The Echini or Sea-urchins, which I believe ought to be ranged with the Kracken, also pass through a metamorphosis.

Finally, it is well known that the Crabs also are subjected to a metamorphosis. Their first condition must be regarded as that of the Worm.

The law is thus universal. The second and third class of every circle traverse after birth the classes, to whose series they belong.

It would even appear that the classes, also of a lower circle, are subject to similar changes; at least many Entozoa, and thus the first class of the third circle, appear to undergo such a metamorphosis, since they first of all resemble an Infusory animal, next change as pupæ into Molluscs, and are then manifested for the first time as Worms.

3299. The Dermatozoa or Sensitive animals range in the following manner, according to the anatomical systems:

A. Intestinal animals.	B. Vascular animals.	C. Respiratory animals.
1. Gastric animals.	4. Venous animals.	7. Reticular animals.
Infusoria.	Mussels.	Worms.
2. Intestinal animals.	5. Arterial animals.	8. Branchial animals.
Polyps.	Snails.	Crustacea.
3. Absorbent animals.	6. Cardiac animals.	9. Tracheal animals.
Acalephæ.	Kracken.	Insects.

3300. According to the developments of the feeling-sense they stand thus:

A. Ovum-animals.	B. Sexual animals.	C. Cutaneous animals.
1. Vitelline animals.	4. Ovarial animals.	7. Papillary animals.
Infusoria.	Mussels.	Worms.
2. Albuminose animals.	5. Orchitic animals.	8. Pedal animals.
Polyps.	Snails.	Crustacea.
3. Involucral animals.	6. Renal animals.	9. Alary animals.
Acalephæ.	Kracken.	Flies.

3301. From these tables the number and kind of relationships or affinities is readily deduced. There is a Relationship of Proximity, as that between Infusoria, Polyps, and Acalephæ. A Relationship of Repetition, as of Infusoria, Mussels, Worms. There is also a Relationship of Series, which takes its rise from the parallelism of the classes. The two last relationships have been confounded together under the term analogy, which at best is but a word of random definition.

Second Province—Sarcozoa, Cephalozoa.

FOURTH CIRCLE—SARCOSE ANIMALS.

3302. The system of motion and sensation, which broke forth so forcibly in the tegument, now passes over into other forms, into the globular form of bone, the fibrous form of flesh or muscle, and the point-form of the nerves.

3303. The nervous system when liberated from the vegetative organs is, or consists of, the myelon (spinal chord) and the encephalon, i. e. brain.

3304. These animals are therefore Osseous, Muscular, and Encephalic animals, whereupon the senses are proportionately developed, and constitute the basis of the highest animal forms—Sarcozoa. The Dermatozoa are asarcose or fleshless animals.

3305. With the sudden appearance of the animal systems all organs of the head, such as tongue, nose, ears, and eyes, are also developed; no asarcous animal has nostrils. Now these parts, especially the nostrils, as marking the terminal extremity of the vertebral column, make up the head—Cephalozoa, or Cephalic animals.

Class 10. Osteozoa, Glossozoa.

3306. The animals, in which the osseous system for the first time makes its appearance are, the Fishes.

The Fish alone has more bones than any other animal. It has dorsal rays, which are wanting in all others.

3307. The animal systems are for the first time slightly separated from each other. Bone, muscle, and nerve are rather a gelatinous mass, which only bids fair for becoming something higher; on this account the bones are frequently but soft cartilages or tendons, the muscle composed of white fibres like those upon the intestine or vessels, the nerve thick, oleaginous, and soft in texture, while the brain is in its constituent parts hardly comparable with that of the Thricozoa.

3308. The muscles of Fishes are not yet perfect in character, since they are devoid of individualization and red colour, and their fibres run mostly parallel with each other without uniting into tendons. Their muscular body is a muscular wall.

3309. The Fishes, ranking upon the first stage of the Sarcozoa, repeat the same stage of the preceding circle, and thus the Infusoria, Mussels, and Worms; or the stomach, veins, and branchial rete or network; furthermore, the vitellus, ovary, and skin, which systems must in them accordingly predominate.

Pelvis.

3310. In Fishes the pelvic organs, sexual parts, and tail predominate. The tail, as being an appurtenance of the pelvis, is larger and stronger than in other animals. It mostly makes up the largest part of the body, and is, properly speaking, its only motor organ.

3311. The sexual parts have still the form of the intestine, and occupy the greatest part of the abdomen. The ovaria are two sacs, like two Infusoria, in whose parietes granules are developed; even the testes are only two such sacs or seminal vesicles wherein the "milt" is contained. Fishes may be termed anorchitic, or animals devoid of testes.

The ova are small, consisting of spawn or roe without shell, but they separate into albumen and vitellus.

3312. External sexual parts are not present. Everything usually opens into a cloaca, which is thus here a true pharyngeal cavity.

3313. With the perfected formation of the head, the animal, so to speak, undergoes a sudden and entire change, and the sexual parts are developed with all their accessory organs, such as the sexual lung. Kidneys are present and mostly an urinary cyst. The kidneys are indeed still so amorphous and soft, that they more resemble coagulated blood than an organ, yet meanwhile agree in this respect with the 'milt' and branchial substance.

Abdomen.

The abdomen is not simply confluent with the sexual body, but thrust by it completely forwards. The anus mostly lies in front of the middle of the body.

3314. The intestine separates for the first time distinctly into cephalic and sexual, or small and

large, intestine.
3315. The vitelline canal is the cœcum. This is very distinct in the Sharks and Rays.

3316. In Birds therefore the vitelline canal is the cœcum also, and the two cœca, falsely so called, which are placed upon the side of the rectum cannot represent the cœcum coli, or else the Bird must have three of these blind appendages or sacs.
3317. The spleen here appears for the first time; while the pancreas divides into a number of cœcal appendages.
Thorax.
3318. The thorax of the Fish is reduced within very narrow limits; it is like the first thoracic formation, and is thus a Mussel's thorax. The branchiæ with their opercula are similarly formed to the branchial plates and shells of the Bivalve Mollusca. The thorax is therefore attached only externally to the body, and the Fish is to be viewed as a Mussel, from between whose shells a monstrous abdomen has grown out.
3319. But this Molluscan thorax is conjoined with animal systems, and has assumed their noble or elevated type of structure. Here therefore the osseous and sarcose system blend together, and the higher formation of the thorax emerges into view.
3320. For the first time an advance is made towards the formation of a trachea, namely, by the branchial framework or skeleton, which opens into the mouth, and corresponds therefore properly to the larynx. Fishes are therefore the first animals which breathe through the mouth. In all the preceding classes the air entered the body, or the water found access to the branchiæ, by other routes.
3321. They may be termed Mouth-breathing animals, for the first formation of the trachea extends no further than to its junction or communication with the mouth; since, for it to be continued into the head and open self-substantially as a nasal organ, constitutes a second step in advance, which cannot be ventured upon in an Abdominal animal. In Fishes everything relates to the abdomen, and this is expressed by the first union of the trachea having taken place with the pharynx or the mouth.
3322. But, at the same time that the trachea is inserted within the mouth, it opens externally and in the lateral direction upon the body, thus letting the water escape from it posteriorly after a mussel-like or sexual fashion. Thus, the trachea is not yet closed inferiorly, nor hence also the thoracic cavity. Between the head and abdomen there are still openings—branchial foramina.
3323. The trachea is itself, however, a thorax upon a small scale, consisting as it does of rings, or as it were ribs. These arches are not yet united by muscles with each other, and the water flows out between them into the apparent thoracic cavity, from which it next escapes beneath the branchial operculum.
3324. These tracheal rings are the branchial arches. The branchial vessels are tracheal and by no means pulmonary vessels.
3325. Thus, upon taking a retrospective glance at the matter, we may fairly conclude that the branchiæ of the Dermatozoa are not equivalent to the lungs of the higher animals, but are only the antetypes of the bronchi, being thus cervical organs.
3326. The trachea is thus formed prior or antecedently to the lungs, but is still membranous and devoid of any continuous connexion between its rings.
3327. The lung is an organ extraneous or foreign to the trachea, and becomes only as if accidentally associated with it.
3328. But a Sarcozoon is not devoid of lungs nor of aerial respiration; for, being the totality of all

lower animals, it consequently unites in itself the respiratory apparatus of the Branchial and Tracheal animals; thus gills and lungs.

3329. In the Fish the first lung appears, that is to say, if we designate or interpret the respiratory organ in Insects only as tracheæ, which do not open into the mouth.

3330. The Fish's lung is the air-or swim-cyst.

3331. The lung is still subordinate to the abdomen and intestine. It is therefore still separated from the trachea or the branchial arches.

3332. Fishes swallow the air and drive it into the pulmonic cyst, where it is analysed or decomposed.

3333. This lung, so soon as it stands opposite to and enters into conflict with a branchia, is what serves to direct or control the action of the heart. It is only when the swim-bladder is allowed to hold good as lung, that the circulation in Fishes admits of being understood. For otherwise venous blood must flow into the heart, from this into the branchiæ, from these directly into the aorta, and so to the different organs without again entering a cardiac cavity, a structure which occurs in no other class of animals, but rather in every instance its reverse. The first heart is arteriose, not venous in all animals, even in the Mussels and Snails and in the embryos. Now, the Fish is still such an embryo, and has only an arteriose heart.

3334. Thus, the following is the "modus operandi" of the parts in question; the swim-bladder is the lung, in which, seeing that it contains air, the blood is oxydized; this oxydized blood flows into the heart and renders it arteriose in character, despite the concomitant influx of venous blood into that organ. Upon this, the blood passes out of the heart through a true aorta, which is called the branchial artery. Instead then of this aorta giving off only some twigs as bronchial vessels to the gills, and next proceeding to pursue its course as a main trunk through the body and along the back, it passes wholly to the branchiæ, i. e. it entirely becomes a bronchial vessel, is slightly oxydized, and then returns upon itself to form the aorta, which should have taken its course directly from the heart.

Senses.

3335. In accordance with the thorax, the corium or true skin is developed. It is mucous in character and muco-secernent like the intestine, from its being constantly in the water. The whole skin is undermined by mucous canals and perforated by their excretory apertures. These foramina of the lateral line are arrested, and metamorphosed branchial foramina, which have only retained the evaporative function of respiration.

3336. The remnants of the tegument's annulation are the scales. They are desiccated air-branchiæ, alary opercula or elytra, and consequently indicate that feature of the Insect, which has continued to work its way into the class of Fishes.

3337. Like the skin, so are the limbs—tegumental members. What is flesh and bone upon them, has kept quite close to the body, and that only which is designed for splitting into digits projects, constituting tegumental digits with cartilages—fins. These fins are something better than the lateral papillæ of the Worms, are furnished at their base with joints, and are in number only four, but crippled or stunted in every possible way.

3338. The fin-rays do not correspond to the digits, but to the nails. They are fibrillated nails, like the wing-feathers of Birds.

3339. Lastly, the head possesses all the organs of sense which belong to a head, but they are still far removed from their perfect condition.

3340. As the nervous system is the first mass, from which the remaining ones have been set free, so also is the nervous sense the first after that of the tegument, which appears as a whole, and serves as a pattern for the subsequent senses. The eye is the sense which is first developed most perfectly, not directly as regards its own completeness, but when considered in reference to the others.

As the sense of feeling has been at once manifested in the vegetative animals as a peripheric nervous sense; so antithetically or vice versâ it is in the higher animals the light-sense, as being the central nervose sense, which appears.

3341. Hitherto both these senses stood altogether alone with each other upon the stage of animal existence, in order, as it were, to play their parts in mutual consent; but so soon as the sense of feeling became individualized into tactile organs, then also the eye made its appearance in a detached or isolated manner.

3342. The Fish's eye is upon the whole composed of the same structures as in the Mammalia; but it is devoid of motion and external coverings.

3343. The ear, as being the sense of motion, has scarcely withdrawn itself from the brain, has not yet become a true external organ, and that portion of it which appears externally, is subservient to inferior systems, such as the branchiæ or gills.

3344. The external auditory organ has become confluent with the branchial aperture, and the auditory ossicles have become parts of the branchial operculum.

3345. In the internal ear only the three semicircular canals have been left. The cochlea is not yet developed.

As true palpebræ or eyelids are wanting to the visual organ, so also are the conchs of the external ear.

3346. The nostrils exist, because a vertebral canal, which terminates in them, is present; stoutly-developed olfactory nerves are also present, so that the sense of smell cannot be wanting. Only this sense has not yet taken the respiratory organ into communication with itself, and both therefore live for themselves in an arrested or stunted condition. This nasal organ does not open into the mouth, admits of neither water nor air passing through it, and therefore serves not as a test-organ to the respiratory process. This is my main point for distinction of Fishes from Reptiles.

3347. Every Sarcozoon, whose nostrils do not open into the mouth, is a Fish. The Siren therefore does not belong to the present class, while the Lepidosiren, having an imperforate nasal organ, does.

3348. The tongue has remained rather an organ of touch and deglutition, than an instrument of taste. The salivary glands are scarcely developed.

3349. But the tongue here appears for the first time in the animal kingdom as a perfect organ provided with muscles, and a lingual bone, as in Man; Fishes are, as regards the development of this sense the Glossozoa, Lingual or Tongue-animals.

3350. As Fishes are the repetition of the intestine and vitellus, so they may be termed abdominal or rather Pelvic animals. They are an abdomen, unto which are appended branchiæ, fins and head.

They are the repetition of the Infusoria, Mussels, and Worms; in them we meet with mucus, branchial opercula, articulated filaments, and labial cirrhi or palpal organs.

Class 11. Myozoa, Rhinozoa.

3351. Those Sarcozoa which for the first time obtain true muscles and a perforated nasal organ are the Reptilia.

3352. True muscles are red, have a definitely circumscribed outline, and are divisible into head, fleshy portion or belly, and tail or tendon. As such they are found for the first time in Reptiles. Thorax.

3353. With the osseous system in Fishes the sexual abdomen was chiefly developed; with the muscular system therefore, as in the present class, the abdomen proper or intestinal abdomen becomes perfectly developed, and the thorax more spacious in calibre.

3354. The thorax is still blended with the abdomen. In Fishes the disposition was already stirring to create for themselves an air-organ; only this was imperfectly attained, since the swim-bladder communicated with the gullet, but not with the branchial arches or larynx, nor with the nasal organ, as in the present class.

3355. Now, if the swim-bladder be symmetrically developed, if it communicates with the branchial larynx and opens through the nose, the aerial respiration is completely attained, and shares its dominion over the body along with the digestive function.

3356. A Fish, respiring through the nose into two swim-bladders, is an Amphibious animal, a Reptile.

3357. In the Reptile, however, the thorax is still subordinated to the abdomen. Its lungs traverse the whole abdominal cavity, and its manner of breathing is still similar in character to the abdominal function. The lungs are simply two membranous bladders, like two intestines, and admit also of being voluntarily filled with air like the intestine is with water. This filling or inflation of the Reptile's lungs takes place also by the abdomen being expanded at the same time that the mouth is closed by the muscles of the fauces, and is therefore a true process of deglutition.

3358. Their inspiration is still therefore fish-like in character, excepting that the air is drawn in by the nostrils; the respiratory process, however, and the respiratory organ is constituted as in the perfect air-breathing animals.

3359. Reptiles may therefore be called Abdominal animals, while the Fishes are Pelvic animals.

3360. The metamorphosis of the branchial arches into larynx is placed beyond doubt in the Reptiles. The anterior branchial arches frequently unite with the lingual bone, whereby the latter obtains several cornua.

3361. The thyroid gland also makes its appearance here for the first time, while the branchial vessels separate from the arches. Fishes have therefore no thyroid gland.

3362. The circulation is complete. The venous and also the arterial blood enters the heart. But both kinds of blood are still mingled with each other as in Fishes. Yet already, through the direction and arrangement of the cardiac orifices, provision seems to be made for their separation.

3363. The rationale of this mixture of the blood appears to reside in the fact, that many of these animals, and probably all of them in the ovum state, breathe by branchiæ. (This proposition, which was announced in the 1st Edition, 1810. § 305, and there based upon the transitional stages traversed by the animal classes during embryonic development, has since then been raised to a state of certainty.)

3364. The mixture of the blood takes place through an opening in the septum separating the two cardiac ventricles, and which corresponds to the foramen ovale of the fœtus. The heart of Reptiles is therefore a persistent fœtal heart.

3365. Without doubt, however, the arterial blood only enters the left heart, while the venous remains in the right, to be thence driven into the lungs.

3366. In other respects the foramen ovale is not a true opening, but a bifurcation of the vena cava, one branch of which passes into the right, the other into the left ventricle, as I have demonstrated in the 'Beyträgen zur Anatomie' (Frankfurt bei Wesche), in treating of the calf's heart.

3367. There are generally no foramina in the body, but only fissures without orifices.

Abdomen.

3368. The digestive organs are pretty similar to those in Fishes, as is exemplified by the intestine, liver, and spleen; the digestive process is, however, more energetic, and frequently aided by venom or poison.

3369. The sexual parts are perfect. The first true testes and ovaria have originated. The tegumental formation has consequently passed over at the extremities into a glandular, because the vascular system preponderates. Even in the highest Fishes, such as the Rays and Sharks, the testes and ovaria are still not developed to the extent which they are in Reptiles.

3370. In the female the oviducts pass together as complete cornua into an uterus. In the male the seminal ducts are still indeed separate, yet the two bodies or crura of the penis are frequently developed, and continue separate in the Serpents and Lizards, but are connate in the Tortoises.

3371. The kidneys are distinct and symmetrical; their ureters usually unite into a large urinary cyst. Thus, the sexual lung is also significantly developed.

3372. Many of these urinary cysts split, as in the Tortoises, into two cornua. This is the form under which the bladder again occurs in Birds, where its two cornua have been incorrectly termed cœca, but the urinary bladder rectum, because the intestine opens into it.

3373. In Reptiles also there is a cloaca or sexual orifice, into which all the sexual apertures meet together.

The ova of Reptiles possess a vitellus and albumen, and are surrounded by a membranous or sometimes by a calcareous shell.

3374. Hence Reptiles have originally also a cœcum, but it has for the most part disappeared; like its antetype the vitelline canal, which is so small in most Fishes and Birds, as to be scarcely distinguishable.

The tail projects from the body, and is rather trailed after the animal, than enabled to assist the body in its motions, as in Fishes.

Senses.

3375. Many Fishes have only a vertebral column without lateral bones; in the Thoracic animals this deficiency can hardly occur any longer, although we are not, on the other hand, to expect any great amount of perfection to be attained by these parts. Reptiles have osseous ramules, whether such be ribs or true feet; the Serpents have indeed only ribs, but these exist in great numbers; in the Frogs, on the contrary, ribs are wanting, but they have limbs.

In no Reptile do fin-rays occur; consequently, there are no peculiar bones as in Fishes.

3376. In Reptiles the limbs themselves are in a varied or undetermined state, although they constitute a class in which the formation of true muscular limbs is decided. In Reptiles the toes appear for the first time.

3377. As regards the tegument, it quite corresponds with that of the Corals, Snails, and Crabs, of which the Reptiles are a repetition. In some cases it is slimy and naked; in others Mussel-shells originate, as in the carapace of the Tortoises; the scales, scutes, and claws of Serpents and Lizards repeat the rings and scutes of Crabs.

3378. The head has separated itself more from the thorax, the eyes and ears being much more perfect; as regards the latter organs the cranium is always perforated, in the former there are eyelids.

3379. The eyes have still no power of free motion, and the eyelids are as yet very imperfect. It is principally the lower eyelid which closes, while in Mammalia the upper one is the most developed.

3380. The external auditory meatus is notwithstanding usually covered by integument, the auditory ossicles are mostly blended into one, which projects above the skull; the cochlea is wanting.

3381. Reptiles produce sounds for the first time through the lungs; they have voice, but as yet no song.

3382. The nose, however, gives the finishing touch to the head. It has not only opened as a vertebral canal in front, but also as a thoracic cavity posteriorly into the mouth, which was not the case in Fishes. The open nose is the æsthetic character of Reptiles, just as the red muscles are the characteristics of the anatomical systems. The Reptiles are Rhinozoa.

3383. The tongue takes a higher rank than in Fishes. It is throughout soft, fleshy, and smooth; but in most species is still slit into two, which reminds us of a similar condition of the penis.

3384. The teeth are in these animals more like digits than in Fishes. While here they consist for the greatest part of front teeth, and are therefore associated with the intestine, in Reptiles they are mostly lateral, and thus true maxillary teeth, which are annexed to the salivary system. With this special dental formation the saliva is also more active; it is a rapidly fatal poison.

3385. The poison-teeth or fangs have likewise a groove, which can be regarded as the continuation of the salivary duct.

Class 12. Neurozoa, Otozoa.

3386. Those Sarcozoa, whose nervous system is for the first time perfectly developed, and whose ear is open externally, are the Birds.

3387. The completion of the nervous system is the brain; now the brain defines or determines the head; the Bird is, properly speaking, the first Encephalic animal.

3388. In the Bird the head has for the first time, and that indeed suddenly, freed itself from the trunk, and been placed upon a long neck, far removed from the thorax. In no class are there such long necks and so many cervical vertebræ to be found as in Birds. They can be therefore called also Cervical animals, as the neck is not simply present, in order to render the head independent or self-substantial.

The caudal vertebræ, on the contrary, are lessened throughout the Birds, to a degree met with in no other class of animals.

3389. The brain separates for the first time completely into cerebrum and cerebellum, begins to exhibit convolutions, and has in general most of the individual parts analogous to those in the Thricozoa. They are therefore Encephalic animals, which the name of Neurozoa will properly enough imply. The nerves are in comparison with the spinal cord much thinner than in Reptiles and Fishes.

In the Bird also, all the spiritual or mental faculties make their appearance for the first time and suddenly, whereas in the preceding classes but slight traces of them were observed. Such for example are their mechanical instincts, varied modes of nidification, powers of imitation, susceptibility to instruction, knowledge of their benefactors, sentiment of joy, wheedling or coaxing manners, and so on. We have no example of Fishes and Reptiles having learnt any artificial tricks.

3390. Birds are the closest repetition of the Insects, but stand remote from the Crabs and Acalephæ, indications of which are reflected by their structure, disposition, mechanical instincts, and in their nests.

Thorax.

As the sexual body with the tail predominates in Fishes, and the abdomen in Reptiles, so does the thorax in Birds. The whole abdomen and sexual body has been subordinated to the thorax.

The ribs are here directed in such a manner for the first time, that the thorax can act as a voluntary pump-organ. Sternum and respiratory muscles are usually large and peculiarly built.

3391. The lung is only a cluster of Insect tracheæ. They are full of foramina, out of which the air can penetrate into the whole body, exactly as in Insects. In Fishes the lung was still an actual intestine, as seen in the single air-bladder; in Reptiles it was nothing more than a double intestine; in Birds this lung is divided after the fashion of an Insect into aerial vessels or ducts. In the Bird the intestine lies in the air, and breathes from it, as in the Insect. Birds are also animals which breathe by the intestine.

3392. The whole Bird is lung. Its body is a thoracic cavity, while in the Fish it was simply a sexual cavity, and in the Reptile had obtained the abdominal form. In Birds therefore there is a number of ribs and a strong ossification, from the air itself penetrating into the bones.

Even the intestine has passed over into a motor organ. In Birds and Insects only do we find a true muscular stomach, wherein the food is crushed.

Members or Limbs.

3393. In these animals the formation of limbs or members must be wholly attained. Everything, that works, must work to the production of limbs. The whole body becomes limb.

3394. With the perfected nervous and respiratory system the bones and muscles make also a perfect appearance. The skeleton is hard, complete, and full of air instead of marrow; the muscles are red and separated, the movements free and complex.

3395. The elevated condition of the motor system is demonstrated in the limbs, which here appear in the greatest diversity. There are only two pairs of limbs, thoracic, and abdominal or sexual members. These two are equivalent, so long only as the sexual and abdominal cavity belonging to the trunk proper predominate, and therefore preserve the equilibrium. But the limbs, which have been given to serve the office of the trunk, are destined for progression or swimming, and are simply terrestrial or aquatic members.

3396. The higher limbs are thoracic respiratory members, which are filled with air and clothed with tracheæ, thus exercising a function conformable to the thorax. The thorax has an aerial character. The highest thoracic limbs must be aerial members.

3397. As the abdominal members move upon the earth or in the water, so do the thoracic in the air. The terrestrial limbs are feet, the aerial limbs, wings.

3398. The wings are in the member-formation the extremes of the thoracic limbs. It does not follow from this that they are the noblest in rank, but only that they are the uttermost unto which bodily motion can attain.

3399. The winged animal is the Bird.

3400. The Bird is an Insect with fleshy limbs.

3401. The wings of Birds repeat the alary appendages of Insects in the flesh.

3402. A Bird's wing is a strange but very instructive composition. It consists, namely, of a Reptile's foot and an Insect's wings.

3403. We saw how the branchiæ of the Insect dried up, separated from the feet, and being liberated as wings, were permeated by tracheæ. In the Bird these wings have remained standing upon the feet and been converted into feathers.

3404. A feather is an Insect's wing.

3405. As the Bird grows out at the thoracic limbs into the Insect's wings, so does it upon the whole body into dried branchial laminæ.
The whole body of the Bird is clothed with branchial laminæ or plates.
3406. The wings of the Insects might be called free tracheæ.
The Bird's feathers are Insect-tracheæ. As in the Insect the wings are a leash or net of tracheæ held together by membranes, so feathers are tracheæ dividing into fibre-like ramules.
3407. The Bird is a Reptilian or Frog's body, beset all over with Insects, like as with parasitic animals.
3408. The highest Insect only attains to the possession of four wings, which in some Moths split again into several feathers. In the Bird a vast number of such wings originates.
3409. An Insect's wing is not more than one feather, and is therefore placed also directly upon the body. These wings must multiply so soon as they occur upon a membered trunk, upon arms. Thus, we need not ask why the Butterfly has four, but the Bird only two wings, seeing that the latter is the nobler animal. The discourse cannot be concerning the wings of the former, for the Butterfly has indeed none, but only four feathers.
3410. What the Bird is, it is by virtue of its feathers. It is throughout a trachea, a pair of bellows. Its bones are hollow, full of air, and stand likewise in communication with the lungs; the feather-quills are also hollow.

Senses.

3411. The wings have all the muscles to themselves; the bone has in them gone to ruin. On the legs, on the contrary, the muscles have declined, and the bone has got the upper hand.
Hence it results, properly speaking, that only the thoracic members are perfected, because the Bird is nothing but thorax. The abdomen has, so to speak, vanished, and through this the abdominal limbs are left only, as thin and dry poles or staffs.
3412. From the same cause the muscular flesh upon the head has disappeared. Neck and head are lean, or like Insect horn, which serves only the nervous system.
3413. Beyond the excess of motion the sense of feeling has been nearly lost. The toes are simply destined for motion, and to be used as scrapers, and the digits have become the supports of feathers.
3414. The bill is an Insect's proboscis. In the Bird no teeth whatever project from the flesh, but the jaws themselves. To such an extent has the flesh been withdrawn. What has been called the cere is the only remnant of the facial flesh. Even nostrils and tongue have suffered ossification.
3415. The tongue is a feather. Saliva is scarcely present.
3416. The ears, as being the sense of motion, are far more perfectly evolved than in all the preceding classes. They have opened outwards, and possess an additional auditory part, the cochlea.
With the limbs the sense of hearing must of necessity grow perfect.
3417. The Bird is thoroughly, or "out-and-out" organized as an animal of song. In it Nature attains unto a definite hearing and speech. The Bird speaketh the language of Nature.
With the Bird, the voice, properly speaking, breaks forth for the first time, and that too in a high grade of perfection, as melody.
3418. The ear is the highest representation of the trachea in muscles and bones. The Bird is the Otozoon.

Sexual Parts.

3419. The kidneys are symmetrically constructed, although not yet a perfectly coherent mass.

They are of very large size. But a strange feature has come to pass in the urinary cyst, which is the sexual lung. Into this, as has been already said, the intestine opens, and thus it here again passes over into the lung, just as it has passed over into a fleshy cardioid stomach, and completely upwards into the feather-like tongue.

Into this sexual lung the seminal ducts also enter, or the penis, when one is present, along with the oviduct.

3420. The ovum consists of two completely separated substances, and these indeed are so distinct, that the vitellus is secreted at an entirely different place to the albumen, the former in the ovarium, the latter in the oviduct—whereas in Fishes both originated together.

The separation cannot extend further, or else the substances would no longer unite with each other, and the vitellus must be first mixed with the albumen, after it has been completely formed, after or when it is an embryo. In the Bird albumen and vitellus come together during their passage out of the parent's body, or in the act of being laid, and thus before the albumen has been converted into the chick. But in the Mammalia they are first completely united, subsequently to their being laid—i. e. during the period of lactation.

3421. The vitellus is directly secreted from the arteries, but the albumen from an enteroidal sac or the oviduct, which is finally converted into mammary glands.

3422. The vitellus is more a product of the thorax, and is therefore formed directly into the intestine and entire embryo.

3423. The albumen is an intestinal or digestive product, a solution of the organic mass into protoplasma or primary mucus. It is not fashioned itself into the embryo, but is only absorbed as fluid nutriment by the latter.

3424. The shell of the ovum is the last bone, which the animal deposits from the sexual blood, as being analogous to the urine. It is an aerial product, or an analogue of the feather's quill.

3425. Even the nest of the Bird is a spiritual repetition of its plumage, for in it the stalks of plants, tracheæ, or feathers are united into one body, which reminds us in the Swallows of the Acalephæ.

FIFTH CIRCLE. AISTHESEOZOA.
Class 13. Aistheseozoa.

3426. An animal with all its organs of sense perfectly developed, is a Thricozoon.

3427. The nervous system emerges at length freely above the other systems, and it is no longer its mass, but its organs which impart character to the animal. The nervous organs are, however, simply the organs of sense. Through these therefore must the present differ from the preceding class of animals.

3428. Now too the senses first make their appearance in a self-substantial manner above the other organs, serving merely their own functions, and only by chance those of others.

3429. As in Birds the whole body was subordinated to the thorax, in Reptiles to the abdomen, in Fishes to the sex, in Insects to the tracheæ, and so on; in like manner is it here subordinated to the system of the senses or the head.

3430. As the higher senses determine or define the anterior part of the head, and are in their state of perfection provided with muscles; so here the face or visage is invested by flesh, whereby, properly speaking, a true, namely, a moveable countenance, first originates. The Aistheseozoa have a fleshy face.

3431. All possess moveable eyes; fleshy noses, which stand open both externally and internally; ears opening outwards, and mostly provided with a moveable flap; a fleshy tongue, free in front, and moveable lips; with at least thoracic limbs and a skin covered with hairs.

3432. In the Bird, Reptile, and Fish the face is merely invested by tegument, nearly devoid of any muscles, and therefore immoveable. They have a tegumental face, which is incapable of producing any expressions.

3433. In the tegumental face the eyes are motionless, and very rarely both directed so forwards that they can view an object together; the nostrils are frequently devoid of a fleshy rim; the tongue often feather-like, cartilaginous, or covered with teeth; true fleshy lips are wanting, as are frequently the teeth, with even the limbs and digits, or these are divided into a number of rays, forming either feathers, or fins; in the Aistheseozoa there are never more than five toes present, and if there be fewer, then the number admits of being referred to some crippling of the normal quantity, five.

3434. It is remarkable, and serves to the discovery of many laws, that the highest sense here appears for the first time in its state of perfection. The eye in the Aistheseozoa is present in a perfect condition throughout the class, with the exception of the eyelids; the other organs of sense are, on the contrary, exhibited in all their gradations of structure.

3435. It appears, as if the whole animal was first perfected, when the eye is present with all its investment or clothing. The eye of the Aistheseozoa has not simply all its internal chambers and all its humours, but also all its muscles; it is moveable and has perfect eyelids, with very few exceptions—Ophthalmozoa.

3436. The ear now begins to suffer arrest. Its completion is indeed the formation of an external concha or flap for receiving the rays of sound, the hand being repeated in the ear, and its skeleton constituted by the auditory ossicles. This auditory hand occurs only in the Aistheseozoa, and might serve as characteristic if it were not wanting in many, while the eyelids are present. As in the Whales, where the auditory passage can be, however, closed, an act which is not possible in any Bird. The Bird must hear whether it will or no.

In all Thricozoa the interior of the ear is perfect, having cochlea, semicircular canals, tympanum, and, as brachial parts, the three conjoined auditory ossicles. The concha of the ear passes besides through all stages of development, from having a simple margin to one with most varied convolutions, lobes and opercula.

3437. Still more than the ear does the nose undergo modifications. In the Whales it seems to be less adapted for the purposes of smell than of respiration. The olfactory nerves are in them very delicate, and the apex of the nose is destitute of motion.

In other beasts, on the contrary, it is elongated into a very muscular proboscis or snout, and is endowed with the power of voluntary motion.

The form also of the nostrils is very varied, they being round, narrow, patulous, and frequently capable of being closed.

3438. The tongue is indeed mostly fleshy and soft; yet in many species it is provided with horny points, in others invested by a dense coriaceous tegument, so that it appears to represent an instrument for deglutition rather than gustation.

3439. The lips also are mostly fleshy and moveable; yet in many they retrograde exceedingly and lose their mobility, as is partly the case in the Ornithorynchus.

3440. In most, however, the limbs, especially the toes, are still subject to variation. Their perfection consists in the number five, and in the difference between the two pairs of limbs, as in Man. In the Apes the posterior feet are also hands, which is an imperfection; in the Marsupials there are hands posteriorly but toes in front; in other respects there are generally toes, sometimes five, then four, finally two perfect and two dew-claws in the Ox, and at length only one in the

Horse, while the posterior extremities are virtually lost in the Whales.

3441. The dental system, as being the set of maxillary claws, is alone present in its state of perfection in the Thricozoa. They alone have, in addition to the incisor teeth, all the five kinds of teeth that differ from each other in form, namely, canine, false molars, laniary, with second and third true molars, corresponding to the five fingers reckoned from the thumb.

3442. In the dentition of those animals which tear their food the greatest amount of completeness and variety is met with, since each tooth has a different form and function.

In the Bears the molars are uniform in character, and so on through the Apes up to Man.
In the Marsupials they are tolerably similar, as also in the Bats and Shrew-mice.
They are still more alike in the Pig and Horse, and the incisor teeth begin to be wanting in the Ruminantia, e. g. the Ox.
In Mice the canines are wanting, in the Sloths the front teeth, and in the Ant-eaters actually all.

3443. In opposition to the perfect eye the general sense of feeling is developed in the tegument. The tegument which is best developed will be that, which represents a self-substantial organ with all its appurtenances, and thus an organ of touch, whose nobility of rank consists in motion. A skin, which is moveable by means of muscles, must take the noblest rank. A skin with tegumental muscles is an organ of feeling, which is already in some degree subjected to the influence of the will. If tegumental muscles do not occur in all these animals, they still do in most of them.

The production of the most perfect coat or covering is the second step, whereby the tegument ascends.

3444. Hitherto the outer covering of animals was pretty inorganic, consisting either of hollow tracheæ, feathers, or semicavernous scales, coat of mail, or lastly, only mucus. All these organs were only constructed after the type of individual vegetable systems, in greatest part only after the respiratory organ, and were therefore partial in aspect or deficient.

The highest covering must also bear the highest meaning. This is that, which grows out of the sanguinary system collectively, out of the capillary vessels. The capillary vessels of the covering are, however, the hairs. The hair is the most perfect covering of the animal.

3445. These animals are the Thricozoa.

Already the highest birds, e. g. the Ostrich and Cassowary, exhibit feathers which merge over into hairs.

3446. An integument clothed with hair, and self-substantially moveable is the perfection of this organ, it is a fur—Furred animals.

3447. The fur is the peripheric combination of the Vegetable and Animal. The hairs are the highest Vegetable, as being the vascular system grown out of or above the animal, and constituting the fundamental or basi-system of the whole body. The tegumental muscles are the lowest Animal. In the fur consequently the whole animal has been represented, but as a limit set between animal and world.

3448. The tegumental covering also varies. The hairs become scanty in respect of number; instead of them spines, horny rings or scales occur, yet there are always hairs on the belly. In many Whales they seem to dwindle into bristles. In some they have coalesced to form a kind of bark, as in the Stellerian Walrus or Sea-cow.

Thorax.

3449. In all the respiration is aerial; in all there is a true costal structure, and a respiration effected through the motion of the ribs or a pumping action; in all the lungs are filled with cells; in all there is a diaphragm, a larynx, a trachea with cartilaginous rings, and a thyroid gland. But the air no longer penetrates from out the lungs into all the cavities of the body, as in Birds. Abdomen.

3450. The two intestines, namely, the large and small, are more distinct from each other than in other animals; the cæcum coli is in most distinct; the stomach expanded and membranous in texture, so that it operates simply through chemical influence. Liver, pancreas, and spleen occur in all.

3451. The sexual parts are in every respect very perfect. The penis is present in all, and all have an uterus, fallopian tubes, and separate oviducts. The penis is in many still retracted into a cloaca, and the testes frequently lie within the abdominal cavity.

3452. As in the tegument the parts separate and each becomes self-substantial, the fibre animal, the tegument vegetal, and so on even in the sexual animal. The ovarium, consisting of two parts, now also separates, since one becomes animal, but the other remains vegetal. The albumen-organ becomes animal, detaches itself from the sexual parts, is developed in the self-substantial tegument and called mamma.

The Aistheseozoa are thus also Mastozoa or Mammalia.

3453. This separation of the sexual animal is one of the first characters. The mammæ could never be wanting, because they indicate an essential stage in the development. The albumen-organ becomes an organ of sensation. If the existence of mammary organs be doubtful, as in the Ornithorynchus, the hairs are in that case perfectly distinct. No Reptile nor Bird can have hairs, because their covering is derived only from a partial system, the respiratory; while hairs are from the general vascular system, which is the foundation of the tegument, of the sense of feeling. The Ornithorynchus is a furred animal, and this would suffice to bring it among the Aistheseozoa, were we even to deny it mammary organs.

3454. Both sexes, male and female, have breasts. In youth they are most readily detected in the male, because their sexual parts then rank nearer to the female. In other respects the mammæ are probably the chiefest organs of absorption for the embryo.

3455. The mammæ are nobler in rank, when, as udders they become self-substantial; and the more they remove from the belly or abdomen, and come, as breasts, to be placed upon the chest.

D. ORDERS AND FAMILIES.

3456. Had the names been given in strict accordance with rank; then the class-divisions, which correspond to the circles, would have been called Orders.

3457. Families would be class-divisions, which correspond to the classes themselves.

3458. In no class therefore could there be more than four orders admitted, or five, if the sensorial organs be reckoned to constitute a special circle.

From the same reason there are no more than thirteen or seventeen families.—These names cannot meanwhile be so strictly adopted, because the classes are not of equal rank, as has been observed; on this account it is necessary to shove in here and there other divisions, which should be termed at one time Cohorts, at another, Alliances.

3459. If the animals of a class differ from each other, it is only possible by their bringing to bear in addition to their characteristic organ some other organ, and consequently becoming similar to an earlier or later class.

3460. Yet nevertheless in this ascent the animals could never outstep the confines of their own circle. There can be no Dermatozoon which could have bones; for in that case it would be an

Osteozoon, and belong to another province.

Every class therefore comprises as many orders only as there are circles with which it comes into contact. Thus in the first circle there is only one order, in the second two, and so on.

Each class therefore includes also as many families only as the circles which are touched by it contain classes. Thus the first to the third class has three, the fourth to the sixth, six; the seventh to the ninth, nine; the tenth and so on, thirteen.

3461. The serial arrangement of animals into families is naturally difficult; but in Physio-philosophy we have to treat not about the execution of the Systematic in detail, but concerning its principles.

FIRST CIRCLE. INTESTINAL, OOZOA—MUCUS-ANIMALS.
First Class.
Gastric, Vitelline Animals—Infusoria.

3462. The Infusoria admit of being reduced to three divisions: the lowest of which, such as the Infusoria proper or Monades, are provided for the most part with cilia; the next in succession, as the Rhizopoda, possess extensible processes, and are mostly covered by a multi-chambered shell; lastly, the most perfect Infusoria have all kinds of internal organs, and especially what has been called rotatory apparatus, as being the dawn of future tentacula.

3463. These three Families obviously correspond to the three classes of the present circle, and that indeed as follows:
To the first class, or Infusoria proper, correspond the Monades.
To the second class, or Polyps, the Rhizopoda.
To the third class, or Acalephæ, the Rotifera.

3464. The Monades are obviously the simplest organized creatures, being mucous vesicles, which move, obtain their food by stirring up vortical currents in the water, and emit what is undigested again by the mouth.

3465. They occur very abundantly in all infusions, and can very well originate, like Fungi, by division of the organic mass, although they are in a condition to propagate themselves, i. e. by spontaneous division.

3466. The Monades are the semen of the animal kingdom, which is dissolved in, or rather produced from, the sea.

3467. The animal body is nothing else than a compound fabric of Monads.

3468. Decomposition is a separation into Monads, a retrogression into the primary mass of the animal kingdom.

3469. All propagation, even that of the sex, commences like the animal kingdom, or with its first family. On that account the embryonic development must be a passage through the animal kingdom.

3470. The Rhizopoda usually adhere together within a multi-chambered calcareous shell, out of which they protrude mucous filaments, and are therefore the antetypes of the Corals or Polyps.

3471. The Rotifera exhibit all kinds of viscera, such as intestine and ovaries; besides what have been called rotatory organs, which remind us of the arms or tentacles of the Acalephæ.

3472. The families of these animals may be therefore aptly named as follows:
Fam. 1. Typical Infusoria—Monades; cilia or vibratile organs.
2. Polypary Infusoria—Rhizopoda; extensible processes.
3. Acalephan Infusoria—Rotifera; intestine and oral organs.
Second Class.

Intestinal, Albuminous Animals—Polypi.

3473. The Polyps also do not admit of being divided into more than three families. The first are only tubes or vesicles with capillary tentacula around the mouth, as the naked Polyps, Tubulariæ, Sertulariæ, and Cellulariæ.

The others have true ciliated tentacular rays surrounding the mouth, and are always condensed inferiorly into a horny, and occasionally stony axis or stem, as is seen in the Gorgoniæ, Alcyoniæ, and Isidiæ.

The third have ordinary tentacula, occurring in great numbers like tassels or tufts around the mouth; their integument is either hardened into stone, or becomes fleshy, as in the Star-corals and Actiniæ.

3474. There is no doubt that the naked Polyps are closely allied to the Infusoria, and among them indeed to the Rotifera, so that they represent but a higher stage of these creatures, characterized by superior size and by tentacular in place of ciliary or vibratile hairs.

3475. The Cellulariæ cannot be more distinctly characterized than by saying that they are cortices or ramules inhabited by Vorticellæ. Thus they are Vorticellæ surrounded by a shell, and may be compared to ova, in whose coriaceous or leathery shell calcareous granules are blended, as in the eggs of the Crocodiles and Tortoises.

3476. They increase by means of ova and by ramification, if the former mode of division is not carried too far.

3477. The tubes of the Tubulariæ appear to be nothing else than the posterior extremity of the Polyp, desiccated or dried. These tubes are not therefore the product of an excretion, but the body itself.

3478. On the contrary, the tubes of the Sertulariæ must be held as being a tegumental excretion, within which the Polyp ramifies and produces ova-cysts. If the naked Polyps resemble ova which are devoid of shells, such as the roe or spawn of Fishes, then the Sertulariæ resemble ova surrounded by a tough skin, like those of Rays and Serpents.

The Radiated Polyps or Horn-corals are invariably ramified and converted towards their interior into a common horny or stony mass; so that the animals themselves appear to have coalesced upon this into a common tegument or bark. They possess a stomach and ovaries surrounding it, which open into the border of the mouth between the rays. They thus increase by ova and ramifications.

They therefore represent the class proper of the Polyps.

The tufted Polyps comprise the proper Corals or Lithozoa, are in form and substance like the Acalephæ but with this difference, that the covering is mostly lithoidal, while in many Acalephans, as the Porpitæ, it only appears as a cartilaginous disc or plate.

3479. These Corals are true ova, having a perfect calcareous shell, like that of the Bird. The gelatinous animal which adheres within a wide-mouthed Madrepore, e. g. Fungia, resembles a vitellus just hatched, and from which the fœtal envelopes have been developed.

3480. The numerous tentacular filaments surrounding the wide mouth resemble the shags of the chorion, which accumulate around the orifice of the umbilical cord to constitute a placenta in the higher animals.

The Corals are brood-eggs inclosed within the uterus of Nature or in the sea. The Coral-animals are an umbilical cord, treasured up for the embryo, while the Tubulariæ are only membranous vitelli, and the Gorgoniæ, ova with desiccated albumen.

3481. The families of Polyps stand therefore in the following order of significance:
Fam. 1. Infusorial Polyps—Tubulariæ.
2. Typical Polyps—Alcyoniæ.
3. Acalephan Polyps—Actiniæ.
Third Class.
Absorbent, Involucral Animals—Acalephæ.
3482. The Acalephæ also can only be brought under three divisions, viz. Röhrenquallen, or Physaliæ, Rippenquallen or Beroes, Hutquallen or Acalephæ.
3483. The Physaliæ are without doubt the lowest, being only giant Infusoria; the Beroes are allied through their simple bodily cavity, to the Gorgoniæ; the Acalephæ, through their form, to the Madrepores, especially the Actiniæ.
3484. They range therefore in the following order of significance:
Fam. 1. Infusorial Acalephæ—Physaliæ.
2. Polypary Acalephæ—Beroes.
3. Typical Acalephæ—Acalephæ.
3485. The first have not yet attained to the unity or single character of the mouth, but imbibe their food through numerous tubes. They are bundles of ramified Vorticellæ, a thoracic duct full of glands and roots, which absorbs nutriment out of the sea, instead of an intestine.
As the first family they are the antetypes of the Mussels and Entozoa, and especially of the Hydatids and Tape-worms, and we shall not be far from hitting the mark, if we compare their air-bladder with the hindermost cystiform member of the body in the Hydatids.
3486. The second family have a single mouth and ribs mostly full of lamellæ, which are probably the antetypes of gills; they are also frequently traversed by nutritive vessels.
They are antetypical of Snails, being in form an abdominal pouch, in substance a liver, and frequently possessing paired tentacula. Higher up, and they indicate the Crabs.
3487. The third have, as a general rule, one central mouth surrounded by four arms and very numerous vessels, which run from the gastric cavity to its margins, and there mostly elongate into filaments or hairs, being thus true lymphatic vessels, which convey the nourishment directly from the stomach through the whole body, and become on its margin tentacula.
3488. As antetypes of the Kracken they particularly manifest the form of the Sepiæ, or Cephalopods, in the strong and frequently papillated arms that surround the mouth. Gland-like nodes are also developed in the margin of the pileus, but their meaning is as yet doubtful. Upon a higher stage, and they are antetypical of Insects.
Their vascular system forms an exceedingly regular, quaternary net, with branches and ramules ranging opposite to each other, which have been believed to be the vitelline network of the ovum when hatched. These vessels terminate likewise in a common marginal vessel, as in the vitellus.
3489. Viewed as a whole they resemble in form, appendages, and substance the fœtal involucra or envelopes. The upper surface may be compared with the convex back of the envelopes, the lower to the concave funnel of the umbilical cord, and the succigerent filaments or hairs to the shags of the chorion. They are probably elongated by injection, like the so-called legs of the Star-fishes.
The secretion of the ova in four ovaries speaks also in favour of their higher grade of development. They lie in four cavities around the stomach, which also open near to the mouth. Their size also is significantly expressive of their higher position.
Finally, they are of separate sexes. In them is the milt for the first time distinctly displayed, and in the same situation indeed where in other animals the ovaries are.

3490. Hence, in these three classes there are found no more than three families, which together make up only one order. The want of a fourth family and so on, proves that the development of these animals does not pass over into the succeeding circle. Their body therefore is only an homogeneous, transparent mass, variously excavated and perforated, but not separated into two cysts, namely, tegument and intestine, and without the other viscera, which are formed by the vascular system, such as the liver, kidneys, and salivary glands.

SECOND CIRCLE. VASCULAR, SEXUAL ANIMALS.

3491. These animals will both traverse the three classes of their circle, as also repeat the preceding three classes, and they consequently divide into two orders and six families.

Fourth Class.

Venous, Ovarial Animals—Mussels.

3492. The Mussels or bivalve Mollusca resolve themselves into two orders, according to the structure of their mantle or respiratory sac. It is either closed in a tubuliform manner, and opens posteriorly into two mostly tubular-shaped respiratory apertures, and in front has an orifice for the passage of the foot—Camacea (Lochmuscheln); or it is slit along its whole length anteriorly; and the pedal aperture is confluent with the anterior and also indeed with the posterior respiratory opening—Ostracea (Spaltmuscheln.)

The first order still represent the cystic form of the Oozoa, and partly by tubuliform calcareous shells placed around the two ordinary shell-valves; but the second, by the freer development of the organs, represent the animals of their own circle.

Order 1. Protozooid Mussels—Lochmuscheln.

3493. These animals repeat the Mucus-animals; are tubular in form, and mostly provided with two long respiratory tubes.

Fam. 1. Infusorial Mussels—Sackmuscheln or Pholades.

Cylindrical, with a tubuliform mantle almost entirely closed and frequently surrounded by a calcareous tube, in addition to the two ordinary shell-valves; the pedal aperture is situated at the oral extremity; the respiratory tubes retractile by means of muscles; the foot is cylindriform.

Of the same kind are the Pfahlmuscheln, Teredines, and Sandmuscheln, as also the Solenidæ. They typify the Nudibranchiate Snails and the Salpæ.

Fam. 2. Polypary Mussels—Klaffmuscheln or Tellinidæ.

Body flat, mostly discoidal; there is a large fissure on the abdominal side of the mantle for the passage of the foot, while posteriorly there are two long respiratory tubes with retractor muscles. Here belong the Tellinidæ and Venusidæ.

I have shown, that they are already to be recognized by the uncinate or hook-shaped groove for the mantle in the shell, and which proceeds from the insertion of the retractor muscles. The foot is usually lancet-shaped. They typify the Patellæ and Ascidiæ.

Fam. 3. Acalephan Mussels—Cardiacea.

Body globular in form, mantle having a pedal slit on the abdominal side; two respiratory apertures without retractor muscles; foot mostly uncinate or strap-shaped.

T have shown that these creatures are to be recognized by a discoid groove in the shell, which simply depends upon the mantle, the retractor muscles being wanting. Here belong the Cardiacea and gigantic Chamidæ, in which last a byssus occurs, and the union also of the two occlusor or sphincter muscles, as in the following order. They antetypify the air-breathing Snails and Cirripedia.

Order 2. Conchozooid Mussels—Spaltmuscheln.

3494. These animals represent their own circle.
The pedal slit in the mantle occupies the whole length of the latter, and is confluent with the anterior or also with the posterior respiratory aperture; so that only one or even no aperture is left, and this when present is besides never elongated in a tubular form; the retractor muscles are therefore wanting also, and the shell simply exhibits a discoid groove for the mantle. The occlusor muscles approximate and unite at their terminations. Both usually coalesce into one.

Fam. 4. Typical Mussels—Mytilaceæ.
Occlusor muscles separate, and mantle slit to such an extent, that only the posterior respiratory aperture remains; foot mostly tongue-shaped or coniform, with occasionally a byssus.
Here belong the fresh-water Mussels and the Mytili proper, of which last many bore into rocks. They typify the Capulidæ and Brachiopoda.
Fam. 5. Snail-Mussels—Arcaceæ.
Have two separate occlusor muscles and a perfectly separate mantle without respiratory aperture; the foot small, mostly cartilaginous. Here belong the Arcæ, Aviculæ, and Pearl-mussels.
They typify the Trochidæ and Pteropoda.
Fam. 6. Kracken-Mussels—Ostracea.
Only one occlusor muscle, mantle entirely open or slit, so that both respiratory apertures are only oblique incisions therein; foot very small, frequently furnished with a byssus. They typify the Buccinidæ and Cuttle-fish.
Fifth Class.
Arteriose, Orchitic Animals—Snails.
3495. The Snails likewise divide into two orders, according to the two circles, preindicated by the Ovum-and Sexual system. Their branchiæ are either ramulate or pectiniform, the sexual parts combined or separated.
The first kind are still frequently gelatinous in texture, transparent and naked; their branchiæ usually stand out freely as filaments, lamellæ, or ramules upon the back, or lie simply as a vascular network within the mantle. All are androgynous. They therefore obviously repeat the Protozoa or Mucus-animals.
The second are invariably covered by the shell and by a mantle, within the cavity of which the branchiæ lie concealed as one or two comb-like bodies. Tentacula and eyes, which are occasionally wanting in the preceding order, are here universally present; the sexes separate.
As in the Snails, the male parts make their appearance for the first time distinct and individualized, and are also a characteristic organ of the class; so also do they serve as bases of division, and the Snails may be divided into those which are Androgynous or bisexual, and those with separate sexes, i. e. Diœcious.
Order 1. Protozooid Snails—Androgyni.
3496. The male and female sexual parts united together in a single individual, branchiæ ramuliform, occurring either as filaments, leaflets or ramules, freely situate upon the naked body, or as a network lodged within the pallial cavity and surrounded by a shell.
The Nudibranchiate species live in the sea, those with hollow branchiæ or the Pulmonea in the air; the former feeding mostly upon animals, the latter upon plants.
Fam. 1. Infusorial Snails—Notobranchiata, or Tritoniæ.
Body gelatinous and membranous, cylindrical and naked, devoid of shell, with branchial filaments or ramules disposed in two rows upon the back.
Here belong the Tritoniæ and Dorides. Their body is muscular; the tentacula not retractile; the

male sexual parts open in company with the female upon the right side of the neck, as in the higher organized Snails. All inhabit the sea. They prefigure or typify the Salpæ.

Fam. 2. Polypary Snails—Pleurobranchiata, or Patellæ.

Body and sexual parts as in the preceding family, the branchiæ, however, occurring as ramules or leaflets upon the sides of the body, are more or less covered.

Here belong the Phyllidiæ, Schüssel-and Schildschnecken. They are the antetypes of the Ascidiæ.

Fam. 3. Acalephan Snails—Dictyobranchiata or Limacidæ.

The branchiæ form a rete or network within the pallial cavity, and respire the moisture of air; mantle and viscera are mostly surrounded by a shell; the body is therefore bipartite, being separated into a splanchnic or visceral body and a foot with head.

Here belong the Air-breathing Snails, both land as well as fresh-water species. They typify the Cirripedia.

The shells are mostly thin and horny, yet nevertheless contain a considerable quantity of calcareous earth, and are mostly devoid of opercula.

Those species, which dwell in fresh water, do not possess introvertible tentacula like the marine Snails, and the eyes are placed at their basis; the sexual orifices are separate.

In the Land-snails, the tentacula are introvertible, and support the eyes upon their apex; the sexual apertures are blended into one. The former, like the marine Snails, lay numerous small ova inclosed within a gelatinous mass in the water; the latter deposit their ova free, covered with a membranous, and occasionally calcareous shell, in the earth. Copulation is effected in all by a reciprocal interchange of the androgynous species.

I have observed that individuals of Limnæa auricularia can continue to propagate for several generations without being impregnated; they must impregnate themselves.

Order 2. Conchozooid Snails—Diœcii.

3497. Branchiæ situated within the cavity of the mantle, and hanging down in the form of a comb; shell for the most part spirally contorted; sexes separate—Pectinibranchiata.

Here belong the Capulidæ, Turbinidæ, Neritæ, Conidæ, Volutidæ, Buccinidæ, Muricidæ and Strombusidæ.

The tentacula are not retractile, and have the eyes mostly seated at their basis; the penis is external, very large, and cannot be drawn within the body, but is only reflected into the pallial cavity; most of them have a protrusile perforating proboscis, and an operculum. They lay numerous small ova, contained within large membranous cases, which frequently hang in rows to each other like a necklace of pearls. The shells are in some instances horny, in others of a stony texture.

Fam. 4. Mussel-Snails, Capulidæ.

Only one branchial comb within the mantle, and covered merely by a flat hood-shaped shell; no operculum.

Here belong the Capulidæ; all the species are marine. They prefigurate the Brachiopoda.

Fam. 5. Typical Snails, Turbinidæ.

Two branchial combs, mantle devoid of a respiratory groove, shell turbinated; mostly furnished with an operculum, and that indeed of a stony texture. Here belong the Turbinidæ, and Trochidæ, such as the Cyclostomata, Paludinæ, Ampullariæ, Janthinæ, Neritæ. Live in the sea and in fresh-water.

Fam. 6. Kracken-Snails, Buccinidæ.
Like the preceding family, but their mantle has a groove, and there is a horny operculum. Here belong the Conidæ and Volutidæ, Buccinidæ, Muricidæ, Strombusidæ. Almost all live in the sea and are of sanguisugal habits. They antetypify the Cuttle-fish.
Sixth Class.
Cardiac, Nephritic Animals—Kracken.
3498. Body cylindrical, without sole or foot; frequently venous hearts together with the arteriose hearts, and a kidney-like organ.
Such are the proper Cylindrical Snails, which are either sessile or fixed, or move themselves by fins and what have been called arms—Sessile and Natant Kracken. Here belong the Salpæ, Cirripedia, Brachiopoda, Pteropoda, Heteropoda, and Cephalopoda. They all live in the sea. Branchiæ and sexual relations very varied.
They divide likewise into two orders, in accordance with the two circles of Mucus-animals and Conchozoa; the former are spathiform, mostly gelatinous, and firmly sessile or fixed; androgynous, without head and rudders or fins. The branchiæ trellis-like and filiform—Ascidiæ, Cirripedia and Brachiopoda.
The others have a kind of head with arms and eyes, or fin-shaped steerage-organs upon the body; branchiæ pectiniform, reticular, and leaf-shaped—Pteropoda, Heteropoda, and Cephalopoda.

Order 1. Protozooid Kracken—Rumpfkracken.
3499. Body sacciform, mostly gelatinous, without head and eyes; species androgynous. They are destitute of independent locomotion, most of them being quite stationary or fixed, and surrounded frequently by a shell; some few float about passively in the sea.
Fam. 1. Infusorial Kracken—Salpæ.
The Salpæ are gelatinous, freely natant cylinders, perforated by an open tube, within which are situated the branchiæ, heart, mouth, intestine, and liver, without any tentacula. They continue to hang for a long time to each other, as though they were in the ovary.
They undergo a remarkable metamorphosis, which traverses two generations, the young meanwhile not resembling the mother but the grandmother.
Body within a sacciform mantle, with two opposite respiratory apertures.
Fam. 2. Polypary Kracken—Ascidiæ.
The Ascidiæ have a sacciform body with two co-approximate respiratory apertures; branchiæ internal and trellis-shaped.
These gelatinous or cartilaginous animals adhere firmly to rocks, with the respiratory apertures directed upwards as in the Mussels, but the branchiæ are not foliiform; the internal cavity is lined with a membranous sac, upon which the branchial vessels are dispersed in a trellis-like manner. At the bottom of and within the sac is the mouth, which is destitute of tentacula. Intestine, liver, and a heart that is simple or undivided. Mode of propagation unknown.
They are frequently connato-sessile, like the Polyps.
Fam. 3. Acalephan Kracken—Cirripedia.
Body sacciform, without branchial openings, oral apertures situated at the inferior extremity of the longitudinal fissure in the mantle, and provided with a kind of maxillary apparatus; above or behind the mouth are annulate or horny filaments representing caudal fins or feet, upon the basal joints of which is a branchial filament. Here belong the Balani and Lepades—Fuss-kracken.
These animals are likewise sessile, at least when full grown, and are surrounded by shell-plates, which remind us of the Echini or Sea-Urchins, whose antetypes they are; this resemblance is

particularly distinct in the Balani, which have similar odontoid valves upon the aperture or mouth of the shell. These animals also undergo a metamorphosis, since in the early periods of their existence they swim about like small Crustacea, and later on become fixed or sedentary, and surrounded by the plates of shell.

The Lepades have the ordinary shells of Mollusca, but each of the two valves consists of two pieces, between which an odd one is interposed down the back.

These five shell-pieces become in the Balani a regular set of teeth, which are adherent within the cylindrical shell, by which the posterior part of the body, or the peduncular and elongated mantle, is surrounded. (Vid. Oken's 'Allgemeine Naturgeschichte,' V, 1835, 509.)

They bear a resemblance to the Crustacea, not simply in the annulate horny feet, but also in the double and ganglionic nervous chord lying along the abdomen; only they are androgynous, which occurs in no Crustacean.

Order 2. Conchozooid Kracken—Kopfkracken.

Tentacula upon the head or fins upon the body. They repeat the Conchozoa or Shell-animals.

Fam. 4. Mussel-Kracken—Brachiopoda.

The body is surrounded by a mantle open above, and by two shells; on the sides of the mouth there are two arms.

These animals look exactly like Bivalve Mollusca; they are firmly adherent to rocks, being frequently supported upon a hollow pedicle, which is a prolongation of the mantle. Branchiæ and sexual parts but little known; species probably androgynous—Armkracken.

3500. In the following families the body is free and sacciform, provided with arms or fins, and mostly with eyes upon a kind of head.

These animals have the mouth situated above or in front of the body, and are distinctly separated into mantle and abdomen, while they are frequently surrounded by a shell. They have a kind of head with tentacles or arms, and frequently furnished with eyes like the Snails, or pretty nearly like the Fishes; the sexual parts united and separated. They all row, paddle, or steer themselves about in the sea.

Fam. 5. Snail-Kracken—Pteropoda.

Body mostly gelatinous, sacciform, closed all round and free; fins upon the sides of the neck, with two tentacula frequently projecting from near the mouth; androgynous, naked, and inclosed within a shell.

These transparent animalcules swim about upright in the sea, and wave the fins in this position as rapidly as a butterfly does its wings. Most of them are covered with a spathiform and likewise transparent shell. The branchiæ are placed externally on the body, but are not very distinct. Several species have an appendage in front of the neck, which is obviously a rudiment of the Snail's pad or sole—Flugelkracken.

The Heteropoda appear to resemble the Plantar or Sole-Snails, but the body is mostly gelatinous, and the foot or sole is only compressed to form a fin, so that they can only swim but not creep with it. Many of them have a shell almost like that of the Argonauta. Pterotracheæ—Ruderkracken.

Fam. 6. Typical Kracken, Cephalopoda or Sepiæ.

Muscular animals inclosed within a sacciform mantle, open in front; head furnished with eyes, and surrounded by more than four strong prehensile arms; laminated or leaf-like branchiæ within the mantle, sexes separate; in the body is a nephroidal or kidney-like gland, which secretes a dark-coloured or inky fluid, on which account they are called by us 'Dintenschnecken'—Sternkracken. They are obviously the highest organized Malacozoa, and already remind us strongly of Fishes, partly from their size, partly by their fleshy body, and in part by their perfect organs of vision.

The body is often as large as the trunk of a man; the head being separated from it by means of a neck, has in its interior a kind of cartilaginous brain-case or skull, with externally a pair of jaws resembling a Bird's beak, and with eyes tolerably similar to those of Fishes. Auditory organs are also present, which consist of a tympanic cavity inclosing an otolithe or ear-ossicle; nostrils are wanting. The so-called arms are perfect organs of locomotion, as also useful for seizing the prey, being provided for this purpose with sucking cups, which adhere by producing a vacuum, when applied. In other respects these arms are nothing else but the Snail's sole divided in front into lobes. The ova resemble berries, and hang attached to one another in branched clusters like a bunch of grapes. The Sepiæ possess a remarkable gland, which is complicated with the liver, and secretes a dark brown juice, which has been called ink or sepia; it probably ranks in the significance of the kidneys.

Many are covered with a shell externally, as the Nautili and Argonautæ; but in the common Sepiæ this shell adheres within the mantle upon the back, and forms a straight lamina or plate, which is sometimes horny, sometimes calcareous in its texture. It has been called the "white fish-bone." What is regarded as the dorsal side of the animal is its ventral side, because upon the former lies the pallial aperture and the anus, while the vitelline sac is also inserted there.

In the form of the body, as well as in the circular position of the cephalic arms, the Sepiæ strikingly resemble their antetypes, the Acalephæ and Cirripedia, and their metatypes, the Asteriadæ and Crustacea.

With these animals the second stage or circle through which the Dermatozoa pass, or that of the Sexual animals, is closed. They require only a slight additional grade of perfection, and they would pass over into another class of animals. Thus were the arms to become horny and articulate, they would be Crustacea; had the head a nose, and the body consequently a myelon or spinal chord, they would then be Fishes.

SECOND CIRCLE. RESPIRATORY, CUTANEOUS ANIMALS—ANCYLIOZOA.

3501. Body annulate.

Here range the Worms, Crustacea, and the proper or volant Insects.

3502. In this circle a remarkable relation is revealed. If, namely, its classes, orders, and families be compared with those of the two preceding circles, it is then distinctly shown that the present one again commences quite from below, ascends parallel with both these, and transcends them in its highest inmates or members.

As an example of the first case, the imperfect condition of some Entozoa, and their great resemblance to Infusoria, is sufficiently expressive.

The Worms evidently pass parallel to the Mucus-animals, and the Entozoa indeed to the Infusoria, the Red-blooded Worms to the Polyps, but the Holothuriæ chiefly to the Acalephæ, near which they are still arranged up to the present day, although they are articulated, and have an intestine and vessels.

Thus this class traverses the three inferior classes, and consequently imparts to them only the value of orders, or must itself be raised only to the rank of an order, a step, however, which, considering the great number of Worms, would not be suitable or just. The names may therefore remain for the sake of uniformity; only it must be borne in mind that they are of unequal value.

The same relation is exhibited in the Crustacea or Crabs. They obviously admit of being parallelized with no other animals than the Malaco-or Conchozoa. The Entomostraca and Crabs evidently repeat the Mussels, the Aselli the Snails; but the Spiders and Scorpions the Kracken or

Cephalopoda. Consequently here also the orders correspond to the preceding classes.

The true or proper Insects transcend the Sexual animals, and therefore conform to the classification of their own circle, namely, they repeat the Worms and Crabs, and finally mount forth upon their own stage.

The inferior or lower organized animals form consequently two series, viz. the Smooth and Annulate, which in their lower members range parallel, and that indeed after the following manner:

A. Smooth.		B. Annulated or Ringed.	
I. Protozoa.		I. Worms.	
a. Stomach,	1. Infusoria.		1. White-blooded Worms.
b. Intestine,	2. Polypi.		2. Red-blooded Worms.
c. Absorbents,	3. Acalephæ.		3. Holothuriæ.
II. Conchozoa.		II. Crustacea.	
d. Veins,	4. Mussels.		4. Crabs.
e. Arteries,	5. Snails.		5. Aselli.
f. Hearts,	6. Kracken.		6. Spiders.
III. Ancyliozoa.		III. Flies.	
g. Retia,	7. Worms.		7. Tracheoptera.
h. Branchiæ,	8. Crabs.		8. Dictyoptera.
i. Tracheæ,	9. Flies.		9. Ceratoptera.

The Worms have simply an annulate body with reticular or filamentary branchiæ, but no feet; the branchiæ are supported by the body.

The Crabs have an annulate body with feet; the branchiæ are supported by limbs.

The Insects have an annulate body with feet and wings; the body is supported by the branchiæ.

Seventh Class.

Reticular, Papillose Animals—Worms.

Protozooid Ancyliozoa.

3503. There are Worms with a soft body and white blood, without proper tentacula; others with red blood, with and without tentacula, having also bristles along the sides of the body; lastly, the rings of the body become hard or muscular, and a circle or ring of osseous maxillæ forms around the mouth, which is usually surrounded by numerous tentacula, as in the Star-fishes and Holothuriæ, in whom the blood is nevertheless colourless.

They divide accordingly into three orders, and each of these again into three families, in accordance with the classes, namely, those of the Protozoa or Mucus-animals, through which they pass.

Order 1. Infusorial Worms—Parasites.

3504. The White-blooded Worms or Entozoa are very imperfect animals, a fact which proves that this circle recommences quite inferiorly or from below. They live for the most part in the interior of other animals, being thus located in darkness and in places where they obtain but little oxygen. Their blood therefore is not simply colourless, but even its vessels are only imperfectly developed. They respire without doubt through the tegument. In many, a distinct or separate intestine being wanting, it is the tegument also which digests; in others the intestine is simply a sac without an anus. The sexual parts too are in many species of a doubtful character, while meantime there are both androgynous and diœcious individuals. In the latter the male parts open invariably at the posterior extremity of the body, as in Insects; the female parts in front of its caudal end, as in the Crustacea. Both are in other respects constructed as in Insects, namely, there are two oviducts or seminal ducts, which unite before reaching their external orifice. They divide into three groups.

In one of these the body is tolerably smooth, and the pharynx commencing from a simple suctorial mouth elongates into a ramified intestine without an anus; they are androgynous—Saugwürmer.

In others the body itself supplies the place of the stomach; it is corrugated, and receives the food through one or several orifices, without separating into a special intestine. They have nearly all a claviform proboscis or rostrum, with which they perforate substantial cavities for themselves; it would seem that they are both androgynous and of separate sexes—Hydatids and Tæniæ.

Others are separated into tegument and intestine like the Mussels and Snails, but without having a distinct vascular system, heart, and liver, though provided with a nervous chord and separate sexual parts—Ascarides.

Fam. 1. Monad-like Worms, Saugwürmer.
Body tolerably smooth, suctorial mouth and sucker, the intestine losing itself in the tissue of the body, and without an anus; androgynous.

They remind us through their small size and structure of the Monades, especially the Cercariæ, and among the Worms typify or prefigurate in particular the Hirudines or Leeches, both by their form and power of suction, as also in the ramification of the intestine.

Many Cercariæ might be metamorphosed into Distomata, and so be or constitute their young. The Planariæ are slightly different from the Liver-flukes or Fasciolæ hepaticæ, since they are white, bloodless, have a ramified intestine without anus, and increase also by fissiparous generation. They belong to the present order, although they live not as parasites, but in a state of freedom.

Fam. 2. Rhizopodoid Worms, Hydatids.
Body annulate or articulate, with a claviform proboscis without an intestine; they appear to have several mouths.

The Cystica develop ova simply upon the internal wall of the tegument; the body's last ring is expanded into a cystic form, and is mostly much larger than the whole body. The Cœnurus cerebralis, or Brain-hydatid, ramifies like the Polyps.

The Cestoidea have in each joint or member of the body an ovarium, and, as it would appear, male parts also, so that they are androgynous. The sexual orifices are at the border or upon the surface of the posterior members. The member can be viewed as an ovarium liberating itself; as in the Lerneæ and Arguli.

The Filariæ or Thread-worms, which also belong here, are in the form of a cylindrical tube, with separate sexual parts, which open posteriorly.

Fam. 3. Rotiferal Worms, Ascarides.
Body cylindrical, intestine free, with mouth and anus; sexual parts separate.

Upon the mouth these Entozoa have some papillæ or points, which may be regarded as tentacula. The larger species have a distinct nervous cord. The Gordius cannot well take any other position than the present, although it does live in a state of freedom.

Order 2. Polypary Worms—Hirudines.
3505. Body cylindrical with branchiæ in or upon the tegument, blood red; two ganglionic nervous threads run along the ventral surface of the body, as in Insects; all are indeed androgynous.

Here belong the Red-blooded Worms, as Hirudines, Lumbrici, Nereides, and Serpulæ.

Fam. 4. Tubularial Worms—Dermatobranchiata.
Body cylindrical. The branchiæ are only a vascular network in the tegument.
Here belong the Hirudines and Lumbrici.
The Hirudines have a perfectly naked body, without filaments and bristles, a perfect network of vessels containing red blood within the tegument, an intestine with an anus, both sexual parts androgynoid; posteriorly they have a pad or sucker; in the mouth there are mostly maxillæ, and upon the head simple eyes.
The Naiades and Lumbrici have bristles in longitudinal rows upon the sides of the body. The latter are androgynous. The Naiades increase by division.
The Thalassemæ have a protuberant white body; red blood-vessels only upon the intestine; mouth rostriform.
These Worms stick in the mud, and are nourished by it. They have here and there bristles, but which do not form any longitudinal rows. They appear to absorb water through the skin, and respire by the intestine. They cannot, by reason of their red blood, range with the Holothuriæ.
Fam. 5. Alcyonioid Worms—Notobranchiata.
Body depressed, with pedal filaments, tentacula and branchiæ in two longitudinal rows upon the back.
These Worms all live in the sea, adhering mostly upright in the earth, and have frequently annulate, rigid tentacula, ocelli, and pairs of maxillæ, like Insects. The gills are not unfrequently covered with pergamentaceous scales, as in many Nereides, and especially in Aphrodite.
Fam. 6. Actinioid Worms—Auchenobranchiata.
Body inclosed within a pergamentaceous or calcareous tube with lateral bristles, branchiæ and tentacula upon the neck or head.
The neck is surrounded by a kind of mantle, almost as in the Snails, so that several animals have been arranged here, which are now known to be veritable Snails. Upon the head many have a horny operculum, whereby they can close the shell.
Here belong the Amphitritæ, Terebellæ, Serpulæ, and Sabellæ.
Order 3. Acalephan Worms—Sternwürmer.
3506. These creatures attain the most perfect structure of the Worms. The blood is white, the form cylindrical, globular and stellate; the mouth surrounded by a ring or wreath of maxillæ. The nerves form a ring around the pharynx, and upon the latter are placed membranous cysts, which spirt water into the tentacula or feet as they have been called, and thereby expand them.
It is impossible for these animals to continue to range with the Acalephæ, although they resemble them in outward form; for they consist of two cysts, the intestine having become freed from the outer one as a special sac; moreover, they have a perfect vascular system, distinct muscles, a mouth with a dental apparatus, which prefigurates a complete skeleton, a peculiar vascular system for the injection of the tentacula or feet, a nervous ring surrounding the pharynx, an ovarium entirely separated, and lastly, a perfectly annulate body.
Fam 7. Physalial Worms, Holothuriæ.
Body cylindrical, very plentifully supplied with muscles. Mouth and anus present, the former of these being surrounded by a dental wreath and ramified branchimorphous tentacula; branchiæ upon the intestine, and branchial aperture at the anal extremity. They correspond to the Physaliæ.
Their tegument is a perfect muscular tunic, consisting of several longitudinal bands, which pursue a downward course upon its internal surface; in other respects being transversely corrugated and full of papillæ, partly replete with hollow podoidal filaments, which by the injection of water into them are elongated; thus, they present a resemblance to the Acalephæ.

Fam. 8. Beroeal Worms, Echini.
Are at bottom Holothuriæ with ossified tegument, or Asteriadæ with shortened rays.
The Echini have an anus, pretty like that of the Holothuriæ, and also similar feet, which are protruded in a similar manner through apertures in the shell. The dental wreath surrounding the mouth is a complete bony framework, which has been called, on account of its form, "Laterna Aristotelis," and bears a considerable resemblance to the valves of the Balani, the antetypes in general of the Echini, just as these are of the Opossum Shrimps.
Fam. 9. Medusal Worms, Asteriadæ.
Body stelliform, consisting of bony rings; contains a free multilobular intestine with blood-vessels and branchiæ; several ovaria; it would appear that they are without male parts.
Here belong the Encrinites, Pentacrinites, and Asteriadæ proper. Around the mouth of the Encrinites and Pentacrinites stand long, ramified, and likewise articulated tentacula, which remind us of the arms of the Acalephæ and Cephalopoda.
Fundamentally too, in the Star-fishes, the disk only is the body proper, and the rays are the tentacula, which, being monstrously developed, surround the mouth; what have been called pedicles upon these, represent the suckers of the Sepiæ.

Eighth Class.
Branchial, Pedal Animals—Crustacea.
Conchozooid Ancyliozoa.
3507. The Crustacea are Worms with horny rings to the body, jointed feet and tentacula, which mostly breathe by branchiæ.
They correspond to the Conchozoa, or Shell-animals, and divide therefore into three orders, each of these including six families.
Here belong the Crabs, Wood-lice, and Spiders.
In the lowest forms the head, thorax, and abdomen are blended together, and the back mostly covered with a great horny scute—Mussel-like Insecta.
In the next place the thorax and abdomen admit of being clearly distinguished, both by their form as well as by their appendages, which are much larger on the thorax, as in the Cray-fish.
Then follow cylindrical forms with numerous feet, appended to uniform rings, but having a head free and distinguished by tentacula, maxillæ, and eyes—Wood-lice.
The Wood-lice bear a resemblance to the air-breathing Insects, although all parts of the body still pass over or blend gradually into each other.
In the Arachnida or Spiders aerial respiration takes place, and their body enters into relations with the more highly organized Insects, since the head becomes small, the abdomen large, thick, and short.
As the Crustacea pass parallel to the Conchozoa or Shell-animals, so do they divide into three cohorts or groups: the Mussel-, Snail-, and Kracken-like Crustacea, or Crabs, Wood-lice, and Spiders.
Cohort I. Mussel-like Crustacea—Crabs.
3508. Head and thorax connate, being mostly covered with a shell or shield, abdomen stunted or caudiform; maxillæ and branchiæ present.
Order 1. Lochmuschelartige, Pfriemenkrebse.
3509. The feet mostly simple and pointed, without large claws or forcipes, and furnished with setæ or lamelliform branchiæ; eyes sessile.
Fam. 1. Pholadoid, Mussel-Insects.

Small, almost microscopical, and slightly annulated animals with an uni-or bi-valvular dorsal shield or testa, stunted maxillæ, and few legs, having attached to them setiform branchiæ. These animals are found in all stagnant waters, in which they are incessantly paddling or rowing themselves about. They remind us of those Infusoria which, like Brachionus, are covered by a scute or shield. They are what have been called Monoculi (Entomostraca or Lophyropoda), e. g. Daphniæ.

Number of feet small, and beyond their appended setæ there are no branchiæ; two eyes frequently blended together; antennæ mostly furcate like the feet; sexes separate; a perfect circulation within a true Mussel-like ventral cavity. They are microscopic Mussels with eyes and feet.

Fam. 2. Tellinoid, Branchiopoda.

Similar animalcules to the above, but with a body strongly annulated, naked or covered with a double shell, and provided with numerous feet, unto which are attached leaflets that serve as branchiæ.

Here belong the Branchiopoda properly so called.

Fam. 3. Cardiaceoid, Shield-Crabs.

Large animals, with strongly annulated body and numerous feet; head and back covered by a shield, whereupon are the eyes—Phyllopoda.

Here range the Trilobites, Apus, and the Molucca or King-Crabs.

Order 2. Spaltmuschelartige, Scheerenkrebse.

3510. Five pairs of thoracic feet, of which the first pair is mostly large and forcipiform; the eyes upon moveable pedicles or foot-stalks.

These animals usually attain a striking size, and are generally the largest among the horny Ancyliozoa.

The sexual parts open mostly upon the posterior legs.

The abdomen or tail usually supports five pairs of stunted feet, to which the ova hang.

The number of maxillæ agrees tolerably well with that of the thoracic feet.

Fam. 4. Mytiloid, Locust-Crabs.

All the feet of pretty equal length, and the forceps stunted in size; the branchial plates free upon the abdominal feet—Squilla.

Fam. 5. Arcaceous, Macroura.

Abdominal feet arrested, branchial comb upon the femora of the thoracic feet and beneath the dorsal shield, tail extended, as in the common Cray-fish.

Fam. 6. Oyster-like Crustacea, Brachyura.

Characters similar, but tail bent under the body—Taschenkrebse.

COHORT II. SNAIL-LIKE CRUSTACEA—ASSELN.

3511. Body mostly cylindrical, annulate, without a true carapax or scute; head free; feet short and simple; branchiæ vesici-or folii-form; mostly placed under the tail; eyes non-pediculated, or even wanting.

Order 3. Androgynoid—Saugasseln.

3512. Body soft and slightly annulate, maxillæ, feet and branchiæ arrested; live by suction as parasites upon other animals, mostly fishes.

Fam. 1. Tritonia-like, Lernæaceæ.

Body soft, without shield, eyes and branchiæ; feet and maxillæ stunted; ova carried in two tubes at the hinder part of the body—Kiemenwürmer or Lerneæ.

Fam. 2. Patella-like, Argulaceæ.
Body distinguished into head, thorax, and abdomen, with few natatory feet, maxillæ moulded to form suctorial organs, mostly eyes; lay also strings of ova; head elongated in many in a scutiform manner. The so-called Fish-lice or Argulaceæ.
Fam. 3. Luftschneckenartige—Pycnogonidæ.
Body short, with four pairs of long feet; eyes, but no branchiæ and maxillæ; abdomen puny in size. The Whale-lice.

Order 4. Kammschneckenartige—Nagasseln.
3513. Body cylindrical, horny, and distinctly annulate, maxillæ, eyes, and mostly seven simple pairs of feet, branchiæ as cysts or leaves.
Fam. 4. Capulidenartige, Walzenasseln.
Body cylindrical, with five or seven pairs of feet, and some branchial vesicles; body and abdomen very puny. The Læmodipoda; Caprella, Cyamus.
Fam. 5. Turbinidenartige, Seitenasseln.
Body horny and distinctly ringed, mostly compressed, with perfect maxillæ; seven pairs of thoracic feet and branchial vesicles; abdominal feet rudder-shaped. They swim usually lying upon the side; many leap—Amphipoda; Flohkrebse or Gammarinæ.
Fam. 6. Buccinidenartige, Sohlenasseln.
Similar to the preceding family, but the body is depressed, and the abdominal feet furnished with branchial plates. The Isopoda; Oniscidæ, to which belongs the genus Armadillo.
COHORT III. KRACKEN-CRUSTACEA—KOBE.
3514. Body not tripartite; spiracula or air-openings; more than three pairs of feet, no wings. Here belong the air-breathing Crustacea; Scolopendræ, Acari, Scorpions, and Spiders.
These animals are abruptly distinguished from the preceding by a conversion of the branchiæ into spiral-shaped tracheæ, which ramify and traverse the whole body. They all therefore live in the air, and if they do also dwell in the water, they still come to the surface to inhale that element. The eyes are only simple points or ocelli, which are accumulated frequently upon the sides of the head.
The most inferior of them are distinguished from the preceding cohort, or the Asseln, by almost nothing save the essential character of their own cohort, the tracheæ. They have mostly a number of feet, and only simple eyes, as the Scolopendræ.

The following have a short body, in which the abdomen predominates; thorax and head connate; never more than four pairs of feet—Acari, Scorpions, and Spiders.
They likewise divide into two orders, like the Kracken.
Order 1. Rumpfkrackenartige—Langkobe.
3515. Body horny, tolerably cylindrical and uniform; feet mostly very numerous.
Fam. 1. Salpoid, Spindelkobe.
Only three pairs of thoracic feet, but still podoidal appendages to the abdomen—Podura, Lepisma.
Fam. 2. Ascidioid, Schnurkobe.
Body cylindrical, with very many feet; sexual parts on the thorax—Juli.
Fam. 3. Cirrhopodoid, Bandkobe.
Body band-shaped and depressed, maxillæ perforate, sexual parts placed posteriorly—Scolopendræ.

Order 2. Kopfkrackenartige—Kurzkobe.
3516. Body thick, mostly globiform; head and thorax blended together; only four pairs of feet.
Fam. 4. Brachiopodoid, Acarides.
Body rounded; all three parts confluent; usually only two simple eyes. The Acari are mostly so small, that their parts can only be distinctly seen through the microscope. Their mouth is always very much arrested, and has maxillæ, which are in some cases adapted for manducation, in others for suction.
Fam. 5. Pteropodoid, Scorpions.
Body tolerably cylindrical, and all three parts connate; palpi very large and forcipiform.
Fam. 6. Sepioid, Spiders.
Body rounded, head and thorax connate, abdomen separate, mostly eight simple eyes.
Their most remarkable organs are the four spinnerets in front of the anus, which probably stand in the signification of the renal organs, just as the material of the Spider-threads does in that of the urine.
The tracheæ are but few in number, and expand into lung-like vesicles.

The sexual parts do not lie posteriorly by the anus, but at the root or base of the abdomen.
It is moreover remarkable, that their mandibles are perforated and pour a poisonous juice into the wounds they inflict. They must be therefore regarded, like the venom-teeth of Serpents, as elongated salivary ducts.
The Crustacean families can also be named according to the orders of their circle, namely, the Worms and Crustacea, e. g. thus:
Order I. Worm-Crabs—Pfriemenkrebse.
Fam. 1. Entozooid—Mussel-Insects.
2. Hirudinous—Branchiopoda.
3. Holothurioid—Shield-Crabs.
Order II. Crustacean Crabs—Scheerenkrebse.
Fam. 4. Typical Crustacea—Sea-Mantes.
5. Asselkrebse "   —Macrura.
6. Arachnoid "   —Brachyura.
Ninth Class.
Tracheal, Alary Animals—Flies.
Typical Ancyliozoa.
3517. Body tripartite, only three pairs of thoracic legs, spiracula and wings. They divide, according to the classes of their circle, into three cohorts, of Worm-, Crab-like, and perfect Flies. As the wings are their characteristic organ, so also must they be divided according to them, and not after the cibarial instruments. Those wings which are developed to the least degree are the homogeneous, transparent wings traversed by few tracheæ or ducts—Tracheoptera, as Flies, Bees, and Butterflies.
Their chrysalis or pupa condition is perfect.
Then follow wings having very numerous veins united by transverse veins so as to form a network—Dictyoptera, as Neuropterous, Orthopterous, and Hemipterous Insects.
The pupa is moveable.
Finally, the anterior and posterior pair of wings become wholly dissimilar to each other, the former being horny, and the latter membranous in texture with reticulated veins, provided likewise with joints like the legs, so that they can be folded up under the anterior pair or elytra—

Ceratoptera, as the Beetles.
Pupa state perfect.
Strictly speaking, it is a matter of indifference whether the names of orders and families be adopted from the first or second parallel series; whether e. g. we speak in the first cohort of Protozooid or Worm-flies. The nearest series will, however, bear the greater amount of resemblance to them. Meanwhile I will in the sequel vary in the choice of names, in order to exhibit different samples of this double parallelism.
COHORT I. WORM-FLIES—TRACHEOPTERA.
3518. Wings membranous with few longitudinal ducts, and almost devoid of transverse ducts, eyes larger than head. Here belong the Flies, Bees, and Butterflies. The abdomen is indeed annulate, but soft; its first ring is frequently set free and unites with the thorax, but supports neither feet nor wings.
The sexual parts always lie at the anal extremity.
The head is almost nothing but eye, and the Insects of this order may very well be termed Megalopidæ, or large-eyed, out of contrast to the succeeding ones. Between the two large compound eyes there are usually found three simple ocular puncta or ocelli, which they have adopted from the preceding class.
The larvæ are either entirely apodal, white and soft like Entozoa, or they have, in addition to the thoracic feet, numerous abdominal feet like the higher Worms.
The Flies with their soft and imperfect body, and the apodal larvæ, repeat the Infusoria and Entozoa; the Bees therefore the Polyps and Red-blooded Worms; the Butterflies with their large farinose or dusty wings and polypodal caterpillars, the Acalephæ and Holothuriæ.
Order 1. Entozooid Flies—Diptera.
3519. All the thoracic rings coalesced, and along with them the first or basi-abdominal ring, unto which are attached the halteres or balancers; only the two posterior wings present; the labium prolonged into a proboscis, which incloses the setiform maxillæ. Larvæ apodal and white.
The Diptera resemble the Entozoa in a striking manner through their apodal, soft and white larvæ, and even through their habitation, which is mostly in fetid and moist animal matters. Added to this, they respire usually through two tubes which open upon the anus.
Many during their metamorphosis do not shed their skin; but it becomes, during the pupa condition, only horny, representing a small case, whose bottom springs up in front like an operculum, and gives exit to the perfect Fly which has been therein developed.
Upon the first abdominal ring of the imago two nodose pedicles stand out, which are called halteres; they are probably the old respiratory tubes.
The maxillæ have changed into bristles, which act like pestels within the groove-shaped labium, puncture and pump in the fluid.
They divide into three families like the White-blooded Worms, or according to the orders of their cohort.
Fam. 1. Typical Diptera, Schmeissen.
Antennæ triarticulate, the last joint being mostly spatulate and furnished with a lateral awn or bristle; only two suctorial setæ inclosed between the terminal valves of a fleshy proboscis, which admits of being retracted within a large cephalic or oral cavity. The Muscidæ, unto whom the Common House-flies belong, and also the Hippoboscidæ or Louse-flies. It is my opinion too, that the Fleas also are to be ranged among the Diptera.
Alliance 1. Klappenmucken. The two punctuating setæ placed between two valves without a proboscis—Flea (Pulex) and Louse-fly (Hippobosca.)

Alliance 2. Acalyptera. Proboscis having thick lips, and retractile into a large cephalic cavity; alulæ or halteral opercula arrested—Hypocera to Dolichocera.

Alliance 3. Dung-flies. Characters similar to the above, but the alulets are of considerable size—Muscidæ.

Fam. 2. Hymenopteroid Diptera, Dasseln.

Antennæ as in preceding family, but the bisetaceous proboscis is thin and horny, with small labia, or else large labia with four setæ—Œstridæ, Conopidæ, and Syrphidæ.

Alliance 4. Parasitic Diptera. Two setæ with or without a lipless proboscis—Œstridæ, Myopariæ, and Conopidæ.

Alliance 5. Syrphidæ. Four setæ within a similar kind of proboscis; the third joint of the antennæ spatulate; the palpi thickened.

Alliance 6. Leptidæ. Four setæ within a short thick-lipped proboscis, the third antennal joint mostly coniform—Therexidæ, Leptidæ, and Dolichopidæ.

Fam. 3. Lepidopteroid Diptera, Gölsen.

Antennæ multiarticulate and stipiform; proboscis with and without labia, mostly four and six setæ for punctuation—Tipulidæ, Tanystomidæ, Stratiomydæ, Tabanidæ.

Alliance 7. Gnats. Antennæ filamentary and multi-articular. Proboscis varied.

Alliance 8. Spiessmucken. Four setæ within a hastate horny proboscis with or without arrested labia; third antennal joint not annulated, with the bristle or awn at the extremity—Tanystomidæ, as Asilidæ, Empidæ, Bombylidæ.

Alliance 9. Stielmucken. Four or six setæ within a thick-lipped proboscis; third antennal joint stipiform and annulated—Stratiomydæ and Tabanidæ.

Order 2. Leech-flies—Hymenoptera.

3520. Four naked veined wings, labium mostly elongated, the two pairs of maxillæ acting above it like spears. Larvæ mostly apodal, or with more abdominal feet or prolegs than in the true Caterpillars.

Most of the larvæ still bear a great resemblance to the Entozoa, though they do not respire like them through the anus, but by lateral apertures or spiracula; the larvæ with abdominal feet repeat the Nereidæ and Aselli. In other respects they live no longer in putrid moisture, fungi, roots, and such like bodies, but in living animals or in cavities specially prepared for their reception by the parents, and even in a state of freedom upon leaves.

And here is particularly worthy of our notice the structure of the cells, which are by many species fabricated quite substantially of wax or wood-shavings, and are to be compared with the webs of Spiders, since both serve as nests for the young. They repeat the Polypidoms.

Others make cases of leaves, and carry into them honey upon which to deposit their ova.

Others again simply bore holes in wood or in the earth, in order that they may in a similar manner provide their young with honey or larvæ. Finally, others simply stick, by means of their ovipositor, the ova into animals or leaves.

The dwellings are fabricated by the maxillæ, which but seldom serve as cibarial instruments, seeing that the labium undertakes this office as a lick-organ.

Another remarkable feature of this order is the arrested development of the sexual parts of the females in certain generations, an occurrence which depends upon the time of the year or the size of the cells, and whereby they are constrained to lead a social life as workers.

They divide, according to the Red-blooded Worms or the orders of their cohort, into three families.

Fam. 1. Dipteroid Hymenoptera, Bees.
Abdomen aculeate; labium elongated in a rostriform manner. They dig or build cells, and carry thither honey to the larvæ, which are apodal.
Alliance 1. Fossorial Bees—Andrenidæ.
2. Carpenter-Bees—Anthophoridæ.
3. Cell-Bees—Apidæ.
Fam. 2. Typical Hymenoptera, Wasps.
Aculeate; labium not elongated—rapacious Hymenoptera, such as Ants, Fossorial, and Cell-wasps, larvæ apodal.

Alliance 1. Hohlenwespen—Formicidæ.
2. Fossorial Wasps—Sphegidæ.
3. Cell-Wasps—Vespidæ.
Fam. 3. Lepidopteroid Hymenoptera, Terebrantia.
Instead of the sting there is an ovipositor, with which the females stick their eggs into other insects, mostly into caterpillars, or else into leaves and wood—Ichneumonidæ, Tenthredinidæ, and Uroceridæ. The larvæ of the first are apodal, those of the second provided with thoracic and abdominal feet like caterpillars, those of the third with thoracic feet only like the larvæ of Coleoptera.
Alliance 1. Stutzwespen. Ovipositor short; wings nearly veinless. Deposit their eggs in small insects; larvæ apodal—Chalcididæ, Oxyuri, Chrysididæ.
Alliance 2. Schlupfwespen. Ovipositor very long, divided into three hairs; wings veined. Deposit their eggs in caterpillars; larvæ apodal—Ichneumonidæ, Evaniidæ.
Alliance 3. Pflanzenwespen. Ovipositor spiral or saw-shaped. Deposit their ova in plants; larvæ mostly furnished with feet, and frequently resembling caterpillars in form—Cynipidæ, Tenthredinidæ, and Uroceridæ.
Order 3. Holothurioid Flies—Lepidoptera.
3521. Four veined wings, covered with small dust-like scales; maxillæ have coalesced into a proboscis; larvæ with thoracic and abdominal legs.
The larvæ or Caterpillars remind us of the Nereides, especially the setaceous Aphrodites, such as the Holothuriæ, and further still, the Aselli and Scolopendræ. There are Caterpillars, which are scarcely to be distinguished from a Wood-louse. They subsist almost throughout upon leaves, and being exposed to the light are therefore variously coloured. They have maxillæ, and in the labium is the orifice of the salivary glands, from which they spin the threads used in weaving their cocoons, or social tents. Where they crawl they leave, like the Spiders, their threads beneath them.
The pupa state is perfect; the perfect insect generally creeps out of the chrysalis by a slit taking place down the back.
The abdomen is almost throughout covered with hair, which is indicative likewise of a strong mucous secretion.
They take in their fluid nourishment by suction, but the mechanism by which this is effected is not yet known, as from the jaws themselves forming the proboscis, there can be no suckers lodged within the latter that might act. It is probably effected by expansion of the abdomen taking place during respiration. Their deglutition would be therefore a respiratory act.
Besides their very non-artistic webs, no artistic instincts are to be observed in this order. The ova are just layed without more ado upon plants, and rarely upon other objects. The Butterflies are

generally related to the plants, and especially to their corollæ, whose colours and forms they carry in themselves.

They divide according to their proximal orders into three families.

Fam. 1. Dipteroid Lepidoptera, Moths.

Antennæ filiform, wings mostly thrown like a mantle round the body; proboscis short.

Small and nocturnal in their habits, proceeding from tolerably apodal Caterpillars, which reside mostly concealed in plants, or make themselves cases of hairs and leaves.

Alliance 1. Typical Moths, Tineidæ.
a. Typical Tineidæ: Alucitidæ.
b. Pyraloid: Tineæ.
c. Tortrix-like: Crambidæ.

Alliance 2. Bombycoid Moths, Pyralidæ.
a. Tinea-like: Aglossæ.
b. Typical: Hydrocampæ.
c. Tortrix-like: Deltoides or Herminiæ.

Alliance 3. Homalopteroid Moths, Tortricidæ.
a. Tinea-like: Fruit-Tortrices.
b. Pyraloid: Heterogeneæ.
c. Typical: Leaf-rolling Tortrices.

Fam. 2. Hymenopteroid Lepidoptera, Silk-Spinners.

Antennæ filiform, wings tile-shaped.

Lepidoptera of considerable size, proceeding out of

Caterpillars, with many feet, and frequently covered with hair, which live mostly in a state of freedom upon plants, and in some cases make large cocoons freely exposed, in others under the earth—Noctuidæ and Bombycidæ.

Alliance 1. Moth-like Spinners, Noctuidæ.
a. Typical: Hadenæ.
b. Pseudobombycoid: Catocalæ.
c. Bombycoid: Erebidæ.

Alliance 2. Typical Spinners, Pseudo-Bombyces.
a. Noctua-like: Psyches, Limacodes.
b. Typical: Notodontæ.
c. Bombycoid: Callimorphæ.

Alliance 3. Homalopteroid Spinners, Bombycidæ.
a. Noctua-like: Bombyces.
b. Pseudo-Bombycoid: Saturniæ.
c. Typical: Hepialidæ.

Fam. 3. Typical Lepidoptera, Homaloptera.

Antennæ varied; wings flat and expanded; proboscis long—Geometridæ, Vespertina, Diurna.

Large Lepidoptera proceeding from free-living, strongly-coloured, and mostly naked Caterpillars.

The Geometridæ have level, tolerably triangular wings and filiform antennæ; they proceed from naked Caterpillars with few abdominal feet.

The Vespertina have level, long, and narrow wings, and spindle-shaped antennæ; they emerge from perfect Caterpillars.

The Diurna have the wings folded upwards in a reverse position, claviform antennæ, and proceed from perfect Caterpillars.
Alliance 1. Moth-like Homaloptera, Geometridæ.
a. Typical: Geometræ.
b. Sphingoid: Aposuræ.
c. Papilionaceous: Uraniidæ.
Alliance 2. Bombycoid Homaloptera, Sphingidæ.
a. Geometroid: Zygænidæ and Sesiadæ.
b. Typical: Sphingidæ.
c. Papilionaceous: Castniæ.
Alliance 3. Typical Homaloptera, Papilionidæ.
a. Geometroid: Hesperiidæ, Lycænidæ and Erycinidæ.
b. Sphingoid: Nymphalidæ and Heliconiidæ.
c. Typical: Pieridæ and Papilionidæ.
COHORT II. CRUSTACEOUS FLIES—DICTYOPTERA.
3522. Four wings, with longitudinal and numerous transverse veins, the anterior pair being mostly leathery in texture; eyes mostly smaller than head.
Here belong the Neuroptera, Orthoptera, and Hemiptera.
The metamorphosis is tolerably imperfect, and the larvæ have never more than three pairs of thoracic feet, and therefore correspond no longer to the Worms, but to the Crabs, where the number of the feet has been already more determined. Many also strikingly simulate the Crustacea.
In all three families also the pupa runs about, eats, and has rudiments of wings.
The abdominal rings are mostly horny and hairless. The spiracula lie upon the upper margin.
The eyes are generally much smaller than the head, and may therefore, in comparison with those of the preceding and subsequent groups, be said to be of middling size; the simple eyes have for the greatest part disappeared. Meanwhile the first family displays large eyes, the second eyes of a middling size, the third small eyes.
They divide like their antetypes, the Crabs, into three orders.
Order 4. Crab-Flies—Neuroptera.
3523. All the wings alike and membranous; maxillæ with mostly large eyes; abdomen soft.
These Insects which, on account of their delicate wings, are known with us by the name of Flohrfliegen, live mostly upon flesh, and many of them by capturing living prey. Many undergo a short pupa stage. Many live in the water and have, too, branchiæ, whereby they strikingly remind us of the Crustacea, especially the Branchiopoda.
Like the Crustacea, or in accordance with the two first cohorts of their class, they divide into two families.
Fam. 1. Tracheopteroid Neuroptera, Nagbolden.
Do not subsist by rapine, but gnaw slowly animal and vegetable matters; many also as flies eat no longer.

Alliance 1. Dipteroid Neuroptera, Kieferläuse. No wings; gnaw skin, feathers and hair, mostly Bird-lice—Philopterus, Liotheum.
Alliance 2. Hymenopteroid Neuroptera, Blumen-and Mulmlaüse, Thrips, Psocus, Termes.
Alliance 3. Lepidopteroid Neuroptera, Phryganeidæ (Wassermotten.)
Fam. 2. Dictyopteroid Neuroptera, Raubbolden.

Alliance 4. Typical Neuroptera, Flohrfliegen, Panorpidæ, Hemerobiidæ, Myrmeleonidæ.
Alliance 5. Orthopteroid Neuroptera, Raphidiidæ, Mantispidæ.
Alliance 6. Hemipteroid Neuroptera, Ephemeridæ, Libellulidæ.
Order 5. Oniscal Flies—Orthoptera.
3524. Maxillæ and dissimilar reticular wings; the anterior, pergamentaceous; abdominal rings, horny; eyes of moderate size; pupa moveable.
Live for the greatest part on plants.
Fam. 1. Tracheopteroid Orthoptera, Saltatoria.
Alliance 1. Dipteroid Orthoptera, Grylli.
Alliance 2. Hymenopteroid Orthoptera, Locustæ.
Have a hard ovipositor, which reminds us of the sting in the Hymenoptera.
Alliance 3. Lepidopteroid Orthoptera, Acrydia.
Fam. 2. Dictyopteroid Orthoptera, Cursoria.
Alliance 4. Neuropteroid Orthoptera, Mantidæ.
Alliance 5. Typical Orthoptera, Blattidæ.
Alliance 6. Hemipteroid Orthoptera, Forficulidæ.
Order 6. Arachnoid Flies—Hemiptera or Bugs.
3525. Horny suctorial proboscis with puncturating setæ; dissimilar reticular wings, the anterior mostly pergamentaceous, the posterior membranous, and mostly capable of being folded up, as in the Chafers; eyes small, pupæ moveable.
The Bugs have much resemblance to the Mites, and Spiders are as it were winged Mites. One party of them suck blood, another vegetable sap. They manifest no artistic instincts.
The Aphides or Leaf-lice require only a single pairing, to enable them to propagate through several generations. During the summer they produce only females, and in autumn the males first appear.
Fam. 1. Tracheopteroid Hemiptera, Leptoptera.
Alliance 1. Dipteroid Hemiptera, Coccidæ, Aphidæ.
Alliance 2. Hymenopteroid Hemiptera, Cicadæ.
Alliance 3. Lepidopteroid Hemiptera, Fulgoridæ.
Fam. 2. Dictyopteroid Hemiptera, Pachyptera.
Alliance 4. Neuropteroid Hemiptera, Water-bugs.
Alliance 5. Orthopteroid Hemiptera, Narrow bugs,
Hydrometræ, Reduvii, Cimicidæ.
Alliance 6. Typical Hemiptera, Broad bugs, Capsidæ,
Lygæidæ, Coreidæ, Scutelleridæ.
COHORT III. PERFECT FLIES—CERATOPTERA, BEETLES.
3526. Maxillæ and dissimilar reticular wings, the anterior being horny, the posterior membranous and susceptible of being folded upon themselves; eyes small; pupæ motionless.
The Chafers or Beetles are the highest Insects, because they possess the greatest variety of organs, namely, two kinds of wings; not to mention the perfect condition of their manducatory apparatus, antennæ, and the first thoracic ring which is free.
The Beetle represents the Flies in their entire perfection; everything is rendered horny in the former even to the lower wings, which are furnished with several joints, almost like the feet. The upper wings are horny and meet together so closely by their inner margins upon the back, that they form a closed suture, like the shells of the bivalve Mollusc. The antennæ are also more perfect than in other families, and mostly jointed like the feet, while at their extremity they are

frequently thickened into laminated moveable clubs, which open when the animal would fly, as if to listen.

Their habitation, mode of living and subsistence is exceedingly complex, and therein also they combine again all the families of this class, and it might be said, all the preceding classes.
The Beetles are also much more numerous than any other order, and could in this point of view even hold good for an entire class, especially, if they represented a special organic system, but which is not the case.
They live upon vegetable saps, and matters, blossoms, leaves, and wood, living animals, putrid flesh, dung, and such like.
They dwell mostly indeed in concealed situations, but also in those that are freely exposed, while many live in water, and are so subjected to its influence that their larvæ actually respire this element through branchiæ.
The larvæ are white and have three pairs of horny thoracic feet. They live concealed.
The pupæ are invested by a transparent tegument, which tears in an irregular manner.
The Beetles divide distinctly into three divisions, which correspond to the three cohorts of this class, or repeat the three cohorts of their circle.
The Beetles have been pretty generally divided according to the number of tarsal joints into those having three, four, five, or an unequal set of these "articuli," whereby, however, the greatest disorder has originated. I have therefore divided them in my 'Naturgeschichte' according to their mode of living into Phytophaga, Sarcophaga, and Rypo- or Coprophaga, which gradually seems to meet with approval. At least a much more natural arrangement comes to light by using these means.
I have also declared the Rhyncophora to be the lowest and the Lamellicornes the uppermost in rank, a view, which at present appears to be generally adopted.
By these means I obtained two firm points, whereby the division or classification of the intermediate members is uncommonly facilitated, although many may still stand in the wrong place. But who can point out a single system in which Insects range correctly, or, what is more, wherein it would not be easy to prove that no single order and family occupies the right place.
In such trumpery systems of classification, one would think, that some respect might be had for principles, or at least a feeling of shame for the blockhead's weakness of intellect. But in vain!
Where the sense for philosophy or for principles is wanting, it is not to be inculcated or driven in. We still see Natural Histories shoot forth, with whose shrub-like ramification we must have compassion. Every thicket of briars is rooted deep, and admits only of being extirpated by wearisome and patient culture.
The Rhyncophora indicate that the Phytophagous beetles are the lowest; the Lamellicornes that the Coprophagous are the highest. The Sarcophagous chafers accordingly take their place between the two.
Now, however, the Rhyncophora are tetrameral. In like manner are the Phytophagous beetles, such as the Borkenkäfer and Holzschröter. They consequently form the main stem, unto which all the true Phytophaga must be annexed. But among them also belong the Pentameral, namely, the Holzböhrer, Schnell-and Prachtkäfer; I have therefore disposed them in this order, and I believe quite correctly.
The Lamellicorn or Dung-beetles are pentameral, and on that account to be co-arranged with the Necrophaga or Carrion-feeders and Sexton-beetles, as well as with the parasitic Lampyridæ, and the predaceous Raub-and Laufkäfer, despite their exceedingly great difference of living and even

anomalous structure. On the contrary, most of the heteromeral kinds agree in their mode of living, and tolerably too in structure, with the Dung-beetles, at least the Mulmfressenden, such as the Meal-beetles or Tenebrionidæ, and indeed the Stenelytra also, while the Cantharides are parasitic at least as larvæ.

It seems also to me, that the fungi must be viewed as flour or dung, and that therefore those very Fungivora, whose structure does not point the way otherwise, can be placed with the Mulm-and Mistkäfern, although they are only tri-and tetrameral, like the Lycoperdinæ and Erotyli. Besides this the antennæ of the Fungivora are usually thickened, and even club shaped, whereby they thus stand nearer to the Necrophaga, such as the Speckkäfern and Sexton-beetles, as being those which likewise live upon putrid substances. It is difficult to separate the Coccinellidæ from them; their external resemblance to the Erotyli is also striking.

The pentameral Raub-Lauf-and Sandkäfern form, as it were, the trunk or main stem of the Zoophagous beetles, to which are annexed likewise the pentameral Malacopterous chafers, namely, the Lampyridæ, as parasites.

But the heteromeral Cantharides have also the same soft wings and mode of life, so that they are not in a natural system, even if devoid of principles, to be separated from the Malacopterous beetles.

The nourishment of the Lamellicorn beetles appears to me to be most perfect in kind. Mould or dung can be regarded as a fully prepared aliment, or, as it were, a minced and cooked meat with greens prepared by Nature, like as Man restores it by art. Thus the lower Thricozoa, e. g. the Mice, eat the crudest vegetable substances, such as roots and seeds; those that stand higher, grass and leaves; then snails, worms, and insects; finally flesh, and last of all fruits, as the Bears and Apes. But Man lets the crude matters ferment or reduces them to rapid decomposition by cooking, whereby a mixed kind of food results, which bears obviously the greatest resemblance to dung which, as just observed, is a food cooked by nature.

According to these considerations I now arrange what have been called Beetles in the following manner into divisions drawn from philosophical principles.

The Beetles again commence, like the whole class of Flies, from below, and the inferior kinds pass therefore parallel to the Tracheo-and Dictyoptera, while the superior project above them, as was the case also in the preceding classes.

In a more remote manner also they repeat the lower classes, namely, the Protozoa, Conchozoa, Worms and Crustacea, a fact which is, properly speaking, self-evident, and which is rendered clear by the following table. We have thus:

Order I. Tracheopteroid Coleoptera—Phytophaga.
Order II. Dictyopteroid Coleoptera—Zoophaga.
Order III. Ceratopteroid Coleoptera—Rypophaga.

It can also be said; the first correspond to the Worms, the second to the Crustacea, the third to the Flies.

Lastly, it may still be said; the first correspond to the Protozoa, the second to the Conchozoa, the third to the Ancyliozoa.

Order 7. Tracheopteroid Beetles—Phytophaga.

3527. Body cylindrical, head mostly long, antennæ setiform, maxillary teeth obtuse, tarsi mostly tetrameral.

They gnaw hard seeds, leaves, and wood, and mostly live concealed. The larvæ almost or entirely apodal.

Fam. 1. Dipteroid Beetles—Rhyncophora.

Head rhynchiform or snout-shaped, tarsi tetrameral.
Fam. 2. Hymenopteroid Beetles, Blattkäfer.
Head tolerably short as well as the antennæ, tarsi tetrameral.
Fam. 3. Lepidopteroid Beetles, Holzkäfer.
Head pretty short, antennæ very long, tarsi tetra-and pentameral—Borkenkäfer, Holzbohrer and Schröter.
Order 8. Dictyopteroid Beetles—Zoophaga.
3528. Body long and depressed, antennæ short, maxillæ large with pointed teeth, feet mostly pentameral and without ungues.
They subsist upon living or dead animals, dwell usually in water or upon dry land, and run very swiftly. The one kind seize upon living beasts, and on that account constantly swarm about them, as the Raub-and Laufkäfer; the others only subsist as larvæ by suction upon other beetles or snails externally, but live, as Flies, upon leaves and flowers, as the Lampyridæ and Cantharidæ.

Fam. 4. Neuropteroid Beetles—Water-beetles.
Like the following, but all three parts of the body closely annexed, and rudder-like fringes of hairs upon the feet—Parnidæ, Hydrophilidæ, Dyticidæ.
Fam. 5. Orthopteroid Beetles—Prædatoria.
Maxillæ very large and pointed, antennæ setiform, wings hard and tolerably flat like the body, tarsi mostly pentameral.—Raub-Lauf-and Sandkäfer.
Fam. 6. Hemipteroid Beetles—Parasitica.
Body pretty cylindrical, head rounded with moderate sized maxillæ, elytra narrow and soft, tarsi penta-and heteromeral.
They hang as larvæ like Mites to other beetles, and suck them; but after the metamorphosis they live upon leaves and flowers—Lampyridæ and Cantharidæ.
Order 9. Ceratopteroid Beetles—Rypophaga.
3529. Body quadripeltate and mostly short, with hard elytra, thorax large, head small, with short claviform antennæ and blunt maxillary teeth; feet mostly furnished with ungues, tarsal joints varied. Their habitation is very varied, being both free and concealed, and they live for the greatest part on decaying excrementitious matters, such as fungi, dry rot, and even animal ordure.
Fam. 7. Phytophagoid—Necrophaga.
Body mostly rounded, with hard and large elytra; antennæ claviform; tarsi tri-tetra-and pentameral—Lycoperdinæ, Erotyli, Coccinellidæ, Byrrhidæ, Dermestidæ, Sylphidæ. They correspond to the Worms and Tracheopterous insects.
Fam. 8. Zoophagoid—Mulmkäfer.
Body elongated and also nearly spherical, mostly of a dark spotted colour, with hard elytra, thorax rounded, head retractile, palpi filiform, tarsal joints heteromeral or unequal in number—Helopidæ, Diaperidæ, Tenebrionidæ, and Pimeliariæ.
They live usually in dark situations, eating meal and dry-rot; have an unpleasant smell. They correspond to the Crabs, Wood-lice, and Spiders, as likewise to the Dictyoptera.

Fam. 9. Perfect Beetles—Erdkäfer or Lamellicornes.
Mostly large, convex and short bodied Beetles with large thorax, broad head, lamellated antennal clubs, spinose tibiæ and pentameral tarsi. Their habitation is in dung or under the earth, where they live in the larval state frequently for years before they change; some Chafers eat also leaves of trees and flowers. They are distinguished by their striking size and singular forms, being

mostly furnished with horns on the thorax or head. I do veritably believe that they may be generally interpreted as being the highest organized beetles and members of their class, although they are generally placed or arranged midway between the lower forms, while the Sand-and Laufkäfer are esteemed the most perfect.

Tabular Co-arrangement.

The Dermatozoa now admit of being co-arranged in the following manner, from which their three Cardinal relations of Proximity, Repetition, and Serial analogy or Parallelism, admit of being distinctly recognized.

PARALLELISM OF THE LOWER ANIMALS.
        IX Cl. Ptilota.
A. PROTOZOA.   VII Cl. Worms.   Co. I. Tracheoptera.
I Cl. Infusoria. O. 1. Weisswürmer. O. 1. Diptera.
F. 1. Monades F. 1. Saugwürmer F. 1. Schmeissen
  2. Rhizopoda   2. Tæniæ   2. Dasseln
  3. Rotifera.   3. Ascarides.   3. Gölsen.
II Cl. Polypi. O. 2. Rothwürmer. O. 2. Hymenoptera.
F. 1. Hydræ   4. Earth-worms F. nbsp; 1. Bees
  2. Alcyonia   5. Nereides   2. Wasps
  3. Actiniæ.   6. Serpulæ.   3. Terebrantia.
III Cl. Acalephæ. O. 3. Sternwürmer. O. 3. Lepidoptera.
F. 1. Physaliæ 7. Holothuriæ F. 1. Moths
  2. Beroeæ   8. Echinidæ   2. Spinner
  3. Acalephæ.   9. Asteridæ.   3. Wanner.
B. CONCHOZOA. VIII Cl. Crustacea. Co. II. Dictyoptera.
IV Cl. Mussels.   Co. 1. Krebse. O. 4. Neuroptera.
O. 1. Röhrenmuscheln.   O. 1. Pfriemenkrebse. F. 1. Nagbolden.
F. 1. Pholades F. 1. Entomostraca A. 1. Kieferläuse
  2. Tellinidæ   2. Branchiopoda   2. Termitidæ
  3. Cardiacea.   3. Phyllopoda.   3. Phryganeidæ.
O. 2. Spaltmuscheln. O. 2. Scheerenkrebse. F. 2. Raubbolden.
  4. Mytilaceæ   4. Squillæ   4. Hemerobiidæ
  5. Arcaceæ   5. Macroura   5. Raphidiidæ
  6. Ostraceæ.   6. Brachyura   6. Libellulidæ.
V Cl. Snails. Co. II. Asseln. O. 5. Orthoptera.
O. 1. Androgyni.   O. 3. Saugasseln.   F. 1. Saltatoria.
F. 1. Tritoniæ F. 1. Lernœæ A. 1. Achetidæ
  2. Patellæ   2. Arguli   2. Gryllidæ
  3. Pulmonea.   3. Pyncnogonides.   3 Locustidæ.
O. 2. Diœcii. O. 4. Nagasseln   F. 2. Cursoria.
  4. Capulidæ 4. Læmodipoda   4. Mantides
  5. Turbinidæ   5. Amphipoda   5. Blattidæ
  6. Buccinidæ.   6. Isopoda.   6. Forficulidæ.
VI Cl. Kracken.   Co. III. Kobe. O. 6. Hemiptera.
O. 1. Rumpfkracken. O. 1. Langkobe.   F. 1. Leptoptera.
F. 1. Salpæ   F. 1. Podurae A. 1. Aphidii

2. Ascidiæ	2. Juli	2. Cicadæ
3. Cirripedia.	3. Scolopendræ.	3. Fulgoridæ.
O. 2. Kopfkracken.	O. 2. Kurzkobe.	F. 2. Pachyptera.
4. Brachiopoda	4. Acaridæ	4. Water-bugs
5. Pteropoda	5. Scorpions	5. Narrow bugs
6. Cephalopoda.	6. Spiders.	6. Broad bugs.
C. ANCYLIOZOA.	IX. Cl. Ptilota. Co. III. Ceratoptera.	
VII CL. Worms	Co. I. Tracheoptera	O. 1. Phytophaga
VIII Cl. Crustacea	II. Dictyoptera 2. Zoophaga	
IX Cl. Ptilota or Flies.	III. Ceratoptera.	3. Moderkäfer.
Co. III. Ceratoptera.	O. 3. Rypophaga.	
A. PROTOZOA.	O. 1. Phytophaga.	F. 7. Necrophaga. F. 7. Erdkäfer.
I Cl. Infusoria.	F. 1. Rhyncophora	A. 1. Fungivora. A. 1. Phyllophaga.
F. 1. Monades	A. 1. Langrüssler	Scaphidiidæ a. Melolonthidæ
2. Rhizopoda	2. Kurzrüssler	Erotyli b. Anoplognathidæ
3. Rotifera.	3. Breitrüssler.	Encidæ. c. Rutelidæ.
II Cl. Polypi.	F. 2. Blattkäfer.	A. 2. Speckfresser. 2. Anthobii.
F. 1. Hydræ	A. 1. Galerucidæ	1. Dermestidæ a. Lepitrichidæ
2. Alcyonia	2. Chrysomelidæ	2. Nitidulidæ b. Dichelidæ
3. Actiniæ.	3. Crioceridæ.	3. Silphidæ. c. Glaphyridæ.
III Cl. Acalephæ.	F. 3. Holzkäfer.	A. 1. Dung-feeders. 3. Melitophaga.
F. 1. Physaliæ	7. Bostrychidæ	1. Byrrhidæ a. Cetoniæ
2. Beroeæ	8. Elateridæ	2. Sphæridiidæ b. Trichii
3. Acalephæ.	9. Cerambycidæ.	3. Histeridæ. c. Goliathi.
B. CONCHOZOA.	O. 2. Zoophaga.	
IV Cl. Mussels.	F. 4. Water-beetles.	F. 8. Mulmkäfer. A. 2. Coprophaga.
O. 1. Röhrenmuscheln.	A. 1. Sumpfkäfer.	A. 1. Trachelidæ. a. Aphodii
F. 1. Pholades	a. Heteroceridæ	a. Lagriæ
2. Tellinidæ	b. Parnidæ	b. Pyrochroæ
3. Cardiacea.	c. Hydrophilidæ.	c. Mordellæ.
O. 2. Spaltmuscheln.	A. 2. Flusskäfer.	Serropalpidæ. b. Copridæ.
4. Mytilaceæ	d. Girinidæ	d. Mycteridæ
5. Arcaceæ	e. Haliplidæ	e. Œdemeræ
6. Ostraceæ.	f. Dytiscidæ.	f. Melandryæ.
V Cl. Snails.	F. 5. Mordkäfer.	A. 2. Stenelytra. Arenicolæ.
O. 1. Androgyni.	A. 1. Raubkäfer	
F. 1. Tritoniæ	a. Pselaphidæ	a. Cistelidæ a. Æglaliæ.
2. Patellæ	b. Scydmænidæ	b. Helopidæ
3. Pulmonea.	c. Staphylinidæ.	c. Cnodalidæ.
O. 2. Diœcii.	A. 2. Laufkäfer.	Taxicornes. b. Trogidæ.
4. Capulidæ	d. Carabidæ	d. Niliondæ
5. Turbinidæ	e. Brachinidæ	e. Cossyhidæ
6. Buccinidæ.	f. Cicindelidæ.	f. Diaperidæ.
VI Cl. Kracken.	F. 6. Schmarotzkäfer.	A. 3. Tenebrionidæ. Geotrupidæ.
O. 1. Rumpfkracken.	A. 1. Leuchtkäfer.	
F. 1. Salpæ	a. Lampyridæ	a. Tenebriones a. Geotrupes.

2. Ascidiæ   b Melyridæ   b. Toxicidæ
3. Cirripedia.   c. Cleridæ.   c. Opatridæ.
O. 2. Kopfkracken.   A. 2. Ziehkäfer.   Melanosomata.   b. Lethri.
4. Brachiopoda   d. Notoxidæ   d. Blapsidæ
5. Pteropoda   e. Horiidæ   e. Pimeladiidæ
6. Cephalopoda.   f. Cantharides.   f. Sepidiæ.
C. ANCYLIOZOA.   O. 3. Moderkäfer.   F. 9. Erdkäfer. A. 3. Lohfresser.
VII CL. Worms   F. 7. Aaskäfer A. 1. Phyllophaga   a. Lucanidæ
VIII Cl. Crustacea   8. Mulmkäfer   2. Coprophaga   b. Passalidæ
IX Cl. Ptilota or Flies.   9. Erdkäfer.   3. Lohfresser.   c. Oryctidæ.
Second Province.
— FOURTH CIRCLE—SARCOZOA.
Pisces, Reptilia, Aves and Thricozoa.
3530. These animals necessarily pass through all 4 circles, and each class therefore divides into 4 orders, or into 5, if the senses be allowed to hold good as constituting a distinct order in themselves.
Two points of departure for classification here admit of being thought of, either wholly from below, or first of all from the commencement of the present circle. In order to be convinced of this, we need only make a slight attempt. The animal series is as follows:
A. Splanchnozoa.
Circle I.   Intestinal animals   Protozoa.
Class   1. Gastric animals   Infusoria.
2. Intestinal   "   Polypi.
3. Absorbent   "   Acalephæ.
Circle II.   Vascular animals   Conchozoa.
4. Venous animals   Mussels.
5. Arteriose   " Snails.
6. Cardiac   " Kracken.
Circle III.   Respiratory animals   Ancyliozoa.
7. Reticular animals   Worms.
8. Branchial   "   Crabs.
9. Tracheal   "Flies.
B. Somatozoa.
Circle IV.   Sarcozoa.
10. Osseous animals   Fishes.
11. Muscular   "   Reptiles.
12. Nervose   "   Birds.
Circle V.   Aistheseozoa.
13. Sense-animals   Thricozoa.
3531. Now it is here conceivable, either that the families of Fishes, Reptiles, Birds, and Thricozoa, pass parallel to all the preceding classes from the Infusoria upwards, or that they first commence with the class of Fishes.
In regard to this question the Birds and Thricozoa give the most clear and decisive answer. If we adopt the last proposition, then the Birds must follow each other thus:
1. Fish-like Birds   Palmipedes.
2. Reptilian Birds   Grallæ.

3. Typical Birds   Gallinæ.
4. Thricozooid Birds   Struthionidæ.
According to this, the Aquatic Birds would occupy the lowest rank, just as they have actually done hitherto in all other systems. The three other families too do not admit of any other position. But what is to be done meanwhile with the large majority of other Birds? They would extend or pass beyond the Thricozoa, and have thus no anatomical system whatever for their basis. Besides, Humming-birds, Sparrows and Linnets would in this way rank higher than the Gallinaceous and Struthious Birds, which some indeed believe to be the case, but without any proof, so that it is scarcely necessary to abide by such an opinion.
The Thricozoa must stand thus:
1. Icthyoid   Cetacea.
2. Herpetoid   Pachydermata.
3. Ornithoid   Ruminantia.
4. Typical   All the remaining Thricozoa.
In the systems now in vogue the Thricozoa certainly follow in this manner, and no scruple whatever is made of placing the Mice, Ant-eaters, Shrews, and Bats above Elephants, Horses, and Deer. Yet apart from all this, the former must pass beyond the Thricozoa, and would thus likewise have no other organ as a foundation than perhaps the organs of sense, of which there are but five, while the families of Thricozoa not yet provided for are much more numerous, namely, besides those named, the Seals, Dogs, Bears, and Apes.
The above methods of arrangement must be consequently cast aside without limitation or reserve, although they may have held good for hundreds of years. But what does or will not hold good in an age that is devoid of principles?

It is thus evident that the smaller Birds stand below the Aquatics, just as the smaller Thricozoa do beneath the Whales, quite apart from organic structure, which in every respect is imperfect. They must correspond, consequently, to the Asarcous animals, wherewith also their great number agrees.
If this holds good of the two highest classes, so also must it hold good of the two lower, namely, Fishes and Reptiles, and we shall thus have:
I.   Protozooid Fishes, Reptiles, Birds, and Thricozoa.
II.   Conchozooid Fishes, &c.
III.   Ancyliozooid   "   &c.
IV.   Icthyoid   "   &c.
V.   Herpetoid   "   &c.
VI.   Thricozooid   "   &c.
VII.   Ornithoid   "   &c.
3532. But this parallelism depends essentially upon the organs, and they are called, in the language of Physio-philosophy, more correctly:
I.   Intestinal Fishes, Reptiles, Birds, and Thricozoa.
II.   Vascular Fishes,   &c.
III.   Pulmonary   "   &c.
IV.   Osseous   "   &c.
V.   Muscular   "   &c.
VI.   Nervose   "   &c.
VII.   Sensorial   "   &c.

But as these names are unused, and give or convey only the anatomical idea, but no external image, the first appellations are to be preferred.

Tenth Class.

Osteozoa, Glossozoa—Pisces.

3533. No one of the upper classes is in such an extensive state of confusion as that of Fishes. Nor does this result simply from the great deviations in their structure from the normal type, these being rather useful than otherwise for the purposes of classification; but mainly from the utter want of any principle which might serve systematists as a guide. Thus at one time we find them having recourse to the nature of the osseous tissue, at another to the fins, now to the teeth or to the scales, ay, even to the fin-rays, and all this because the characteristic or typical organ has never been sought after, nor the presence of such an organ as indispensable been so much as known.

3534. The characteristic organ of Fishes is the Osseous system, which is consequently the principle also of their division.

The physical nature, form, position, and number of the osseous parts must therefore be principally considered, and hence, above all, the substance or texture of the bones; the limbs also, and the maxillæ with their teeth, as well as the teeth upon the palate, upon the tongue and the branchial arches.

With regard to the component substance or texture of the bones we encounter a great difficulty. The Cartilaginous Fishes appear to belong to each other, and are also usually arranged together. Yet amongst them we find those species, such as the Lampreys, which obviously occupy the lowest grade of all Fishes, while the Sharks and Rays remind us of the Reptilia and Thricozoa, as well by their external structure as the development of their sexual parts, since they possess perfect testes, and ovaria separate or distinct from the oviducts, while they no longer deposit roe, but large ova inclosed in leathery shells, like the higher Reptilia. Now, if we separate these Fishes from the Lampreys, with whom in the scaleless tegument, the branchial foramina, and even the external form they have many points of resemblance, nothing else remains to be done than assign them the uppermost place, and so parallelize them with the Thricozoa. But one is next constrained to unite into one family the Pikes and Herrings, which perhaps admits of being done.

There belong namely to the upper Fishes without doubt the Abdominales, which are divisible into five families: the Carps, Pikes, Shads, Salmons, and Herrings. Now, if the Sheat-fishes be placed inferiorly on account of their scaleless body and amorphous maxillæ, four families will still be left, which should correspond to the Fishes, Reptiles, Birds, and Thricozoa, so that no place is left remaining for the Sharks. Now, however, the Salmons correspond decidedly to the Reptilia; and the Flying Fishes, which are ranged below the Pikes, probably to the Birds. If we unite them with the Herrings, then the Sharks may occupy the place of the Thricozoa.

This being preassumed, we can now attempt the classification. The substance or texture of the bones is of such importance, that notwithstanding the separation of the Sharks and their congeners, the other Cartilaginous Fishes must be left along with them, and range upon the lowest stage, so that they thus correspond to the Intestinal animals.

The next great distinction in the osseous system is the regular and irregular form which it imparts to the body, so that the Regular-shaped can be separated in a tolerably "tranchant" manner from the Irregular Fishes.

The regular form of a Fish is obviously the ellipse, as we find to be the case in our fresh-water Fishes, namely, the Perches, Salmons and Carps. They are collectively covered with large scales, which is therefore also a sign of their regularity.

The Irregular Fishes are cylindrical, fusi-clavi-spheri-and tubuliform, usually destitute of scales or covered with plates, scutes, and spines. Thus, since the Cartilaginous Fishes are collectively irregular in form, the Irregular Osseous Fishes must be allowed to follow them.

The greatest variety of the osseous system is shown in the limbs, especially the posterior pair, which in the other classes also are generally imperfect and make their appearance the last. In Fishes they are not divided into digits, but only into rays, which probably correspond only to the digital ungues or to feathers. An animal which has fin-rays is assuredly a Fish, for fin-rays do not occur in any Reptile.

The posterior fins change even their situation. Those Fishes in whom they are placed near to the anus, are obviously the more perfect, as the Abdominales.

In others they advance to the rear of the thoracic fins, and are even attached to what has been called the "girdle" or humerus—Thoracici.

In others they even get in front of the thoracic fins on the throat—Jugulares.

Lastly, they are actually wanting—Apodales.

In the Lampreys there are neither thoracic nor ventral fins.

The skeleton of Fishes is not simply divided to a greater extent than in other animals, but has actually a greater number of bones, such as the rays in the perpendicular fins, which are wanting in all animals, even in the Reptilia. An animal with dorsal rays must surely be placed among the Fishes, and consequently the Lepidosiren also.

The misshapen Fishes will therefore occupy the lower stages, the regular the upper, not directly by reason of their form alone; but because the other organs are also more imperfectly developed, the bones being cartilaginous, the tegument asquamous, mucous, or covered with spines, scutes, and plates; the fins wanting or abortive, or displaced from their proper situation; the head disproportionate in size to the body, the mouth wide or narrow, the eyes placed superiorly or upon the forehead.

A lower character is afforded also by very long dorsal fins, such namely as extend from the head to the tail. In the Abdominales, Sturgeons, Sharks, and Rays, the perpendicular fins are small; in the Thoracici, Jugulares, and Apodes they are, on the contrary, mostly very long.

Moreover, a lower character is a very long coccygeal or caudal fin, which denotes that the anus lies far forwards, and therefore that the tail has a great preponderance over the trunk. In the Abdominales and the Sharks the tail is short; in the Rays thin and terminated abruptly, as in Reptiles and Thricozoa. Thus the higher the animals ascend the more does the tail diminish in length.

Regard being had to all these relations, the Irregular Fishes must be viewed as those which correspond to the lower classes of animals, and the Cartilaginous Fishes will indeed make the commencement; to these are annexed the Irregular Osseous Fishes, and nearest to them indeed those with arrested ventral fins, whether wanting or placed on the throat; then come the Regular Fishes, and of these first of all the Thoracici, and next the Abdominales. In this manner we obtain four divisions.

A. Body irregular.
I. Pisces Cartilaginei.
II. Stummelflosser—Apodales and Jugulares.
B. Body regular.
III. Thoracici—Tunnies, Breams, and Perch.

IV. Abdominales—Carps, Pikes, Salmons and Herrings, Sharks.
Now, these divisions, having been discovered by a simple analysis or testing of facts, are to be arranged according to philosophical principles, and further subdivided, whereby the ground and legality of their existence will be recognized.
A. IRREGULAR FISHES.
3535. Body deviating from the elliptical form, devoid of scales, or covered with spines, scutes, and plates; head and tail disproportionate; fins mostly arrested.
They correspond to the unarticulate Proto-and Conchozoa; their irregular-shaped mucous or mailed body agreeing perfectly with these animals.
Order 1. Intestinal, Protozooid Fishes.
3536. Mouth round and without maxillæ, or disproportionately narrow and wide.
There can be no doubt that the Lampreys are the lowest Fishes, since they remind us in every respect of the Worms by their naked, mucous, and lineiform body, with indistinct head, almost devoid of bones and true teeth, having a circular mouth, obliterated nostrils, puny eyes, and finally branchial cysts, which occupy a higher rank only from their opening into the pharynx. They pass therefore parallel to the Infusoria, or rather to the commencements of the second animal series, namely, the Etozoa Ancyliozoa.

Now, although much dispute prevails concerning the division of Fishes which should be associated with the Lampreys, yet it appears to me that no others but the narrow-mouthed Fistularidæ and Pipe-fishes can follow, since they resemble them not only in their cartilaginous bones, but in the structure also of their branchiæ. They will thus truly occupy the place of the second family. Whether the narrow-mouthed Globe-fishes are likewise to be united with them or to be set up as a third family, may seem to be matter of doubt. I adopt the first course, and arrange at present the wide-mouthed or Frog-fishes and Shads in the third family.
Fam. 1. Infusorial Fishes, Lampreys.
Body vermiform, naked, and slimy, without membral fins. Ex. Branchiostoma or Amphioxys, Myxine, Petromyzon.
In these Fishes the mouth is quite in front and round, being without maxillæ and adapted for the purposes of suction; only one nostril, but mostly several branchial foramina, which lead to cysts provided with reticular branchiæ, but without opercula.
The Branchiostomata are the smallest Fishes, not much above 1" in length, almost devoid of head, yet with traces of eyes and a nostril. The Myxinæ crawl even into the rectum of other Fishes, and live therein like Entozoa. The river or lesser Lampreys stick in the mud; the Lampreys cling fast by suction to stones, and do not draw the water in through the mouth, but through the branchial foramina themselves, like the lower animals.
Fam. 2. Polypary Fishes, Narrow-mouthed.
Body cartilaginous, mouth having maxillæ, but unusually narrow, only one branchial foramen with immoveable operculum—Fistularidæ, Pipe-fishes and Globe-fishes.
In this family we still meet with species entirely naked, but covered also with plates, scutes, nails, and spines. The corymbiform or tufted branchiæ of the Syngnathi or the Lophobranchii still remind us strongly of the cystiform reticular branchiæ of the first family.
Fam. 3. Acalephoid Fishes, Wide-mouthed.
Body naked or covered with plates; mouth in front and mostly unusually wide.
I here arrange in doubt the Frog-fishes and the Shads, although the last are Abdominales. But they deviate from the Carps and Pikes by their asymmetrical, naked or mailed body, the large

transverse mouth, the eyes staring upwards, and the arrested branchial opercula.

Order 2. Vascular, Conchozooid Fishes—Stummelflosser.

3537. Asymmetrical Osseous Fishes, Apodales, and Jugulares.

Among the true Osseous Fishes the Eels must undoubtedly range the lowest on account of their vermiform and asquamous body, and the want of ventral fins. To them are allied the Jugulares, namely, the Blennii and Gadidæ, as well as the Plaice; lastly, the asymmetrical Thoracici without scales, being quite naked or covered with scutes, as the Gobii and Triglæ.

Fam. 4. Mussel-Fishes, Eels.

Body naked and serpentiform, without ventral fins.

The Eels, from their naked, cylindrical or riband-shaped body, the long dorsal and coccygeal fin, the small branchial foramina occasionally confluent beneath the neck, and by their dwelling in the mud, rank among the imperfect Fishes. The one set have soft, the other hard, fin-rays, and on that account they have not simply been separated, but even far removed from each other. This difference alone is not so great as to justify their being arranged, when the structure of the body agrees in other respects, into distinct orders. The influence of the dorsal spines upon the life and natatory or waving movement of Fishes is so slight, and its value generally, in comparison with other parts, to be taken so little into account, that a natural arrangement can never result from these appendages, which do not deserve to be called organs, and viewed as principal characters.

Fam. 5. Snail-Fishes, Haddocks.

Jugulares without distinct scales and spines on the branchial operculum, and hard fin-rays.

These Fishes are allied to the Eels by the aborted ventral fins, the tolerably cylindrical, naked, or small-scaled body and soft dorsal fins. The viviparous Blennius bears with us its name of Aalmutter not in vain, for it resembles a shortened, slimy Eel. The Gadidæ are indeed less slimy and have in part scales, but, by reason of their lengthened form and their fins, cannot be removed far from the Eels; the same holds good of the Plaice.

Fam. 6. Kracken-Fishes, Grundeln.

Asymmetrical Jugulares and Thoracici, with naked and mailed body, in addition to hard fin-rays.

Here belong the Gobiidæ and Triglidæ, the first whereof are usually naked, the second mailed, with roughnesses upon the head and spines on the opercula; in all the eyes placed high up.

B. REGULAR FISHES.

3538. Body elliptical, mostly covered with scales, Thoracici and Abdominales; eyes placed sideways.

Order 3. Pulmonary, Ancyliozooid Fishes.

3539. Regular Thoracici.

The position of the abdominal fins immediately behind the thoracic is obviously a step further in the perfection of structure, and these Fishes must be therefore placed above the preceding kinds, in whom, apart from the asymmetrical form of body, the position of the fins is mostly upon the neck or advanced very close to it. The anus is still situated far forwards, and the tail is therefore mostly larger than the trunk. The dorsal fins are still very predominant. Among them we still meet with naked or microlepidal species, which in their abnormal form also remind us of the irregular Fishes, as the Tunnies and Haberdines or Stock-fish. They are therefore to be regarded as the lowest.

As is the case here in Fishes, so also in the Ancyliozoa the truly regular or bilateral body originates in them for the first time, or at least constitutes a persistent character.

Fam. 7. Worm-Fishes, Tunnies.

Tolerably naked or microlepidal, cylindriform or very much compressed, with small head and mouth; the teeth very feeble, only like a brush; the branchial opercula without spines. Here belong the proper Tunnies and Haberdines: all of them marine Fishes.

Fam. 8. Crustacean Fishes, Brassen.

Body perfectly regular, covered with great scales; mouth small with strong teeth; branchial opercula unarmed. Here belong the Labridæ or Lipped Fishes, Seabream and Osphromanus.

Fam. 9. Ptilotoid Fishes, Perch.

Body quite regular with large scales, mouth of moderate size with scythe-shaped teeth, branchial opercula armed. Here belong the Scianoidæ and Percoidæ. Dwell in the sea and rivers.

Order 4. Sarcose Fishes.

3540. Abdominales, dorsal fins small, with soft ramified rays, mostly placed far behind. Plainly regular Fishes with large scales. The head is regular, with the eyes upon its sides; the set of teeth varied; the trunk large, tail small, as are also the dorsal fins, which proceed more and more backwards to the sacrum or even the tail, a fact indicating their gradual disappearance, and therefore a sign also of greater perfection. Added to this, they are generally distributed over the whole earth in rivers and seas. Finally, it is they that yield the most nutriment to Man, which is also a constant sign of greater perfection; as is evidenced in the vegetable kingdom by the Fruit-trees or plants; in the animal kingdom by the Oysters, Snails, Sepiæ, Holothuriæ, Geese, Fowls, Cattle, &c.

Fam. 10. Typical Fishes, Carps.

Body of the Carps covered with large scales, mostly but one dorsal fin placed pretty far back, mouth nearly edentate, supra-maxillary bones arrested, large teeth on the posterior branchial arches, or what have been called pharyngeal bones. They are for the greatest part fresh-water Fishes, and those which are most used as articles of food.

Fam. 11. Reptilian Fishes, Salmons.

Mouth with strong teeth in the superior and intermaxillary bone; behind the radial dorsal fin there is still a fatty fin and no large scales. Here belongs the Lizard-like fish (Saurus), so called on account of its resemblance to the Lizards. Live in sea and fresh-water.

Fam. 12. Ornithic Fishes, Herrings and Pikes.

The Herrings have teeth in the superior and intermaxillary bone; only one dorsal fin; mostly large scales.

Body of the Pike slightly scaled, furnished mostly with a small dorsal fin situated very far back, mouth full of teeth, but none of these in the rudimental intermaxillary bone. Dwellers in the sea and in fresh water.

Among the Pikes is placed the Exocœtus or Flying fish.

Order 5. Sensorial Fishes.

3541. Bones cartilaginous, mouth opening transversely under the snout.

Fam. 13. Thricozooid Fishes, Sharks.

Abdominales; bones cartilaginous, mouth opening transversely under the projecting snout; mostly several pairs of separate branchial apertures.

Here belong the Chimæræ, Sturgeons, Sharks, and Rays. The last ought to be held as higher in rank, partly on account of their slender tail, partly because the huge Rays, which are called Cephalopterus, have the anterior thoracic rays free and so moveable that they can seize their prey with them as with hands. All lay, with the exception of the Sturgeons, large and leathery ova, and in this approximate pretty closely to the Reptilia.

It has been already remarked, that the large Cartilaginei would not pass correctly into the others, and obviously seem to claim the highest post. At some later period the principle may probably be discovered whereupon their union with the higher Osseous fishes depends. Both are at all events Abdominales.

Eleventh Class.

Myozoa, Rhinozoa—Reptilia.

3542. Body entirely naked or covered with scales, with distinctly separate and red-coloured muscles; two nostrils permeable throughout.

As the Muscular system is here the characteristic organ, it must be regarded principally in the division of the present class; the limbs also exhibit nowhere so great a variety as in this class, since they are in some cases wanting, while in others two only, in some four, are present. In place of fin-rays true toes have, however, made their appearance, these again indicating the greatest variety in the number of the joints; but meanwhile there are in no instance more than five toes.

The osseous system is constructed after the pattern of that of the Thricozoa, and is never furnished with dorsal rays as in Fishes.

The dental formula begins also to be regular. The teeth stand usually in the superior intermaxillary bone, and are sometimes pointed, at others obtuse; in the Ichthyosauri and Crocodiles they are even inserted by gomphosis, as in the Thricozoa. In many Serpents additional kinds of teeth occur, namely, the curved poison-teeth or fangs, which have an involuted groove traversing their concavity. In most Reptiles teeth also occur upon the palatal, but there are no longer any upon the lingual bone. In the Chelonia or Tortoises the teeth are entirely wanting, and they are also scarcely indicated in the Asquamous Salamanders and Frogs. The os quadratum is found as in Birds; in Serpents, however, the mastoid bone has been freed, and hence the capability possessed by these creatures of expanding the mouth.

Viewed in accordance with the perfection of the limbs, the Salamanders and Frogs ought to be regarded as the highest in rank; but their scaleless tegument, their development out of spawn in the water, as well as branchiæ brings them near to the Fishes. In other respects their position is determined by the dental formula, which, as I have already shown, belongs to the limbs and thus to the motor system.

The position of the nostrils is now, throughout the class, in front of and upon the snout, being no longer situated almost on the forehead or vertex, as in Fishes. Their relation to the scales is likewise of importance for the purposes of division.

The naked Salamanders and Frogs will occupy the lowest place. By their form, absence of teeth, and mode of life, the Chelonia are allied to them.

Then follow the Serpents and Lizards with a perfect set of teeth. The Serpents are distinguished from the Lizards by their want of feet and the long bifid tongue inclosed within a sheath.

Among the Lizards apodal species occur, as the Blind-worms; but they have under the skin some pedal bones, and are, in addition to this, sufficiently distinguished from the Serpents by the short and sheathless tongue.

Among the Lizards with perfect feet there are some with small visual organs, like as in the Serpents; others with unusually large eyes, e. g. the Chameleon, Gecko, Ichthyosauri, and Crocodiles. All other Lizards, the Serpents, Chelonia, Salamanders and Frogs, have small eyes and consequently follow each other.

The Crocodile has teeth articulated by gomphosis, and consequently resembles the Thricozoa. I divide the present class therefore into the following groups.

A. DERMAL REPTILES—SMALL-EYED.
Order 1. Protozooid Reptilia-Kröten.
3543. Body slimy and scaleless, or maxillæ devoid of teeth.

Fam. 1. Infusorial Reptiles, Petromyzoid—Caudate Batrachia, Salamanders.
Body naked and furnished with a tail.
These cylindrical animals bear the greatest resemblance to the Lampreys, or are Petromyzoid, being provided with setaceous and scarcely discernible teeth, frequently only one pair of feet; they are developed also out of spawn in the water, and many retain the branchiæ during the whole of life.
Fam. 2. Polypary Reptilia, Kugelfischartige—Stutzkröten, Frogs.
Body thick and naked and without a tail, but with four feet.
The Frogs and Toads proper are likewise developed out of spawn, but soon lose their branchiæ. In their form and even the structure of their mouth they are Kugelfischartige, i. e. remind us strikingly of the Globe-fishes, among the Engmäulern or Plectognathi.
Fam. 3. Acalephan Reptilia, Welsartige—Chelonia.
Body thick and coated with scales, maxillæ quite edentulous.
The Chelonia appear certainly to occupy a pretty high rank; they lay large ova covered with a calcareous shell, but frequently live or dwell in the water, and have, in their mode of living as well as form, a striking resemblance to the naked Batrachia. Through their want of teeth they range among all the succeeding families.
Order 2. Conchozooid Reptilia—Ophidia.
3544. Body cylindriform and scaly, teeth acuminate, tongue longitudinally bifid and inclosed within a sheath; no feet and eyelids.
The Serpents stand without doubt below the Lizards; yet it is difficult to arrange them properly into families. The poison-teeth appear to indicate a lower character, because the cranium is thereby removed or recedes from the usual type of structure, while the superior maxilla becomes very much arrested. But as there are moveable and immoveable poison-teeth, and the latter gradually pass over into the ordinary kinds of teeth, while externally also, no character has been found to distinguish the venomous from the non-venomous Serpents, it is best at present to discontinue this separation.
Then the Serpents can be brought according to the structure of their scales into 3 families. They are either of equal and small size, around the whole body; or there are plates upon the belly and tail; or finally, the plates under the latter are divided into two tablets.
Fam. 4. Mussel-Reptiles, Eel-like—Schuppenschlangen.
All the scales around the body small and of equal size; upon the belly only being somewhat larger. Here belong the venomous Hydridæ or Sea-snakes, and huge Boas or Pythons. If once the venomous Serpents could be distinguished as a distinct family, they will indeed be placed higher, and the non-venomous be separated from them. Then probably the Boas may be raised, together with their allies, as the highest family.
Fam. 5. Snail-Reptiles, Haddock-like—Tafelschlangen.
Plates under the belly, those beneath the tail being halved or divided into two.
Here belong our Colubri, but also the venomous Adders.
Fam. 6. Kracken-Reptilia, Grundelartige—Schienenschlangen.
Entire plates upon belly and tail, as in the Rattle-snakes.
Order 3. Ancyliozooid Reptilia—Lizards.

3545. Scales, the usual kind of teeth and tongue; inferior maxillæ anchylosed in front, mastoid bone not freed, mostly feet and eyelids.
They divide likewise into three families.
Fam. 7. Worm-Reptiles, Tunny-like—Schleichen.
Body round and serpentiform, with small scales, feet rudimental, or even wanting.
Here belong the Cæciliæ, Blind-worms, and Scinci. I have united these animals, which elsewhere have stood dispersed among the Serpents and Lizards as well as the Salamanders, and placed them as a distinct family between the two former. By reason of their rudimental feet they keep simply upon the earth and bore themselves passages therein. Their motion is serpent-like, because they lie upon the belly and cannot assist themselves with the feet.
Fam. 8. Crustacean Reptiles, Bream-like—Schuppenechsen.
Four perfect feet, body covered all round with small granular scales, tongue short and hardly slit. Here belong the Flying Lizards, the Iguanæ and Basilisks. They usually climb about on trees and seek for beetles and berries.
Fam. 9. Ptilotoid Reptilia, Perch-like—Schienenechsen.
Four perfect feet, but plates upon the belly and tail, tongue thin and bifid. Here belong the common Lizards and the "Sauvegards" or Monitors. Their body is usually depressed. They cannot climb, but run about briskly upon the ground and eat beetles, and even the higher animals. Many are a fathom in length.
B. SARCOSE REPTILIA—LARGE-EYED.
3546. These animals have tubercles, spines and plates, with four perfect feet, as in the higher Lizards, but the toes are of pretty equal length; there are no palatal teeth, and the tongue is not fissured. They are of varied size, lead a sluggish and mostly nocturnal mode of life, and have a slow pace, occurring too only in warm climates.
Order 4. Sarcose Reptiles.
3547. Feet abnormal, being fin-and wing-shaped, adapted for climbing or clinging. They correspond quite closely to the three next animal classes.
Fam. 10. Ichthyoid Reptilia, Herring-like—Ichthyosauri.
All four feet converted into fins.
Here belong clearly the monstrous extinct animals, which formerly lived in the sea. They had gomphotic teeth almost like the Crocodile.—Ichthyosaurus, Plesiosaurus.
Fam. 11. Typical Reptiles, Salmon-like—Geckos.
Climbing feet, or retractile claws as in the Cats, and perpendicular laminæ or plates under the soles, by means of which they cling to walls. They usually dwell in houses and catch beetles—Chamæleo, Gecko.
Fam. 12. Ornithoid Reptilia, Pike-like—Dragons.
Body short and bird-like, with very long neck and a long digit, unto which an alary membrane has probably been attached.
Here belong plainly the fossilized species of a single genus known by the name of Pterodactylus. They have been hitherto discovered in different parts of Europe, and probably flew about, like the Bats, by night.
Order 5. Sensorial Reptiles.
3548. Teeth wedged into the jaws, toes regular.
Fam. 13. Thricozooid Reptilia, Shark-like—Crocodiles.
Feet and toes of equal length, with swimming webs between them.
The Crocodiles must be without doubt regarded as the most perfect Reptilia, on account of their

notched teeth and regular feet. In many respects they remind us, like the Sharks do among Fishes, of the Thricozoa.

Twelfth Class.

Neurozoa, Otozoa—Aves.

3549. In these animals it is again shown in a striking manner, that the characteristic organ is the principle of classification. Without recourse being had to the varied structure of the head, especially of the beak, the division of Birds would not be possible, although the feet frequently yield good characters.

3550. The development of the young makes an essential distinction upon a large scale, since one great body of Birds must be fed as nestlings, whilst the other, when scarcely emerged from the egg, runs about and seeks its own nourishment. Although in the lower animals the young do not require the assistance of the parents, and, on the other hand, those of the Thricozoa are suckled for a longer time by the mother; yet still those Birds which can, upon issuing from the egg, at once nourish themselves, such as the Fowls, Geese, &c., are probably the most perfect, for they pass parallel to the upper classes of animals, and for the Ostrich to be a Thricozoon, there is indeed as little wanting as to the Crocodile.

3551. There is no doubt whatever that the Natatores in every respect, both in the structure of the feet and head as also in their mode of living and feeding, repeat the Fishes. This opinion the empirical zoologists have already adopted from the Physio-philosophy.

The same may be said of the Grallæ or Wading Birds in reference to the Reptiles. Their feet, neck, and beak are serpentiform, and associated with a frog-like body. Their mode of living and feeding is likewise amphibious. But these birds pass over so directly into the Fowls, that the latter could not be arranged elsewhere, apart even from their displaying by their domestication to Man the higher grade of understanding, which is manifested for the first time in the class of Birds.

The Bustards and Ostriches are, finally, the highest stage of Birds, and form the closest alliance with the Thricozoa.

3552. This point being once settled, it is self-evident that the Birds, which do not stand in need of being fed, occupy the uppermost place, and consequently the nestling or parent-fed Birds the lowest, i. e. the former correspond to the Sarcozoa, the latter to the Dermatozoa.

A. SPLANCHNIC BIRDS.—NESTHÖCKER.

3553. Remain, after exclusion from the egg, in the nest and are fed; neck and feet short, toes four in number and ununited, beak pointed.

By their short neck, which rarely appears longer than the head, these Birds approximate the Reptiles and Fishes, in whom also the head has scarcely separated itself from the thorax.

The uniformity of the feet and toes is likewise an inferior character, as it is found in the Dermal Reptiles and Fishes; while in the Sarcose Reptilia and Fishes the feet and fins make their appearance in the greatest variety, both as regards the form and length of the feet themselves, as the structure of the toes; and such is the case too in the higher organized Birds.

The same holds good also of the form of the head, and especially of the maxillæ and teeth, which are very uniform in the lower Fishes and Reptiles, but occur under very varied conditions in the upper kinds, exactly as in the Natatores and Grallæ, in the Gallinæ and Struthionidæ,

Order 1. Protozooid Birds—Tenuirostres.

3554. Bill awl-shaped.

These Birds present a resemblance to each other, not simply in their mode of feeding, since they are collectively insectivorous, but also in the slenderness of their form, and in the dazzling, varied, sharply defined, and very striking colours of their plumage, as well as in their habits, for they employ their feet and tail usually for the purposes of support, and so climb about the upright stems and branches of trees.

Among them also occur the smallest sized Birds, a fact which, compared with a similar one in the Thricozoa, indicates likewise their lower grade or rank.

Fam. 1. Infusorial Birds, Tree-runners or Creepers.

Bill awl-shaped, three toes in front—Humming-birds, Tree-creepers.

The small size of the Humming-birds seems to render them the lowest in rank of the class, and by this means the system obtains a point of departure, unto which similar forms may be annexed. Their manner of feeding is rather a process of lapping, than an actual snapping with the bill; their food also, which consists of small Beetles and their larvæ, requires scarcely any operation of the bill, so that here the cibarial instruments obviously rank upon the lowest stage, and remind us of the proboscis in Flies, Butterflies, and Bugs.

Fam. 2. Polypary Birds, Woodpeckers.

Bill straight and chisel-shaped, two of the toes directed forwards and two backwards.

The Woodpeckers stand obviously a step higher, because their bill is specially active in seeking out larvæ, and their body is held securely by the toes and stiff tail.

Fam. 3. Acalephan Birds, Cuckoos.

Bill rounded, slightly arched and obtuse; feet scansorial.

These Birds are less animated or lively than the preceding ones, keep themselves more concealed, and live mostly upon caterpillars in warmer countries. Some of them lay their eggs in the nests of other Birds.

Order 2. Conchozooid Birds—Conirostres.

3555. Bill short, thick, and straight, without a notch, three toes directed forwards.

These Birds usually perch upon branches, without, however, being able to walk upon them, since they usually hop, and must therefore make an auxiliary use of their wings. They are all of them granivorous, and in a condition both to crush the seeds with their strong bill, as well as pound them with their fleshy stomach or gizzard.

Their instruments of manducation and digestion are therefore perfect, added to which they are so allied to the subsequent order of Rapaces, that they could not be arranged lower; probably they ought even to rank higher.

Fam. 4. Mussel-Birds, Sparrows.

Bill short and coniform, without bristles; eat simply grains.

Fam. 5. Snail-Birds, Crows.

Bill long and coniform; eat grains, beetles and flesh, and pass gradually over into the Rapaces.

Fam. 6. Kracken-Vögel, Parrots.

Bill very thick and curved. The food consists of grains and fruits. Their spiritual energies or capacities are also more strongly developed.

Order 3. Ancyliozooid Birds—Dentirostres.

3556. Bill with a notch; food consists of worms and flesh.

Fam. 7. Worm-Birds, Cantores or Songsters.

Bill tolerably long and straight; eat worms and berries.

Fam. 8. Crustacean Birds, Fly-catchers.

Bill straight, with sharp edges superiorly, point de-curved or bent downwards. Their aliment consists of insects, which they mostly snap at during flight—Fly-catchers, Shrikes, Swallows.

Fam. 9. Ptilotoid Birds—Rapaces.

Bill unciform; seize upon the Sarcozoa with their claws.

B. SARCOSE BIRDS—NESTFLÜCHTER.

3557. Run about soon after being hatched, and nourish themselves. Bill and feet very varied, the former being mostly obtuse.

These Birds do not fly much nor hop, but walk, run, or swim. In them are found all the diversities of bill and feet; the latter are mostly placed far back, so that the body is usually directed upwards.

Their food also is very varied, consisting of seeds, grass, fruits, worms, and flesh.

Order 4. Sarcose Birds.

3558. Neck long, i. e. much longer than the head and bill, frequently longer than the body.

Fam. 10. Fish-Birds, Natatores.

Natatory feet short.

Mostly large Birds, which live upon fishes, worms, and many of them even upon herbs. Their bills are so varied, that they could represent several families, a fact which likewise speaks in favour of their higher position. It may be regarded as an instructive hint, that animals occupy a higher rank, if in them a richness or variety of forms is to be perceived. This is the case throughout the Thricozoa. The Natatores, through the structure and posterad insertion of their feet, through their closely-set plumage, which frequently presents short and scale-like feathers, by their swimming and diving, and lastly, through their fish-catching, approximate as closely to the structure and mode of living of Fishes as it is possible for a Bird to do, while still retaining the characters of its class. Many row even with the wings, and consequently use them as veritable fins.

Fam. 11. Reptilian Birds, Grallæ or Waders.

Legs, neck, and bill very long.

These Birds are a lively image of, or a composition from, the Reptilia, having a Frog's body with its long feet, and a Serpent's neck with a Tortoise's head.

They wade about in marshes to catch worms and fish; have also bills of very varied structure, yet generally very long and slender, being in some cases naked, in others covered with a skin.

Fam. 12. Typical Birds, Gallinæ.

Bill and legs shorter, the former being arched, the latter armed with strong toes for scraping.

Feed usually upon grains and worms, and live in dry situations. The Marsh-hens pass at once into the true Fowls, and these are attached, like as is no other family, through their capacity for domestication, to Man.

Order 5. Sensorial Birds.

3559. Run only, cannot fly.

Fam. 13. Thricozooid Birds, Bustards.

Fowls with long legs and mostly diminished toes—Bustard, Cassowary, and Ostrich.

These Birds have mostly such shortened wings, that they cannot elevate themselves from the earth. In the Cassowary we find, in place of the primary feathers upon the wings, five barbless quills like so many claws. The Ostrich has a pelvis closed in front like the Mammalia. It has not incorrectly been compared with the Camel, since the Ruminant animals in general have many resemblances in common with Birds, especially in the development of the horny substance, which obviously agrees with the feathers.

Thirteenth Class.

Aistheseozoa, Opthalmozoa—Thricozoa.

3560. All the senses perfect, five digits; the face covered with skin and flesh, the body with hairs;

mammæ present.

3561. The Thricozoa combine in themselves all the animal classes, and indeed, so far as the development by grades makes no distinction, in equal proportion.

The differences are in no class therefore so numerous and so strongly pronounced as in this; and yet one is in more doubt almost concerning the rank of the families than in the former classes. It turns out, therefore, that they are not treated according to sound or solid principles, but that these, after the manner in which Natural History is still prosecuted, are despised.

Upon viewing superficially the groups of Thricozoa, it is certainly difficult to determine, which are the lowest families, although the uppermost ones are very well known; so that here matters are the reverse of what they were in Plants. In the case of Birds the empirical inquirer into nature knows neither which are the upper, nor which the lower groups; while in the Reptiles and Fishes he fares still worse.

It is pretty generally conceded that the Whales are the lowest in point of rank, because their posterior feet are wanting, and in this they certainly do depart the most from the usual or normal form of the Thricozoa; yet still it is just these animals which form the proper starting-point for the present class. Even the empirical naturalists here begin to employ physio-philosophical language, and designate these animals by the title of Fish-like Thricozoa. But, if they repeat the Fishes, then there must be Thricozoa, which stand or rank below them, and pass parallel to the Dermatozoa. Which these latter are, is not indeed to be declared until the other groups have been separated and brought into their proper station or place.

If once there are Ichthyoidal there must also be Herpetoidal or Reptile-like Thricozoa. Now, if the Whales correspond to the former, then the serial arrangement of the latter kinds is readily enough decided. To the Whales no other animals admit of being annexed but the Swine-like genera, such as the Hippopotamus, Pig, and Elephant, which, through their muscular mass, almost naked skin, and residence in marshy situations, agree very closely with the Reptilia.

In the next place come the Ruminantia, to take their site opposite the class of Birds, with whom they correspond in their susceptibility to domestication, large ears, fine sense of hearing, and timid disposition. The horns must be regarded as the obscure metatype (nachregung) of the feathers. These three families may be aptly termed Ungulata or Hoofed-animals.

Were matters to fare simply thus, then the center-building of this class would stand firm, and it would consequently not be difficult to say which animals stood below and which above it. For to the Apes are allied the Bears, to these the Dogs, Cats, and Martens, and to the latter again the Seals; all would rank above the Ungulata or Hoofed animals, and represent the proper or typical class of Thricozoa.

But the Thricozoa are Æsthetic or Sense-animals, and consequently the upper families must pass parallel to the organs of sense, if the lower correspond to the anatomical systems.

Now, if the eye has attained its maximum development in Man, the same must be said of the ear in the Apes, of the nose in the Bears, of the tongue in the Carnivora, and of the skin in the Seals. Having so disposed of these, the only remaining Thricozoa are the Bats, Shrews, Moles, Marsupials, Sloths, and Rodentia.

In common parlance we compare the Bats with the Birds on account of their wings, unto which may be further added the large size of their ears. Their close relation, however, with the Shrews and Rodents assigns them a lower rank. They must be placed parallel with the Ptilota or Flies.

The dental formula of the Bats resembles too so strikingly that of the Shrews that a rusty-grown prejudice can alone place the former in the neighbourhood of the Apes. They have obtained this post, without doubt, simply because they have but two mammæ, and these placed upon the chest.

The Moles cannot be removed from the Shrews.

The choice of position now remains between the Marsupials and Rodents. And to determine this point cannot prove difficult, for every one will place the former above the latter on account of their size, more perfect dental formula, and the hands, upon whose model the hind-feet have been in many species formed. To the Marsupials are annexed the Sloths, because several of them still possess marsupial or pouch-bones.

Viewed in this and in every other respect, the Rodentia stand or rank the lowest; and since the members of this family are much more numerous than those of any other, we may conclude that they fill up several families of the asarcose animals.

3562. The series would accordingly be as follows; first of all Rodents, then Sloths and Marsupials; with Moles, Shrews, and Bats, all as the repetition of the Dermatozoa. They are all Myoidal or Mouse-like in character. Next come the Whales, Pigs, and Ruminants as the repetitions of the Fishes, Reptiles, and Birds.

The Carnivora with the Bears, Apes, and Man, as being the proper representatives of the senses, conclude or wind up the list. They alone have a regular dental formula. We have thus—

I. Splanchno-Thricozoa; Mice, Edentata, Marsupialia, Shrews and Bats.
II. Sarco-Thricozoa; Whales, Pachyderms, Ruminants.
III. Æsthesio-Thricozoa; Carnivora, Seals, Bears, Apes and Man.

3563. It is here shown, just as distinctly as in the series of the classes, that no simple scale exists in the history of development, and consequently in the arrangement of animals. The Mouse-like species stand off from the rest, and then follow the entirely different Ungulata with the Pigs and Ruminants, which once again diverge in like manner, and make room for the development of the Seals, which then proceed through the Dogs, &c., in a less interrupted series up to Man.

He who marvels at this, let him take and set the table of the class-series before his eyes, and he must give utterance with us to the following words; namely, that the lower animals diverge or turn aside, and the entirely different Fishes, Reptiles, and Birds follow, which, once again diverging, make room for the development of the Thricozoa, or, in other words, the "Compendium Animalium." A perfect parallelism is thus found to exist between the classes of animals generally and the families of Thricozoa; but no linear or continuously progressive connexion is discoverable between one set and the other, but an appearance by fits or starts of new forms, just as the systems and organs also are not gradually evolved metamorphoses of one system, but sudden productions "en avant" with new tissues, forms, and functions. The animal system is a multifariously-constructed temple, with its nave, choir, chapels, and towers, while these again are present with the whole diversity of forms, which belongs to them in their several characters or bearings.

A. SPLANCHNO-THRICOZOA—PFOTENTHIERE, MAUSARTIGE.

3564. Small animals with irregular set of teeth; four feet with claws. The regular set of teeth has included all kinds of teeth, and with them four or six incisors.

A set of teeth is irregular, which has more or less than the ordinary number of incisors, in which moreover one or the other kind of tooth is wanting, or if it is separated by a breach or interspace. The small Thricozoa divide into three orders.

The one have blunt uniform molars, two rodent or gnawing-teeth, and no canines—Rodentia.
The others have, one might say, a wholly aberrant and confused dental formula, there being at one time too few, at another too many teeth; molars uniform, with perfectly irregular incisors and canines—Sloths, Marsupials.
Others, lastly, have a tolerably regular set of teeth, presenting quadriacuminate molars, mostly

small canines, and rodent-like incisors—Shrews and Bats.
Order 1. Gedärm-, Eyer, Schleimthier-Haarthiere—Rodentia.
3565. Rodent teeth, without canines.
The dental formula of the gnawing Rats and Mice is so varied, and so devoid of influence upon their bodily structure and mode of life, that all attempts to arrange these animals in accordance with it have failed.
3566. The lower position of the Rodentia admits of being easily proved by taking to our aid the meaning of the dental system. The splanchnic or visceral teeth obviously rank lower than the membral teeth. Now, it so happens that the anterior teeth are in the Rodentia the principal organ, not only of the dental formula, but of the entire animal. Without gnawing-teeth the Mice could not maintain their existence, much less their character. They thus depend wholly and solely upon the visceral teeth, and are consequently the lowest Thricozoa.
In addition to this comes their small size, which is by no means a character to be despised, and one to which all naturalists pay regard, even while they keep it, upon the other hand, in the background; or else they must place the Elephant below the Field-mice.
3567. I divide them, as well according to structure as habit, into three groups. The one has the front and hind feet of equal length, with blunt claws for digging.
The other has similar feet, with sharp claws for climbing.
Lastly, the third has the hind feet longer than the fore for leaping.
Fam. 1. Infusorial Thricozoa—Wühlmäuse.
Eyes and auditory conchæ feebly developed, feet for scraping, tail lax; live always under the ground, and eat roots and grains—Spalax, Rat, Common and Field-mouse, Beaver; mostly only three molars.
Fam. 2. Polypary Thricozoa—Klettermäuse.
Eyes, ears, and tail large, the latter being stiff and hairy, claws pointed; live mostly upon trees, eating nut-kernels and fruit; usually four molars—Squirrels.
Fam. 3. Acalephan Thricozoa—Laufmäuse.
Eyes and ears large, tail hairy, hind feet longest, claws obtuse, lamellar or plicated teeth; live upon the surface of the earth and eat grass—Hares and Guinea-pigs; mostly more than four molar teeth.
Order 2. Ader-, Geschlechts, Schalthier-Haarthiere—Kaumäuse.
3568. Teeth and toes deviating completely from those of every other order; nor in a less degree the method of propagation.
In some instances all the teeth are wanting, in some they exceed the ordinary number, and are quite uniform; sometimes the lateral teeth only are similar, but in this case there are no canines and incisors; or else there are canines with more than six incisors, or also with rodent teeth.
In like manner the toes are irregular; the one kind being disproportionately large and the other absurdly small; some are for the most part wholly connate; in many cases there are hands posteriorly. The claws also are in some instances obtuse, in others sharp for climbing, or else asymmetrically large and unciform. Hands adapted for swimming or flight also occur.
Finally, the sexual parts are quite aberrant, being mostly very large and singularly formed; the mammæ frequently lodged in a pouch, or at least furnished with marsupial bones. The young are born as naked and immoveable embryos, and suckled for a very long time.
All this reminds us of the Conchozoa or Shell-animals as doth also their unusual covering of belts, scales, spines and long hairs.
Fam. 4. Mussel-like Thricozoa, Sloths.

Lateral and canine teeth equal in size and obtuse, incisors mostly wanting, and occasionally all the teeth; claws very large and curved—Ornithorynchi, Ant-eaters, Armadillos and Sloths.
Fam. 5. Snail-like Thricozoa, Herbivorous Marsupials.
Rodent teeth, usually with stunted proximal and canine teeth, lateral teeth level; toes mostly connate and very unequal; they live in the Old World upon roots, grass, and fruit—Wombat, Dasyure, Kangaroo, Opossum.
Fam. 6. Kracken-Haarthiere, Carnivorous Marsupials.
Mostly more than six incisors, triangular molars and large canines; live in the New World and in Australia, eating worms, insects, eggs and flesh—Vulpine Phalanger, Phascogale, Beutelratze. The abnormal structure of the sexual parts reminds us of the same relation in the Snails and Kracken.
Order 3. Lungen-, Fell-, Ringelthier-Haarthiere—Raubmäuse.
3569. Molar teeth mostly quadriacuminate, with a break in the series, canines and pointed incisors, or rodent teeth with lateral incisors, five toes; live upon worms and insects.
Fam. 7. Worm-Thricozoa, Moles.
Claws, sharp incisors or rodent teeth, with lateral incisors or false molars; live exclusively under the earth, and cast up the mould.
Fam. 8. Crustaceous Thricozoa, Shrews.
Paws, rodent teeth, with small lateral incisors and canine teeth. Many burrow passages without throwing up the soil.
Fam. 9. Ptilotal Thricozoa, Bats.
Alary membrane between the feet and anterior digits; pointed canine and incisor teeth.
Order 4. Sarco-Thricozoa—Ungulata.
3570. Body large and heavy; teeth stunted, molars uniform, tolerably obtuse; feet fin-or hoof-like; mostly udders, rarely mammæ.
Fam. 10. Ichthyoid Thricozoa, Whales.
Skin naked; no hind feet; two horizontal caudal fins; toes of the anterior feet surrounded by a common skin; no auditory conchs; posteriorly two udders. All live in the sea.
It is hardly necessary, in speaking of the Whales, to direct attention to their monstrously-developed osseous system and large fleshy tongue, as also to their correspondence with Fishes in the entire form, manner of living, and imperfect nose. Their head is still confluent with the neck, the teeth are horny plates of coalesced hairs, or uniform simple points, as in Fishes. Most of them have even dorsal fins; the two udders are hardly separated from the sexual parts.
Fam. 11. Reptilian Thricozoa, Pigs.
Four feet with hoofs; canines and mostly also incisors; stomach simple, do not ruminate—Nylghau, Pig, Elephant, Rhinoceros, Horse. These animals love the marshes, and are through their mode of living, as also their form, similar to the larger Reptiles; or, in other words, through their colossal skeleton, with preponderating muscular mass, they are Myozoa, through the proboscis or snout, Rhinozoa.
Fam. 12. Ornithic Thricozoa, Ruminants.
Toes bifid, surrounded by a hoof; above there are rarely incisor and canine teeth; udders behind; stomach fourfold; they ruminate. The horn-formation indicates a relationship with feathers; the want of incisor teeth, large ears, and timidity of disposition that of the family with the class of Birds.
Both families enter into connexion with the Whales through the size of their body, the structure of their feet, form of the head and disposition towards water and mud. They are quadrupedal

Whales, which have come out of the water, and adopted a manner of living like the Amphibious Reptiles and Grallatorial Birds.

Order 5. Æsthesio-Thricozoa—Unguiculata.

3571. Here for the first time an equiponderance of the sensorial organs makes its appearance and along with it therefore a resemblance between the animals, which is no more interrupted by such strange or odd forms, as in the preceding orders.

All of them have divided toes with claws or nails, and all kinds of teeth with multiacuminate enamelled molars.

3572. No doubt can exist about the animals belonging to the present order; they are the Apes, Bears, and Carnivora generally, as Seals, Cats, Dogs, Martens, and such like beasts.

I also cherish no doubt concerning the rank of the two first families, namely, the Apes and Bears, although they have been separated in a strange manner by the interposition of the Bats and likewise the Shrews. At some future time one will not believe that the Bats and Shrews were once placed next to the Apes.

3573. Doubt, however, may exist concerning the rank of the Carnivorous or Rapacious animals; so that here the principles of our philosophy must be brought to bear in our behalf. I regard then these animals as the highest representatives of the sensorial organs. By this step three families at once take their proper positions, viz.:

Man upon the station or rank of the Eye.
The Apes on that of the Ear.
The Bears upon the post of the Nose.

Difficulties are consequently presented in regard only to the Rapacious animals, but which are removed, so soon as the three families just named are parallelized with the others thus:

Bone, Tongue, Whales.
Muscles, Nose, Pigs—Bears.
Nerves, Ear, Cattle—Apes.
Senses, Eye—Man.

Now the Seals range of themselves next the Whales.

There are thus left the Rapacious animals proper, which as Dermatozoa or Sentient animals must consequently correspond to the Splanchnozoa. In favour of this, evidence is afforded not only by the particular use made of their feet, but their great number also, which can admit of comparison with no individual family or order.

Now, however, there are three orders of Murine animals.
1. Rodent Mice—the Rodentia proper.
2. Chewing Mice—the Sloths and Marsupialia.
3. Rapacious Mice—the Moles, Shrews, and Flitter-mice, or Bats.

In like manner do the rapacious Carnivora divide into three groups.
1. The mostly sneaking and scansorial Martens and Viverræ.
2. The sneaking and fossorial Skunks, Gluttons, and Badgers, with soles and blunt claws.
3. The high-legged Digitigrades, as Dogs, Hyænæ, and Cats.

As I have given up the dental formula in the Rodentia as a means of division, so now it seems to me that it must be abandoned also in the Beasts of Prey. The whole appearance of them and their mode of living, which is still the main point in view, obviously directs us more towards consideration of the feet, than of the dental formula. The Marten or Weasel kind were formerly compared with the Mice, and called on that account Mustela. They cannot be regarded otherwise than as the lowest in rank.

Unto them are obviously annexed the short-legged Civets, despite their cunoidal set of teeth. Many have half soles or pads under the feet. With these again the Fox-like animals, notwithstanding their viverrine dental formula, enter into alliance. I believe that I have rightly parted the Badger from the Bears, and rightly done it too in this place.

The highest are without doubt the Dogs, Hyænas, and Cats, with their long and upright legs, not to speak of their mental faculties. We accordingly obtain the following arrangement.

Fam. 13. Dermal Thricozoa—Carnaria.

All kinds of teeth included; six broad incisors, a longer canine, two to three small false molars, a large carnivorous tooth and large Querzahn, and frequently also the Kornzahn. Mostly five separated toes, with nails resting upwards, and either sharp or obtuse feet, occasionally soled. They eat flesh, mostly that of warm-blooded animals, and kill their prey themselves.

Alliance 1. Rodent-like Unguiculate Animals—Schleicher.

Feet short and oblique, mostly sharp claws without entire soles.

1. Mole-like Animals—Martens.

Sharp claws without soles, and no Kornzahn—Martens and Otters.

2. Squirrel-like—Civets.

Sharp, curved, and mostly retractile claws, superiorly the Kornzahn—Civet, Paradoxurus.

3. Leporine or Hare-like—Ichneumons.

Straight claws with half soles—Ichneumon, Ryzæna.

Alliance 2. Kaumausartige—Fossores.

Large and straight claws with soles, no Kornzahn.

1. Sloth-like Animals—Mephitic Animals.

Dental formula, like that of the Marten—Skunk, Mydaus.

2. Wombat-like—Gluttons.

Dental formula, like that of Marten, but behind the carnivorous is a small tubercular tooth which is wider, being broader than long.

3. Opossum-like—Badgers.

The tuberculous tooth larger than the carnivorous tooth, and nearly quadrangular.

Alliance 3. Raubmausartige—Digitigrada.

Legs high and upright; no soles.

1. Mole-like—Dogs.

Claws blunt, the Kornzahn both in upper and lower jaw.

2. Shrew-like—Hyænas.

Claws blunt; no Kornzahn but a small Querzahn.

3. Bat-like—Cats.

Claws sharp, curved, and retractile; no lower tubercular tooth, but a small upper one.

Fam. 14. Lingual Thricozoa, Seals.

Feet fin-shaped, the hinder pair stretched out, dental formula complete, but the lateral teeth tolerably even, and six or four incisors; tongue mostly somewhat slit. They correspond to the Whales.

Fam. 15. Nasal Thricozoa, Bears.

Nose elongated into a snout, walk upon soles; all kinds of teeth, of which, however, the carnivorous or tearing tooth is similar to the grinders, six incisors.

The slower gait, originating from their walking on the soles of the feet, with the less pointed molars, assign to these animals a less rapacious habit of living. They therefore kill no large

animals, and are satisfied with worms and even roots, fruit, and honey. They repeat the Pigs.
Fam. 16. Nasal Thricozoa, Apes.
The ears begin to acquire the human form, as do even the teeth; never more than four incisors, but a longer canine; hands both fore and aft.
These animals live upon fruits and beetles, and are by their scansorial feet destined to live upon trees. Their varied, piping, and sonorous cry is a property of the sense of hearing, and along with it the larynx also usually obtains a stronger amount of development.
3574. It seems that every family of Thricozoa contains five genera, and that these accord with the organs of sense. In many families this relation is at least striking, e. g.
Among the Pigs the Elephant is obviously characterized by the nose, the Hippopotamus by the skin, the Pig by the dental system; and thus by the sense of taste, the Rhinoceros by large ears, the Horse by the eyes, thus
1. Skin Hippopotamus.
2. Tongue   Pig.
3. Nose   Elephant.
4. Ear Rhinoceros.
5. Eye Horse.
In like manner do the Ruminantia or Cattle become distinctly marshalled according to the five senses. The Camel recedes from the others by its simply uplying hoofs, the Musk-deer by its monstrous canine; the Goat is distinguished by its fine sense of smell, and the Giraffe by its large ears, the Ox by the large and beautiful eye, thus
1. Hautrind   Camel.
2. Zungen-   Musk-deer.
3. Nasen-   Goat.
4. Ohren-   Giraffe.
5. Augen-   Ox.
In the Bears also we are confronted by this principle of classification.
The Skin-bear is the Common Bear.
The Tongue- " is the Raccoon.
The Nose- " is the Nasua.
The Ear-   " is the Arctitis.
The Eye-   " is the Cercoleptes.
After this principle had been so strikingly verified in several families, I proceeded with confidence also to those which were more difficult, and it resulted that each consists only of five genera severally distinguished by the predominance of some one organ of sense. This method has been carried out throughout my "Allgemeine Naturgeschichte."
Fam. 17. Ophthalmozoa, Man.
Superiorly or in front hands, inferiorly or behind soles.
3575. Here all the senses enter for the first time into a state of perfect equiponderance or proportion. Skin naked, and therefore a perfect organ of feeling; feet and hands differently constructed for progression and manipulation; tongue and lips fleshy, while the latter have hitherto been only tegumental; all the kinds of teeth different, but still very similar, being of equal height and nearly equal size; nose elevated by its whole length from the face, and fleshy; ears oval, laid close against the head and having regular windings or convolutions; eyes directed forwards, with perfect eyelids, and moveable in all directions.
3576. Man by the upright walk obtains his character, namely, that of bodily freedom, for his hind

feet take the place of all the four feet of other animals, by which means the hands become free and can achieve all other offices, the feet alone serving to support the body.

He is the only animal that surveys with the axes of the eyes borne parallel the most extensive horizon. All animals whose eyes look higher up or above the ground, as the Horse, Elephant, Ostrich, and such like creatures, have eyes directed sideways.

3577. With the freedom of the body has been granted also the freedom of the mind. Man sees everything, the whole universe, while the animals can only view individual parts thereof, two of these even invariably appearing different, so that the images seen by them are never reduced to unity.

3578. There is only one human family, only one human genus, and only one species; and this just because Man is the whole Animal Kingdom.

3579. But yet there are five kinds or varieties of Men, according with the development of the sensorial organs.

1. The Skin-Man   is the Black, African.
2. The Tongue- "   is the Brown, Australian—Malayan.
3. The Nose-   "   is the Red, American.
4. The Ear-   "   is the Yellow, Asiatic—Mongolian.
5. The Eye-   "   is the White, European.

Co-arrangement.

3580. The classes of Sarcozoa pass or rank parallel to each other in the following manner:

	Class X. Anatomical Systems.	Class XI. Osteozoa, Fishes.	Class XII. Myozoa, Reptiles.	Neurozoa, Birds.
I. Intestinal.		Ord. I.	Ord. I.	Ord. I.
Stomach—Infusoria		F.1. Lampreys	F.1. Salamanders	F.1. Tree Creepers
Intestine—Polypi		2. Plectognathi	2. Frogs	2. Woodpeckers
Absorbents—Acalephæ.		3. Shads.	3. Tortoises.	3. Cuckoos.
II. Vascular.		Ord. II.	Ord. II.	Ord. II.
Veins—Mussels		4. Eels	4. Schuppen-	4. Sparrows
Arteries—Snails		5. Haddocks	5. Tafel-	5. Crows
Heart—Krackenchlangen.		6. Grundeln.	6. Schienens-	6. Parrots.
III. Pulmonary.		Ord. III.	Ord. III.	Ord. III.
Skin—Worms.		7. Tunnies	7. Schleichen-	7. Singing-birds
Branchiæ—Crabs		8. Breams	8. Schuppen-	8. Fly-catchers
Tracheæ—Flies.		9. Perch.	9. Schienen-Echsen.	9. Hawks.
IV. Flesh.		Ord. IV.	Ord. IV.	Ord. IV.
Bones—Fishes		10. Herrings	10. Icthyosauri	10. Geese
Muscles—Reptiles		11. Salmons	11. Geckos	11. Herons
Nerves—Birds.		12. Pike.	12. Pterodactyli.	12. Fowls.
V. Senses.		Ord. V.	Ord. V.	Ord. V.
Thricozoa.		13. Sharks.	13. Crocodiles.	13. Bustards.

Class XIII.

Anatomical Systems.	Sense-animals, Thricozoa.	Aisthesio-Thricozoa. 13. Mausartige	Sense-Organs. 1. Tegument.	
I. Intestinal.	Ord. I.	1. Wühlmäuse	I. Ovum.	
Stomach—Infusoria	F.1. Wühlmäuse		F.1. Martens	Vitellus—Infusoria
Intestine—Polypi	2. Kletter- "		2. Civets	Albumen—Polypi
Absorbents—Acalephæ.	3. Lauf- "		3. Ichneumons.	Envelopes—Acalephæ.
II. Vascular.	Ord. II.	2. Kaumäuse	II. Sex.	
Veins—Mussels		4. Sloths	4. Skunks	Spawn—Mussels
Arteries—Snails		5. Beutelhasen	5. Gluttons	Milt—Snails
Heart—Kracken.		6. Beutelmarder.	6. Badgers.	Kidneys—Kracken.
III. Pulmonary.	Ord. III.		3. Raubmäuse	III. Limbs.
Skin—Worms.		7. Scheermäuse	7. Dogs	Papillæ—Worms
Branchiæ—Crabs		8. Shrews	8. Hyænas	Feet—Crabs
Tracheæ—Flies.		9. Bats.	9. Cats.	Wings—Flies.
IV. Flesh.	Ord. IV.		IV. Cephalic Senses.	
Bones—Fishes		10. Whales	14. Seals	2. Tongue—Fishes
Muscles—Reptiles		11. Pigs	15. Bears	3. Nose—Reptiles
Nerves—Birds.		12. Ruminants.	16. Apes.	4. Ears—Birds
V. Senses. Thricozoa.	Ord. V.	13. Aisthesio-Thricozoa.	17. Man.	5. Eyes—Thricozoa.

3581. By the accompanying table the parallelism of the different families is recognized, as well as their remaining relationships, both among themselves as also with the Asarcous animals, if their table at p. 614 be compared with it, and which could not for want of room be inserted here.

3582. It is, moreover, proved from this table that the classes stand one above the other, but yet that each recommences from below, so that the lower animals of a higher class are more stunted or rudimental than the upper ones of a lower class. Thus the Salamanders are more rudimental, that is, they have organs more imperfect than the Sharks; the Tree-creepers are more rudimental than the Crocodiles, the Mice than the Fowls and Bustards.

Nevertheless these stunted animals stand higher than those of the lower classes, because they are characterized by a higher organ.

What holds good of the classes holds good again of the orders and families also. The lower animal of a consecutive family is again more rudimental than the upper one of the antecedent family. Thus the Ornithorynchus is more rudimental than the Beaver, the Shrew-mouse more rudimental than the Opossum and such like creatures.

In the highest families the equiponderance is first restored, and the lowest Man is still higher than the uppermost Ape.

IV. PSYCHOLOGY.
FUNCTIONS OF THE ANIMALS.

3583. The present section treats concerning the functions of the Whole animal, just as the preceding one did of those belonging to individual organs. It is at bottom the psychological part of Physio-philosophy. The functions are so numerous and difficult also to arrange, that I place this section here, to point out rather its situation than to follow out or trace its development.

3584. All the functions of an entire animal are spiritual or sensorial functions; at least they are conditionated by the senses, and I will also speak of them only in this respect. The mechanical and chemical functions have been already comprehended in the physiological part. The senses only make their appearance gradually in animals, and with them also the spiritual functions.

A. FUNCTIONS OF THE DERMATOZOA.

1. Enterozoa or Oozoa.

3585. These animals are governed chiefly by a passive sense of feeling, from their consisting for the most part of a naked, homogeneous, and gelatinous integument, and living in water, an element wherein the other senses can be but slightly active.

3586. Their sense of feeling stands upon the lowest grade, since it is only the sensibility of the tegument, there being no articulated organs of touch, so that it consists only in the discrimination of an opposing object.

The ability or power to discriminate is not yet consciousness; for unto this a reflexion upon the object discriminated is necessary.

The Infusoria, Polypi, and Acalephæ, simply feel that something else is there, but they are so completely imprisoned in this feeling that they are unable to submit the same to an internal process of comparison.

3587. By reason of this inability to compare their own feelings, not a trace is left unto them of internal change; so that these creatures are truly devoid of memory or recollection.

The Infusoria have only sensation, nothing else; they are therefore in ceaseless motion. They are actually capable of nothing but moving and eating. Of all other spiritual functions they are utterly devoid.

3588. Their spiritual life is in some degree a mesmeric condition. Destitute of the senses of seeing, hearing, smelling and tasting, they feel every thing, or, properly speaking, perform all these functions at one and the same time, and by one organ, the visceral mass. By mesmerism they find their food, perceive the light, and become transparent unto themselves, just as they are really in a physical point of view. For they are only viscera or visceral nerves.

Development of the Mind.

3589. The mind, just as the body, must be developed out of these animals. The human body has been formed by an extreme separation of the neuro-protoplasmic or mucous mass. So must the human mind be a separation, a memberment of infusorial sensation.

3590. The highest mind is an anatomized or dismembered mesmerism, each member whereof has been constituted independent in itself.

The skeleton of this dissected mind, when scientifically represented, would be the science of the mind, i. e. Philosophy, properly so called.

Pneumato-philosophy is the likeness of Physio-philosophy. For spirit is only the tension of nature, and nature only the spirit set in motion.

The philosophy of spirit must develop itself out of the philosophy of nature, as doth the flower out of the stem. For nature is the spirit analyzed and at rest, which we can handle at our pleasure. It does not appear only for an instant; but as stone, air and such like entities, abideth always, as if to solicit and preserve us for its investigation.

A Philosophy or Ethicks apart from Physio-philosophy is a nonentity, a bare contradiction, just as a flower without a stem is a non-existent thing.

3591. As many essential members as Physio-philosophy hath, into so many must Pneumato-philosophy also divide, and this too so exactly that the two shall cover each other.

The reason why one has hitherto rambled about in Pneumato-philosophy without ballast and without compass, depends solely upon the disregard which has been paid to the science or knowledge of Nature. It is in fact not difficult to understand how impossible it must be, from observations made upon the rapidly evanescent phenomena of the spirit, to thence abstract a system of the laws in conformity wherewith this spirit manifests itself or acts. Spirit is nothing different from Nature, but simply her purest outbirth or offspring, and therefore her symbol, her language. With such a basis as this, we shall no longer pursue the ignes fatui of the mind, but first of all endeavour to banish them into the provinces or realms of Nature, and there co-ordinate them in conformity with her laws; then for the first time shall we recognize the flaming lights of the mind and the divine voices, which all matter proclaimeth through the speech of Man.

He, who were once in a condition to reveal or disclose this conformity of Nature's phenomena with those of Spirit, will have learnt the philosophy of the latter.

2. Functions of the Vascular or Sexual Animals.

3592. These animals are no longer merely sentient, 'clairvoyant' Acalephæ, because, in addition to the nervous mass, they are provided with other systems, such as the sexual and vascular, with the special organs of digestion and taste.

These three or four organs must also resolve themselves into three spiritual functions; the vascular system furnishes special organs of feeling and therewith a voluntary sense of feeling; the intestine and chiefly the liver is now the cardinal organ, and will therefore execute the mesmerically percipient functions.

3593. In the liver the faculty of anticipation and foresight, with melancholy, choleric passion, and anger, appear to reside. Encephalic thought is reflected in it.

The liver is the soul in a state of sleep, the brain is the soul active and awakening. In it the spirit broods unconsciously for years, and then breaks forth fearfully, as capriciousness, tyranny and sorrow, but also as earnestness and strength.

Circumspection and foresight appear to be the thoughts of the Bivalve Mollusca and Snails. Gazing upon a Snail, one believes that he finds the prophesying goddess sitting upon the tripod. What majesty is in a creeping Snail, what reflection, what earnestness, what timidity and yet at the same time what firm confidence! Surely a Snail is an exalted symbol of mind slumbering deeply within itself.

The old artists must have felt this signification, as in many of their representations they have introduced a Snail. One can hardly think, that in so doing they wished to express such common and lascivious ideas, as are at present manifested openly or secretly by our daily enjoyments.

3594. The intestine must moreover be concerned with the sense of taste. Taste, however, leads to voracity, gluttony, daintiness, sluggishness and drowsiness.

3595. Taste in union with the sexual function is the expression of venery or lust.

This is indicated by the secretion of slime, by the monstrous size too of the sexual organs, and by their androgynism, which enables either individual during copulation to enjoy the delicious feeling, belonging to the male and female, either at once or alternately. Their food also appears to be selected from a feeling of desire.

Circumspection in feeling, dainty voracity, and immoderate lust appear to constitute the spiritual character of the Malacozoa, especially of the Snails.

3. Functions of the Respiratory or Arthric Animals.

3596. The Insect is mainly an aerial and motor organ, and therefore its spirit is also of an aerial and motor kind.

The respiratory process produces strength, and this again courage, both which are the distinguishing properties of Insects. The Insect is the strongest and boldest animal upon the earth.

Health, plenitude of life, generosity, nobleness, and heroism dwell in the thorax.

3597. But besides these virtues the spirit of the thorax is also that of smell. Insects have an excellent sense of smell, the spirit of which is cunningness and treachery, wherein no animal will easily surpass them.

3598. The Insect has moreover a spirit of motion or versatility of the tactile sense, which is displayed in the representation of symmetrical figures. This faculty proceeds especially from the creative sexual functions—as mechanical or artistic instinct.

All spirit of motion launches out into mechanical instinct. It disappears in all classes of animals, which chiefly correspond to the sex and belly, as, e. g. in Fishes and Reptiles. On the other hand, in the moveable thoracic animals or Birds, the mechanical instincts at once re-appear.

Mechanical instinct and dexterity of limbs run parallel together.

The dexterity of the limbs taken up into the spirit is an art-sense.

B. FUNCTIONS OF THE CEPHALOZOA.

3599. Here the head is for the first time placed in a perfect condition, and hence an antagonism arises for the first time between head and trunk.

The Cephalozoon no longer distinguishes nature and self only like the acephalous and amnemonic animals; but it distinguishes even its body from its head, because the Fish has begun to be a double animal.

3600. The Cephalozoon hath consciousness; consciousness of its condition, of its body, but not of its head and the operations therein. It has no self-consciousness.

3601. As soon as an animal contemplates a part of its body, of its world, and hath consciousness in a general sense, it has also memory. For memory is a repetition of its own condition, not the reiterated feeling of a foreign object.

The Acephalous or anencephalic animals have therefore no memory, because they live only in opposition to the world, but never in antagonism to themselves. Every perception is therefore a new one for them, because it is always an actual object which excites them. Whether Insects have memory, has not yet been made out.

3602. The brainless animals have no ideas, and naturally so, because they have no consciousness. It would appear likewise that they do not feel pain.

The Cephalozoa have ideas, and quite certainly pains, because they become partly an object unto themselves.

4. Functions of the Osteo-or Glossozoa.

3603. The Fish's head is the lowest, and therefore its mind also will manifest only the first function, that ranks above the mind of the Acephalozoa, the memory.

With this memory, however, all the spiritual functions, exhibited by the preceding classes, but chiefly mesmerism, have been bestowed.

3604. Fishes are again provident, zealous animals, that, drawn together by mysterious bands, make the longest voyages, wherein they ascend and descend rivers, knowing how to find their prey over miles in extent.

All the mechanical instincts are, on the contrary, obliterated in them, as being fingerless, finned animals. Their principal business is propagation—Pelvic animals.

3605. Gluttony is the principal character of the Glossozoa, in so far as taste stands upon the lowest stage. Touch and taste are only motion and deglutition.

Smell becomes evidently stronger.

The ear still ranks upon the lowest stage, but yet they hear closely. In other respects they are

mute, and exhibit all the consequences of aphony.
3606. Fishes are of the Phlegmatic temperament.
5. Functions of the Myo-or Rhinozoa.
3607. Reptiles are Cephalozoa with well-developed lungs and nose, and are therefore gifted with voice.
3608. To memory comes the art of ambuscade or lying in wait, a property of the olfactory sense, the application of memory. The circumspection of the Snails passes over into ambuscade and surprise.
With this character the higher perfection of the belly or of the digestive system, along with the poisoning saliva, is in parallel accordance. To take by surprise and to poison are acts of one series.
3609. Reptiles appear enabled to reflect, i. e. several reminiscences are at their bidding. Now, the comparison of these constitutes reflection. They are therefore more sagacious than Fishes, docile and in some measure susceptible of instruction.
3610. The courage, which they have probably obtained through their aerial respiration, passes over more into impudence and sauciness.
3611. Reptiles are of the Melancholic crasis.
6. Functions of the Neuro-or Otozoa.
3612. The spirit of the thorax and limbs is here predominant, whence comes the restlessness and mechanical instinct of the Bird.
The motor sense, the ear, is the prominent one. The ear, however, is partly the sense for the Indefinite, partly for the demolition of matter. The auditory spirit is fear.
But it passes over into joy, passion, levity, if it perfectly perceives tones.
3613. With the ear and the moveable organs of voice originates a kind of language, which is in a condition to express a multitude of sentiments. The language of Birds hath not a few tones, and expresses not a few passions.

3614. The Bird knits or associates for the first time with some completeness a sense, or definite sensation, to a simple tone. The Bird hath for the first time signs, or symbols, which are not the things themselves, but only signify or mean them.
The Bird understands the relation of the spiritual expression to the organ or matter. It comprehends a connexion where none is materially, but which is only imparted by the idea.
The capacity to understand the thing in the image or idea, I call imagination, and of this Birds furnish us with very definite proofs.
Birds can therefore dream.
3615. Reptiles and Fishes appear to have no ideas because they have no signs, or tones, indicating the resolution of the organ into mind. For the tone is none other than the ghost of the organ or animal. Fishes and Reptiles do not indeed dream.
3616. The Bird, however, appears to get no further than to mere images or ideas. The conception is wanting unto them.
It has therefore no sense of shame. But possesses in a full degree circumspection, desire of imitation, comparison.
3617. To the Bird it is not simply the sensation of its body that becomes, like a foreign product, objective; but its own product, its voice, as something already distinct from its own mind.
3618. It is clear, that if all the sensorial functions were to become objective to the Bird, it would be self-apparent and resolve itself into self-consciousness. Thus does self-consciousness sprout

forth gradually with the sensorial functions.
3619. Birds are of the Sanguineous temperament.
7. Functions of the Æsthesio-or Ophthalmozoa.
3620. All the faculties hitherto mentioned occur here. The soul of the eye is still associated with them, and therewith the faculties of perception, understanding and conception appear to be bestowed.

One cannot refuse understanding to the Thricozoa. The actions of the Dog, Horse and Elephant, do not admit otherwise of being conceived; nor also the shame and pride, fidelity, animosity, desire of revenge, and yet many other properties exhibited by these animals.
But it is an understanding without self-consciousness, if we may so venture to express ourselves; an understanding of many signs, but devoid of any combination and separation of these signs; in a word, there is no faculty of judgment.
The Thricozoa are Choleric animals.
8. Functions of the Panæsthetic Animal.
3621. If finally all its organs become objective to the animal, through contemplation of the universe, through hearkening of the animal restored into symbols; it thus contemplates itself, is apparent unto itself, and is quite conscious of itself.
Then is the animal equivalent to the whole animal kingdom and to the universe.
3622. The universal spirit is Man.
In the human race the world has become individual. Man is the entire image or likeness of the world. His language is the spirit of the world. All the functions of animals have attained unto unity, unto self-consciousness, in Man.
3623. The thorough penetration of all the animal's symbols, the comparison of all the world's symbols, and thus free comparison is Reason.
The understanding compares only the symbols of sounds, of men; but the reason compares also the symbols of light, of the world.
Reason is world-understanding; the understanding is animal reason.
All the mental functions of animals have become ratiocinative in Man.
Feeling is in him consciousness, consciousness is self-consciousness, the understanding is reason, the passion, freedom, the mechanical instinct artistic sense, the comparison science.

3624. The spirits of the senses are Art-spirits, the spirit of reason is the spirit of Science.
Art.
3625. Art is the representation of the senses in Nature.
3626. The sense is, however, the last or ultimate design of Nature.
Art is consequently the representation of Nature's design. (Ed. 1st, 1811. § 3517.)
3627. That is beautiful which represents the will of Nature.
But non-beautiful is that which represents real Nature by means of art.
3628. Art is one universal business. Beautiful is that, which represents the world within a fragment or portion of the world.
3629. There is also one natural beauty—unconscious formation of the world's laws.
3630. The highest natural beauty is the universal portion of nature, i. e. Man.
Man expresses the ultimate goal or purpose of Nature's design.
3631. The terminus or goal of Nature is, in Man to revert again into herself. The human countenance most perfectly repeats the trunk, and again reverts wholly and actually into the

trunk. That human countenance is beautiful, in which the vertebral column runs back again parallel with the vertebral column of the trunk. The facial vertebral column is the nose.
3632. The face is beautiful, whose nose runs parallel to the spine.
No human face has grown unto this estate, but every nose makes an acute angle with the spine. The facial angle is, as is well known, 80°.
What as yet no Man has remarked, and what is not to be remarked either without our view of the cranial signification, the old artists have felt through inspiration. They have not only made the facial angle a right angle, but have even stepped beyond this, the Romans going up to 96°, the Greeks even to 100°.
Whence comes it, that this unnatural face of the Grecian works of art, is still more beautiful than that of the Roman, when the latter comes nearer unto Nature? The reason thereof resides in the fact of the Grecian's artistic face representing Nature's design more than that of the Roman; for in the former the nose is placed quite perpendicular, or parallel to the spinal cord, and thus returns whither it has been derived.
3633. He who paints or otherwise copies Nature in a purely mechanical manner, is consequently a bungler; he is devoid of ideas, and imitates her no better than a bird does song, or an ape the postures of the human body. The province of Art is alas! not yet understood.
3634. In Man all the beauties of nature are associated or combined.
3635. Thus, Nature can be beautiful, in so far as she represents the individual ideas of Man.
3636. There are only two art-senses, the eye and the ear; and thus but two departments of art, the Plastic and Sonant, or that of form and of motion.
3637. The province of form represents the material universe in its ideas, its design, and thus in its freedom.
3638. The representation of the bodily universe in the ideas is the Architectural art. (Ed. 1st 1811, § 3533.)
3639. The representation of heaven in the Plastic, is the church architecture.
The temple or church is the art of heaven.
3640. The representation of the planet in the Plastic is the house.
The house is the art of the planet.
The architectural art is the cosmical art.
3641. The representation of the Individual is the Sculptor's art.
The sculptural art represents the Terrestrial, and in its highest estate Man. It is the Heroic art.
3642. This art rendered manifest in matter, or repeated in light, is Painting.

Painting represents the symbol, the naught of the world, the heavenly, and also in its lowest estate a Spiritual.
Painting is the art of religion, the Sacred art.
Painting is the art of heathens, whose deities are men; Painting is the art of Christians, whose Men, being sacred or holy, are gods.
God can be painted, but not formed or sculptured.
3643. The art of motion represents the material and the spiritual motion.
3644. The representation of the world's material laws of motion is the Dance.
3645.The representation of the motion of individuals is the Histrionic art.
3646. The representation of the world's spiritual laws of motion, of the laws of the dance, is Music.
3647. The spiritual representation of the Histrionic is the Poetic art.

Science.
3648. The representation of the rational world is Science.
3649. The first science is the Science of Language, the architecture of science, the earth.
3650. The second science is the Art of Rhetoric, the sculpture of science, the river.
3651. The third science is Philosophy, the painting of science, the respiration or breath.
This, like painting, divides into a number of branches, whereof the art of government is the highest.
3652. The fourth science is the Art of War, the art of motion, histrionism, music, poetic art of science, the light.
As in the art of poetry all arts have been blended, so in the art of war have all sciences and all arts.
The art of War is the highest, most exalted art; the art of freedom and of right, of the blessed condition of Man and of humanity—the Principle of Peace.
THE END.

ERRATA AND CORRIGENDA.

Page 4, § 30, read upon the primary proposition of mathematics or the axiom.
5,   § 35, for magnitudes read quantities.
8,   § 39, for Manifold read Different.
11, § 49, l. 10, for elevating themselves in power, read becoming suppressed. § 50, l. 4, dele the second —.
12, l. 1, for magnitudes read quantities; l. 5, for — read 0. § 52, for emerged read been evolved out of nothing, or been produced from it by addition; but it is, &c. § 54, for removing read suppressing.
14, § 59, l. 8, for posits itself read itself posits.
16, for presentation read passim, representation.
32, § 134, for Disintegrated read Dissevered.
32, § 137, for never read nowhere.
37, § 161, l. 7, for it read there.
38, § 164, for inventive read passim throughout Part I. postulate.
39, l. 1, for and the axiom is read the axiom being.
44, § 195, l. 15, for that it may shine read upon which it shines.
50, § 214, last line, for darker read dark.
59, § 253, read heavy and material.
61, § 269, 1 and 2, for caloric read hydrogen, and for hydrogen read carbon.
80, § 388, for nitrogen read hydrogen.
112, l. 6, for Terpentin read Serpentin; l. 7 from bottom, for Calces read Calcareous earths.
124, § 562, for the recent read new.
139,> § 633-4, for disintegrations read dispersions.
159, § 761, for metallic read magnetic.
186, § 914, for heat of blood read blood-heat.
209, § 1060, read constantly circulate therein.
213, § 1077, l. 6, for but before read and whereas; l. 7, for have read previously, and dele it has then, and read the appearance is now.

228, § 1161, read has been reproduced after trunk, and dele being.
249, § 1314, for umbilicus read hilum, and for cause read basis.
307, § 1679, instead of stipaceous read hypogynous.
308, § 1687, l. 2, read Stielblumen or Hypogynes.
320, § 1763, after light-æther read from the dead mass, &c.
327, § 1800, for substance, &c., read contents become, &c.
328-9, for Myxozoa read Protozoa, and for Acalephæ read Acalephæ here and passim.
330 l. 2, after the word radiating read æther.
332, § 1824, for substance read contents, and § 1827, for substance read mass.
334, § 1837, read upon the middle rate of oxydation.
327, l. 5, transpose the words matter and spirit.
345, § 1928, read elemental matter.
346, § 1931, read at certain points, after each other.
354, § 1992, read through the medium of the gills, the water through that of the intestine, &c.
355, § 2006, for along with read which belong to the category of.
361, § 2039, read plexus-forming ramules; § 2046, for is imparted read furnishes us, and for by read with.
362, § 2050, for œsophageal read pharyngeal, passim, as also, pharynx for pharynx.
372, § 2114, for enter into the composition of read appertain to; and for Mammalia read Thricozoa.
376, § 2141, read the bone, the fluid which has been secreted from it and rigidified.
381, § 2176, for the last word bladder read cyst, hic et passim.
386, § 2211, read bivalve Mollusca, passim throughout Zoogeny.
393, l. 1, read exsecernent.
398, § 2313, read Snail-type of organization.
405, § 2371, read, or the integument after respiratory organ.
408, § 2387, for carnivorous read false molars; for incisor read laniary molar; for premolar read second, and for tuberculous read third true molar.
493, § 3054, for testaceous Mollusca read Conchozoa or Shell-animals.
515, § 3170, for stock read trunk.
544, § 3301, after analogy read which at best is but a word of random definition.
532, § 3256, read pharyngeal or pneumogastric nerves, to end of paragraph. For the words evolution and evolved, read passim in the Botanical and Zoological parts, perfection and perfected, as the text may require.

www.ingramcontent.com/pod-product-compliance
Lightning Source LLC
Chambersburg PA
CBHW051850170526
45168CB00001B/51